Ecology of the Marine Fishes of Cuba

Ecology of
the Marine Fishes
of Cuba

EDITED BY RODOLFO CLARO,
KENYON C. LINDEMAN,
AND LYNNE R. PARENTI

SMITHSONIAN INSTITUTION PRESS
Washington and London

© 2001 by the Smithsonian Institution

Publication of this book was supported by a grant from the Atherton Seidell Fund.

Based, in part, on a translation by Georgina Bustamante

Copy editor: Fran Aitkens
Production editor: E. Anne Bolen

Library of Congress Cataloging-in-Publication Data

Ecologia de los peces marinos de Cuba. English
    Ecology of the marine fishes of Cuba / edited by Rodolfo Claro,
Kenyon C. Lindeman, and Lynne R. Parenti.
        p. cm.
    Includes bibliographical references (p.).
    ISBN 1-56098-985-8 (alk. paper)
    1. Marine fishes—Ecology—Cuba. I. Claro, Rodolfo. II. Lindeman,
Kenyon C. III. Parenti, Lynne R. IV. Title.
    QL631.C9 E2613 2001
    597.177'365—dc21   2001020344

British Library Cataloguing-in-Publication Data available

Manufactured in the United States of America

08 07 06 05 04 03 02 01   5 4 3 2 1

⊗ The paper used in this publication meets the minimum requirements of the American National Standard for Information Sciences—Permanence of Paper for Printed Library Materials ANSI Z39.48-1984.

For permission to reproduce illustrations appearing in this book, please correspond directly with Rodolfo Claro, Instituto de Oceanología, Ave. 1ra. No. 18406 entre 184 y 186, Reparto Flores, Playa, Habana 21, Ciudad de La Habana, Cuba C.P. 12100. The Smithsonian Institution Press does not retain reproduction rights for these illustrations individually, or maintain a file of addresses for figure sources.

CONTENTS

List of Illustrations   vii

List of Tables and Appendixes   x

Preface   xiii

Contributors   xv

1. PHYSICAL ATTRIBUTES OF COASTAL CUBA   1
RODOLFO CLARO, YURI S. RESHETNIKOV, AND PEDRO M.
ALCOLADO

1.1 Geographic Characteristics   1

1.2 Shelf Geology and Geomorphology   3

   1.2.1 Structure, Origin, and Evolution   3

   1.2.2 Sediments   7

1.3 Climate   8

1.4 Hydrological Characteristics   8

1.5 Hydrochemical Characteristics   11

1.6 Hydrobiological Characteristics   12

   1.6.1 Primary Production   12

   1.6.2 Benthos   13

1.7 Major Habitats   13

   1.7.1 Coral Reefs   14

   1.7.2 Seagrass Beds and Softbottom Areas   17

   1.7.3 Mangroves   18

   1.7.4 Brackish Lagoons   18

   1.7.5 Pelagic Environment   18

1.8 Anthropogenic Effects   18

1.9 Summary   20

2. THE MARINE ICHTHYOFAUNA OF CUBA   21
RODOLFO CLARO AND LYNNE R. PARENTI

2.1 Introduction   21

2.2 Faunal Composition   21

2.3 Distribution and Habitat   22

   2.3.1 Coral Reef Fish Fauna   22

   2.3.2 Mangrove Fish Fauna   26

   2.3.3 Fish Fauna of Seagrass Beds and Euhaline
       Lagoons   27

   2.3.4 Fish Fauna of Estuaries and Low-Salinity
       Lagoons   28

   2.3.5 Oceanic Fish Fauna   29

   2.3.6 Bathyal Fish Fauna   30

**2.4 Changes in Population and Community
Structure**   30

2.5 Additional Information on Fish Fauna   31

   2.5.1 Fish Size   31

   2.5.2 Biotoxicity   32

2.6 Summary   32

3. BEHAVIOR OF MARINE FISHES OF THE CUBAN
SHELF   58
EMILIO VALDÉS-MUÑOZ AND ANDREI D. MOCHEK

3.1 Introduction   58

3.2 Diurnal Fishes   58

   3.2.1 Inshore Pelagic Fishes   58

   3.2.2 Epibenthic Pomacentrids   59

   3.2.3 Suprabenthic Fishes   60

   3.2.4 Territorial Benthic Fishes   64

3.3 Nocturnal Fishes   67

   3.3.1 Inshore Pelagic Fishes   67

   3.3.2 Epibenthic Apogonids   67

   3.3.3 Suprabenthic Fishes   67

3.4 Crepuscular Demersal Fishes   69

**3.5 Behavioral Mechanisms of Community
Differentiation**   69

3.6 Summary   71

4. REPRODUCTIVE PATTERNS OF FISHES OF THE
CUBAN SHELF   73
ALIDA GARCÍA-CAGIDE, RODOLFO CLARO, AND
BORIS V. KOSHELEV

4.1 Introduction   73

4.2 Sexuality   73

   4.2.1 Hermaphroditism   73

   4.2.2 Sexual Dimorphism   77

   4.2.3 Sex Ratio   79

**4.3 Gonad Development and Spawning
Patterns**   79

   4.3.1 Sexual Differentiation and Attainment of
       Sexual Maturity   79

   4.3.2 Gametogenesis   81

   4.3.3 Ecological Patterns of Spawning   84

   4.3.4 The Resorption Process   89

   4.3.5 Gonadosomatic Index   89

4.4 Annual Reproductive Cycle   92

4.5 Fecundity   95

4.6 Reproductive Strategy   98

4.6.1 Spawning at the Shelf Edge 99
4.6.2 Spawning on the Inner Shelf 101
4.6.3 Other Spawning Strategies 101
4.7 Summary 102

5. TROPHIC BIOLOGY OF THE MARINE FISHES OF CUBA 115
LUIS M. SIERRA, RODOLFO CLARO, AND OLGA A. POPOVA
5.1 Introduction 115
5.2 Major Trophic Groups 115
5.3 Trophic Relationships of the Cuban Marine Fish Fauna 117
5.3.1 Food Composition 117
5.3.2 Seasonal Variations in Feeding Habits 119
5.3.3 Ontogenetic Variations in Feeding Habits 120
5.4 Feeding Intensity 122
5.4.1 Diurnal Cycle 122
5.4.2 Digestion Rate 126
5.4.3 Food Ration 127
5.4.4 Ontogenetic Patterns in Feeding Intensity 129
5.4.5 Seasonal Variations in Feeding Intensity 130
5.5 Predator–Prey Size Relationships 130
5.6 Trophic Comparisons with other Habitats and Regions 131
5.6.1 Trophic Structure among Habitats 131
5.6.2 Trophic Patterns among Regions 132
5.7 Summary 134

6. GROWTH PATTERNS OF FISHES OF THE CUBAN SHELF 149
RODOLFO CLARO AND JUAN P. GARCÍA-ARTEAGA
6.1 Introduction 149
6.2 Growth Mark Formation 149
6.2.1 Annual Mark Formation 149
6.2.2 Juvenile Marks 152
6.3 Seasonal Variations in Growth Rate 152
6.4 Relationships among Different Size Measures 154
6.5 Age and Growth Rate 154
6.5.1 Methods Used to Evaluate Growth Rate 154
6.5.2 Growth Parameters of Species of the Wider Caribbean 154
6.5.3 General Patterns of Growth 163
6.6 Methodological Considerations in Growth Investigations 165
6.7 Summary 166

7. ECOPHYSIOLOGY OF CUBAN FISHES 179
GEORGINA BUSTAMANTE, RODOLFO CLARO, AND MIJAIL I. SHATUNOVSKY
7.1 Introduction 179
7.2 Fish Fat Content 179
7.3 Seasonal Patterns of Physiological and Biochemical Indicators 180
7.3.1 Juvenile Stages 180
7.3.2 Adult Stages 182
7.4 Ontogenetic Dynamics of Physiological and Biochemical Indicators 186
7.5 Summary 191

8. CUBAN FISHERIES: HISTORICAL TRENDS AND CURRENT STATUS 194
RODOLFO CLARO, JULIO A. BAISRE, KENYON C. LINDEMAN, AND JUAN P. GARCÍA-ARTEAGA
8.1 Introduction 194
8.2 Fishery Resources of the Cuban Exclusive Economic Zone 195
8.3 Ecological Subsystems 198
8.4 Fishing Gear 198
8.4.1 Nets 198
8.4.2 Fish Traps 200
8.4.3 Longlines 202
8.5 Fisheries Infrastructure 202
8.6 Catches by Species and Region 205
8.6.1 Snappers 205
8.6.2 Groupers 209
8.6.3 Grunts 210
8.6.4 Mojarras 210
8.6.5 Sardines and Herring 211
8.6.6 Mullets 211
8.6.7 Jacks 212
8.6.8 Tunas 213
8.6.9 Mackerels 214
8.6.10 Billfishes 214
8.6.11 Sharks 215
8.6.12 Other Fishes 215
8.6.13 Bycatch 216
8.7 Fishery Productivity 217
8.8 Use of Artificial Reefs 217
8.9 Conservation and Management 218
8.10 Summary 218

REFERENCES 220
INDEX 245

ILLUSTRATIONS

Fig. 1.1. Map of Cuba. 2

Fig. 1.2. Bathymetric profiles from the shore to the outer shelf at different areas of the western Cuban shelf. 3

Fig. 1.3. Northwest and southwest areas of the Cuban shelf. 4

Fig. 1.4. North-central shelf of Cuba (Archipiélago Sabana-Camagüey). 5

Fig. 1.5. The Archipiélago Jardines de la Reina, southeast coast of Cuba. 6

Fig. 1.6. Distribution of sediments on the Cuban shelf. 7

Fig. 1.7. Coarse schematic of large-scale oceanic circulation around Cuba. 9

Fig. 1.8. Variable flow directions off southwestern Cuba, measured by satellite-tracked drifters. 10

Fig. 1.9. Daily fluctuations of surface water temperature on the northwest coast near La Habana. 12

Fig. 1.10. Maximum biomass of benthic invertebrates recorded in the Golfo de Batabanó. 14

Fig. 1.11. Distribution of habitats of the Golfo de Batabanó, including estimates of seagrass biomass and sediment color. 15

Fig. 1.12. Cross-shelf habitat profiles at different locations in Cuba. 16

Fig. 2.1. Comparison of fish density and biomass between 1988/1989 and 2000 at slope reef sites across the Archipiélago Sabana-Camagüey. 26

Fig. 3.1. Species diversity and similarity of the fish fauna of different habitats at Punta del Este, Isla de la Juventud. 59

Fig. 3.2. Behavioral characteristics of the sergeant major, parrotfishes, and doctorfish in several habitats. 61

Fig. 3.3. Behavioral characteristics of the ocean surgeon, blue tang, bluehead, slippery dick, blackear wrasse, and dusky damselfish in reefs and seagrass beds. 63

Fig. 3.4. Behavioral characteristics of the foureye, banded, and spotfin butterflyfishes, and beaugregory, yellowtail damselfish, and French grunt in several habitats. 65

Fig. 3.5. Intensity of the attacks of the dusky damselfish on an aggressor relative to the distance of the aggressor from the shelter. 66

Fig. 3.6. Behavioral characteristics of the bluestriped grunt, dog snapper, schoolmaster, and longspine and dusky squirrelfishes in reefs and mangroves. 68

Fig. 3.7. Diurnal–nocturnal activity of fishes in mangroves, seagrass beds, and reefs. 71

Fig. 4.1. Size and frequency distributions of juveniles, and adult females and males of Golfo de Batabanó fish populations. 78

Fig. 4.2. Size and age distribution of adult males and females of Golfo de Batabanó fish populations. 80

Fig. 4.3. Schematic representation of primary spawning patterns of fishes, based on field and laboratory studies of gametogenesis. 85

Fig. 4.4. Primary features of synchronous and asynchronous vitellogenesis in the species studied. 86

Fig. 4.5. Oocyte size distribution of three ovaries of lane snapper, characterized by synchronic vitellogenesis and intermittent spawning. 88

Fig. 4.6. Monthly variation in the gonadosomatic index of different sexual stages in several species from the eastern Golfo de Batabanó. 92

Fig. 4.7. Monthly variation of average gonadosomatic index in spawners and the proportion of individuals in different stages of maturation for various Cuban coastal fishes. 94

Fig. 4.8. Spawning periods of marine fishes inhabiting the Cuban shelf. 96

Fig. 4.9. Temporal patterns of spawning in selected species of the Cuban shelf. 97

Fig. 4.10. Schematic of mutton snapper migrations to form spawning aggregations at Corona San Carlos, northwestern Cuba. 100

Fig. 4.11. Potential larval recruitment pathways to shallow-water areas of the eastern Golfo de Batabanó. 101

Plate 4.1. Photomicrographs of gonad histological sections. 76

Plate 4.2. Photomicrographs of representative stages of oocyte development.   83

Plate 4.3. Histological sections of ovaries of species with different spawning patterns.   87

Plate 4.4. Representative sections of ovaries in which resorption of the empty follicle and the mature oocyte are observed.   90

Fig. 5.1. Ontogenetic variation in food consumption in lane, mutton, gray, and yellowtail snappers, and Nassau grouper.   121

Fig. 5.2. Ontogenetic variation of food consumption in margate, bar jack, yellow jack, redear sardine, and false pilchard.   123

Fig. 5.3. Ontogenetic and interannual variation in food consumption of the bar jack, 1978–1980.   124

Fig. 5.4. Diurnal variation in the index of gastric repletion in various Cuban fish species.   125

Fig. 5.5. Digestion rates of various fishes in winter and summer while feeding on fishes or crabs.   126

Fig. 5.6. Digestion rate of lane snapper fed dwarf herring, redear sardine, or crabs.   127

Fig. 5.7. Ontogenetic variation in feeding intensity in three species.   128

Fig. 5.8. Seasonal variations of feeding intensity in various fish species from Cuba.   128

Fig. 5.9. Monthly variation of food composition and feeding intensity of three different size classes of bar jack, 1978–1980.   129

Fig. 5.10. Monthly variation of mean daily food ration in mutton snapper under experimental conditions in aquaria.   130

Fig. 5.11. Size relationships among predators and prey in four reef fish species found in Cuba.   132

Fig. 6.1. Periods of mark formation on skeletal structures of fishes studied in Cuba.   150

Fig. 6.2. Seasonal variation in the width of the spacing between the last increment and the outer margin of the skeletal structure for various species.   153

Fig. 6.3. Correlation between asymptotic fish length and ϕ', a growth performance index, for selected marine fishes occurring in Cuba.   155

Fig. 6.4. Back-calculated values for fish length, mean annual growth, and theoretical growth curves for lane and mutton snappers in southwestern Cuba.   157

Fig. 6.5. Back-calculated values for fish length, mean annual growth, and theoretical growth curves for gray and yellowtail snappers in southwestern Cuba.   159

Fig. 6.6. Back-calculated values for fish length, mean annual growth, and theoretical growth curves for dog snapper and margate in southwestern Cuba.   162

Fig. 6.7. Back-calculated values for fish length, mean annual growth, and theoretical growth curves for bluestriped and white grunts in southwestern Cuba.   164

Fig. 6.8. Back-calculated values for fish length, mean annual growth, and theoretical growth curves for bar jack and Nassau grouper in southwestern Cuba.   164

Fig. 6.9. Back-calculated values for fish length, mean annual growth, and theoretical growth curves for tiger grouper and hogfish in southwestern Cuba.   165

Fig. 6.10. Theoretical growth curves of mutton, yellowtail, gray, and lane snappers in southwestern Cuba.   166

Fig. 7.1. Seasonal variations of physiological and biochemical parameters in juvenile bar jack of southwestern Cuba.   181

Fig. 7.2. Mesenteric fat index of spawning lane snapper during the spawning peak of May 1972 in southwestern Cuba.   183

Fig. 7.3. Seasonal variations of different indicators of fat content, and the gonadosomatic index of adult females of various species of southwestern Cuba.   184

Fig. 7.4. Seasonal variations of the hepatosomatic index, and lipid, protein, and water content in liver tissues of lane snapper in southwestern Cuba.   187

Fig. 7.5. Seasonal variations of the liver weight, and its protein, lipid, and water content in the adult bar jack in southwestern Cuba.   188

Fig. 7.6. Seasonal relationships among habitats, coastal water temperatures, feeding intensity, and food composition of 0+ bar jack, and mesenteric fat content and growth for cohorts born in the first and second spawning peaks.   189

Fig. 7.7. Ontogenetic changes of the mesenteric fat index with size in male and female fishes of several species.   190

Fig. 7.8. Ontogenetic variation in annual growth increments, feeding intensity, food composition, and physiological and biochemical indicators in the bar jack.   191

Fig. 7.9. Seasonal and size-specific variations in the relative percentage of lipids, proteins, and water content in the muscle of redear sardine at the beginning and end of the reproductive season.   192

Fig. 8.1.   Summary of fisheries landing data from the Cuban Exclusive Economic Zone, 1959–1998.   195

Fig. 8.2.   Composition of average catch from 1986–1990: country total and four statistical reporting zones.   197

Fig. 8.3.   Four steps in the deployment of a small trawl (*chinchorro de boliche*).   199

Fig. 8.4.   Important set net types used in four different areas of the Cuban shelf.   200

Fig. 8.5.   Important trap types used in four different areas of the Cuban shelf.   201

Fig. 8.6.   Types of longlines used in billfish, marlin, swordfish, and shark fisheries.   202

Fig. 8.7.   Location of fishing associations and their components.   203

Fig. 8.8.   Schematic distribution of fishing activities, habitats, and life stages of fishes in several regions of the Cuban shelf.   204

Fig. 8.9.   Regional and total catches for all finfishes in Cuba, 1959–1998.   205

Fig. 8.10.   Regional and total catches of lane snapper in Cuba, 1957–1998.   207

Fig. 8.11.   Regional and total catches of mutton snapper in Cuba, 1962–1998.   207

Fig. 8.12.   Regional and total catches of gray and cubera snappers in Cuba, 1962–1998.   208

Fig. 8.13.   Regional and total catches of yellowtail snapper in Cuba, 1962–1998.   208

Fig. 8.14.   Regional and total catches of Nassau grouper in Cuba, 1962–1998.   209

Fig. 8.15.   Regional and total catches of grunts in Cuba, 1962–1998.   210

Fig. 8.16.   Regional and total catches of mojarras in Cuba, 1959–1998.   211

Fig. 8.17.   Regional and total catches of mullets in Cuba, 1959–1998.   212

Fig. 8.18.   Regional and total catches of jacks in Cuba, 1959–1998.   212

Fig. 8.19.   Regional and total catches of skipjack and blackfin tunas in Cuba, 1959–1998.   213

Fig. 8.20.   Regional and total catches of mackerels in Cuba, 1959–1998.   214

Fig. 8.21.   Regional and total catches of swordfishes and billfishes in Cuba, 1959–1998.   214

Fig. 8.22.   Regional and total catches of sharks in Cuba, 1959–1998.   215

TABLES AND APPENDIXES

Table 2.1. Number of species, diversity index, richness index, evenness index, density, and fish biomass of different reef habitats and regions   23

Table 2.2. Abundant fishes and the F/a−b index for various habitats in different regions   24

Table 2.3. Fish assemblages on artificial reefs at several locations in the Golfo de Batabanó   29

Table 3.1. Behavioral categories of selected coastal fishes in Cuba   59

Table 4.1. Hermaphroditic species occurring on the Cuban shelf   74

Table 4.2. Description of the gonad stages used in this book   81

Table 4.3. Microscopic characteristics of the different phases of fish oocyte development   84

Table 4.4. Gonadosomatic index of ripe females of different species of the Cuban ichthyofauna   91

Table 4.5. Fecundity of fishes in gonad Stage IV of several species with asynchronous vitellogenesis and multi-batch spawning (Type B)   98

Table 5.1. Primary food items of 365 fish species studied in Cuba   116

Table 5.2. Interannual variations in the diets of four common fish species of the Cuban shelf   120

Table 5.3. Relationships between predator and prey sizes in some Cuban marine fishes   131

Table 5.4. Trophic structure of fish communities in different habitats and regions   133

Table 5.5. Trophic composition of the ichthyofauna of different regions   134

Table 5.6. Biological characteristics of predatory fishes of different latitudes   134

Table 6.1. Length and growth parameter estimates for the lane snapper, *Lutjanus synagris,* by different authors   156

Table 6.2. Length and growth parameter estimates for the mutton snapper, *Lutjanus analis,* by different authors   158

Table 6.3. Length and growth parameter estimates for the gray snapper, *Lutjanus griseus,* by different authors   160

Table 6.4. Relationships between diet composition and growth of the gray snapper, *Lutjanus griseus,* in different latitudes   160

Table 6.5. Length and growth parameter estimates for the yellowtail snapper, *Ocyurus chrysurus,* by different authors   161

Table 6.6. Length and growth parameter estimates for three grunt species of various regions by different authors   163

Table 7.1. Physiological and biochemical indicators in fe male bar jack, *Caranx ruber,* at two stages of sexual development   180

Table 7.2. Gonadosomatic index, condition factor, fat content in the body cavity or mesenteric fat index, and the percentage of lipids in muscles of adult bar jack during the reproductive cycle   182

Table 7.3. Protein, lipid, and water content in ovaries and testes of adult bar jack, *Caranx ruber,* and margate, *Haemulon album,* at different gonadal stages   183

Table 7.4. Composition of ripe ovaries of lane snapper, *Lutjanus synagris,* bar jack, *Caranx ruber,* and margate, *Haemulon album,* during the reproductive season   185

Table 8.1. Mean annual catches and proportion of total catch for major groups of marine organisms in the Cuban Exclusive Economic Zone during three five-year periods   196

Table 8.2. Mean annual catches and proportion of total catch for major finfish groups during three five-year periods   206

Table 8.3. Composition of shrimp trawl bycatch in the Golfo de Guacanayabo and Golfo de Ana Maria   216

Appendix 2.1. Fish species recorded from Cuba.   33

Appendix 4.1. Sizes at sexual differentiation and maturation of some marine fishes inhabiting Cuban waters.   103

Appendix 4.2. Types of vitellogenesis, spawning characteristics, and duration in some fish species inhabiting the Cuban shelf.   107

Appendix 4.3. Spawning seasonality of coastal fish species from studies in Cuba.   109

Appendix 4.4. Individual fecundity and relative fecundity of some marine fishes inhabiting Cuban waters.   112

Appendix 5.1. Feeding habits and main dietary components of most marine fish species occurring in Cuba.   136

Appendix 5.2. Dietary patterns of marine fishes studied in Cuba.   145

Appendix 6.1. Length parameters that can be used to estimate weight from growth studies of marine fishes of Cuba.   167

Appendix 6.2. Growth parameters of some fishes that occur on the Cuban shelf.   173

PREFACE

For many people interested in the ecology, biogeography, evolution, and management of tropical island faunas, Cuba represents a fundamental gap in our understanding. For those specializing in the Caribbean, knowledge of the Cuban fauna is crucial because of its unparalleled size, positioning, and biodiversity. Cuba is of particular interest to fishery researchers and managers because its massive reef tracts–several of which equal or exceed the Florida Keys in length—are spawning areas for many important fish species. Knowledge and informed management of the Cuban marine fish fauna may be of as much scientific value to its neighbors as to Cuba itself, due to its close proximity to both North and Central America.

From the early 1970s through the mid-1980s, Cuban ichthyologists collaborated extensively with colleagues from the former Soviet Union and elsewhere to compile and analyze a wealth of information on the ecology of marine fishes from the southwest and northwest coasts of Cuba. Subsequently, the Russian book *Ekologia pib Shelfa Kubi* (Sokolov and García 1985) summarized many results obtained through 1984. The Cuban team continued working in the same regions and also on the north-central shelf of Cuba (the Archipiélago Sabana-Camagüey).

Much of the new and earlier work was summarized relative to the broader Caribbean literature in *Ecología de los Peces Marinos de Cuba* (1994) published in Spanish by the Centro de Investigaciones de Quintana Roo and the Instituto de Oceanología de Cuba. Rodolfo Claro, Institute of Oceanology, Cuban Ministry of Science, Technology and Environment, edited that book and authored or coauthored seven of its eight chapters. The book contained more than 150 pages of detailed tables summarizing ecological parameters for important species occurring in Cuba and the wider Caribbean and was "by far the most comprehensive guide to the ecology of marine fishes of Cuba" (Lindeman 1996). Another unique aspect was the perspective brought to the collection and analysis of data by authors largely trained in the ichthyology program of G. V. Nikolsky at the Institute of Evolutionary Morphology and Animal Ecology in Moscow, a program steeped in the rich European ichthyological literature. Many concepts and methods were inspired by well-known Russian ichthyologists such as Nikolsky, D. V. Radakov, B. P. Manteifel, and others. This program emphasized hands-on data collection at the level of the organism. An extremely logical approach was taken: specimens were simultaneously processed by a team of

workers, each focusing on a particular demographic attribute (growth, reproduction, feeding). Many of the resulting databases are unparalled for the region, or, in combination with extensive data from Jamaica (Munro 1983d), constitute a still underexploited opportunity to compare island to island variations in key demographic parameters.

Unfortunately, distribution of most of the primary Cuban journal literature and the book itself was limited in North America and many other regions. In addition, production of this work involved significant obstacles. Most of the text was composed without word processors. Original manuscripts were not intensively copyedited. The most basic tools (e.g., paper, pencils, and drafting supplies) were of poor quality or unavailable. Many calculations were done by hand or using hand calculators and graphics were hand drawn. Correspondence with the publisher and with colleagues was slow and erratic, as was access to research literature. Independent technical review of the manuscript was lacking. These problems explain the extensive typographical errors, repetitious passages, and layout problems throughout the 1994 book. These and other problems that were beyond the authors' control detracted from the book's utility, but not from its scientific value.

Because of the potential value of this earlier work to both English- and Spanish-speaking researchers, the incomplete status of its original production, and the explosion of new literature in the 1990s, Claro, in consultation with Ken Lindeman and Lynne Parenti, initiated this revised and updated English version. The Cuban authors of these chapters have generated many recent references cited within this new book. In addition, this book has been updated with more than 350 new English and Spanish references from both the most recent and older literatures. Retaining the original approaches and the Cuban perspective was paramount in interpreting issues of either substance or style. The majority of text additions were made by the Cuban authors. All deletions from the original Spanish version were approved by the Cuban authors and typically were suggested by them. The literature treatment is not absolutely comprehensive and the feasibility of achieving such a goal across the breadth of topics within this book is questionable. Yet, we know of no books that summarize new and earlier components of the Spanish and English literature across such a wide diversity of topics related to reef fish biology. We hope this work will serve to introduce workers of both languages to the rich ideas and scientific traditions within each.

The original text was translated from Spanish to English by Georgina Bustamante of The Nature Conservancy. She was a member of the original team in Cuba and is senior author of the chapter on ecophysiology. Juan Pablo García-Arteaga, Alida García-Cagide, Emilio Valdes, Julio Baisre, and Luis Sierra, also members of the original team in Cuba, provided much new information and editorial assistance during preparation of this volume.

John Munro (International Center for Living Aquatic Marine Resources) reviewed the entire book draft and R. Grant Gilmore, Jr. (Dynamac Corporation), reviewed the first four chapters. Their extremely detailed reviews greatly contributed to the final version. A variety of experts reviewed individual chapters: Tom Lee and Doug Wilson (Chapter 1), Anne-Marie Eklund and L. Allan Collins (Chapter 4), Doug Weaver (Chapter 5), Charles Manooch, Mike Burton, and Jennifer Potts (Chapter 6), and Rich Appeldoorn (Chapter 8). Jeffrey Clayton, National Museum of Natural History (NMNH), provided first-class technical and editorial assistance with Appendix 2.1. William D. Anderson, Jr. (College of Charleston), Karsten Hartel (Museum of Comparative Zoology, Harvard University), Carole C. Baldwin, David G. Smith, and Jeffrey T. Williams (all NMNH), Bruce B. Collette and Tom Munroe, National Marine Fisheries Service (NMFS), C. Richard Robins (University of Kansas), William F. Smith-Vaniz (U.S. Geological Survey), Joseph S. Nelson (University of Alberta), and William N. Eschmeyer and John McCosker (both California Academy of Sciences) reviewed and provided valuable technical comments on Appendix 2.1.

Vince Burke, Anne Bolen, Peter Cannell, Nicole Sloan, and many other staff at the Smithsonian Institution Press patiently guided us through the manuscript preparation and publishing details. Fran Aitkens copyedited the text with remarkable skill and patience. Michael L. Smith (Conservation International), then with the Center for Marine Conservation (CMC), organized a joint Cuban Academy of Sciences–National Museum of Natural History field expedition in 1994 that accumulated a large reference collection of Cuban shorefishes now housed at the NMNH in Washington, D.C. That field expedition is one of many Cuban and U.S. scientific exchanges Michael Smith catalyzed.

Production of this book was supported by two generous grants from the Smithsonian Institution's Atherton Seidell Fund, and additional support from the office of Ross B. Simons, National Museum of Natural History. Coastal Research and Education, Inc., Miami, Florida, financially assisted many components of the project over its six-year duration, as did Environmental Defense, New York, during the final 18 months. Financial assistance was also provided by the Bacardi Family Foundation, Arlington, Virginia, and Jay Venable, St. Petersburg, Florida.

Bruce Collette (NMFS), Susan L. Jewett and Marsha Sitnik (both NMNH) contributed useful ideas and encouragement as the translation and editing project was initially conceived. Tina Ramoy located information for us with her usual aplomb. The following interns or contractors at Coastal Research and Education also aided this project: Calan Leyendecker, Helena Molina, Molly Munro, Jennifer Somers, and Monica Valle. Fernando Bretos (CMC) assisted with logistical questions. Doug Rader, Joe Roumelis, and Jocelyn Karazsia (Environmental Defense) provided well-timed support.

The background literatures that fuel our research should be shared. This product demonstrates the value of increased collaboration between Cuban and U.S. scientists, and on a larger scale, between researchers from different cultures and of different languages. Many scientific questions and management problems involving northern Caribbean marine ecosystems will not be resolved without increased collaboration among Cuban and North American researchers.

Rodolfo Claro
Kenyon C. Lindeman
Lynne R. Parenti

CONTRIBUTORS

Pedro M. Alcolado, Instituto de Oceanología, CITMA, Avenida 1ra., No. 18406, Playa, La Habana, Cuba.

Julio A. Baisre, Ministerio de la Industria Pesquera, Barlovento, Sante Fe 19 100, Playa, La Habana, Cuba.

Georgina Bustamante, The Nature Conservancy, 4245 North Fairfax Drive, Suite 100, Arlington, Virginia, USA 22203 (formerly of the Instituto de Oceanología, La Habana, Cuba).

Rodolfo Claro, Instituto de Oceanología, CITMA, Avenida 1ra., No. 18406, Playa, La Habana, Cuba.

Juan P. García-Arteaga, Instituto de Oceanología, CITMA, Avenida 1ra., No. 18406, Playa, La Habana, Cuba.

Alida García-Cagide, Instituto de Oceanología, CITMA, Avenida 1ra., No. 18406, Playa, La Habana, Cuba.

Boris V. Koshelev, A. N. Severtzov Institute of Ecology and Evolution, Russian Academy of Sciences, Moscow, Russia (deceased).

Kenyon C. Lindeman, Environmental Defense, 14630 SW 144 Terrace, Miami, Florida, USA 33186.

Andrei D. Mochek, A. N. Severtzov Institute of Ecology and Evolution, Russian Academy of Sciences, Leninsky Prospect, 33, 117071 Moscow, Russia.

Lynne R. Parenti, Department of Vertebrate Zoology, National Museum of Natural History, Smithsonian Institution, 10th and Constitution Avenue NW, Washington, D.C., USA 20560-0109.

Olga A. Popova, A. N. Severtzov Institute of Ecology and Evolution, Russian Academy of Sciences, Leninsky Prospect, 33, 117071 Moscow, Russia.

Yuri S. Reshetnikov, A. N. Severtzov Institute of Ecology and Evolution, Russian Academy of Sciences, Leninsky Prospect, 33, 117071 Moscow, Russia.

Mijail I. Shatunovsky, A. N. Severtzov Institute of Ecology and Evolution, Russian Academy of Sciences, Leninsky Prospect, 33, 117071 Moscow, Russia.

Luis M. Sierra, Estacion de Biologia Marina, Escuela de Ciencias Biologicas, Universidad Nacional, Apdo Postal 126, Puntarenas, Costa Rica (formerly of the Instituto de Oceanología, La Habana, Cuba).

Emilio Valdés-Muñoz, Instituto de Oceanología, CITMA, Avenida 1ra., No. 18406, Playa, La Habana, Cuba.

1

Physical Attributes of Coastal Cuba

RODOLFO CLARO, YURI S. RESHETNIKOV, AND PEDRO M. ALCOLADO

1.1 Geographic Characteristics

Cuba is situated on the Tropic of Cancer, between 23° 10′ 34″ and 19° 9′ 32″ N, and between 74° 7′ 55″ and 84° 57′ 11″ W. The Republic of Cuba consists of the main island and many smaller islands, including Isla de la Juventud and 4,195 islets and keys that range in area from a few hundred square meters to approximately 150 km² (Núñez Jiménez 1982). All the islands are located on the insular shelf, many of them at the edge (Fig. 1.1). The main island is about 1,200 km long. The maximum width is 191 km in the eastern part, the minimum is 31 km between Bahía de Mariel and the Golfo de Batabanó (Academia de Ciencias de Cuba and Academia de Ciencias de la URSS 1970). The area of the main island is 105,007 km²; the total for the Republic is 110,922 km² if keys and islets are included (Academia de Ciencias de Cuba 1965). Coastal mangroves and mangrove islands occupy about 6% of the overall territory (Ionin et al. 1977). Cuba is surrounded by the deep basins and trenches of the Caribbean Sea, the Gulf of Mexico, the Straits of Florida, and the Bahamas.

The coastline of the main island is 5,746 km long. The surrounding shelf resembles a submerged plain with an area of 67,831 km² (Núñez Jiménez 1982). Most of the shelf is bordered by extensive reefs. The shelf drop-off is near the outer border of the keys and fringing reefs and drops steeply to 400 m or more (Fig. 1.2). Observations from the research submersibles *Argus* and *Johnson Sea-Link II* show that along the entire northwest coast, from Bahía de Matanzas to Cabo San Antonio and then to Punta de Maisí along the south coast, an almost vertical slope (70–90 degrees) extends from the shelf drop-off (at 20–50 m) to 100–150 m. Beyond this incline is a second, less steep incline (60–70 degrees) that continues with little change for more than 600 m. These drop-offs constitute the geographic limits of Cuba and are evidence that Pleistocene sea levels were 120–150 m lower than at present.

According to Ionin et al. (1977) the Cuban shelf has several distinctive features that distinguish it from a typical platform: (1) it is a shallow insular shelf; the largest region, the Golfo de Batabanó, averages only 6–7 m deep; (2) near the shelf-edge are lines of fringing keys and reefs; and (3) the shelf edge consists of a series of narrow submerged terraces and the outer border is an almost vertical slope. From a geomorphological perspective, the platform can be divided into four areas of wide shelves separated by areas of narrow shelves (Fig. 1.1). Two wide-shelf areas occur on the north coast (the Archipiélago Sabana-Camagüey and a smaller area on the northwest coast, the Archipiélago Los Colorados) and two on the south coast (the Golfo de Batabanó and the area including Golfo de Ana María and Golfo de Guacanayabo).

Along the northwest coast, from Cabo San Antonio to Punta Hicacos (Fig. 1.3), the shelf is widest from Cabo San Antonio to Bahía Honda, an area of approximately 3949 km². The Archipiélago Los Colorados, formed by approximately 160 keys and islets, extends for 225 km. A reef of the same name fringes the shelf and Banco de Sancho Pardo. Oceanic pelagic bony fishes, sharks, and some reef fishes are the major fisheries resources of this region. The easternmost part of the Archipiélago Los Colorados differs from the Golfo de Guanahacabibes by having a more complex bottom relief, more embayments and keys fringed by mangroves, and more numerous reef crests. This shelf region is relatively broad at the Golfo de Guanahacabibes, and gradually narrows from Cayo Buenavista toward the east. The narrowest part of the northeast shelf has a submerged terrace of complex relief that stretches eastward from Bahía Honda to Punta Hicacos (Ionin et al. 1977) (Fig. 1.3). This narrow shelf area includes several large bays, including Bahía de la Habana.

The Archipiélago Sabana-Camagüey extends along the north-central coast from Punta Hicacos to Cayo Sabinal (Fig. 1.4). The archipelago ranges from 6–35 km wide and has an area of 10,115 km². There are several macrolagoon systems with many islands, islets, and keys (2,517 in total). Some islands are large, such as Cayo Coco, Cayo Romano, and Cayo Sabinal, each more than 100 km² in area

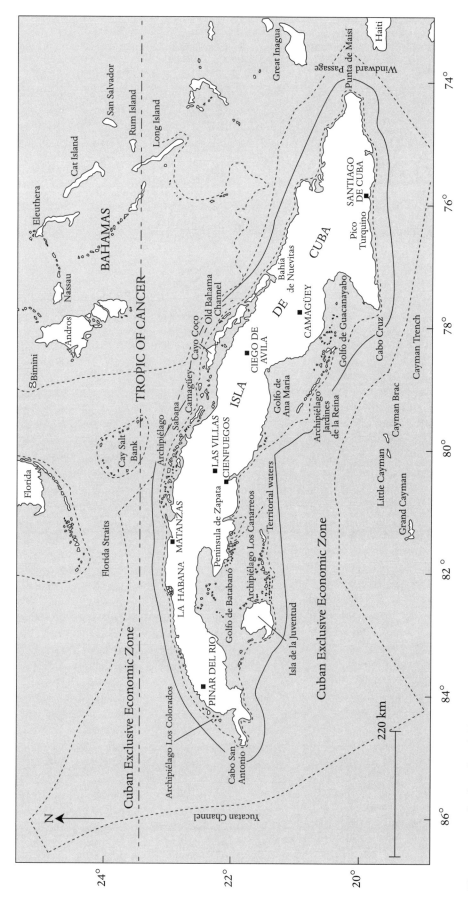

Fig. 1.1. Map of Cuba. The solid line represents the boundary of territorial waters, the inner dashed line is the shelf boundary, and the outer dashed line is the boundary of the Cuban Exclusive Economic Zone.

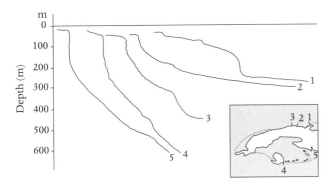

Fig. 1.2. Bathymetric profiles from the shore to the outer shelf at different areas of the western Cuban shelf. 1: Punta Hicacos; 2: Santa Cruz del Norte; 3: Playas del Este, Habana; 4: Punta del Este, Isla de la Juventud; 5: Golfo de Cazones.

(Núñez Jiménez 1982). The keys fringe the shelf for more than 460 km, forming a series of lagoons that includes Bahía de Cárdenas, Bahía Santa Clara, Bahía de Sagua, Bahía de Nazabal, Bahía de San Juan de los Remedios, Bahía de Buenavista, Bahía Los Perros, Bahía de Jigüey, and Bahía de La Gloria (Fig. 1.4). These bays have a maximum depth of 2–3 m and many shoals are emergent at low tide; therefore only small boats can navigate this region. Water exchange with the ocean is limited and terrestrial runoff is low because few major rivers, except for Río Sagua la Grande near La Isabella, are present and many others have been dammed. North of the keys, a narrow belt of rocky bottom with sandy areas, sparse seagrass beds, and numerous reef crests and fore reefs extends to the shelf edge.

The sections from Cayo Sabinal to Punta de Maisí, along the northeast coast, and from Punta de Maisí to Cabo Cruz along the southeast coast, lack substantial shelves (Figs. 1.1 and 1.4). Coastal land is of tectonic origin with many submarine terraces along the entire coastline (Ionin et al. 1977) and large bays of erosive origin. Estuarine regions are present in some areas. The shelf edge is located close to the coast, just a few hundred meters offshore in some places. Off southeastern Cuba from Punta de Maisí to Cabo Cruz, a steep slope that reach a maximum depth of 7,041 m delineates the northern border of the Cayman Trench, the second deepest trench in the Atlantic Ocean. The total relief from Pico Turquino, near the southeast Cuban coast, to the bottom of the adjacent Cayman Trench exceeds the height of Mount Everest.

From Cabo Cruz to Casilda is a long, wide shelf area fringed by the Archipiélago Jardines de la Reina, which encompasses the Golfo de Ana María and Golfo de Guacanayabo. These two gulfs are partly separated by a shallow area of many keys and shoals (Fig. 1.5). This wide-shelf region has an area of 17,992 km² and an average depth of 15 m (Emilson and Tápanes 1971). The greatest depths (to 28 m) of inshore waters of the Cuban shelf are

located in these two gulfs. The gulfs are bordered by or include more than 840 islands, islets, and keys (Núñez Jiménez 1982).

The Golfo de Guacanayabo is divided into two interior basins by a shallow and relatively large bank, the Gran Banco de Buena Esperanza. Unlike other wide-shelf areas, the outer border is not rimmed by keys, sand banks, and reefs, but by a steep slope at the shelf edge. The Golfo de Ana María is delineated by an array of keys and reef crests that separate it from the open sea (Fig. 1.5). Along the coastal margin, both gulfs are fringed extensively by lagoons and estuaries fed by several rivers. Shrimp (*Penaeus schmitti* and *P. notialis*) are the economic basis of coastal fisheries in this region (see Pérez Farfante and Kensley 1997). High catches of estuarine and reef fishes are also obtained here.

The coast from the Golfo de Batabanó to Golfo de Ana María has no substantial shelf area (Fig. 1.1). The Golfo de Batabanó, the largest submerged area of the Cuban shelf, is 90–140 km wide and has an area of approximately 20,870 km² (Fig. 1.3). Depths reach 15 m in channels, but typically average 3–6 m. This massive lagoon is fringed by or includes many hundreds of islands, keys, and islets that make up the Archipiélago Los Canarreos to the east of Isla de la Juventud and Cayo Los Indios and Cayo San Felipe to the west.

The Golfo de Batabanó is also divided by a series of keys and shoals that stretch from the mainland to the Isla de la Juventud (Fig. 1.3). The gulf is bordered on the east by a series of large reef crests that do not fully impede high-volume exchanges with the oceanic waters of the deep Golfo de Cazones. The Golfo de Batabanó is Cuba's major lobster fishery ground. Some demersal fishes of this area, such as grunts, snappers, and porgies and a few shallow-water pelagic species, such as sardines and jacks, also constitute important fishery resources. To the west, between the Golfo de Batabanó and Cabo San Antonio, the shelf is narrow (Fig. 1.3).

1.2 Shelf Geology and Geomorphology

1.2.1 Structure, Origin, and Evolution

Cuba is a continental island for which various hypotheses of geological origin and evolution have been proposed. According to Corral (1940), Cuba was once connected to South America. Dietz and Holden (1970) support the theory that the bedrock of Cuba was derived from a region in the Gulf of Mexico, whereas Malfait and Dinkelman (1972) propose that it was once farther south as part of Central America. Our understanding of the geological history of the Caribbean Plate, Cuba, and the rest of the Greater Antilles grows with each revision of plate tectonics. The current consensus is that most of Cuba was once part of Central America, although eastern Cuba is of Cenozoic tectonic origin. The following summary is based on the works of

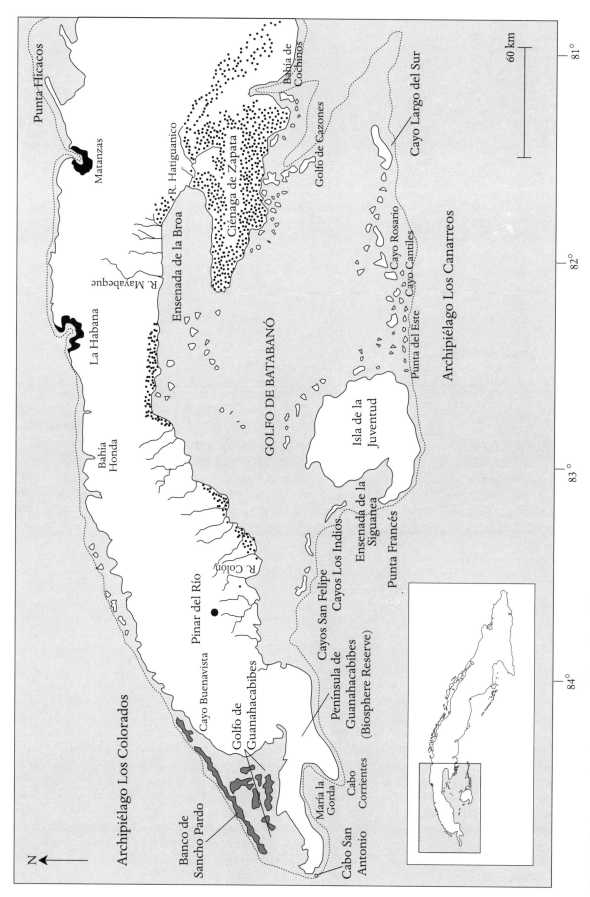

Fig. 1.3. Northwest and southwest areas of the Cuban shelf. The shelf boundary is represented by a dashed line.

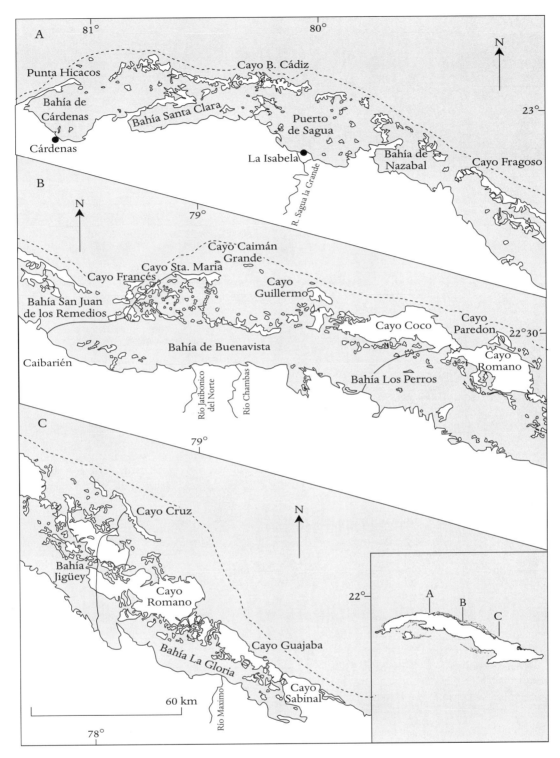

Fig. 1.4. North-central shelf of Cuba (Archipiélago Sabana-Camagüey). The lines across the lagoons in panels B and C are recently constructed causeways. The shelf boundary is represented by a dashed line.

many authors (e.g., Furrazola-Bermúdez et al. 1964; Meyerhoff and Hatten 1968; Iturralde-Vinent 1972, 1975, 1977, 1981, 1982; Rosen 1975, 1985; Sykes et al. 1982; Durham 1985; Lewis et al. 1990; Donnelly 1988; Briggs 1995; and Roughgarden 1995).

The North American and South American tectonic plates began separating during the late Jurassic and moved westward relative to the Caribbean Plate, which passed between them with Cuba on its leading edge, moving relatively northeastward. The proto–Greater Antilles formed

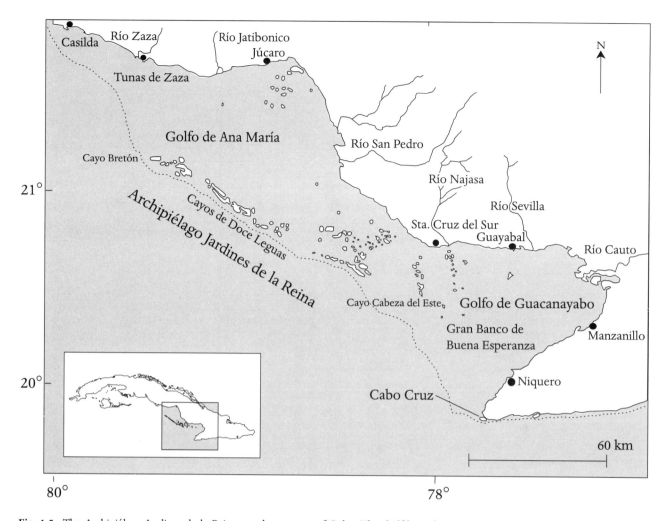

Fig. 1.5. The Archipiélago Jardines de la Reina, southeast coast of Cuba. The shelf boundary is represented by a dashed line.

in the eastern Pacific during the early Cretaceous; joining with lithosphere, it became part of the Caribbean Plate. Cuba has occupied its present position relative to the Bahamas since the Eocene when the leading edge of the Caribbean Plate collided with the Bahama Bank, which is part of the North American Plate. During the Eocene, the relative movement of the Caribbean Plate changed from northeastward to eastward, such that Cuba occupied the northern margin of the plate and the Lesser Antilles occupied the leading edge (Fig. 2 in Durham 1985). After collision with the Bahama Bank, Cuba comprised an archipelago of uplifted blocks. Continued uplift during the Miocene resulted in the modern outline of the Cuban archipelago by early Pliocene times.

Biogeographic data suggest a correspondingly old, continental origin for the terrestrial Cuban biota that supports a relatively high number of endemic genera (i.e., Rosen 1975; Roughgarden 1995). Iturralde-Vinent (1982) argues that some marine habitats might have formed from ancestral habitats situated within the limits of proto-Cuba during the Eocene, whereas other biotopes are

undoubtedly much older. Tropical seagrasses were worldwide in distribution by the Cretaceous according to den Hartog (1970), and it is generally accepted that corals, mangroves, and seagrasses were associated by the Cretaceous (McCoy and Heck 1976; Phillips and Menez 1988; Veron 1995). Complex, large-scale movements of the Caribbean Plate have affected underlying biogeographic patterns throughout the region (Rosen 1985). In the western Pacific, marine shorefish distributions have been shown to coincide with plate boundaries (Springer 1982). Our understanding of the biogeography of Caribbean marine fishes will benefit from the identification of areas of endemism of marine taxa with consideration of plate boundaries, and the inference of historical relationships among such areas from the phylogenetic history of the taxa (e.g., Rosen 1985; Humphries and Parenti 1999). One might predict that a hierarchy or series of the biogeographic patterns as inferred from both freshwater and marine fishes will be correlated with a hierarchy or series of area relationships throughout geological time (Rosen 1985). In addition, many paleoclimatic or ecological factors operating over more recent time scales might have substantially influenced present-day distributions of fishes of Cuba and the northern Caribbean (Robins 1971; Colin 1973; Gilmore 1995).

1.2.2 Sediments

Knowledge of sediment types and distributions is critical to understanding coastal fish ecology because sediments influence the habitat landscape available to organisms across the shelf. Sediments from the Cuban shelf are composed of complicated facies caused by the intervention of various factors during formation. The three main factors are chemical formation of calcium carbonate in sea water, biological production of carbonate-reducing organisms whose skeletons are sometimes the main component of the sediments, and terrestrial runoff. These processes can occur separately or in combination, which can result in substantial heterogeneity in the distribution of skeletal remains. It is possible to find areas with a high percentage of biocomponents (> 50%), whereas nearby areas have a low organic content. Sedimentary characteristics of the four primary regions of the Cuban shelf are summarized below, after Ionin et al. (1977).

The Golfo de Guanahacabibes, in the northwest region, is divided into three main sediment subzones (Fig. 1.6): (1) inshore areas are low in carbonate sediments and have pure aleuritic and clay mud with mollusk valves, partly derived from terrestrial runoff and partly from organic matter from littoral vegetation; (2) the middle gulf is largely carbonate biogenic sediments composed of muddy sand, including abundant remains of calcareous algae (*Halimeda*) and mollusk valves; and (3) the outer areas near the coral reefs that fringe the gulf have a strip of carbonate debris and coralline conch sand formed by storm waves pounding the

reefs. A more detailed description can be found in Ionin et al. (1977).

The narrow shelf area extending from the Golfo de Guanahacabibes eastward is covered typically by sandy mud. The proportion of mud increases toward the shore, whereas sand predominates near reefs. The section from Bahía Honda to Matanzas along the northwest coast has practically no shelf, but rather a series of submerged terraces with predominately rock and coral bottoms, with sand patches.

On the north-central coast, the outer shelf that stretches from La Isabela to Cayo Coco (Fig. 1.4) is covered by coarse carbonate sand, but toward the shore, the bottom is muddy sand (Fig. 1.6). The region including the bays of Buenavista, Los Perros, Jigüey, and La Gloria constitutes an almost closed sedimentary basin with little exchange with the open sea. The bottom is covered by a relatively thick layer (0.8–1.0 m) of aleurite with mollusk shells and large amounts of sulfur compounds (Ionin et al. 1977; Avello 1979).

In the Golfo de Batabanó on the southwest coast, five main sedimentary regions can be identified (Fig. 1.6): (1) the eastern region, with dominant oolitic sand and high levels of sedimentation; (2) the central region, with muddy sand carbonate sediments, including biogenic components; (3) the western region, with a complex of muddy-sand carbonate sediments of terrestrial origin; (4) Ensenada de la Broa, where a slow process of clay carbonate sedimentation is occurring; and (5) the northern and northwestern part of Isla de la Juventud with a fringe of terrestrial silicate sediments that stretches up the Ensenada Siguanea.

In the southeast region, in the Golfo de Ana María, the sediment layer is notably uniform (Fig. 1.6). Gray muddy sand, typically containing mollusk valves, covers most of the central area. Farther from shore, the mud is lighter in color because of increased calcium carbonate content. Nearshore,

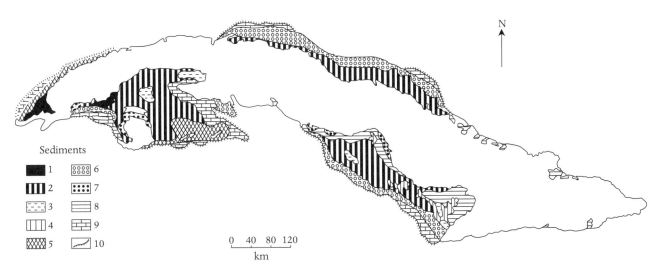

Fig. 1.6. Distribution of sediments on the Cuban shelf (from Avello 1979). 1: sand-aleuritic *Halimeda* mud; 2: sand-aleuritic detrital mud; 3: carbonate mud with mollusks; 4: sand-mud detrital sediments; 5: oolitic sand with foraminifera, 6: mollusk-shell detritic sand; 7: terrestrial sands; 8: terrestrial clay mud; 9: lack of unconsolidated sediments; 10: submerged terrace at the shelf drop-off.

off the mouth of the Zaza and Manatí rivers, are fine-grained sediments (aleuritic and clay). The outer part of the shelf is typically covered by organogenic sands (with fragments of coral, mollusk, and other organic remains). The shallow water area that joins the gulfs of Ana María and Guacanayabo has many keys, and the bottom consists of aleuritic-clay bluish-gray mud that contains small mollusks. The Golfo de Guacanayabo is mostly covered by an aleuritic-clay gray or greenish-gray mud, which is lighter in color at the river mouths. The outer border of the shelf typically has sandy organogenic sediments containing coral, mollusk, and *Halimeda* fragments.

1.3   Climate

The Cuban climate can be generally classified as grassland-wet, semicontinental with low-intensity winters (Núñez Jiménez 1965). The two main seasons are warm-wet (May to October) and cool-dry (November to April). The warm water masses that surround the Cuban archipelago contribute heat and humidity. The anticyclonic area over the North Atlantic is a major factor that influences the seasonal fluctuation of climate, generating high temperatures. The expansion and contraction of the anticyclonic area determines the extent of temperature and humidity changes. Summer weather is influenced by the upper stream of equatorial or tropical neritic, wet, hot, air blowing from east to west (Núñez Jiménez 1965). During winter, north and northwest winds influence the weather with continental highs and cold fronts. Because of the proximity to North America, low temperatures can occur during the winter.

Two main winds influence the climate of Cuba: trade winds and mainland breezes. Cuba is located in the zone of the northeast trade winds. Trade winds blow throughout the year from high-pressure to low-pressure areas or equatorial doldrums. The circulation of these winds can be interrupted by local atmospheric systems that generate breezes and terrestrial winds that occur most intensively near the coast and whose influence can be substantial. On the north coast, daytime breezes appear as northeast winds, reinforcing the trade winds; on the south coast, daytime breezes blow from the southeast. Nighttime mainland breezes are weaker than those in the daytime. On the northern coast these breezes occur as southeast winds and on the south coast as northeast winds (Núñez Jiménez 1965).

Annual rainfall has two clearly defined seasons: the cool-dry season from November to April with an average rainfall of 316 mm and the warm-wet season from May to October with rainfall reaching 1,059 mm. Many factors influence the geographic distribution of precipitation during dry and wet seasons. Hurricanes (with winds up to 300 km/hour) and changes in direction and velocity of the prevailing trade winds are among the most important factors. Hurricanes are most frequent in September and October (Academia de Ciencias de Cuba and Academia de Ciencias de la URSS

1970). Hurricanes can influence the long-term structural integrity of many habitats used by fishes in the northern Caribbean (Woodley 1992; Lirman and Fong 1995; Wulff 1995).

1.4   Hydrological Characteristics

Considerable information is available on the hydrology of the Cuban shelf and its oceanic waters (Suárez-Caabro and Duarte-Bello 1962; Emilson 1968; Kabanova et al. 1968; Elizarov and Gómez 1971; Tápanes 1972a, 1972b; Ramírez 1974; Gómez 1979; García Díaz 1981; Blázquez-Echandi and Romeu 1982; Victoria del Río et al. 1997; Victoria del Río and Penié 1998). Fine- and meso-scale current dynamics are influenced by many hydrological and meteorological factors that vary spatially and temporally. Mean flows off the south and north coasts are westerly (Fig. 1.7), but the component current systems are complex, with considerable variations in flow directions. Oceanic circulation off Cuba's southwest coast is largely influenced by interactions between the westerly Caribbean Current and the easterly Cuban Counter-current (Sukhovey et al. 1980). This countercurrent, originating in the Yucatán Straits, penetrates the Cayman Sea in an approximate east-southeast direction and has been detected at a longitude of 81.5°W (at the westernmost margin of the Cayman Islands).

Our summary primarily follows the work of Victoria del Río and Penié (1998). These authors found evidence of a complicated system of cyclonic and anticyclonic circulation, variable in location and intensity, in this area. At the southeast end of Isla de la Juventud, an anticyclonic eddy has been observed. To the west of Isla de la Juventud, south of Cayos San Felipe, there is evidence of cyclonic circulation and complex interactions with the Cuban Countercurrent (Fig. 1.7). Southeast of the Golfo de Cazones, flow directions at the Fosa de Jagua (Jagua Trough) are variable and depend on the intensity of currents from adjacent areas. Similar evidence was obtained by García Díaz et al. (1991).

Flow directions have recently been estimated for oceanic waters off southwestern Cuba based on satellite-tracked drifter paths. This information is from drifters drogued at depths of 10–20 m and represents a variety of potential recirculation mechanisms (Lindeman et al. 2001). Fig. 1.8 shows the tracks of four buoys that entered southwestern Cuban waters in 1998 and 1999. Buoy 23472 was released on Julian day 291/98 at 10.7°N 76.7°W. It made one cyclonic circle of the Panama-Colombia gyre in approximately 60 days and entered southwestern Cuban waters in March 1999. Circulation was then influenced by variable eddies with diameters of 10–100 km (Fig. 1.8). Buoy 15992 was released on 043/99 off southwestern Cuba. It stayed in 10- to 50-km, largely cyclonic eddies for approximately 30 days in February and March 1999 (Fig. 1.8). Buoy 30659 was released on 151/98 at 10.9°N 76.7°W. It moved immediately to the north and

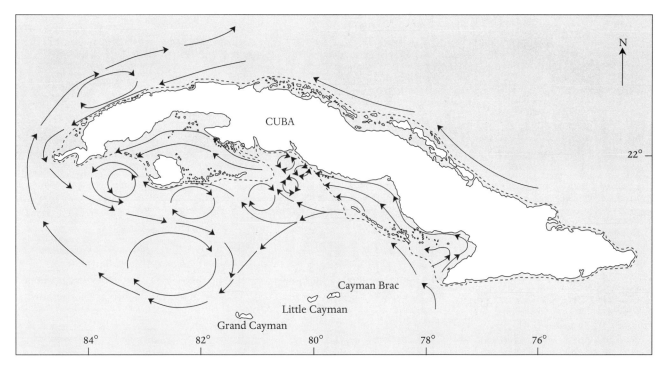

Fig. 1.7. Coarse schematic of large-scale oceanic circulation around Cuba (data from Victoria del Río and Penié 1998) and shelf circulation (data from Emilson and Tápanes 1971; Blázquez-Echandi and Romeu 1982).

entered the region from the west, south of Cabo San Antonio, in July 1998. It spent eight months in various eddies before leaving the region (Fig. 1.8). Buoy 30660 was released on 150/98 at 11.2°N 77.4°W. It made two cyclonic loops of the Panama–Colombia gyre before entering the region in October 1998. It made one large anticyclonic loop with smaller cyclonic eddies before exiting the region to the west in March (Fig. 1.8) and grounding in Belize in April 1999.

The occurrence of such a diverse system of eddies off the southern end of the Cuban archipelago is important for understanding fish recruitment processes because recirculation systems can provide favorable conditions for the retention of eggs, larvae, and nutrients. Many species spawn near the shelf edge, and eddies or gyres might foster self-recruitment of local populations by retaining larvae near spawning areas. Despite the presence of these retention mechanisms, the mean westerly flow and the sheer number of eggs and larvae produced along the hundreds of kilometers of outer reefs of the southern Cuban shelf must also contribute larvae to downstream areas in Florida. This is particularly applicable to lobsters and other species with larval periods of several months or more. However, a sizable anticyclonic system might also be present in the eastern part of the Yucatán Channel during periods of the year (Lindeman et al. 2001). Coupled with the short larval duration of many reef species, these recirculation systems off Cuba and similar systems off the Florida Keys (Lee and Williams 1999) provide alternatives to the hypothesis that long-distance dispersal

from southern Cuba to Florida is a primary recruitment pathway operating between the two regions.

Information on shelf circulation in Cuba is somewhat limited. Emilson and Tápanes (1971) proposed that circulation over the southern shelf was most influenced by tidal regimes, currents in the adjacent ocean, and winds. They also suggested that, because of low periodicity and height, tides do not directly affect long-term shelf circulation. Nevertheless, the effect of tidal currents on vertical mixing is significant. Because of the great extent of the southern lagoons and the narrow passages between many keys, currents can reach high velocities in some channels. Such currents might play an important role in the transport of fish and invertebrate eggs and larvae and, therefore, in the reproductive strategies and recruitment processes of commercial species.

A primary factor that influences water circulation on the southern shelf is wind. Shallow water depths insure that wind energy is distributed throughout the water column and variations in water movement correlate well with fluctuations in wind direction and velocity. Wind displaces water masses to the west at 3–5 nautical miles per day on average (Emilson and Tápanes 1971). Therefore, oceanic water that penetrates the eastern Golfo de Batabanó through the Golfo de Cazones might pass fully across the gulf in about 30 days. The circulation scheme proposed by Emilson and Tápanes (1971) has been supported by work using drift bottles (Blázquez-Echandi and Romeu 1982) and by current measurements (Blázquez-Echandi et al. 1988).

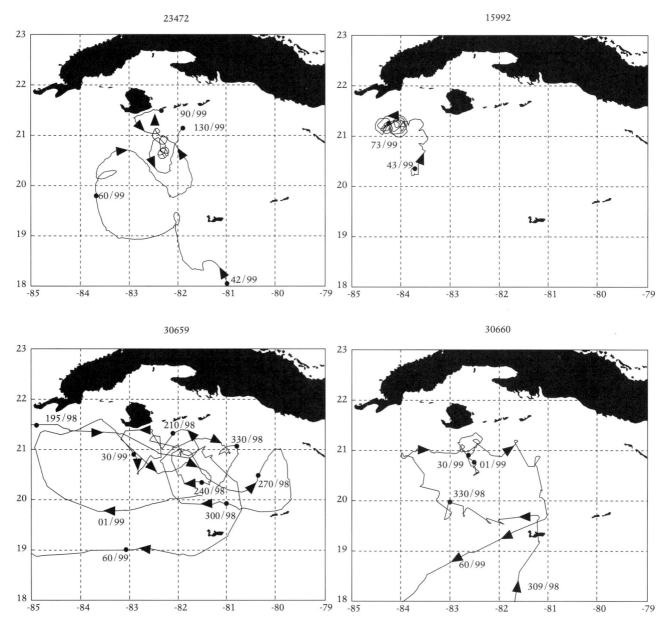

Fig. 1.8. Variable flow directions off southwestern Cuba, measured by satellite-tracked drifters drogued at depths of 10 to 20 m. The numbers above each panel are the individual buoy numbers; numbers on the plots are Julian calendar days (from Lindeman et al. 2001).

Little is known about circulation on the southeast shelf. Factors that determine water movement in this region (tides, winds, and the influence of oceanic water) may be similar to the Golfo de Batabanó (Emilson and Tápanes 1971). However, the bottom topography of the Golfo de Ana María and Golfo de Guacanayabo is more irregular than that of the Golfo de Batabanó, contributing to variable current speeds and directions that can be detailed only through continuous observations using a network of stations.

Knowledge of the circulation on the northwest shelf is also limited. Shallow depths probably make this area highly dependent on prevailing winds. In the western part of this area, circulation seems to be influenced by a clockwise gyre

of the Yucatán current that flows to the east on the open sea side and to the west on the shelf slope side during much of the year (Gómez 1979; García Díaz et al. 1991; Victoria del Río and Penié 1998). Closer to the coast, currents might be influenced more by tides and winds. North of La Habana, the dominant movement is toward the east (Rodríguez-Portal and Nadal-Llosa 1983), but inshore inversions of this flow can occur, depending on the tide flow.

Off the north-central coast, circulation patterns are determined mainly by tidal currents and, to a lesser extent, by prevailing winds. Tidal currents are stronger in the channels and passages between the keys fringing the lagoons where the mean tidal amplitude is 40–60 cm and the maximum is 120 cm. Close to the mainland coast, such currents are weaker and the highest tidal amplitude does not exceed 10–20 cm. In the inner waters, tidal influence induces complex current patterns (CUB/92/G31 1999).

Sinking of surface waters and upwelling events in the vicinity of Cabo San Antonio were recorded by Rossov and Santana (1966). Siam-Lahera (1983) reported upwellings to the south of Isla de la Juventud and Cienfuegos and other upwellings occurring in the fall near the southeastern provinces. At the eastern extreme of the island, Gallegos et al. (1998) recorded an upwelling event at the slope off the western margin of the Windward Passage.

1.5   Hydrochemical Characteristics

Comprehensive hydrochemical surveys on the Cuban shelf have been conducted by Lluis-Riera (1972, 1974, 1977, 1981a, 1981b, 1983a, 1983b, 1983c, 1984). Several important conclusions from this work are summarized in this section. Because of shallow depths and wide shelves, nearshore waters show substantial spatial variation, mainly influenced by seasonal changes in evaporation and terrestrial runoff. During the dry season, the Golfo de Batabanó has large areas of high salinity (> 36‰); low values (32.5–34.0‰) were recorded only at the Ensenada de la Broa (Lluis-Riera 1972, 1983b, 1983c). In the wet season, however, almost half of the gulf includes areas of 35‰ or less. This is due to major river drainage from the Habana and Pinar del Río provinces and the Península de Zapata, especially the latter. Tropical storms can lead to marked variations in salinity, especially in nearshore areas. In this eastern part of the gulf, the influence of summer and early fall rains is often evident until January or February. The dissolved oxygen content is about 4 ml/l or higher throughout the year. By the end of the dry season (March to April), large areas of high salinity and density and relatively homogenous pH and oxygen content are dominant in the gulf. Substantial vertical mixing takes place in the Golfo de Batabanó primarily because of its shallow depth.

Compared with the southwest coast, the southeastern lagoons inside the Archipiélago Jardines de la Reina do not show such marked seasonal variations, largely because they are deeper and receive moderate drainage from rivers (Batista 1974). Oceanic influence on the hydrological systems of the southeastern region is also relatively limited. Restricted exchange with the open ocean is particularly apparent in the Golfo de Ana María. In comparison with the Golfo de Guacanayabo (Fig. 1.5), the former is characterized by higher salinities, a wider seasonality in nearshore waters, and greater salinity and density gradients at the shelf edge.

On the northwestern Cuban shelf, seasonal changes in the hydrological systems are less notable because the shelf is narrower and has more oceanic water exchange. Average salinity ranges from 25.5–36.3‰. Nutrient concentration values are extremely low during the dry season, but increase during the summer because of terrestrial runoff (Lluis-Riera 1983a).

For the north-central region, we summarize information from a report on biodiversity and sustainable development in the Sabana-Camagüey region (CUB/92/G31 1999). The hydrological regime is quite variable; fast and sporadic local changes are prompted by meteorological events because of limited exchange with the ocean, many geographic obstacles (keys, points, shoals), and depths that barely exceed 4 m (Lluis-Riera 1981a, 1984; Rodríguez-Portal et al. 1983; and Fernández-Vila et al. 1990). Other significant changes have occurred after construction of causeways and other human modifications to lagoons of the area. In addition, precipitation has been below the historical average since 1973. Average rainfall over the Archipiélago Sabana-Camagüey from 1964 to 1995 was 800–1,000 mm/year, whereas the annual mean evaporation was 2,100–2,200 mm/year. With some exceptions, inshore salinity values have tended to be higher than in offshore waters. In large areas the salinity has exceeded 40‰. In Bahía Los Perros and Bahía Jigüey, maximum values exceeding 85‰ were recorded in 1995 after construction of a causeway between Cayo Coco and Cayo Romano that greatly constrained circulation and the interruption of runoff caused by the Estero-Socorro dike.

Seasonal temperature variations are less than in higher latitudes, but they vary enough to influence the seasonal dynamics of biological cycles in fishes (see Chapter 7). Minimum and maximum water temperatures recorded in Cuban shelf waters are 19°C and 31°C, respectively (Lluis-Riera 1972, 1977, 1981a, 1981b, 1983c), although maximum values in semienclosed water bodies might exceed 31°C. Data from the Siboney Hydrological Station (at the Instituto de Oceanología, La Habana) showed a 6°C range in average monthly values (Fig. 1.9). The most abrupt changes observed at this station were in November to December and March to April, which mark, respectively, the beginning and end of the dry season. Seasonal temperature patterns are also seen offshore to depths of at least 50–60 m (García Díaz 1981). Water temperatures are fairly stable in winter, but they can undergo relatively abrupt shifts (2°C or more overnight) caused by the arrival of northern cold fronts (Fig. 1.9). These changes can be even more pronounced in shallow areas with reduced circulation.

Nutrient concentrations in shelf waters are typically low (Lluis-Riera 1983c). In the wet season, however, nutrient concentrations rise in the vicinity of river mouths, reaching values comparable to those in relatively fertile regions, such as the Banco de Campeche (Bessonov and González 1971; Bessonov et al. 1971). In these areas, high nutrient levels are typically accompanied by high turbidity that affects primary production. In some regions, increases in water column nutrients are caused by resuspension of nutrients within and above the muddy bottom by increases in water motion due to climatic events. Compared to shelf waters, the physical and chemical properties of oceanic waters are more uniform

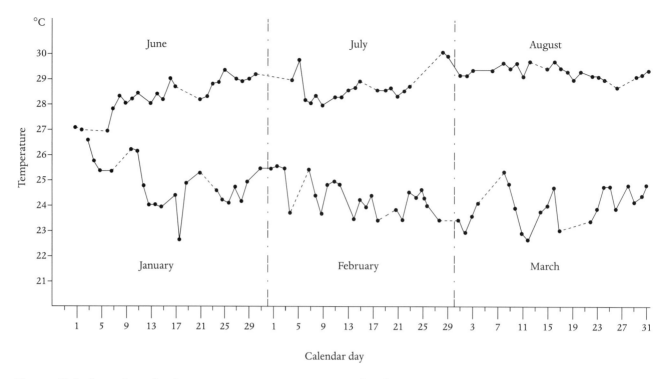

Fig. 1.9. Daily fluctuations of surface water temperature, January to March and June to August 1983, on the northwest coast near La Habana (from Institute of Oceanology, Siboney Station).

spatially and less variable seasonally. The influence of oceanic water over the shelf waters, therefore, is less variable. The extent of that influence depends on the intensity of water exchange and the topography of the shelf, as well as circulation patterns within both areas.

1.6   Hydrobiological Characteristics

1.6.1   Primary Production

Primary production in Cuban shelf waters is from two main sources: phytoplankton and phytobenthos. Kondratieva and Sosa (1967) reported a low mean annual production of phytoplankton in shallow waters (3–5 m) of the northwest region (78 mg $C/m^3/day$). In deep waters (up to 100 m) close to the shelf border, primary production can reach 160 mg $C/m^3/day$. The overall values ranged from 150 to 1,150 mg $C/m^3/day$ (mean, 330). Other studies, however, have found high values in surface waters of the shallow-water areas in the northwest and northeast regions (Kabanova and López-Baluja 1973). Values decreased abruptly outside the shelf boundary (from 200 mg $C/m^3/day$ to 1–19 mg $C/m^3/day$ in the northwest region and from 660 to 100–200 mg $C/m^3/day$ in the northeast region), and there

was great spatial variability in primary production. Evidently, patterns of horizontal distribution of primary production in shelf waters are determined by hydrological systems, which in turn are highly influenced by winds, oceanic currents, and terrestrial runoff. Changes in any of these factors, or other factors associated with them, could increase variation in production. A more detailed review of primary production in Cuban coastal waters can be found in Claro and Reshetnikov (1994).

The macrophyte benthic community is mainly composed of marine phanerogams. Among these, turtle grass (*Thalassia testudinum*) could be the main element of primary productivity on the Cuban shelf. The contribution from microphytobenthos and periphyton is low. A large part of the foliar biomass of marine phanerogams becomes detritus after decomposition. This process plays an intermediate role between primary production and the secondary consumers, suggesting an important role for heterotrophic bacteria in the shelf's aquatic ecosystem. The mangroves bordering almost all the coasts of the island and keys can supply large amounts of organic matter in the form of leaves, branches, and roots, which become detritus. This represents an important energy source in the large estuarine areas of the Cuban archipelago (González-Sansón and Lalana-Rueda 1982).

1.6.2 Benthos

Seasonal and spatial distribution patterns of benthic organisms determine the main food supplies of coastal fishes in Cuba. In the northwest region, mean biomass and density sampled with benthic grabs were 45.3 g/m^2 and 524 individuals/m^2, respectively (Murina et al. 1969). Maximum values (69.1 g/m^2 and 616 individuals/m^2) occurred in seagrass beds and minimum values (0.001 g/m^2) on sandy bottoms. Polychaetes (211 individuals/m^2) and crustaceans (127 individuals/m^2) were the most numerous, although sponges (36.9%) and mollusks (34.7%) prevailed by biomass. The estimated benthic biomass of the region was 8.95 g/m^2, lower than that reported by Formoso (1975) for the eastern region of the Banco de Campeche (12.3 g/m^2).

In the Golfo de Batabanó, means of 35.9 g/m^2 benthic biomass and 550 individuals/m^2 density have been reported (Gómez et al. 1980). Mollusks and polychaetes were most numerous, followed by sponges and echinoderms. Later detailed studies of benthos in this same region (Alcolado 1990; Alcolado et al. 1990a, 1990b; Corvea et al. 1990; Ibarzábal-Bombalier 1990; Martínez-Iglesias and Alcolado 1990; Martínez-Estalella and Alcolado 1990) found that megabenthic organisms were 85–92% (mean values) of the total dry biomass. Most of this biomass, however, was composed of organisms such as sponges and some corals (often gorgonians) that are not typically used as food by most fishes. Subtracting the weight of these organisms, the potential benthic food supply for fishes (Fig. 1.10A) ranged from means of 7.0 to 23.2 g/m^2 (Alcolado 1990). Accounting for the difference between wet and dry weight (the wet weight of these organisms is five times the dry weight), these biomass values are notably higher than those reported in earlier studies (Murina et al. 1969 and Gómez et al. 1980). However, differences in sampling gear limit definitive comparisons among these studies.

Among the macroinvertebrates, echinoderms (typically sea urchins, mainly *Lytechinus variegatus*) were most abundant (Alcolado 1990; Corvea et al. 1990), followed by mollusks. Sea urchins were most common on vegetation-rich bottoms and near reefs, but were scarce on muddy bottoms. Mollusks, which constitute a major food supply for fish, were most abundant (2.5–20 g/m^2) on white sandy-muddy bottoms, where organic content and sea-grasses are limited. Although small crustaceans are the food items most frequently found in fish stomachs (see Section 5.3), they were found in low proportions in these benthic samples and were probably underestimated by the sampling methods.

In the Archipiélago Sabana-Camagüey, maximum densities of megabenthic organisms are associated with ranges of 37.5–41.5‰ in salinity, 5–15% in particulate organic matter, and 30–50% in silt content in sediments (Jiménez and Ibarzábal 1982). As these proportions increase, maximum densities of benthic animals tend to decrease. The best conditions for high species diversity and organism density are often found where circulation or water exchange with the open sea is high (see Alcolado et al. 1996).

Macroinfaunal biomass in the Golfo de Batabanó ranged from 0.34 to 14.0 g/m^2 (Ibarzábal-Bombalier 1990). The higher values were recorded south of Ciénaga de Zapata and northwest of Isla de la Juventud (Fig. 1.10), where the bottom is covered by medium- to high-density seagrass beds. In both areas, echinoderms (mainly ophiuroids) made up the greatest biomass, and polychaetes accounted for the highest density (45% of the total fauna and 619–780 individuals/m^2, respectively). Polychaetes played an important role in the macroinfauna, accounting for 30–50% of the dry biomass at most of the surveyed stations. The lowest macrobenthic biomass levels (2–4 g/m^2) were recorded over areas of the Golfo de Batabanó where seagrass is sparse or absent (Fig. 1.10B). In these regions, polychaetes (85%) or crustaceans (55%) were most numerous.

1.7 Major Habitats

Habitat distributions on the Cuban shelf are influenced by many factors, which result in a wide variety of environmental conditions at several spatial scales. The main habitats of Cuban coastal fishes include coral reefs, seagrass beds, mangroves, softbottom areas, hardbottom, and pelagic open-water environments. Factors influencing habitat distributions include distribution of surface sediments; relief (mainly hard bottom, coral reefs, and other natural or artificial substrates); hydrochemical and hydrological systems, which can be strongly associated with terrestrial runoff or exchange with the open ocean; and nutrient sources and aquatic vegetation, which are a main source of primary production and also provide refuge to a great variety of organisms. Fig. 1.11 shows basic habitat distributions across the Golfo de Batabanó as an example of the complex factors influencing habitat.

Habitats are often arranged in cross-shelf geomorphic gradients from the shoreline toward the ocean (Fig. 1.12). Structurally similar or identical habitat types can be used by very different fish assemblages, depending on their position across the shelf (Valdés-Muñoz et al. 1990; Lindeman et al. 1998).

Approximately 60% of the mainland shore is fringed by mangroves and shallow lagoons are often connected with the open shelf through narrow channels and creeks. The sand content of sediments increases gradually offshore as prevailing conditions become more oceanic. Here, the bottom is more sandy or rocky, and coral heads approximately 0.5–4 m in diameter and patch reefs of 100–1,000 m^2 occur. Farther seaward, reef crests often occur close to the fore reef, the latter with slopes of 30–45 degrees (Fig. 1.12). The shelf edge is often followed by an escarpment that falls from 25–40 to 100–150 m. Where the shelf is narrow, no coastal lagoons occur. In high-energy areas, the shoreline can consist of long sections of sand or hard bottom. This is the case along substantial parts of the

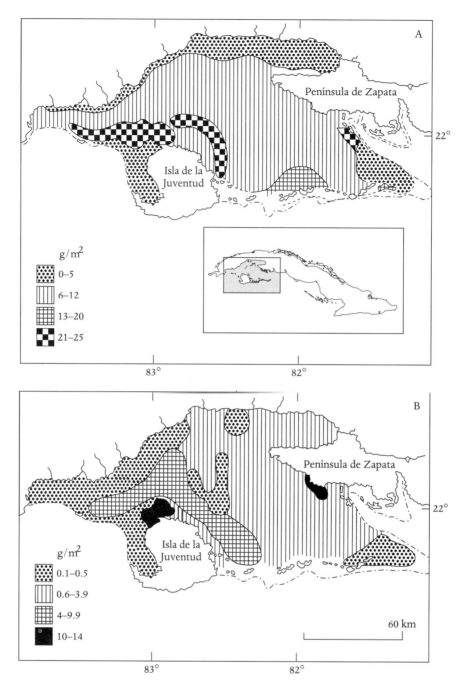

Fig. 1.10. Maximum biomass of benthic invertebrates recorded in the Golfo de Batabanó (from Alcolado 1990). (A) Potential food for fishes (all organisms larger than 4 mm, exclusive of sponges, corals, and gorgonians); (B) Macroinfauna only. Biomass estimates are based on dry weights.

north coast near La Habana, Matanzas, and the north and south coasts of eastern Cuba.

This combination of habitat features forms what Baisre (1985) terms a seagrass–reef complex. This zonation can be interrupted by an array of keys that border much of the shelf. The leeward sides of the keys are fringed by mangroves with a fish fauna that can differ from the mainland mangroves. Taking into account the behavioral, trophic, and habitat relationships of coastal fishes, we suggest that Baisre's (1985) formulation should include mangroves, thus expanding the seagrass–reef complex to a seagrass–reef–mangrove complex. Such a complex has also been suggested by Marshall (1980).

1.7.1   Coral Reefs
The outer borders of the shelf are typically delimited by submerged reefs or arrays of keys and islets fringed by reefs.

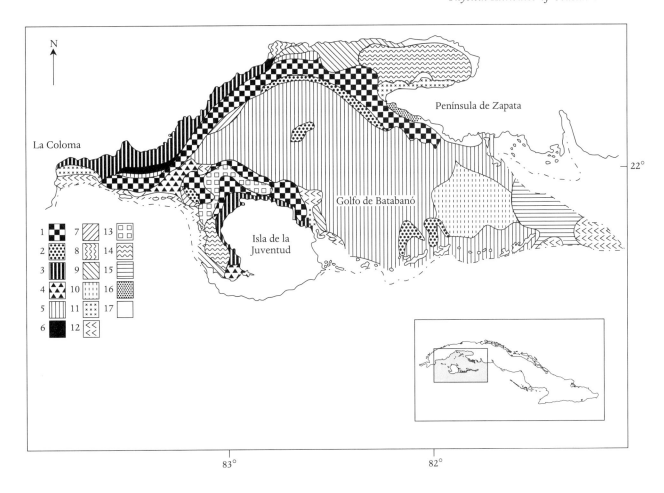

Fig. 1.11. Distribution of habitats of the Golfo de Batabanó (after Alcolado 1990), including estimates of seagrass biomass and sediment color. The latter is a primary indicator of the quantity of particulate organic matter (POM), terrigenous influence, and the balance between sedimentation and transport processes. 1: Intermediate to high seagrass biomass, gray sediments, high POM; 2: Intermediate to high seagrass biomass, whitish sediments, low POM; 3: Intermediate to high seagrass biomass, dark gray sediments, high POM; 4: Hardbottom with seagrass patches of variable biomass, brownish gray sediments; 5: Low seagrass biomass, whitish sediments, low POM; 6: Low seagrass biomass, whitish sediments, low POM; 7: Low seagrass biomass, brownish gray sediments, high POM; 8: Very low seagrass biomass, dark gray sediments, high POM; 9: Low or very low seagrass biomass, medium gray sediments, low POM; 10: Very low seagrass biomass, whitish sediments, low POM; 11: Hardbottom with seagrass patches of variable density, medium gray sediments, low POM; 12: Hardbottom with seagrass patches of low density, whitish sediments, low POM; 13: Hardbottom with seagrass patches of variable density, brownish gray sediments, low POM; 14: No vegetation; muddy, medium gray sediments, low POM; 15: Bottom with little vegetation; muddy, whitish sediments, very low POM; 16: Peat bottom, no vegetation; 17: Seagrass of intermediate biomass, pale gray sediments.

The shelf edge extends 3,966 km; 2,150 km on the north coast and 1,816 km on the south coast (Alcolado et al. 1997). The most well-developed reefs are distant from the mainland and are not easily accessed by people. This isolation has contributed to maintaining the relatively high quality of many Cuban reefs. Recent studies of the status of Cuban coral reefs (Alcolado et al. 1997, 2000; Woodley et al. 2000) and several particular reef areas (Alcolado et al. 1999a, 1999b) were used in the following summary.

The Cuban shelf edge is almost continuously fringed by deep fore reefs. These fore reefs exhibit great variation in their profiles and ecological zonation. Usually, after a shallow limestone terrace (poorly covered by corals), a 10- to 15-m

coral-rich slope descends to a deeper rocky-sandy terrace, often with some dispersed patch reefs. Farther out on this terrace a spur and groove system often occurs. Sometimes on the deep terrace just before the drop-off, a transverse sand corridor is developed in front of an elevated coralline threshold. Examples include fore reefs sheltered from terrestrial runoff by the Cayos Los Indios and Cayos San Felipe (western Golfo de Batabanó); Cayo Diego Pérez (eastern Golfo de Batabanó); Cayo Rosario (south of Golfo de Batabanó); Los Colorados and Cayo Levisa (northwestern Cuba) (Alcolado et al. 1999c).

Reef crests tend to be more abundant at the edges of the four broadest sections of the Cuban shelf—the Golfo de

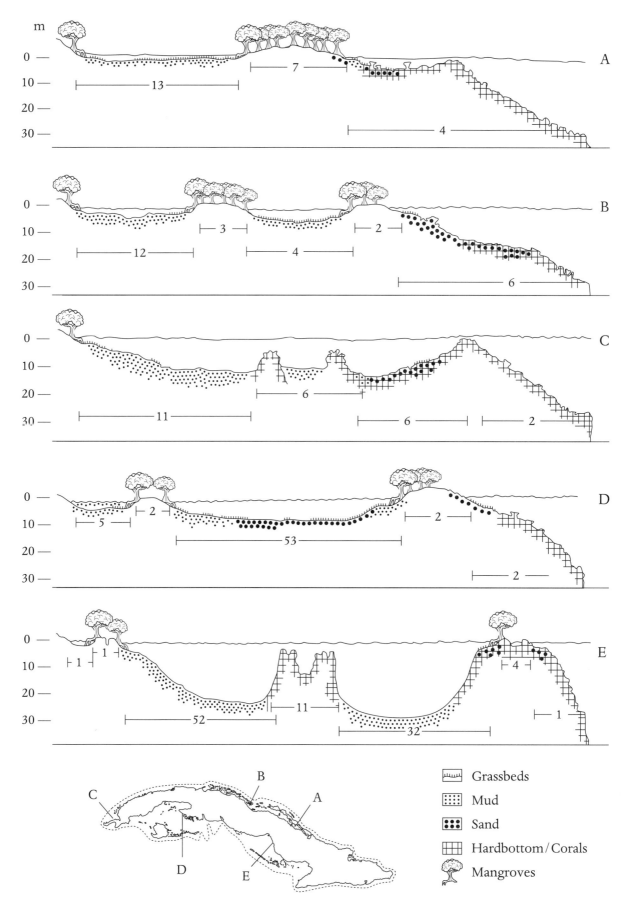

Fig. 1.12. Cross-shelf habitat profiles at different locations in Cuba. The numbers below each habitat indicate the approximate width in kilometers; the *y*-axis indicates depth in meters.

Guanahacabibes (northwest), Archipiélago Sabana-Camagüey (north-central), Golfo de Ana María-Guacanayabo (southeast), and Golfo de Batabanó (southwest). However, along the narrow shelf of northeastern Cuba, reef crests can be widespread and quite long. Reef crests can be linear or can occur as a series of patch reefs. The linear types are more common and can differ greatly in species composition and zonation. Patchily distributed reef crests have an irregular to semicircular outline and can be characterized by dense fringes of elkhorn coral (*Acropora palmata*). Patchy reefs are common along the border of the Gulf of Guanahacabibes (Colorados reefs) and along the eastern edge of the Golfo de Batabanó (e.g., Arrecifes de María de Médano Vizcaino). Reefs on muddy substrates were found by Zlatarsky and Martínez-Estalella (1980) in the Golfo de Guacanayabo. These structures can be 25 m high and are covered mainly by *Oculina* spp. and *Cladocora arbuscula*, as well as many sponges that give the reef a gelatinous appearance.

At least 41 species of stony corals are represented in Cuba (40 scleractinians and one hydrocoral species with three forms), following the Zlatarsky and Martínez-Estalella (1980) classification. This figure rises to 58 if some forms are considered to be different species. Important examples include *Acropora cervicornis* (*cervicornis* and *prolifera*), *Agaricia agaricites* (*agaricites, fragilis, grahamae, tenuifolia, undata,* and *lamarcki*), *Scolymia lacera* (*lacera, cubensis,* and *wellsi*), *Mycetophyllia lamarckiana* (*lamarckiana, aliciae, ferox,* and *danaana*), *Montastrea annularis* (*annularis, faveolata,* and *franksi*), *Siderastrea radians* (*radians* and *siderea*), *Porites porites* (*porites, furcata,* and *divaricata*), *Madracis decactis* (*decactis* and *mirabilis*) and *Isophyllia sinuosa* (*sinuosa* and *rigida*) (Alcolado et al. 1999c).

Acropora palmata and *Millepora complanata* are the most dominant and common species on reef crests. Plate-shaped *Montastrea annularis* and *Agaricia agaricites* (f. *unifasciata*) usually dominate at slopes below a depth of 25 m. Herrera-Moreno and Martínez-Estalella (1987) observed small *Siderastrea radians* (forms *siderea* and *radians*) to be abundant near the polluted reefs of La Habana. In polluted areas with large amounts of suspended particles, other common corals include *Dichocoenia stokesii, Stephanocoenia intersepta, Porites astreoides,* and *Montastrea cavernosa*. *Dichocoenia stokesii* was once found to be a locally dominant species at 10 and 20 m depths at Cayo Francés reef, in the north-central part of Cuba (Alcolado et al. 1999c).

Fifty-five gorgonian species recorded in Cuba are reef dwellers. In shallow reef crest areas, *Plexaura homomalla, Eunicea mammosa, E. tourneforti, E. flexuosa, Muricea muricata, Gorgonia flabellum,* and *Briareum asbestinum* are among the typically common species. Between 5 m and 20 m the most common species include *Plexaura kuekenthali, E. mammosa, E. tourneforti, E. flexuosa, E. calyculata* (f. *coronata*), *Plexaurella dichotoma,* and *Pseudopterogorgia americana*. Below 25 m the most common are *Iciligorgia schrammi, Elluella elongata* and

E. barbadensis (Guitart Manday 1959; Behety 1975; Alcolado 1981; García-Parrado and Alcolado 1998).

Approximately 160 sponge species have been collected from Cuban reefs. Of these, 144 have been identified to species. In shallow reef zones, down to 7 m, the most common and abundant species are *Aplysina fistularis, Clathria virgultosa, Cliona caribbea* f. *aprica, Chondrilla nucula, Scopalina ruetzleri, Cliona vesparia,* and *Spirastrella coccinea*. In deep reefs (7–35 m), the most abundant sponges are *Aplysina cauliformis, Cliona caribbea* f. *aprica, Ectyoplasia ferox, Iotrochota birotulata, Mycale laevis, Niphates amorpha, Aiolochroia crassa,* and *Scopalina ruetzleri*. Other common but not often dominant sponges are *Aplysina fistularis, Callyspongia vaginalis, Chondrilla nucula,* and *Niphates digitalis*. At greater depths (30–35 m), the most common and dominant species are *Aplysina cauliformis, Ircinia felix,* and *Ectyoplasia ferox* (sometimes 10% or more of the total number of individuals).

At least 526 species of marine macroalgae (Chlorophyta, Phaeophyta, and Rhodophyta) have been recorded from Cuba (A. M. Suárez, personal communication) and approximately 60% of these have been found on coral reefs (C. Jiménez, B. Martínez-Daranas, A. M. Suárez, J. Trelles, and D. Zúñiga, unpublished data). Unexpectedly high algal biomass (0.1–1.5 kg/m² dry biomass; 0.3–3.1 kg/m² wet biomass) was found in research conducted in reef areas of the Archipiélago Sabana-Camagüey (at 37 stations distributed along 10 reef profiles from the back reefs to 20 m deep, April to May 1994). The highest dry biomass values were found most frequently at 5 m, where about 80% of the stations displayed values of 0.69 kg/m² or higher. This value or higher was also found at 60% of the stations at 10 m and 20% at 20 m. Unfortunately, no earlier values are available for intrasite comparisons. Phosphate concentrations were fairly high and exceeded threshold values for excessive algal proliferation as defined by Lapointe et al. (1992). This may be the result of inputs from the nutrient-rich inshore water bodies. *Cladophora catenata, Microdyction marinum, Dictyota* spp., and *Lobophora variegata* were the dominant species (Alcolado et al. 1999c; Woodley et al. 2000). High algal biomass was also unexpectedly found during May 1999 on reefs near Cayo Largo (southwest Cuba) in areas with high nutrient levels of uncertain origin.

1.7.2 Seagrass Beds and Softbottom Areas

Large areas of the Cuban shelf are covered by marine phanerogams, typically turtle grass, and to a lesser extent, other grasses or algae. These beds of grasses and algae, known in Cuba as *seibadal*, constitute a major habitat that provides refuge for benthic organisms, which in turn play an important role in fish food webs. Mixed areas of grass and sand are common in regions such as northeastern Isla de la Juventud and the southeastern part of the Golfo de Batabanó (Fig. 1.11), where precipitation of calcium carbonate forms oolitic sand

(Ionin et al. 1977). Alcolado (1990) and Jiménez-Dominguez and Alcolado (1990) detail benthic habitats of the Golfo de Batabanó, estimate the biomass and distribution of marine phanerogams, evaluate sediment color as an indicator of the content of particulate organic matter, and suggest vectors of sediment transportation.

Seagrass beds are a primary habitat of the spiny lobster (*Panulirus argus*), the most important fishery resource of the Cuban shelf. Juveniles of many coastal fishes also find shelter and food among seagrass blades. Starfish (*Oreaster reticulatus*), several sea urchins (*Lytechinus variegatus, Meoma ventricosa*, and others), and large mollusks (*Strombus gigas, S. costatus, Cassis flammea, C. tuberosa*) also dwell among seagrass blades during the day.

Although seagrass beds and sandy areas often have low fish biomass, in many places the areas are interrupted by patch reefs or coral heads surrounded by large numbers of fishes that evidently feed in the sandy and seagrass areas. For example, coastal lobster and multispecies finfish fisheries exist in complexes of seagrasses, mangroves, and patch reefs. In those areas of the Golfo de Batabanó where patch reefs are sparse, many fishermen have established artificial refuges (known as *pesqueros*) for attracting finfish and lobsters (Claro and García-Arteaga 1999).

1.7.3 Mangroves

Most coastal areas of the Cuban archipelago are fringed by mangroves, as well as by coastal lagoons, and estuaries. All of the mangrove community types are outlined in Lugo and Snedaker (1974). The mangrove formations of the southeast region are some of the largest in the Antilles (Díaz-Piferrer 1972). Throughout Cuba, these habitats are formed mainly by red mangrove (*Rhizophora mangle*), with incidental groups of black mangrove (*Avicennia nitida*) and white mangrove (*Laguncularia racemosa*) (González-Sansón and Lalana-Rueda 1982). The roots of red mangroves provide habitat for many fishes and invertebrates (see Chapter 2). The dominant invertebrates are crustaceans (mainly Brachyura of the families Grapsidae and Ocypodidae) and mollusks (mostly *Crassostrea rhizophorae, Isognomon alatus*, and *I. radiatus*). Also, the barnacles *Balanus eburneus* and *Chthanalus* sp. and various algae and sponges live attached to the mangroves and are in turn hosts to numerous organisms, such as colonial ascidians and coelenterates, principally of the order Hydroida (Ortiz 1976).

The composition of mangrove fish faunas depends in part on hydrological features. In estuaries and lagoons, the influence of terrestrial runoff is critical. In keys, distant from the mainland, oceanic conditions can favor different faunas. Species particularly tolerant of changes in salinity, such as gray snapper (*Lutjanus griseus*) and several species of mojarra (family Gerreidae), can live among mangrove roots in many different cross-shelf regimes.

1.7.4 Brackish Lagoons

Brackish lagoons in Cuba are typically shallow water bodies (less than 2 m deep) with little seawater exchange. These systems can include softbottom, mangrove, and seagrass habitats. Tidal flow is often through channels and creeks. These lagoons usually receive considerable terrestrial runoff that can create high levels of biological productivity. The main sediments are dark-colored muds, but close to river mouths, rocky substrates can occur. Salinity characteristics of the brackish lagoons and estuaries of Cuba are varied, with several systems represented: (1) those dominated by salt wedges (strongly stratified), (2) those with strong estuarine freshwater input throughout most of the year, (3) hypersaline systems, and (4) isohaline systems (near 36‰ for most of the year). Nevertheless, most lagoons endure abrupt intra-annual salinity changes. In addition, damming of many Cuban rivers has increased salinity over large areas of the Golfo de Ana María and Golfo de Guacanayabo, as well as several bays within the Archipiélago Sabana-Camagüey.

Temperature, dissolved oxygen, pH, and nutrient concentrations are more seasonally variable in brackish lagoons and estuaries than in outer shelf and oceanic waters. Therefore, these lagoons are considered ecologically variable. The fishes inhabiting them are typically tolerant of abrupt environmental changes.

1.7.5 Pelagic Environment

The hydrological and hydrochemical conditions of the open sea are often relatively stable and predictable. Currents can influence the distribution and migration of pelagic fishes and many neritic fishes move to the open sea to spawn. *Sargassum* and other floating plants constitute a mobile substrate where some invertebrates and fishes, or their juveniles, find refuge and food (Dooley 1972). A large number of pelagic fishes inhabit the oceanic waters surrounding the Cuban shelf and are connected in various ways to shelf resources. Several coastal pelagic fishes are found in shallow shelf waters during substantial periods of their life cycle (e.g., Clupeidae, Engraulidae, Carangidae, and Belonidae).

1.8   Anthropogenic Effects

Direct and indirect human effects on fishes and their habitats fall into dozens of categories. It can be difficult to separate natural and anthropogenic mechanisms of habitat or stock degradation (e.g., Lessios 1988; Butler et al. 1995; Hughes and Connell 1999). Nonetheless, a hierarchy of potential anthropogenic stressors can be identified within two broad categories: (1) effects from coastal construction, agricultural, and recreational activities; and (2) effects through commercial fishing and gear damage to habitats. This section briefly summarizes some information on the former effects, Chapter 8 addresses the latter effects.

Limited empirical information is available on the direct effects of nonfishing activities on fish populations and habitats in Cuba. Industrial wastes and their derivatives from the sugar industry are an important source of water pollution on the Cuban shelf (see Vales et al. 1998, for a recent review of this and other anthropogenic effects). Approximately 150 sugar factories currently operate during the sugar cane season and many of them channel their effluents into the coastal zone (González 1995; Díaz-Briquets and Pérez-López 2000). These wastes typically consist of organic matter whose principal effect is the depletion of dissolved oxygen (Basu et al. 1975b). Suárez et al. (1983) studied the effluents of 47 sugar factories and reported that 12% had biochemical oxygen demand (BOD) levels lower than 400 mg/l, 69% had BOD levels of 400–3,000 mg/l, and 12% had levels higher than 3,000 mg/l. Suárez and colleagues found a ratio between BOD and chemical oxygen demand of 0.82, which suggests these wastes are biodegradable over time. Since these factories are active only three to six months a year, there might be limited recovery periods for affected fishes and habitats.

Sugar industry effluents can cause local fish kills and other effects, particularly in the coastal lagoons of the north-central and southeast regions, and in the Ensenada de la Broa, where little exchange with outside waters delays the diffusion of pollutants (Lluis-Riera 1972, 1983a; Basu et al. 1975c). Distilleries, breweries, and yeast factories processing the final molasses, a byproduct of the sugar production process, are adjacent to many sugar factories. Such industries generate highly pollutant, but largely biodegradable, discharges (Basu et al. 1975a). Nevertheless, these effluents have caused considerable deterioration of littoral areas of Bahía de Cárdenas, Bahía de la Habana, Bahía de Nipe, Santa Cruz del Norte (east of La Habana), and Ensenada de la Broa.

Domestic sewage wastes are also an important source of pollution. Many urban centers in Cuba are close to the coast and drain sewage into coastal areas. Such wastes, frequently untreated or only poorly treated, can severely affect water quality, especially in the bays (Díaz-Briquets and Pérez-López 2000). Food, fish processing, textile, and tannery industries are also major sources of pollution, although some of these industries have partial or total treatment processes. On the other hand, much agricultural waste (from activities such as livestock farming and land drainage) contains pesticides, herbicides, and fertilizers, and reaches coastal waters untreated. In the 1970s, the Golfo de Batabanó was estimated to be moderately polluted because BOD, chemical oxygen demand, and total phosphorus levels were above those reported for Japan and the United States (Basu et al. 1975c). Unfortunately, no water quality standards have been established for Cuban shelf waters, nor are data available on the capacity for ecosystem recovery.

Although generally in fairly good condition, Cuban coral reefs are suffering variable levels of deterioration. Watersheds have been extensively deforested and coastal human populations have grown, both of which increase sediment and pollutant runoff to the sea. Many outer shelf reefs are separated from the main island, often by broad lagoons bordered by offshore keys. In these regions, the effects of terrestrial activities on coral reefs can be buffered. Land-based sedimentation from deforestation is estimated to occur at approximately 20% of the reefs that fringe the shelf; substantial reforestation efforts have been underway since the 1960s (Alcolado et al. 1997, 1999c). Organic and chemical pollution has also affected the coral reefs of Ciudad de la Habana, as well as other urban centers. Examples include Moa (sedimentation and heavy metals from mines), east of Bahía de Mariel (thermal pollution by a power plant), and Rio Mosquito (pollution by fiber wastes from an agave fiber-processing factory).

Less than 3% of the reef habitat at the shelf border is estimated to be affected by severe pollution (Alcolado et al. 1994, 1997), but many of these areas are being affected by some degree of eutrophication and algal proliferation (e.g., Archipiélago Sabana-Camagüey, Cayo Largo, Cayo Juan García, and north of Pinar del Río Province). This could be exacerbated by the massive mortality of the sea urchin *Diadema antillarum*. Extensive purple to black cyanobacterial mats have been observed during the spring at the coral reefs of Cayo Largo in 1998 and 1999 (Alcolado et al. 1999b) and by Elena de la Guardia (personal communication) in Punta Francés in 2000 (both areas off the southwest coast). The mats can blanket algae, corals, and sponges (Alcolado et al. 2000). Coral reef diseases are also becoming more common, causing extensive acroporid mortality, often by "white band" disease. Another significant disease is aspergillosis in gorgonians. To a lesser extent, massive corals are affected by "white plague." Other pathogenic diseases occasionally found are "yellow band," "black band," and "dark spot" (Alcolado et al. 2000). Coral bleaching events often take place during El Niño–Southern Oscillation (ENSO) events, mostly along the north coast (Carrodeguas et al., in press), but a substantial coral bleaching event extended along the north and south coasts of Cuba during the late summer of 1998 during La Niña. Alcolado et al. (2000) subsequently found evidence of recovery, with the probable exception of the Archipiélago Sabana-Camagüey, where much less coral cover was observed in 2000 than in 1994. The observed increase of ciguatoxic fishes could be a consequence of the deterioration of reefs off northwest Cuba (Alcolado et al. 1997).

Declines in carnivorous and herbivorous fishery resources from habitat deterioration and commercial, subsistence, and illegal overfishing are significant issues (see Chapter 8). In part because large predators are scarce, it is possible to find large populations of species that feed on or otherwise damage corals and gorgonians (*Coralliophylla* spp., *Cyphoma gibbosum,* and *Stegastes* spp.). Illegal collection of gorgonians for the manufacture of jewelry and handcrafts led to the devastation of gorgonian gardens in the shallow reef zones of Ciudad de la Habana, Playa de Varadero, and other sites (Alcolado et al. 1997, 1999c). SCUBA diving tourism in

Cuba, in spite of its high potential, is still underdeveloped and has apparently had a limited effect on coral reefs. More than 35 diving centers exist or are proposed, with more than 400 diving sites. Measures are being considered and applied by relevant agencies to minimize additional damage. Anchoring on coral reefs continues to be used in fishing and diving activities. At present, mooring buoys are used or deployed at some reef dive sites (Alcolado et al. 1997), and efforts are underway to increase the use of mooring buoys.

The legal infrastructure for controlling pollution sources in Cuba is still developing, as are enforcement mechanisms. Several recent laws may encourage efforts to more effectively regulate major sources of contaminants and to protect biodiversity. For example, Resolution 111/96, dated October 14, 1996, contains the Regulations on Biodiversity that establish guidelines for management strategies to promote sustainable resource use and biodiversity conservation in Cuba. A broad new environmental law passed in 1997, Ley 81, establishes a new framework for protecting Cuban natural resources and places considerable authority under the Ministry of Science, Technology and Environment (Speir 1999). Despite government efforts, lack of public awareness about coral reef issues is still a fundamental problem (Alcolado et al. 1997, 2000). Until recently, the management of coral reefs has not been administratively designed as a component of integrated coastal zone management, a situation that is now improving in some instances (Woodley et al. 2000).

1.9 Summary

Nearshore and midshelf areas of the Cuban shelf often have the characteristics of a flat and shallow submerged plain. Four wide shelf areas are bordered by long reef lines, forming several massive coastal lagoons. These lagoons are separated by stretches with a narrow shelf and a steep drop-off. These large lagoons, each with distinctive features, may function fairly independently of neighboring areas. In some regions, nearshore areas are greatly influenced by terrestrial runoff. The steep insular slopes of Cuba are probably an ecological barrier for adults of typical continental shelf demersal fishes.

Because of some isolation from oceanic waters, shallow depths, and the influence of river discharges, two climatic seasons are clearly defined in shelf waters. The dry season, November to April, is characterized by high evaporation rates, high salinity, and low nutrients. The rainy season, May to October, induces high terrestrial runoff and lower salinity

in coastal waters. Shelf water temperature extremes range from 19°C to 32°C in different areas, but monthly average temperatures range from 23.5°C to 29.5°C. The most abrupt fluctuations occur in winter after the passage of northern cold fronts, when temperatures can drop 2.5°C in less than 24 hours. Temperature variations, although relatively low, can influence the seasonal patterns of biological processes in fishes.

The oceanic waters surrounding Cuba are characterized by their relative stability and low nutrient levels. In most regions, oceanic waters might contribute less to coastal productivity than terrestrial runoff. Sediment composition is an important determinant of habitat distributions. On the shelf, shallow-water primary production is driven by seagrasses and algae. Because herbivorous species are relatively scarce in some areas, much biomass is probably converted to detritus through decomposers. Turtle grass biomass is higher in warm periods than in the dry season when temperatures are lower. Benthic biomass follows a similar pattern. Biomass of grasses and algae greatly influences not only the feeding habits, but also shelter use by many invertebrates and small fishes.

The major habitats for marine fishes on the Cuban shelf are coral reefs, seagrass beds and softbottom areas, mangroves, and oceanic waters, representing a gradient of environmental conditions. Fishes are distributed widely across these habitats. Reefs contain the most diverse fish fauna; the mangroves of offshore keys, highly influenced by oceanic waters, are also used by many reef fishes, whereas mainland mangroves can have quite different fish faunas. Seagrass beds and softbottom areas are important food sources for many species inhabiting reefs and mangroves, and they also serve as nursery grounds for many species. Estuarine regions have low species diversity but high biological productivity. Oceanic areas adjacent to the shelf are dominated by migratory fishes.

Agricultural and industrial development have led to a gradual increase in coastal pollution, causing local damage to the flora and fauna of some estuarine and coastal regions. Although many pollutants are organic and biodegradable (e.g., fertilizers, sugar wastes, sewage), the scope of their negative influence has increased. Other human factors, such as tourism development (e.g., hotel and causeway construction), fisheries activity, mining, and the damming of rivers, have also affected the coastal resources of Cuba. New research initiatives, education outreach programs, and regulatory measures, if properly integrated, can foster more sustainable development activities.

2

The Marine Ichthyofauna of Cuba

RODOLFO CLARO AND LYNNE R. PARENTI

2.1 Introduction

Extensive collection and study of the Cuban fish fauna by Felipe Poey y Aloy (1799–1891) in the nineteenth century produced the primary basis for marine ichthyology in Cuba (viz. Poey Aloy, 1851–1861). Felipe Poey's major work, the 30-volume *Ictiología Cubana,* was not published until 117 years after completion (Poey Aloy 2000). Poey described and illustrated 542 marine and freshwater species, of which approximately 104 species among those previously described and published are currently accepted as valid (see Appendix 2.1). The status of taxa whose descriptions were first published in Poey Aloy (2000) remains to be evaluated. The Cuban ichthyofauna was not reviewed again as a whole for another century until Duarte-Bello (1959) compiled a catalog of the known species, later revised by Duarte-Bello and Buesa (1973). The multivolume *Sinopsis de los Peces Marinos de Cuba* (Guitart 1974–1978) was published soon after. This work summarized information on the systematics of Cuban marine fishes and provided keys and illustrations for species identification. More recently, Rodríguez et al. (1984) updated the list of marine fish species, assigned a Cuban common name to each, included new records and changes in nomenclature, and arranged the checklist following the classification of Robins et al. (1980).

2.2 Faunal Composition

The most recent previous compilation of the Cuban marine ichthyofauna, by Rodríguez et al. (1984), listed 776 species in 22 orders, 139 families, and 387 genera. Duarte-Bello and Buesa (1973) listed 133 species separately because their presence in Cuban waters was considered probable but had not been verified. Appendix 2.1 lists the 1,030 species, including subspecies, currently recognized as valid, combining and updating the nomenclature in both catalogues and adding new records from subsequent publications and museum records. This list represents our modern knowledge of the Cuban ichthyofauna. Undoubtedly, the list will be revised and updated with continued collecting and revision of poorly known taxa. More than 20 exotic species, such as the tilapia (*Oreochromis aureus*), are not included. The genus- and species-level classification largely follows Eschmeyer (1990, 1998), whereas higher classification largely follows Nelson (1994). The local English common names follow Robins et al. (1991), if a name is in use, or the online database FishBase (2000) maintained by ICLARM (International Center for Living Aquatic Resources Management). Additional taxonomic information and common names are from Böhlke and Chaplin (1993) for the Bahamas, McEachran and Fechhelm (1998) for the Gulf of Mexico, and Smith-Vaniz et al. (1999) for Bermuda.

Of the strictly marine species listed in Appendix 2.1, 910 are teleosts and 80 are elasmobranchs. Many of the species inhabit shallow waters (less than 10 m) and occupy reefs, mangroves, seagrass beds, and other coastal habitats. One hundred and forty neritic species have commercial value or are accepted as edible in Cuba. At least 50 species have economic significance.

The majority of fishes belong to the order Perciformes: 62 families, 245 genera, and approximately 485 species. The families Serranidae, Carangidae, Lutjanidae, Labridae, Labrisomidae, Gobiidae, and Haemulidae are particularly speciose. Among the remaining orders, those with the greatest numbers of species are Anguilliformes (59 species), Carcharhiniformes (31), Tetraodontiformes (41), Myctophiformes (34), Aulopiformes (28), and Pleuronectiformes (26). All of the totals at the species, genus, subfamily, and family level will change as our knowledge of the fish fauna increases.

Appendix 2.1 includes the 38 species classified as freshwater fishes by Vergara (1992) and subsequent records (viz. Watson 2000) because some of these species can tolerate saltwater and can be found in coastal habitats during at least part of their life cycle. Twenty-one of these species are endemic to Cuban waters. Notable among the endemics are fishes of the order Cyprinodontiformes, including species in the families Rivulidae, Fundulidae, Cyprinodontidae, and Poeciliidae. These fishes are tolerant of a wide range of temperatures and salinities and some are common in the hypersaline habitats along Cuba's north coast.

2.3   Distribution and Habitat

The distribution and abundance of habitats for shelter and food in tropical coastal areas is correlated with fish diversity and biomass. Fish feeding habits and shelter associations are strongly correlated, in part because of specializations in both morphology and behavior that allow individuals to use resources available within certain habitats. Combinations of feeding relationships, defense mechanisms, and morphological and behavioral adaptations allow resource sharing, which may permit the coexistence of a large number of species. Consequently, attributes of the food supply and the topography of each habitat might, to some extent, determine the abundance and diversity of reef fish assemblages (Jones 1991; Caley et al. 1996). However, considerable evidence also indicates that variable planktonic recruitment and physical disturbance events are of great importance in shaping assemblage composition (Sale 1991; Doherty and Fowler 1994).

2.3.1   Coral Reef Fish Fauna

Most demersal marine fishes of Cuba (at least 350 species) associate with coral reefs during part or all of their life history. Reef fishes are either residents or transients and their relationship with the reef community is complex. Only a relatively small number of species are found both frequently and in high abundance because many species are cryptic and are seldom visible to humans.

Among the resident species, many are territorial (see Chapter 3), whereas others roam across the reef. Resident species are typically small fishes that use the reef for shelter and food and, in some cases, for breeding. Many of these fishes are primary consumers; they feed on epiphytic algae on corals or nearby seagrasses. Various species of Acanthuridae and Scaridae often make up much of the fish biomass on reef crest habitats in Cuba (Claro et al. 1990b; Claro and García-Arteaga 1994b). A large number of resident reef fishes maintain close contact with their microhabitat, most are primary carnivores or omnivores that eat epiphytic algae and small invertebrates. Among the most abundant are the pomacentrids (*Stegastes* spp., *Abudefduf saxatilis*), some sciaenids (*Equetus* spp.), and chaetodontids (*Chaetodon* spp.). Other species roam over the reefs, apparently without sheltering in habitats, and consume algae, sponges, tunicates, zoantharians, zooplankton, and other benthic organisms.

Families showing this pattern include the Pomacanthidae, Ephippidae, and Labridae.

Invertebrate consumers that feed on nearby seagrass beds and sand areas are also abundant on reefs. Among these, grunts (Haemulidae) are typically most numerous. The nocturnal feeding migrations of grunts are an important source of energy exchange between adjacent habitats and can measurably increase coral growth rates (Meyer et al. 1983). Other intermediate-sized reef fishes that consume invertebrates have diurnal feeding habits; these include squirrelfishes (Holocentridae) that live in or close to reef crevices, and goatfishes (Mullidae) and hogfishes (*Bodianus rufus* and *Lachnolaimus maximus*) that roam over the reef and its vicinity.

Snappers (Lutjanidae) are prominent reef fishes in Cuba. These coastal species often inhabit coral heads or patch reefs during the day. Their association with reefs appears to be lower than that of grunts, perhaps because of their larger size. The lane snapper (*Lutjanus synagris*), mutton snapper (*L. analis*), gray snapper (*L. griseus*), schoolmaster (*L. apodus*), and cubera snapper (*L. cyanopterus*) are the most abundant snappers on the Cuban shelf. Lane snappers are numerous in many locations. They often form large groups in the Golfo de Batabanó and can display defensive schooling behaviors similar to gregarious pelagic species (Claro 1976, 1981a).

Some piscivores are shelter-associated, such as moray eels (Muraenidae) and some grouper species (Serranidae). Adult Nassau groupers (*Epinephelus striatus*) and *Mycteroperca* species use mid- or outer-shelf reef habitats for protection and to ambush prey (Silva Lee 1977). Other piscivores cruise the water column over the reef in search of food; they do not use the reef as shelter. Among them, the great barracuda (*Sphyraena barracuda*) and the jacks (*Caranx ruber* and *C. bartholomaei*) are particularly abundant and are often found around reefs in large schools. Some of the most frequent transients on the reefs are scombrids (*Scomberomorus regalis*, *S. maculatus*, etc.), several sharks (*Carcharhinus* spp., *Sphyrna* spp.), and, in shallow waters, needlefishes (*Strongylura notata*, *Tylosurus crocodilus crocodilus*, *T. acus acus*, and others) (Claro et al. 1990b).

Among tropical marine habitats, coral reefs can have exceptionally high fish biomass because of their high productivity. Most investigations of reef fish community structure and productivity have been limited to areas outside of Cuba and only a handful of studies have been conducted in Cuba (Claro et al. 1990a, 1990b; Claro and García-Arteaga 1994b; Claro et al. 2000). The association of species with habitat types might follow similar patterns across different regions of the Cuban shelf. A cluster analysis of three subzones of the Archipiélago Sabana-Camagüey (Bray Curtis dissimilarity index: Boesch 1977) showed a clear affinity among the ichthyofauna of the same reef type (Claro and García-Arteaga 1994b). Three clusters were defined: slope reefs, shallow reefs (patch reefs, reef crests, and bank reefs),

Table 2.1. Number of species, diversity index (H′), richness index (R₁), evenness index (J), density, and fish biomass of different reef habitats and regions

Region, Date	Habitat[a]	Area (m²)	No. Spp.	No. Families	H	R₁	J	Density (ind/m²)	Biomass (g/m²)
SW Cuba, Archipiélago Los Canarreos, 1984–1985									
	Reef crests	11,895	113	40	4.58	7.66	0.67	2.132	114.39
	Patch reefs*	55,668	129	36	3.26	6.86	0.47	7.459	582.41
	Slope reefs	30,370	136	39	4.14	8.48	0.58	2.040	111.31
	Mangroves	21,580	82	29	3.22	4.72	0.51	6.789	154.28
North-central Cuba, Archipiélago Sabana-Camagüey, 1987–1989									
	Reef crests	22,470	125	41	4.29	7.89	0.62	2.385	117.10
	Patch reefs*	16,498	131	38	4.11	8.36	0.58	2.907	138.50
	Bank reefs	7,940	92	29	4.55	6.85	0.70	1.083	84.66
	Slope reefs	20,610	132	33	3.95	8.54	0.56	2.019	139.90
	Mangroves	39,705	87	30	1.82	4.44	0.28	19.899	165.47
SE Cuba, Archipiélago Jardines de la Reina, 1997									
	Reef crests	4,200	103	32	4.37	7.67	0.65	2.397	108.82
	Slope reefs	11,340	136	38	3.28	8.61	0.46	4.617	203.73
	Mangroves	5,200	58	24	1.55	3.63	0.26	10.232	141.67
Guadeloupe, 1991									
	Reef crests	1,200	38	16	3.26	3.24	0.62	2.300	90.41
	Patch reefs*	250	34	—	1.95	3.13	0.38	5.928	70.60
	Slope reefs	3,800	76	28	3.39	5.60	0.54	2.841	63.55
Martinique, 1991									
	Slope reefs	5,400	89	30	2.57	5.63	0.40	9.438	62.96
Key West, Florida, 1995									
Sand Key	Slope reefs	1,800	81	26	4.04	6.46	0.64	2.959	395.41
W. Sambo	Slope reefs	1,800	56	18	3.40	4.78	0.59	1.608	50.41
Total Key West	Slope reefs	3,600	90	28	4.24	6.84	0.65	2.284	222.93

Sources: Claro and García-Arteaga 1994b; Claro et al. 1998; Claro and García-Arteaga, unpublished data.

[a] Most transects were 100 m by 3 m (300 m²); * indicates shorter transects.

and mangroves. Similar results were obtained using the Sorensen index of similarity. Approximately 45% of the species were common to the four types of reefs; 19% were found in just one reef type. Slope reefs showed the highest species richness (Table 2.1). The diversity on patch reefs was similar to the slope reefs, perhaps because of their proximity to the shelf border. Bank (very shallow mid-shelf) reefs in the Archipiélago Sabana-Camagüey showed lower density and biomass, perhaps because they have less structural complexity.

In the Golfo de Batabanó, where patch reefs are abundant far from the shelf slope, diversity was usually much lower, but the abundance of some species was much higher (Claro et al. 1990a) (Table 2.1). The reefs in the Archipiélago Jardines de la Reina in southeastern Cuba had the highest number of species, although the survey area was the smallest. This area is characterized by higher fish density and biomass, perhaps because of the influence of nutrients from terrestrial runoff over the inner shelf that may result in high biological productivity. This area also has the greatest fisheries productivity on the Cuban shelf (see Chapter 8).

An interdecade comparison of fish density and biomass among 11 slope-reef and 9 reef-crest stations in the Archipiélago Sabana-Camagüey was made using identical survey methods (Claro et al. 2000). Between 1988/1989 and 2000, fish density and biomass decreased threefold (Fig. 2.1), and there was a notable decrease in species richness. At least three factors could be involved (Claro et al. 2000): (1) the massive coral bleaching that occurred during the El Niño–Southern Oscillation (ENSO) events of 1995, 1997, and 1998 (see Section 1.8), which greatly reduced coral cover and subsequently increased algal cover (by 60–80% at most stations); (2) eutrophication from degraded inner shelf areas; and (3) overfishing caused by increased subsistence fishing since the early 1990s (Claro et al. 2000; see Chapter 8).

Tables 2.1 and 2.2 show diversity measures by habitat and an index of frequency, abundance, and biomass on slope reefs and mangroves at sites in Cuba and in the Caribbean (Martinique, Guadeloupe, and Key West, Florida). Despite fewer surveys in Martinique, Guadeloupe, and Key West, preliminary comparisons were developed to stimulate discussion of potential differences in community structure

Table 2.2. Abundant fishes and the F/a−b index for various habitats in different regions. The dominant species in a habitat (in bold) are those with a frequency > 67% with density or biomass > 1% (A/1 or A/2), or a frequency > 33% with density or biomass > 10% (B/1) (Claro et al. 1998).

Family Species	SW Cuba: Golfo de Batabanó					North-Central Cuba: Archipiélago Sabana-Camagüey					Martinique	Guadeloupe	Florida
	Slope Reefs	Patch Reefs	Reef Crests	Artificial Reefs	Mangroves	Slope Reefs	Patch Reefs	Reef Crests	Nearshore Reefs	Mangroves	Slope Reefs	Slope Reefs	Slope Reefs
Clupeidae													
Jenkinsia lamprotaenia	—	—	—	—	**B/1–3**	—	C/3–4	—	—	**A/1–2**	—	—	—
Holocentridae													
Holocentrus rufus	A/3–3	—	A/3–2	C/4–4	C/4–4	A/3–3	A/3–3	A/2–2	A/3–2	—	A/4–2	A/3–2	B/3–3
Myripristis jacobus	C/4–4	—	C/4–4	—	—	—	C/4–4	B/3–3	C/3–3	—	A/3–2	**A/2–2**	C/4–4
Serranidae													
Cephalopholis fulva	B/3–3	—	C/4–4	C/4–4	—	—	B/4–3	B/3–3	B/2–2	—	B/4–3	A/3–2	—
Grammatidae													
Gramma loreto	**A/2–3**	C/4–4	C/4–4	—	—	A/2–4	B/3–4	C/4–4	B/3–4	—	C/4–4	—	—
Lutjanidae													
Lutjanus apodus	A/3–2	A/3–2	B/4–2	C/4–4	**A/2–1**	B/3–3	A/3–2	**A/2–2**	B/2–2	**A/2–1**	—	C/4–3	B/3–2
Lutjanus griseus	C/3–3	**A/2–2**	C/3–3	**A/1–1**	**A/2–1**	—	C/3–2	C/4–4	—	A/3–1	C/4–2	—	B/3–3
Ocyurus chrysurus	**A/2–2**	B/4–4	A/3–2	B/3–3	A/3–3	**A/2–4**	A/3–2	A/3–3	**B/1–1**	C/4–4	B/4–2	B/3–2	**A/2–1**
Gerreidae													
Eucinostomus spp.	—	—	—	—	—	—	—	—	—	**A/2–3**	—	—	—
Haemulidae													
Haemulon aurolineatum	C/4–4	C/3–2	C/3–3	B/1–2	B/2–2	—	C/1–2	C/4–4	—	C/3–3	C/4–3	B/2–2	—
Haemulon flavolineatum	A/3–3	A/3–3	A/3–2	B/2–2	B/3–2	B/2–3	**A/1–2**	A/1–1	A/2–2	A/3–2	B/4–3	A/3–2	B/2–2

Continued on next page

F/a−b Index[a]

Table 2.2. continued

Family / Species	C1	C2	C3	C4	C5	C6	C7	C8	C9	C10	C11
Haemulon plumieri	A/3-3	C/4-3	B/4-2	B/3-3	A/2-3	A/2-2	A/1-1	A/2-2	B/3-2	A/1-1	A/3-2
Haemulon sciurus	B/2-2	C/4-3	C/3-2	A/2-1	A/3-2	A/2-2	A/2-2	A/2-2	A/2-1	A/2-2	A/3-2
Pomacentridae											
Abudefduf saxatilis	B/3-3	B/4-3	C/4-2	A/3-3	A/2-2	A/2-2	C/3-3	C/4-4	A/3-3	B/4-4	C/3-3
Chromis cyanea	A/2-3	A/1-2	A/2-2	—	B/2-3	A/1-3	B/2-3	A/1-3	—	—	A/1-2
Chromis multilineata	B/2-3	A/1-2	A/1-1	—	—	B/3-3	B/3-4	—	—	—	B/2-3
Microspathodon chrysurus	B/4-4	B/3-3	C/4-4	C/4-4	B/2-2	A/2-2	B/3-4	C/4-4	—	—	C/4-4
Stegastes adustus	B/4-4	C/4-4	—	C/4-4	A/2-3	A/2-3	B/3-4	C/4-4	C/4-4	—	B/3-4
Stegastes partitus	A/1-3	A/1-2	A/1-2	C/4-4	B/2-4	A/2-3	A/2-4	A/2-3	—	C/4-4	A/2-3
Labridae											
Bodianus rufus	B/3-4	B/4-4	B/4-3	C/4-4	A/2-3	A/3-3	B/3-3	A/3-3	C/4-4	C/4-4	B/4-3
Clepticus parrae	B/2-2	A/2-1	B/2-3	—	—	C/4-4	C/4-3	A/2-3	—	—	A/1-2
Halichoeres bivittatus	B/3-4	C/4-4	—	C/4-4	A/2-3	C/4-4	A/2-3	A/2-3	B/4-4	C/4-4	C/4-4
Halichoeres garnoti	A/2-3	A/2-3	A/3-2	—	A/3-4	A/3-3	A/2-3	A/2-3	C/4-4	C/4-4	A/2-3
Thalassoma bifasciatum	A/1-2	A/2-3	A/3-3	C/4-4	A/1-3	A/1-2	A/2-3	A/2-3	C/4-4	A/3-4	A/1-2
Scaridae											
Scarus iserti	A/2-3	A/2-2	A/3-2	B/4-4	A/2-3	A/2-3	A/2-3	A/2-3	B/3-3	C/3-4	A/2-2
Sparisoma chrysopterum	A/3-3	A/2-2	A/3-3	B/4-3	A/3-2	A/3-3	A/2-2	A/3-3	B/4-3	B/2-2	A/3-3
Sparisoma viride	A/3-3	A/3-2	A/3-2	B/4-4	A/2-2	A/2-2	A/3-2	A/2-2	B/4-3	B/3-3	A/3-2
Gobiidae											
Coryphopterus spp.	B/1-4	A/1-3	C/2-4	—	—	—	—	—	—	—	C/2-4
Acanthuridae											
Acanthurus bahianus	A/3-3	B/4-3	B/4-3	A/2-2	A/2-2	A/2-2	—	A/3-2	C/4-4	C/4-3	B/3-3
Acanthurus chirurgus	B/3-3	—	A/3-3	A/3-3	A/2-2	A/3-3	—	A/3-3	C/4-4	A/3-3	B/4-3
Acanthurus coeruleus	A/2-3	B/4-3	A/4-3	A/2-2	A/2-2	A/2-3	—	A/3-2	B/4-3	B/4-4	A/3-2

Sources: Data summaries from Claro and García-Arteaga 1994b; Claro et al. 1998 (includes details of the F/a−b index).

a F: frequency; a: abundance; b: biomass. For frequency (F), A: > 67%, B: 33–67%, C: < 33%. For abundance (a) and biomass (b), 1: > 10%, 2: 1–10%, 3: 0.1–1.0%, and 4: < 0.1%.

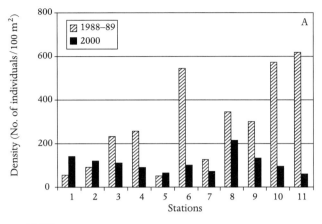

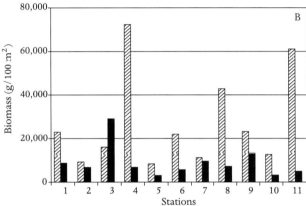

Fig. 2.1. Comparison of fish density (A) and biomass (B) between 1988/1989 and 2000 at 11 slope reef sites (20–25 m depths) across the Archipiélago Sabana-Camagüey (Claro et al. 2000a).

within the Caribbean (Claro et al. 1998). Definitive comparisons among all of these sites will require frequent sampling over extended time periods. Even with exhaustive and standardized sampling it will be difficult to compare habitat-use data from different regions because of uncontrollable differences in biological and physical characteristics among the widely separated sites.

Some earlier surveys suggested a lower biomass in certain areas of the Lesser Antilles than in Cuba and Florida (Claro and García-Arteaga 1994b; Claro et al. 1998). This could have been due to the limited number of large- and midsized fishes on Lesser Antilles reefs, a possible result of growth or ecosystem overfishing (Gobert 1990; Mahon 1993). In the samples, planktivorous and small omnivorous fishes (Pomacentridae, Labridae) were dominant in biomass, whereas mid-sized fishes prevailed in Cuba and Florida (Claro et al. 1998). On Cuban slope reefs, small zooplanktivores dominate the number of individuals; fishes and invertebrate predators dominate the biomass. Many of the latter species, including several species of snappers and grunts, reside on shallow reefs, seagrass beds, and mangroves during their early life stages. Other species use the slope reefs as juveniles (e.g., groupers).

On Cuban reef crests and patch reefs, grunts were dominant (Claro et al. 1990a, Claro and García-Arteaga 1994b; García-Arteaga and Claro 1998). The density and biomass of grunts and gray snapper were much higher on patch reefs located in seagrass beds, far from the slope reefs (as in the Golfo de Batabanó), than on other studied reefs. Herbivorous fishes showed the greatest biomass on reef crests. On all reef types, herbivores were more abundant in Cuba than in Martinique and Guadeloupe (Claro et al. 1998; see Table 5.4).

2.3.2   Mangrove Fish Fauna

Mangrove habitats can occur in many physiographic regimes across the shelf (Lugo and Snedaker 1974). By looking at community structure, trophic relationships, and habitat conditions, the fishes that associate with mangroves can be divided into two basic groups: those in mangroves of more turbid, lower salinity estuaries and those in mangroves associated with fringing islands, keys, and islets on the open shelf under more oceanic conditions. Ley et al. (1999) detailed shifts in fish assemblages across a gradient from estuarine to open shelf waters that many mangrove habitats can occupy. Because of the differing characteristics of estuarine areas, their fish fauna is addressed separately (see Section 2.3.4).

In visual censuses of the mangroves of Punta del Este, Isla de la Juventud in the Golfo de Batabanó, 55 fish species were noted, with an additional 12 species in the vicinity of this habitat (Valdés-Muñoz 1981). Only eight of these species were recorded from softbottom habitats near mangroves in the coastal lagoons of south-central Cuba (González-Sansón et al. 1978). A total of 78 species were recorded in 18 euhaline mangrove stations in the Golfo de Batabanó, although only 21 to 51 species (mean, 46) were present in each (Valdés-Muñoz et al. 1990). In the Archipiélago Sabana-Camagüey, a total of 85 species were observed in 63 surveys of mangrove habitats in 1988/1989 (Claro and García-Arteaga 1993). High overall densities (Table 2.1) were due in part to large schools of dwarf herring (*Jenkinsia lamprotaenia*).

The most abundant species in mangroves were the dwarf herring, the hardhead silverside (*Atherinomorus stipes*), and *Harengula* spp. These planktivorous species accounted for 80–90% of all fishes surveyed in this habitat, although they were low in biomass (2–8%). The dominant species according to biomass and frequency were bluestriped grunt (*Haemulon sciurus*), gray snapper, and schoolmaster. Other species also common in this habitat included sergeant major (*Abudefduf saxatilis*), foureye butterflyfish (*Chaetodon capistratus*), beaugregory (*Stegastes leucostictus*), great barracuda juveniles, dog snapper (*Lutjanus jocu*), doctorfish (*Acanthurus chirurgus*), and the French grunt (*Haemulon flavolineatum*) (Claro and García-Arteaga 1993).

Ten of the same mangrove stations in the Archipiélago Sabana-Camagüey were resampled in 2000 using the same methods. Total fish density was four times lower than in 1988/1989 and fish biomass was 1.6 times lower. The difference in density was due to a decrease in schools of dwarf herring, hardhead silverside, and *Harengula* spp.; the reduced biomass was due to a decrease in grunts and snappers. These declines in fish density and biomass could be the result of degradation of the inner lagoons of the Archipiélago Sabana-Camagüey (Claro et al. 2000; and see CUB/92/G31 1999).

Although most of the individuals are planktivores, most species using euhaline mangroves are carnivorous, mainly invertebrate feeders. Piscivores are also present. The number of herbivorous species is relatively small (Sierra et al. 1990; Claro and García-Arteaga 1993). Many carnivores (such as snappers and grunts) feed at night in nearby seagrass beds and use the mangrove roots during the day, although some predators also feed in the mangroves.

Many relatively abundant species maintain a less direct association with mangroves. Small mojarras (*Eucinostomus* spp. and *Gerres cinereus*) frequently move between mangrove and seagrass habitats. Parrotfish and surgeonfish show similar behaviors. Typically at sunset planktivores, as well as fishes and invertebrate consumers, spread out through the nearby seagrass beds to forage (Rooker and Dennis 1991). Various predators, such as great barracuda, needlefishes (*Strongylura notata*, *Tylosurus* spp.), and small groups of bar jacks (*Caranx ruber*), move near or under the mangroves to seize dwarf herring and sardines.

These euhaline mangrove fishes include many species also living on coral reefs, but of earlier life stages and in different proportions and biomass. Relatively similar species compositions have been found in high-salinity, low-turbidity mangrove habitats of southeast Florida (Ley et al. 1999) and Puerto Rico (Rooker and Dennis 1991; Dennis 1992a).

2.3.3 Fish Fauna of Seagrass Beds and Euhaline Lagoons

Large areas of the shelf, influenced by oceanic waters, are covered by seagrass beds or bare sandy bottoms inhabited by a substantial number of species, including herbivores, detritivores, and benthic carnivores. Among the 98 species recorded from seagrasses of the Golfo de Batabanó, the most common are batrachoidids (mainly *Opsanus phobetron*), chaenopsids (*Chaenopsis ocellata*), gobiids (*Ctenogobius boleosoma*), and ophichthids (*Ophichthus gomesii, O. ophis, Myrichthys breviceps*) (E. Valdés-Muñoz, unpublished data). Many of these species burrow into the sediment during the day and forage at night. Some diodontids (mainly *Diodon hystrix* and *D. holocanthus*), ostraciids (*Acanthostracion quadricornis, Lactophrys trigonus,* and *L. bicaudalis*), and tetraodontids (*Sphoeroides* spp.) are also

common. Close to mangroves, small mojarras (*Eucinostomus spp.* and *Gerres cinereus*) are abundant. Stingrays, particularly the southern stingray (*Dasyatis americana*), are common in seagrass beds, sandy areas, and muddy bottoms.

Despite the abundance of food for fishes in sandy seagrass beds (Murina et al. 1969; Ibarzábal-Bombalier 1982), this fish fauna is often low in biomass and diversity in areas located far from reefs. An absence of heterogenous and high-relief habitats over wide areas of the shelf probably explains the low abundance of larger demersal and pelagic fishes in some localities. In areas closer to the shelf border, seagrass beds alternate with patch reefs and coral heads, creating a complex mosaic of habitats. In these areas, seagrass beds and, to a lesser extent, mud or sand areas constitute the principal feeding areas for many fishes. Through nocturnal feeding migrations, seagrass beds and nearby softbottom areas are important sources of energy for fishes inhabiting reefs and mangroves (Parrish 1989).

Nearshore seagrass beds (close to the mainland and to the keys as well) of the Golfo de Batabanó are also important nursery areas for early stages of many reef fishes, particularly snappers and grunts that subsequently migrate to deeper reef areas as they grow (García-Arteaga et al. 1990). Juvenile predators, such as the Nassau grouper, can also be found among the seagrass blades. On larger temporal scales than diurnal feeding migrations, ontogenetic habitat shifts across the shelf into deeper waters are shown by many snapper, grouper, grunt, and other species. The temporal dynamics and spatial complexity of interactions among these habitats and the many dozens of fish species and life stages suggest that many of these taxa have considerable plasticity in the use of trophic and habitat resources.

In seagrass and sand areas of the Golfo de Batabanó, where patch reefs are limited, fishermen build artificial refuges for concentrating fishes. These structures, called *pesqueros,* are built with mangrove branches, tires, scrap metal, and other disposable objects (Silva Lee 1975a). They are a major factor in the high productivity of trawl fisheries in that region (Bustamante et al. 1982; Silva Lee and Valdés-Muñoz 1985; Giménez et al. 1986; Claro and Giménez 1989; Claro et al. 1990a; Claro and García-Arteaga 1999). They also appear to be an important structural element in the aggregation of fish communities in regions that lack natural shelters. The fish species that aggregate at the *pesqueros* are similar to those on patch reefs, although the community is less diverse (see previous references). An important factor that affects diversity and density of fishes using artificial habitats is that the *pesqueros* are usually placed far from reefs and in densities of 0.2 to 0.5 (mean, 0.4) per square kilometer (Giménez et al. 1986); patch reefs are rarely so isolated.

Recent surveys of seagrass beds and mud bottoms of the inner part of the Archipiélago Sabana-Camagüey (north-central coast) recorded 127 species (Claro et al. 2001a). Differences between the Archipiélago Sabana-Camagüey and the Golfo de Batabanó faunas were evident. In the former, the dominant species by number of individuals were mojarras (*Eucinostomus* spp., *Eugerres* spp., and *Gerres cinereus*) and some other fishes associated with estuarine seagrass beds: *Archosargus rhomboidalis, Elops saurus, Megalops atlanticus, Albula vulpes, Centropomus undecimalis, Caranx ruber, C. bartholomaei, C. hippos, Opsanus beta,* and *O. phobetron.* Various monacanthids (*Aluterus schoepfi, Monacanthus ciliatus,* and *Stephanolepis setifer*) were also abundant. In this ecosystem, invertebrate consumers were most common.

In the Archipiélago Sabana-Camagüey, Bahía de Buenavista (Fig. 1.4) had the highest number of species. When standardized for area surveyed, however, species richness was higher in Puerto de Sagua where there is more water exchange with the open ocean (Claro et al. 2001a). Otter trawl surveys yielded a higher fish density in Bahía de Buenavista and Puerto de Sagua than in other locations, with smaller species predominating, mainly *Eucinostomus gula* and *E. jonesii* (62% of all individuals).

Bahía Los Perros and Bahía La Gloria showed the lowest species richness. In 1995 surveys, only six fish species were present in Bahía Los Perros (salinities were 67–75‰). This is a result of artificially elevated salinities caused by river damming over the last 30 years and, more recently, the disruption of previously euryhaline coastal lagoons by the construction of long, impermeable causeways to join the main island with new tourist resorts on offshore islands (see Fig. 1.4). Mojarras (*Eucinostomus* spp.) and the sheepshead minnow (*Cyprinodon variegatus*) are most abundant. Salinity was lower in 1996 and 1998 (43–51‰), and 15–16 species were recorded (Claro et al. 2000, in press a). Similar environmental conditions have been observed in Bahía La Gloria and Bahía Jigüey. The commercial fauna has almost disappeared from these lagoons; Bahía de Buenavista has been seriously affected as well. Bahía de Buenavista appears to be an important nursery area for some fishes, where, as adults migrate west to Bahía San Juan de los Remedios, they are fished intensively. Bahía San Juan de los Remedios and Puerto de Sagua had the highest fish density and biomass.

Stingrays, mainly the southern stingray (*Dasyatis americana*), constitute one of the major fishery targets in the inner water bodies of the Archipiélago Sabana-Camagüey. In less affected bays, such as San Juan de los Remedios, Nazabal, and Puerto de Sagua, snappers (mainly lane, mutton, and gray) and jacks (*Caranx* spp., *Selar crumenophthalmus,* and others) were the most abundant commercial fishes. They appear to move out when salinity exceeds 40–44‰. When salinity decreased in 1996, the number of predators increased and the density of small fishes decreased (Claro et al. 2000, in press a).

2.3.4 Fish Fauna of Estuaries and Low-Salinity Lagoons

Most species in low-salinity areas are extremely tolerant of environmental variations. However, spatial and seasonal variation in hydrology can affect fish community structure as less tolerant species move in response to physiological stressors.

In the Tunas de Zaza lagoon system along the south-central Cuban coast (Fig. 1.5), 58 species have been reported (González-Sansón et al. 1978). Using the same sampling methods, a similar number was found in the estuarine systems of the Río Cauto and Ciénaga Litoral del Sur, east of Tunas de Zaza (González-Sansón and Aguilar Betancourt 1983). Similar fish diversities and evenness were estimated (by the Shannon-Weaver index) in these three areas. Four trophic groups have been defined:

- Herbivores, represented exclusively by tilapia (*Oreochromis aureus*), were introduced in the dammed reservoirs close to the Río Zaza, arrived at the lagoons in 1976, and became the main fishery resource.
- Detritivores, such as mullets (*Mugil liza, M. curema, M. trichodon,* and *M. incilis*) and some cyprinodontiform fishes (e.g., *Cyprinodon variegatus* and *Limia vittata*) are common, with mullets constituting an important fishery resource.
- Small invertebrate consumers were common before the introduction of tilapia. In the inner areas of the Tunas de Zaza and Cauto lagoon systems, this group includes the mojarras *Eugerres brasilianus* (33% of all individuals) and *Diapterus rhombeus* (34%); in Ciénaga Litoral del Sur, sea bream (*Archosargus rhomboidalis,* 48%) and yellowfin mojarra (*Gerres cinereus,* 13%) were more abundant. The species *Anchovia clupeoides* and *Eucinostomus* spp. also belong to this group.
- Consumers of small fishes and decapods, among which *Lutjanus griseus, Centropomus ensiferus,* and *Bairdiella ronchus* were most abundant. *Elops saurus, Megalops atlanticus* (juveniles), *Centropomus undecimalis,* and *Sphyraena barracuda* were also abundant. In Ciénaga Litoral del Sur, *Albula vulpes* and *Haemulon bonariense* were also relatively abundant.

The low abundance of natural predators on primary species (mojarras, tilapia, and mullet) has been suggested as a major factor influencing the structure of fish assemblages in these lagoons (González-Sansón and Aguilar Betancourt 1983).

Mangrove habitats of nearshore, high-turbidity, low-salinity areas differ in structural and trophic attributes from euhaline, clear-water mangroves farther out on the shelf (Lugo and Snedaker 1974). Limited information suggests a gradient of differing ichthyofaunal assemblages between these physiographic zones. Thayer et al. (1987) found a higher percentage of estuarine species in nearshore, high-turbidity mangrove habitats in south Florida (by rotenone sampling). This pattern was also observed by Ley et al. (1999)

in mangrove habitats of both high- and low-salinity areas (by visual surveys and traps).

2.3.5 Oceanic Fish Fauna

At least 300 fish species are known to inhabit the oceanic waters adjacent to the Cuban shelf, most notably pelagic species and those associated with the deep bottom, bathyal, and abyssal zones. Large migratory predators are the best known of the pelagic–oceanic species: tuna (Scombridae), swordfish (*Xiphias gladius*), billfishes (Istiophoridae), and sharks. Skipjack tuna (*Katsuwonus pelamis*) and blackfin tuna (*Thunnus atlanticus*) are abundant and of high commercial significance. They occur together in large mixed schools that frequently approach the coastal zone. Like other scombrids, they live in the water column above the thermocline, which in Cuba is located at depths of 100–150 m. These species are frequently found close to the surface where they are caught by fishermen. Blue marlin (*Makaira nigricans*), white marlin (*Tetrapturus albidus*), and swordfish prevail in the commercial and billfish/swordfish recreational fisheries (Guitart Manday 1975). The blue and white marlin runs occur in the summer, as the fishes move toward their spawning grounds in the Caribbean Sea and Gulf of Mexico. Swordfish runs occur in winter and early spring (Guitart Manday 1975).

Many oceanic shark species almost never enter shelf waters (Alopiidae, Lamnidae, and some Carcharhinidae). Similarly, species that live at great depths (Hexanchidae,

Scyliorhinidae, Squalidae, and some other species of carcharinids) are seldom found in shelf waters. Some other sharks, such as sphyrnids and *Galeocerdo cuvier*, can be found occasionally in both shelf and oceanic waters. Some sharks are territorial, although the territorial boundaries of each individual depend on its migratory nature (Guitart Manday 1983). Most of these species are carnivorous and feed on small fishes and squid.

Biological communities aggregating under and within *Sargassum* macroalgae and other drifting objects may play an important role in ocean ecosystems. Large numbers of fishes, such as jacks, dolphin, and small tuna, can be found associated with drifting objects (Hunter 1968). One inventory found 100 invertebrates and 72 fish species associated with Atlantic *Sargassum* (Dooley 1972). Similar patterns were found by Settle (1993). Forty-two species (24 neritic, 12 oceanic, and 6 deep-water fishes) were reported from *Sargassum* collected southwest of Bermuda (Fedoriako 1982). Species frequently recorded include *Histrio histrio*, *Syngnathus pelagicus*, *Coryphaena hippurus*, *C. equiselis*, *Diodon holocanthus*, and *Kyphosus sectatrix*, all of which inhabit Cuban waters.

An estimated 8,400 individual fishes were found associated with a 3.5-metric ton (t) mat of floating *Sargassum* collected in the Florida Straits. The fishes belonged to 8 orders, 23 families, 36 genera, and 54 species (Dooley 1972). The families Carangidae (14 species), Balistidae and Monacanthidae (14 species), and Antennariidae (1 species) accounted for 90% of the total number of individuals. Of all

Table 2.3. Fish assemblages on artificial reefs at several locations in the Golfo de Batabanó

Species or Family	Relative Percentage of Individuals					
	Cayo Tablones		Flamenco-Traviesa		West Cayo Sigua	
	1974[a]	1984[b]	1978/80[c]	1984[b]	1979/80[c]	1984[b]
Lane snapper (*Lutjanus synagris*)	32.8	1.3	2.3	2.2	7.7	2.2
Gray snapper (*Lutjanus griseus*)	26.4	23.8	8.2	10.2	9.1	14.9
Other snappers	2.2	0.5	12.2	0.4	6.8	0.5
Total snappers	61.4	25.6	22.5	12.8	23.6	17.6
Bluestriped grunt (*Haemulon sciurus*)	12.7	8.0	6.8	1.1	2.7	1.4
White grunt (*Haemulon plumieri*)	14.4	28.6	45.9	11.0	23.1	24.9
Tomtate (*Haemulon aurolineatum*)	0	2.4	1.4	63.2	19.2	48.1
Other grunts	0	42.2	1.4	0.2	1.2	0.8
Total grunts	27.1	17.6	54.1	75.5	46.2	75.2
Jacks (Carangidae)	8.7	5.0	1.9	8.1	6.5	4.5
Parrotfishes (Scaridae)	0.05	0.1	3.8	0.1	11.5	0.2
Porgies (Sparidae)	0	0.1	12.0	0.2	8.1	1.3
Hogfish (*Lachnolaimus maximus*)	<0.1	0.4	0.4	0.4	0.2	0.2
Sphyraenidae	2.3	2.6	0.4	0.2	0.2	0
Other fishes	0.9	6.4	5.3	2.6	3.4	0.6

Sources: [a] Silva Lee 1975a; [b] Claro et al. 1990b; [c] Bustamante et al. 1982.

fishes, 74% were triggerfishes and jacks. Triggerfishes are a particularly important food target for dolphin and tuna (Gibbs and Collette 1959; Lewis 1967).

Many factors are involved in the aggregation of *Sargassum* and the fish communities beneath. Several hypotheses were examined by Fedoriako (1982) who emphasized hydrological factors, mainly Langmuir circulation. *Sargassum* often collect along narrow convergence zones associated with Langmuir circulation cells. Mechanical effects of moving water masses can also advect organisms toward the *Sargassum,* and aggregations are then maintained through visual and tactile mechanisms.

The visual orientation that objects provide in such a uniform optical medium is a primary factor in the formation of drifting communities (Senta 1962; Hunter and Mitchell 1967). Thigmotropism can also play a role in species associations with the algae (Breder and Nigrelli 1938). Several authors suggest that *Sargassum* provides abundant food resources (Pratt 1935; Ida et al. 1967; Dooley 1972). Some species (several balistids, for example), consume floating plants (Soemarto 1958; Gooding and Magnuson 1967). Other species use the shadows provided by objects as a background for easier sighting of planktonic prey (Damant 1921). Some fishes might use floating objects as a refuge from predators (Soemarto 1958; Hunter and Mitchell 1967; Ida et al. 1967; Dooley 1972) and as mechanical protection against surge (Becebnov 1960). The aggregates might also serve as cleaning stations for ectoparasites of many fishes (Gooding and Magnuson 1967).

2.3.6 Bathyal Fish Fauna

More than 130 fish species have been recorded from the bathyal zone around Cuba. A survey of several slope and bathyal regions off the southern Cuban coast was performed using the *Johnson Sea-Link II* submersible in December 1997 in collaboration with R. Grant Gilmore Jr., C. Richard Robins, and John McCosker. At depths of 100–600 m, more than 30 species were collected from Cuban waters for the first time (Claro et al. 2000b). Most of these were small species, typically those associated with topographic irregularities (rocks, crevices, etc.). They included representatives of the families Serranidae, Lutjanidae, Bythitidae, Symphysanodontidae, Grammatidae, Epigonidae, and Percophidae (see Appendix 2.1). In addition, three potential new species in the families Percophidae, Gobiidae, and Bythitidae were collected.

At these depths, large invertebrate and fish predators are prominent, including the sixgill shark (*Hexanchus griseus*), the silk (*Lutjanus vivanus*) and blackfin (*L. buccanella*) snappers, and the misty grouper (*Epinephelus mystacinus*). All of these species are commercially significant. Many deep-water taxa perform notable vertical migrations for feeding (e.g., Sternoptychidae, Myctophidae). Other taxa represented in deeper pelagic waters off Cuba include the Chauliodonti-

nae, Melanostomiinae, Malacosteinae, Paralepididae, Trachipteridae, Bramidae, Gempylidae, Stromateidae, and others.

2.4 Changes in Population and Community Structure

The species composition of reef fish assemblages changes over time as a result of the dynamics between mortality and recolonization (Doherty and Williams 1988; Sale 1991; Caley et al. 1996). Assemblage structure typically changes when the relative proportions of one or more components change. This can result from shifts in environmental resources, recruitment variation, increases or decreases in the survivorship of a certain generation, or population declines caused by disturbance events, epidemics, and changes in food supply. Many of these changes can be initiated, directly or indirectly, by human activities that degrade habitat (e.g., coastal construction, trawling) or promote overfishing (e.g., fish traps, unregulated fishing of spawning aggregations).

The most observable human-induced changes in species compositions are related to fisheries activity, the first notable manifestation of which is a variation in the age structure of fish stocks (Nikolsky 1974a). One example of such a change in Cuba occurred in the lane snapper population of the Golfo de Batabanó in the 1970s and 1980s. Until 1976, the lane snapper was the dominant species in that region (more than 40% of total finfish catch) (Claro 1991; see Section 8.6). Although no fish surveys were conducted in the 1970s, our underwater observations of reefs and seagrasses (during other fisheries investigations in the area) allowed us to frequently observe large schools of lane snapper (much larger schools than any other species). Later fish surveys conducted in 1984/1985 (Claro et al. 1990a) found an almost complete absence of lane snapper from its previous habitats and a dominance of some grunt species (Table 2.3). On patch reefs, once holding the highest lane snapper concentrations of all habitats, the number of individuals decreased to less than 1%, whereas grunts (mainly tomtate, *Haemulon aurolineatum*) increased to 75% of the total number of individuals.

The greatest evidence of changes in species composition is from data collected in artificial habitats before and after the collapse of the lane snapper fishery. Table 2.3 shows data obtained by Silva Lee (1975a) during the period of highest catch, Bustamante et al. (1982) in 1978–1980 soon after the fisheries decline, and Claro et al. (1990a) in 1984/1985. In 1978–1980, a dramatic reduction in lane snapper was already evident. In 1974 snapper species accounted for more than 60% of total individuals; 10 years later, they represented 10–25%, being more abundant on inner shelf areas (Cayo Tablones). In 1984/1985, snappers reached 4–10% on patch reefs and 38% in mangroves (Claro et al. 1990a; Valdés-Muñoz et al. 1990). Grunts, unlike lane snappers, increased in artificial refuges of the eastern part of the Golfo de Batabanó: in 1974 they represented only 27.1% of the individuals and in 1984/1985 they represented 66–75%.

Lane snapper overfishing—particularly the disturbance of reproduction by the use of set nets across migratory routes that catch concentrations of migrating spawners—might foster its replacement by other species. In addition, the absence or reduction in the number of new recruits during one or more years might have contributed to the replacement of lane snapper in the ecosystem by species with longer reproductive seasons, such as grunts (see Chapter 4). Although the relative abundance of gray snapper was low, biomass was high and catches of gray snapper increased as it replaced lane snapper as the main fishery in the region. In addition, since 1977 fishermen stopped fishing small grunts because of their low commercial value, which may also have contributed to the increase of grunt populations. This could explain why, despite protective measures implemented since 1978 (such as partial or total closures during spawning, limited fishing quotas), lane snapper stocks started to show signs of recovery only 20 years later, and their fishery potential remains at 50% of the earlier estimates. A related type of population alteration could be occurring in the Nassau grouper, in which young individuals are currently prevalent (Ministerio de la Industria Pesquera 1980; Leonel Espinosa, personal communication).

Ecosystem overfishing resulting from chronic over-exploitation can provoke substantial changes in fish community structure (Pauly 1979; Pauly et al. 1998). Increased overfishing has been documented for serranids and lutjanids in the Florida Keys (Ault et al. 1998). As a result of excessive fishing effort, the abundance of large carnivores (snappers and groupers) has also diminished markedly in northern Jamaica (Munro 1983b; Koslow et al. 1988), Martinique, Guadeloupe, and St. Lucia (Gobert 1990). The Jamaican study (Munro 1983b) demonstrated that the collapse of these resources led to dramatic decreases in fisheries yield (to less than 100 g/trap-day) and catch quality (small reef fishes became dominant in catches).

In other Caribbean locations, several authors have also documented the limited abundance of large groupers on reefs and in fisheries. A comparative study of four areas (Florida Keys, central Bahamas, Bahía Guantánamo in southeast Cuba, and Parque Nacional del Este in southeast Dominican Republic) revealed significant variation in grouper composition, density, and size, attributable to different levels of protection (Chiappone et al. 1998). The abundance of Nassau grouper (*Epinephelus striatus*), the main target species, was notably reduced in intensively exploited locations, and smaller groupers (*Cephalopholis fulva* and *C. cruentata*) were numerous (Sluka et al. 1994; Sluka and Sullivan 1996). Although no historical data are available, reef fish surveys in southeast Dominican Republic showed that yellowtail snapper (*Ocyurus chrysurus*) was the most numer-

ous snapper, whereas mutton and gray snappers were much less abundant, and large groupers were essentially absent (Bustamante et al. 1998). Fisheries landings in the area reflected the same pattern.

Overfishing of the oldest individuals of hermaphroditic groupers might alter sex ratios in a variety of detrimental ways (Bannerot et al. 1987; Huntsman and Schaaf 1994; Koenig et al. 1996; Coleman et al. 1999). Overfishing affects not only the quantity and quality of fisheries resources, but also the quality of the ecosystem. Unfortunately, information on fish community structure before and during the growth of Cuba's fisheries industries is limited, making it impossible to detect such changes in other areas. Nevertheless, the information related above emphasizes the need to examine ecosystem resources with an integrated approach to facilitate sound management. The cases provided here are only a few examples of the many anthropogenic changes generated within Cuba and similar marine ecosystems.

2.5 Additional Information on Fish Fauna

2.5.1 Fish Size
More than 70% of the fish species inhabiting Cuban waters do not reach a maximum length of 50 cm. Of 398 neritic species, 42.2% have a maximum length of less than 20 cm and 36.7% have a maximum length of 21–50 cm (Claro 1994b). The many smaller species might allow the fish fauna to maximize shelter use in reef areas, as well as to exploit a variety of food resources. About 88% of the small species (maximum length < 20 cm) are demersal, the remainder are pelagic. Among demersal habitats, reef and nearshore seagrass beds are used most commonly. Smaller species usually live in reef crevices and fissures, and their feeding and reproductive movements are limited. Fishes of the families Blenniidae, Labrisomidae, Tripterygiidae, Chaenopsidae, Gobiidae, and others, whose maximum length is typically less than 10 cm, forage within several centimeters or a few meters of their refuges; they usually are solitary or live in small groups. Among fishes that permanently inhabit seagrass beds, species with a maximum length less than 20 cm prevail.

Among fishes with a maximum length of 10 to 20 cm, the grunts *Haemulon flavolineatum* and *H. aurolineatum* are particularly abundant in number and biomass in a variety of habitats. In areas of the wider Caribbean, individuals of these species can reach much larger sizes than in Cuba (Manooch and Barans 1982; García-Arteaga 1983). These differences could result from latitudinal temperature differences, or perhaps high population densities in Cuban waters result in lower growth rates.

The majority of commercially significant neritic species with relatively large populations have intermediate lengths, ranging from 20 to 70 cm. Among the most abundant are snappers (lane, gray, mutton, schoolmaster, yellowtail, and others), grunts (*Haemulon plumieri, H. sciurus*), jacks (*Caranx latus, C. crysos, C. ruber, C. bartholomaei, Selar crumenophthalmus*), and groupers (*Epinephelus striatus, Mycteroperca* spp.). A small group of mid-sized species, including snappers, groupers, and bigeyes (Priacanthidae), live on the insular slope. Other than *Sphyraena barracuda,* sharks and rays, billfishes and swordfishes, mackerels (*Scomberomorus* spp.), moray eels (*Gymnothorax*), and the cobia (*Rachycentron canadum*), few species exceed 1 m in length.

Among pelagic species, fishes larger than 1 m make up 35% of the oceanic species, but only 6.5% of the neritic species (calculated from Appendix 2.1; Claro 1994). Small pelagic species are the most numerous fishes in oceanic waters in terms of abundance. Because of the relative absence of upwelling zones around Cuba (see Chapter 1), large offshore concentrations of taxa such as engraulids are not an important fishery resource in Cuba. Nevertheless, flyingfishes and other small pelagics are relatively abundant. In addition, small ubiquitous mesopelagic fishes (gonostomatids, myctophids) migrate to surface waters each night, thus spending half of their lives in near-surface waters (R. Grant Gilmore, Jr., personal communication).

2.5.2   Biotoxicity

Ichthyosarcotoxins have affected human consumers of fishes at many latitudes, but the effects seem to be more frequent and more dangerous among tropical marine fishes. Because different toxins can cause similar symptoms, the pathology is not always well defined and can be mistaken for other kinds of poisoning. According to Halstead (1967) and Claro (1994), 107 fish species of the Cuban ichthyofauna are potentially toxic. Relevant information on ciguatera in coastal fishes of Florida has been reviewed by De Sylva (1994).

The majority of fish poisoning cases reported in Cuba have probably been caused by ciguatoxin. Poey Aloy (1866) recognized that predatory fishes on the northern coast of Cuba had a greater propensity to acquire ciguatoxin, whereas those living on the south coast could be eaten with much less risk. Among five of the six species examined from northwest Cuba, Valdés-Muñoz (1980) found that 17–44% of the individuals were toxic. However, in our surveys throughout the Golfo de Batabanó (southeast Cuba) and the Archipiélago Sabana-Camagüey (north-central Cuba), only one ciguatoxic fish specimen was found, a black grouper (*Mycteroperca bonaci*). Valdés-Muñoz (1980) obtained the highest percentage of toxic individuals among *M. bonaci, M. tigris, Caranx lugubris,* and *Sphyraena barracuda*. A higher occurrence of toxic fishes in northwest Cuba could be caused by the greater abundance of dead corals in that area, the result of strong swells generated by winter storms. The data do not rule out the presence of ciguatoxic fishes on the south and northeast coasts; there have been several anecdotal reports from both regions.

2.6   Summary

The marine ichthyofauna of Cuba is characterized by a high species diversity, with 1,030 recognized species including subspecies (Appendix 2.1), that is associated with a great variety of nearshore and oceanic habitats, a large shelf area, and well-developed reefs. Systematic revisions of many taxa are necessary, numerous species require some clarification, and occurrences in Cuban waters still require confirmation. Habitat use and trophic relationships are key determinants of the assemblage structure of fishes in Cuba. In addition to habitat variables, differences in fish community structure within and among different reef types are influenced by recruitment and physical disturbance events. Presumably because of high predation pressure, defensive mechanisms of many species are associated with refuge use.

Slope reef habitats have the highest fish species diversity. The many reefs along the shelf edge ensure a great variety of shelters, foods, and spawning areas. Fish density and biomass is higher on patch reefs than on other reef types, but can be even higher on artificial habitats in the extensive shallow seagrass beds, otherwise void of hard structure, in the Golfo de Batabanó. High fish biomass in these habitats is possibly enhanced by energy transfer from adjacent seagrass beds.

A decrease in fish species richness has been observed in the last 10 years in the north-central region, as well as a threefold decrease in fish density and biomass on the fore reefs, reef crests, and mangroves. In addition to fishing pressure, this may be the result of coral cover decreases and algal cover increases associated with increased water temperatures during ENSO events, amplified by nutrient loading from terrestrial sources.

Although seagrass beds are highly productive habitats, the relative lack of topographic complexity for larger species results in low fish biomass per unit area. Close to offshore keys, where seagrasses are dense, juveniles of commercially significant fishes (such as snappers and grunts) often find refuge and food. Euhaline mangrove fish assemblages can be similar in composition to shallow-reef fish assemblages, although differing from assemblages near the mainland. Patch reefs, mangroves, and seagrass beds constitute a habitat complex used by many species, particularly in their early life stages. The availability of food and refuge can limit fish abundance and diversity in many habitats, as well as variations in larval recruitment.

Estuarine fish assemblages of Cuba can tolerate abrupt environmental changes and are characterized by low diversity and high biomass. Detritivores, herbivores, and small invertebrate consumers are most numerous; large piscivorous predators are less abundant. Among fishes inhabiting Cuban oceanic waters, large migratory predators such as tuna and billfishes prevail, as well as sharks and some snappers and groupers associated with deep bottom habitats. The lack of notable upwelling areas may limit the abundance of pelagic fishes.

Appendix 2.1. Fish species recorded from Cuba. Classification largely follows Nelson (1994) at the family and subfamily level and above and Eschmeyer (1990, 1998) at the genus and species level. For some species (*), presence in Cuba or species identification requires confirmation.

Scientific Name	English Name	Cuban Name
SUBPHYLUM VERTEBRATA		
SUPERCLASS AGNATHA		
CLASS MYXINI		
Order Myxiniformes		
Myxinidae		
Eptatretus minor Fernholm & Hubbs, 1981*		
Eptatretus springeri (Bigelow & Schroeder, 1952)*	Gulf hagfish	
SUPERCLASS GNATHOSTOMATA		
CLASS CHONDRICHTHYES		
SUBCLASS ELASMOBRANCHII		
Order Hexanchiformes		
Hexanchidae		
Heptranchias perlo (Bonnaterre, 1788)	sharpnose sevengill shark	tiburón de 7 branchias
Hexanchus griseus (Bonnaterre, 1788)	bluntnose sixgill shark	marrajo
Hexanchus vitulus Springer & Waller, 1969	bigeye sixgill shark	marrajo ojigrande
Order Orectolobiformes		
Rhincodontidae		
Rhincodon typus Smith, 1828	whale shark	dámero
Ginglymostomatidae		
Ginglymostoma cirratum (Bonnaterre, 1788)	nurse shark	tiburón gata
Order Lamniformes		
Odontaspididae		
Carcharias taurus Rafinesque, 1810	sand tiger	tigre arenero
Odontaspis ferox (Risso, 1810)[1]	ragged-tooth shark	
Lamnidae		
Carcharodon carcharias (Linnaeus, 1758)	white shark	jaquetón de ley
Isurus oxyrinchus Rafinesque, 1810	shortfin mako	dientuso azul
Isurus paucus Guitart Manday, 1966	longfin mako	dientuso prieto
Cetorhinidae		
Cetorhinus maximus (Gunnerus, 1765)	basking shark	pez elefante
Alopiidae		
Alopias superciliosus (Lowe, 1841)	bigeye thresher	zorro ojón
Alopias vulpinus (Bonnaterre, 1788)	thresher shark	zorro
Order Carcharhiniformes		
Scyliorhinidae		
Apristurus riveri Bigelow & Schroeder, 1944	broadgill catshark	gatica afilada
Galeus arae (Nichols, 1927)	marbled catshark	gatica manchada
Scyliorhinus boa Goode & Bean, 1896	boa catshark	gatica
Scyliorhinus torrei Howell Rivero, 1936	dwarf catshark	gatica prieta
Pseudotriakidae		
Pseudotriakis microdon Capello, 1867	false catshark	falsa gata
Triakidae		
Allomycter dissutus Guitart Manday, 1972*		tiburoncito narigón
Mustelus canis insularis Heemstra, 1997	smooth dogfish	boca dulce
Mustelus norrisi Springer, 1939	Florida smooth-hound	boca dulce
Proscylliidae		
Eridacnis barbouri (Bigelow & Schroeder, 1944)	Cuban ribbontail catshark	tiburón turquino
Carcharhinidae		
Carcharhinus acronotus (Poey, 1860)	blacknose shark	tiburón limón
Carcharhinus altimus (Springer, 1950)	bignose shark	baboso

Continued on next page

Appendix 2.1. continued

Carcharhinus brevipinna (Müller & Henle, 1839)	spinner shark	tiburón de arrecife
Carcharhinus falciformis (Müller & Henle, 1839)	silky shark	jaquetón
Carcharhinus isodon (Müller & Henle, 1839)	finetooth shark	galano dientefino
Carcharhinus leucas (Müller & Henle, 1839)	bull shark	cabeza de batea
Carcharhinus limbatus (Müller & Henle, 1839)	blacktip shark	balicero
Carcharhinus longimanus (Poey, 1861)	oceanic whitetip shark	galano
Carcharhinus obscurus (Lesueur, 1818)	dusky shark	tiburón amarillo
Carcharhinus perezii (Poey, 1876)	reef shark	cabeza dura
Carcharhinus plumbeus (Nardo, 1827)	sandbar shark	arenero
Carcharhinus signatus (Poey, 1868)	night shark	tiburón de noche
Galeocerdo cuvier (Péron & Lesueur, 1822)	tiger shark	tiburón tigre
Negaprion brevirostris (Poey, 1868)	lemon shark	galano de ley
Prionace glauca (Linnaeus, 1758)	blue shark	tiburón azul
Rhizoprionodon porosus (Poey, 1861)	Caribbean sharpnose shark	cazón de ley
Rhizoprionodon terraenovae (Richardson, 1836)	Atlantic sharpnose shark	cazón de playa
Sphyrnidae		
Sphyrna lewini (Griffith & Smith, 1834)	scalloped hammerhead	cornuda
Sphyrna mokarran (Rüppell, 1837)	great hammerhead	cornuda de ley
Sphyrna tiburo (Linnaeus, 1758)	bonnethead	cornuda de corona
Sphyrna tudes (Valenciennes, 1822)	smalleye	cornuda tudes
Sphyrna zygaena (Linnaeus, 1758)	smooth hammerhead	cabeza de martillo
Order Squaliformes		
Dalatiidae		
Dalatiinae		
Dalatias licha (Bonnaterre, 1788) [1]	kitefin shark	
Somniosinae		
Centroscymnus coelolepis Bocage & Capello, 1864	Portuguese shark	tiburón portugués
Somniosus pacificus Bigelow & Schroeder, 1944[2][*]	Pacific sleeper shark	
Somniosus rostratus (Risso, 1827) [1]	little sleeper shark	
Squalidae		
Centrophorus granulosus (Bloch & Schneider, 1801)	gulper shark	galludo manchado
Centrophorus uyato (Rafinesque, 1810) [3][*]	little gulper shark	
Etmopterus hillianus (Poey, 1861)	blackbelly dogfish	galludo enano
Squalus acanthias Linnaeus, 1758 [1]	spiny dogfish	galludo espinoso
Squalus asper Merrett, 1973 [1]	roughskin dogfish	
Squalus cubensis Howell Rivero, 1936	Cuban dogfish	galludo cubano
Order Pristiformes		
Pristidae		
Pristis pectinata Latham, 1794	smalltooth sawfish	pez sierra
Order Torpediniformes		
Torpedinidae		
Torpedo nobiliana Bonaparte, 1835	Atlantic torpedo	torpedo de lo alto
Narcinidae		
Benthobatis marcida Bean & Weed, 1909	blind torpedo	torpedo
Narcine brasiliensis (Olfers, 1831)	lesser electric ray	torpedo brasileño
Order Rajiformes		
Anacanthobatidae		
Anacanthobatis longirostris Bigelow & Schroeder 1962		
Rhinobatidae		
Rhinobatos percellens (Walbaum, 1792)	guitarfish	guitarra
Rajidae		
Breviraja atripinna Bigelow & Schroeder, 1950	blackfin skate	raya aletinegra
Breviraja colesi Bigelow & Schroeder, 1948	lightnose skate	
Breviraja cubensis Bigelow & Schroeder, 1950	Cuban pygmy skate	
Cruriraja atlantis Bigelow & Schroeder, 1948	Atlantic legskate	

Continued on next page

Appendix 2.1. continued

Cruriraja poeyi Bigelow & Schroeder, 1948	Cuban legskate	
Fenestraja ishiyamai (Bigelow & Schroeder, 1962)	plain pygmy skate	
Fenestraja plutonia (Garman, 1881)	Pluto skate	
Fenestraja sinusmexicanus (Bigelow & Schroeder, 1950)	Gulf of Mexico pygmy skate	
Dasyatidae		
Dasyatis americana Hildebrand & Schroeder, 1928	southern stingray	raya americana
Dasyatis guttata (Bloch & Schneider, 1801)	longnose stingray	raya máxima
Dasyatis sabina (Lesueur, 1824)	Atlantic stingray	raya enana
Dasyatis say (Lesueur, 1817)	bluntnose stingray	raya mediana
Himantura schmardae (Werner, 1904)	Chupare stingray	lebisa
Myliobatidae		
Aetobatus narinari (Euphrasen, 1790)	spotted eagle ray	obispo
Manta birostris (Walbaum, 1792)	manta	manta
Mobula hypostoma (Bancroft, 1831)	devil ray	manta enana
Mobula mobular (Bonnaterre, 1788)	giant devil ray	manta mobula
Rhinoptera bonasus (Mitchill, 1815)	cownose ray	cara de vaca
Urolophidae		
Urolophus jamaicensis (Cuvier, 1816)	yellow stingray	tembladera
SUBCLASS HOLOCEPHALII		
Order Chimaeriformes		
Chimaeridae		
Chimaera cubana Howell Rivero, 1936	chimaera	quimera
CLASS ACTINOPTERYGII		
Order Lepisosteiformes		
Lepisosteidae		
Atractosteus tristoechus (Bloch & Schneider, 1801)[e]	Cuban gar	manjuarí
DIVISION TELEOSTEI		
Order Elopiformes		
Elopidae		
Elops saurus Linnaeus, 1766	ladyfish	banano
Megalops atlanticus Valenciennes, 1847	tarpon	sábalo
Order Albuliformes		
Albulidae		
Albula vulpes (Linnaeus, 1758)	bonefish	macabí
Order Notocanthiformes		
Halosauridae		
Aldrovandia affinis (Günther, 1877)	halosaur	
Aldrovandia gracilis Goode & Bean, 1896		
Aldrovandia phalacra (Vaillant, 1888)		
Halosaurus ovenii Johnson, 1864		
Order Anguilliformes		
Anguilloidei		
Anguillidae		
Anguilla rostrata (Lesueur, 1817)	American eel	anguila
Heterenchelyidae		
Pythonichthys sanguineus Poey, 1868		morena sanguínea
Moringuidae		
Moringua edwardsi (Jordan & Bollman, 1889)	spaghetti eel	morenita de arena
Neoconger mucronatus Girard, 1858	ridged eel	morenita acanalada
Muraenoidei		
Chlopsidae		
Chilorhinus suensonii Lütken, 1852	seagrass eel	anguililla bembona
Chlopsis dentatus (Seale, 1917)	mottled false moray	anguililla moteada

Continued on next page

Appendix 2.1. continued

Kaupichthys hyoproroides (Strömman, 1896)	false moray	anguililla de arrecife
Kaupichthys nuchalis Böhlke, 1967	collared eel	anguililla de collar
Muraenidae		
Uropterygiinae		
Anarchias similis (Lea, 1913)	pygmy moray	morena pigmea
Channomuraena vittata (Richardson, 1845)*	broadbanded moray	morena rara
Muraeninae		
Echidna catenata (Bloch, 1795)	chain moray	morena jaspeada
Enchelycore carychroa Böhlke & Böhlke, 1976	chestnut moray	morena parda
Enchelycore nigricans (Bonnaterre, 1788)	viper moray	morena mulata
Gymnothorax conspersus Poey, 1867	saddled moray	
Gymnothorax funebris Ranzani, 1840	green moray	morena verde
Gymnothorax hubbsi Böhlke & Böhlke, 1977	lichen moray	morena liquen
Gymnothorax maderensis (Johnson, 1862)	sharktooth moray	
Gymnothorax miliaris (Kaup, 1856)	goldentail moray	morena de cola dorada
Gymnothorax moringa (Cuvier, 1829)	spotted moray	morena manchada
Gymnothorax ocellatus Agassiz, 1831	Caribbean ocellated moray	morena ocelada
Gymnothorax polygonius Poey, 1875	polygon moray	morena de polígonos
Gymnothorax saxicola Jordan & Davis, 1891	honeycomb moray	morena punteada
Gymnothorax vicinus (Castelnau, 1855)	purplemouth moray	morena de boca púrpura
Congroidei		
Synaphobranchidae		
Atractodenchelys phrix Robins & Robins, 1970[4]	arrowtooth eel	
Synaphobranchus affinis Günther, 1877		
Synaphobranchus kaupii Johnson, 1862	northern cutthroat eel	
Ophichthidae		
Ophichthinae		
Aplatophis chauliodus Böhlke, 1956[5]	tusky eel	
Aprognathodon platyventris Böhlke, 1967[6]	stripe eel	
Bascanichthys scuticaris (Goode & Bean, 1880)	whip eel	safio pardo
Echiophis intertinctus (Richardson, 1848)	spotted spoon-nose eel	safio dentudo lunado
Echiophis punctifer (Kaup, 1860)	snapper eel	safio dentudo punteado
Ichthyapus ophioneus (Evermann & Marsh, 1900)	surf eel	safio amarillo
Myrichthys breviceps (Richardson, 1848)	sharptail eel	safio de manchas blancas
Myrichthys ocellatus (Lesueur, 1825)	goldspotted eel	safio ocelado
Ophichthus cylindroideus (Ranzani, 1840)	dusky snake eel	safio
Ophichthus gomesii (Castelnau, 1855)	shrimp eel	safio prieto
Ophichthus ophis (Linnaeus, 1758)	spotted snake eel	safio de manchas negras
Ophichthus spinicauda (Norman, 1922)	banded snake eel	safio bandeado
Quassiremus ascensionis (Studer, 1889)	blackspotted snake eel	
Myrophinae		
Ahlia egmontis (Jordan, 1884)	key worm eel	safio de llave
Myrophis platyrhynchus Breder, 1927[4]	broadnose worm eel	
Myrophis punctatus Lütken, 1852	speckled worm eel	safio pecoso
Congridae		
Bathymyrinae		
Ariosoma anale (Poey, 1860)	longtrunk conger	congrio estirado
Ariosoma balearicum (Delaroche, 1809)	bandtooth conger	congrio algino
Paraconger caudilimbatus (Poey, 1867)	margintail conger	congrio
Congrinae		
Bathyuroconger vicinus (Vaillant, 1888)	large-toothed conger	
Conger esculentus Poey, 1861	Antillean conger	congrio común
Conger triporiceps Kanazawa, 1958	manytooth conger	congrio dentudo
Rhechias thysanochila (Reid, 1934)	conger eel	congrina
Rhynchoconger flavus (Goode & Bean, 1896)	yellow conger	
Rhynchoconger gracilior (Ginsburg, 1951)	whiptail conger	congrina filamentosa
Heterocongrinae		
Heteroconger longissimus Günther, 1870	brown garden eel	congrio de jardín

Continued on next page

Appendix 2.1. continued

Muraenesocidae
Cynoponticus savanna (Bancroft, 1831)   sapphire eel   morena arenera
Nemichthyidae
Avocettina infans (Günther, 1878)   snipe eel
Nemichthys scolopaceus Richardson, 1848   slender snipe eel
Serrivomeridae
Serrivomer beanii Gill & Ryder, 1883   sawtooth eel
Nettastomatidae
Hoplunnis tenuis Ginsburg, 1951   spotted pike-conger   congrio colilargo
Nettastoma syntresis Smith & Böhlke, 1981
Nettenchelys exoria Böhlke & Smith, 1981[4]

Order Clupeiformes
Pristigasteridae
Chirocentrodon bleekerianus (Poey, 1867)   dogtooth herring   anchoa pelada
Neoopisthopterus cubanus Hildebrand, 1948[e]   Cuban longfin herring   sardina cubana
Clupeidae
Dussumieriinae
Etrumeus teres (DeKay, 1842)   round herring   sardina canalera
Jenkinsia lamprotaenia (Gosse, 1851)   dwarf herring   manjúa
Jenkinsia majua Whitehead, 1963   little-eye herring   sardineta
Clupeinae
Harengula clupeola (Cuvier, 1829)   false pilchard   sardina escamuda
Harengula humeralis (Cuvier, 1829)   redear sardine   sardina de ley
Harengula jaguana Poey, 1865   scaled sardine   sardina escamudina
Opisthonema oglinum (Lesueur, 1818)   Atlantic thread herring   machuelo
Sardinella aurita Valenciennes, 1847   Spanish sardine   sardina de España
Sardinella janeiro (Eigenmann, 1894)   orangespot sardine   sardina orejinaranja
Engraulidae
Anchoa cayorum (Fowler, 1906)   key anchovy   manjúa de los cayos
Anchoa cubana (Poey, 1868)   Cuban anchovy   manjúa cubana
Anchoa filifera (Fowler, 1915)
Anchoa hepsetus (Linnaeus, 1758)   striped anchovy   manjúa listada
Anchoa lamprotaenia Hildebrand, 1943   bigeye anchovy   manjúa ojuda
Anchoa lyolepis (Evermann & Marsh, 1900)   dusky anchovy   manjúa mulata
Anchoa parva (Meek & Hildebrand, 1923)   little anchovy   manjúa parva
Anchovia clupeoides (Swainson, 1839)   Zabaleta anchovy   hachudo
Anchoviella perfasciata (Poey, 1860)   flat anchovy   manjúa chata
Cetengraulis edentulus (Cuvier, 1829)   Atlantic anchoveta   bocón

Order Siluriformes
Ariidae
Bagre marinus (Mitchill, 1815)   gafftopsail catfish   bagre

Order Osmeriformes
Argentinoidea
Argentinidae
Argentina georgei Cohen & Atsaides, 1969
Argentina striata Goode & Bean, 1896   striated argentine   argentina
Glossanodon pygmaeus Cohen, 1958   pygmy argentine
Bathylagidae
Bathylagus bericoides (Borodin, 1929)   deep-sea smelt
Alepocephaloidea
Alepocephalidae
Conocara macroptera (Vaillant, 1888)   long-fin smooth-head
Leptoderma macrops Vaillant, 1886   grenadier smooth-head
Rouleina maderensis Maul, 1948   Madeiran smooth-head

Order Stomiiformes
Gonostomatidae
Bonapartia pedaliota Goode & Bean, 1896

Continued on next page

Appendix 2.1. continued

Cyclothone braueri Jespersen & Tåning, 1926	Brauer's bristlemouth	
Cyclothone microdon (Günther, 1878)	veiled bristlemouth	
Cyclothone obscura Brauer, 1902		
Cyclothone pallida Brauer, 1902	bicolored bristlemouth	luciérnaga bicolor
Cyclothone pseudopallida Mukhacheva, 1964	slender bristlemouth	luciérnaga palida
Gonostoma atlanticum Norman, 1930	Atlantic fangjaw	luciérnaga atlántica
Gonostoma bathyphilum (Vaillant, 1884)		
Gonostoma elongatum Günther, 1878	longtooth anglemouth	luciérnaga dentada
Manducus maderensis (Johnson, 1890)		
Margrethia obtusirostra Jespersen & Tåning, 1919		
Sonoda paucilampa Grey, 1960[4]		
Triplophos hemingi (McArdle, 1901)[4]		
Sternoptychidae		
Maurolicinae		
Maurolicus weitzmani Parin & Kobyliansky, 1993	pearlsides	luciérnaga perlada
Sternoptychinae		
Argyripnus atlanticus Maul, 1952		
Argyropelecus aculeatus Valenciennes, 1850	marine hatchetfish	pez hacha plateado
Argyropelecus affinis Garman, 1899	Pacific hatchetfish	pez hacha laminado
Argyropelecus hemigymnus Cocco, 1829	half-naked hatchetfish	pez hacha ganchudo
Argyropelecus olfersii (Cuvier, 1829)		pez hacha luminoso
Polyipnus asteroides Schultz, 1938		pez hacha aquillado
Polyipnus clarus Harold, 1994		
Polyipnus laternatus Garman, 1899		pez hacha linterna
Sternoptyx diaphana Hermann, 1781	diaphanous hatchetfish	pez hacha transparente
Phosichthyidae		
Ichthyococcus ovatus (Cocco, 1838)	lightfish	
Pollichthys mauli (Poll, 1953)	stareye lightfish	luciérnaga amarilla
Polymetme thaeocoryla Parin & Borodulina, 1990		luciérnaga musculosa
Vinciguerria nimbaria (Jordan & Williams, 1895)	oceanic lightfish	luciérnaga orbitaria
Yarrella blackfordi Goode & Bean, 1896		luciérnaga negra
Stomiidae		
Chauliodontinae		
Chauliodus danae Regan & Trewavas, 1929	viperfish	víbora oscura
Chauliodus sloani Bloch & Schneider, 1801	Sloane's viperfish	víbora marina
Stomiinae		
Stomias affinis Günther, 1887	scaly dragonfish	
Astronesthinae		
Astronesthes leucopogon Regan & Trewavas, 1929		
Astronesthes macropogon Goodyear & Gibbs, 1970		
Astronesthes micropogon Goodyear & Gibbs, 1970		
Astronesthes niger Richardson, 1845		
Astronesthes richardsoni (Poey, 1852)	snaggletooth	tachonado
Astronesthes similus Parr, 1927		
Melanostomiinae		
Bathophilus digitatus (Welsh, 1923)	scaleless black dragonfish	
Bathophilus longipinnis (Pappenheim, 1914)		pez dragón
Bathophilus nigerrimus Giglioli, 1882	scaleless dragonfish	
Bathophilus schizochirus Regan & Trewavas, 1930		pez dragón gris
Echiostoma barbatum Lowe, 1843		pez dragón barbudo
Eustomias brevibarbatus Parr, 1927		pez dragón pelado
Eustomias fissibarbis (Pappenheim, 1914)		pez dragón ramudo
Eustomias leptobolus Regan & Trewavas, 1930		pez dragón negro
Photonectes margarita (Goode & Bean, 1896)		
Malacosteinae		
Aristostomias grimaldii Zugmayer, 1913		mandibulón barbudo
Aristostomias polydactylus Regan & Trewavas, 1930		

Continued on next page

Appendix 2.1. continued

Aristostomias tittmanni Welsh, 1923	loosejaw	mandibulón delgado
Aristostomias xenostoma Regan & Trewavas, 1930		mandibulón torpedo
Malacosteus niger Ayres, 1848[7]		
Photostomias guernei Collett, 1889		mandibulón negro
Idiacanthinae		
Idiacanthus fasciola Peters, 1877	black dragonfish	dragón negro
Order Ateleopodiformes		
Ateleopodidae		
Ijimaia antillarum Howell Rivero, 1935[e]	jellynose	
Order Aulopiformes		
Aulopoidei		
Aulopidae		
Aulopus nanae Mead, 1958	yellowfin aulopus	aulopus amarillento
Chlorophthalmoidei		
Chlorophthalmidae		
Chlorophthalmus agassizi Bonaparte, 1840	shortnose greeneye	ojiverde ñato
Parasudis truculenta (Goode & Bean, 1896)	longnose greeneye	ojiverde espátula
Ipnopidae		
Bathypterois atricolor phenax Parr, 1928	blackfin spiderfish	
Bathypterois quadrifilis Günther, 1878		
Bathypterois viridensis (Roule, 1916)		
Bathytyphlops marionae Mead, 1958		
Notosudidae		
Scopelosaurus mauli Bertelsen, Krefft & Marshall, 1976		
Scopelosaurus smithii Bean, 1925		
Scopelarchidae		
Scopelarchoides nicholsi Parr, 1929		
Scopelarchus analis (Brauer, 1902)		
Scopelarchus michaelsarsi Koefoed, 1955		
Alepisauroidei		
Synodontidae		
Synodontinae		
Synodus foetens (Linnaeus, 1766)	inshore lizardfish	lagarto máximo
Synodus intermedius (Spix & Agassiz, 1829)	sand diver	lagarto manchado
Synodus poeyi Jordan, 1887	offshore lizardfish	lagarto barbado
Synodus synodus (Linnaeus, 1758)	red lizardfish	lagarto rojizo
Trachinocephalus myops (Forster, 1801)	snakefish	lagarto ñato
Harpadontinae		
Saurida brasiliensis Norman, 1935	largescale lizardfish	lagarto brasileño
Saurida caribbaea Breder, 1927	smallescale lizardfish	lagarto caribeño
Saurida normani Longley, 1935	shortjaw lizardfish	lagarto espinoso
Paralepididae		
Lestidiops affinis (Ege, 1930)	barracudina	barracudina incolora
Lestidiops mirabilis (Ege, 1933)	strange pike smelt	
Lestidium atlanticum Borodin, 1928	Atlantic barracudina	barracudina listada
Lestidium gracile (Ege, 1953)		barracudina traslúcida
Lestrolepis intermedia (Poey, 1868)		barracudina antifaz
Paralepis atlantica Krøyer, 1868	duckbill barracudina	barracudina pequeña
Stemonosudis intermedia (Ege, 1933)		
Alepisauridae		
Alepisaurus ferox Lowe, 1833	longnose lancetfish	conejo de lo alto
Order Myctophiformes		
Myctophidae		
Benthosema suborbitale (Gilbert, 1913)	smallfin lanternfish	pez linterna suborbital
Bolinichthys photothorax (Parr, 1928)		pez linterna alilargo
Centrobranchus nigroocellatus (Günther, 1873)		pez linterna ojinegro
Ceratoscopelus maderensis (Lowe, 1839)		

Continued on next page

Appendix 2.1. continued

Ceratoscopelus townsendi (Eigenmann & Eigenmann, 1889)	dogtooth lampfish	
Ceratoscopelus warmingii (Lütken, 1892)	Warming's lanternfish	pez linterna mayor
Diaphus bertelseni Nafpaktitis, 1966		
Diaphus brachycephalus Tåning, 1928	short-headed lanternfish	
Diaphus dumerilii (Bleeker, 1856)		pez linterna coliapagado
Diaphus fragilis Tåning, 1928	fragile lanternfish	
Diaphus garmani Gilbert, 1906		
Diaphus lucidus (Goode & Bean, 1896)		
Diaphus luetkeni (Brauer, 1904)		
Diaphus minax Nafpaktitis, 1968		
Diaphus mollis Tåning, 1928		
Diaphus perspicillatus (Ogilby, 1898)	transparent lanternfish	
Diaphus problematicus Parr, 1928		pez linterna problemático
Diaphus splendidus (Brauer, 1904)		
Diaphus termophilus Tåning, 1928	Taaning's lanternfish	
Gonichthys cocco (Cocco, 1829)		pez linterna de cocco
Hygophum hygomii (Lütken, 1892)		
Hygophum taaningi Becker, 1965		
Lepidophanes guentheri (Goode & Bean, 1896)		
Lobianchia gemellarii (Cocco, 1838)		
Myctophum affine (Lütken, 1892)	metallic lanternfish	pez linterna
Myctophum asperum Richardson, 1845		
Myctophum brachygnathum (Bleeker, 1856)*		
Myctophum nitidulum Garman, 1899*		
Myctophum obtusirostre Tåning, 1928		
Myctophum selenops Tåning, 1928		
Myctophum spinosum (Steindachner, 1867)		
Notolychnus valdiviae (Brauer, 1904)		pez linterna de Valdivia
Neoscopelidae		
Neoscopelus macrolepidotus Johnson, 1863	blackchin	
Neoscopelus microchir Matsubara, 1943		

Order Lampriformes (=Lampridiformes)
Lamproidei
Lampridae (=Lamprididae)

Lampris guttatus (Brünnich, 1788)	opah	pez mariposa

Trachipteroidei
Trachipteridae

Desmodema polystictum (Ogilby, 1898)	polka-dot ribbonfish	
Trachipterus trachyurus Poey, 1861*		cinta gris
Zu cristatus (Bonelli, 1819)	scalloped ribbonfish	cinta moñuda

Order Polymixiiformes
Polymixiidae

Polymixia lowei Günther, 1859	beardfish	barbudo de lo alto
Polymixia nobilis Lowe, 1838	stout beardfish	

Order Gadiformes
Bregmacerotidae

Bregmaceros atlanticus Goode & Bean, 1886	antenna codlet	ciliado
Bregmaceros mcclellandii Thompson, 1840	spotted codlet	

Macrouridae

Bathygadus favosus Goode & Bean, 1886		
Bathygadus macrops Goode & Bean, 1885		
Bathygadus melanobranchus Vaillant, 1888		
Coelorinchus caribbaeus (Goode & Bean, 1885)	blackfin grenadier	granadero alinegro
Coelorinchus coelorhinchus (Risso, 1810)	saddled grenadier	

Continued on next page

Appendix 2.1. continued

Coelorinchus occa (Goode & Bean, 1885)	swordsnout grenadier	
Coryphaenoides armatus (Héctor, 1875)	abyssal grenadier	granadero máximo
Gadomus arcuatus (Goode & Bean, 1886)	doublethread grenadier	
Gadomus longifilis (Goode & Bean, 1885)		
Lionurus carapinus (Goode & Bean, 1883)	carapine grenadier	granadero
Nezumia aequalis (Günther,1878)	common Atlantic grenadier	granadero espinoso
Nezumia bairdii (Goode & Bean, 1877)	marlin-spike	granadero narizón
Moridae		
Physiculus kaupi Poey, 1865	morid cod	
Steindachneriidae		
Steindachneria argentea Goode & Bean, 1896	luminous hake	granadero luminoso
Order Ophidiiformes		
Ophidiidae		
Brotulinae		
Brotula barbata (Bloch & Schneider, 1801)	bearded brotula	brótula
Neobythitinae		
Benthocometes robustus (Goode & Bean, 1886)		brótula robusta
Grammonus claudei (Torre & Huerta, 1930)	reef-cave brotula	brótula parda
Monomitopus magnus Carter & Cohen, 1985		
Neobythites gilli Goode & Bean, 1885		brótula amarillenta
Neobythites unicolor Nielsen & Retzer, 1994[4]		
Ophidiinae		
Lepophidium brevibarbe (Cuvier, 1829)	blackedge cusk-eel	clarín
Lepophidium kallion Robins, 1959[4]		
Lepophidium marmoratum (Goode & Bean, 1885)		clarín manchado
Lepophidium profundorum (Gill, 1863)	fawn cusk-eel	clarín
Carapidae		
Carapus bermudensis (Jones, 1874)	pearlfish	fierásfer
Encheliophis dubius (Putnam, 1874)	Pacific pearlfish	
Encheliophis homei (Richardson, 1846)	silver pearlfish	
Bythitidae		
Bythitinae		
Calamopteryx goslinei Böhlke & Cohen, 1966[4,7]	longarm brotula	
Calamopteryx robinsorum Cohen, 1973		
Stygnobrotula latebricola Böhlke, 1957	black brotula	brótula oscura
Brosmophycinae		
Lucifuga dentatus Poey, 1858[e]	toothed Cuban cusk-eel	
Lucifuga simile Nalbant, 1981[e]		
Lucifuga subterraneus Poey, 1858[e]	Cuban cusk-eel	
Lucifuga teresinarum Diaz Perez, 1988[e]		
Ogilbia cayorum Evermann & Kendall, 1898	key brotula	brótula rojiza
Order Batrachoidiformes		
Batrachoididae		
Batrachoidinae		
Opsanus beta (Goode & Bean, 1880)	gulf toadfish	sapo de boca blanca
Opsanus pardus (Goode & Bean, 1880)[8]	leopard toadfish	sapo leopardo
Opsanus phobetron Walters & Robins, 1961	scarecrow toadfish	
Porichthyinae		
Porichthys plectrodon Jordan & Gilbert, 1882	Atlantic midshipman	sapo narinero
Order Lophiiformes		
Antennarioidei		
Antennariidae		
Antennarius multiocellatus (Valenciennes, 1837)	longlure frogfish	pescador caña larga
Antennarius ocellatus (Bloch & Schneider, 1801)	ocellated frogfish	pescador ocelado
Antennarius pauciradiatus Schultz, 1957	dwarf frogfish	
Antennarius striatus (Shaw in Shaw & Nodder, 1794)	striated frogfish	pescador
Histrio histrio (Linnaeus, 1758)	sargassumfish	pez sargazo

Continued on next page

Appendix 2.1. continued

Chaunacoidei
Chaunacidae
Chaunax pictus Lowe, 1846   redeye   bostezador rosado
Chaunax stigmaeus Fowler, 1946[4]   redeye gaper
Chaunax suttkusi Caruso, 1989
Lophioidei
Lophiidae
Lophius gastrophysus Miranda-Ribeiro, 1915   blackfin goosefish
Ogcocephaloidei
Ogcocephalidae
Halieutichthys aculeatus (Mitchill, 1818)   pancake batfish   diablito espinoso
Halieutichthys caribbaeus Garman, 1896
Halieutichthys smithii Evermann & Marsh, 1900   diablito
Ogcocephalus corniger Bradbury, 1980   longnose batfish   pez diablo narizón
Ogcocephalus nasutus (Cuvier, 1829)   shortnose batfish   pez diablo ñato
Ogcocephalus parvus Longley & Hildebrand, 1940[4]   roughback batfish
Ogcocephalus radiatus (Mitchill, 1818)   polka-dot batfish
Ogcocephalus vespertilio (Linnaeus, 1758)   pez diablo narizón
Ceratioidei
Ceratiidae
Ceratias uranoscopus Murray in Thompson, 1877
Cryptopsaras couesii Gill, 1883   triplewart seadevil   diablo marino
Gigantactinidae
Gigantactis vanhoeffeni Brauer, 1902
Linophrynidae
Linophryne coronata Parr 1927
Melanocetidae
Melanocetus johnsonii Günther, 1864   negro
Melanocetus murrayi Günther, 1887

SERIES ATHERINOMORPHA
Order Atheriniformes
Atherinidae
Atherininae
Alepidomus evermanni (Eigenmann, 1903)[e]
Atherinomorus stipes (Müller & Troschel, 1848)   hardhead silverside   cabezote
Hypoatherina harringtonensis (Goode, 1877)   reef silverside   catacuche
Menidiinae
Melanorhinus microps (Poey, 1860)   Querimana silverside   cabezote rey

Order Cyprinodontiformes
Aplocheiloidei
Rivulidae
Rivulus cylindraceus Poey, 1860   Cuban rivulus   rivulo cubano
Rivulus garciai de la Cruz & Dubitsky, 1976 [e]   Matanzas rivulus   rivulo matancero
Rivulus insulaepinorum de la Cruz   Isle of Pines rivulus   rivulo pinero
    & Dubitsky, 1976[e]
Rivulus marmoratus Poey, 1880   mangrove rivulus   rivulo de mangle
Cyprinodontoidei
Cyprinodontidae
Cyprinodontinae
Cyprinodon laciniatus Hubbs & Miller,1942   Bahama pupfish   guajacón
Cyprinodon variegatus Lacepède, 1803   sheepshead minnow   pipón
Cubanichthyinae
Cubanichthys cubensis (Eigenmann, 1903)[e]   Cuban killifish   guajacón cubano
Fundulidae
Fundulus grandis saguanus Rivas, 1948[e]   Cuban gulf killifish   guasábalo

Continued on next page

Appendix 2.1. continued

Poeciliidae

Gambusia rhizophorae Rivas, 1969	mangrove gambusia	gambusia de mangle
Gambusia punctata Poey, 1854[e]	Cuban gambusia	gambusia cubana
Gambusia puncticulata Poey, 1854	Caribbean gambusia	gambusia caribeña
Girardinus creolus Garman, 1895[e]	Creole topminnow	guajacón creol
Girardinus cubensis (Eigenmann, 1903)[e]	Cuban topminnow	gambusino
Girardinus denticulatus Garman, 1895[e]	toothy topminnow	guajacón dentado
Girardinus falcatus (Eigenmann, 1903)[e]	goldbelly topminnow	guajacón dorado
Girardinus metallicus Poey, 1854[e]	metallic topminnow	guajacón metálico
Girardinus microdactylus Rivas, 1944[e]	smallfinger topminnow	guajacón
Girardinus uninotatus Poey, 1860[e]	singlespot topminnow	guajacón parchado
Limia vittata (Guichenot, 1853)[e]	Cuban limia	fanguillo
Quintana atrizona Hubbs, 1934[e]	barred topminnow	guajacón barreado

Order Beloniformes

Scomberesocidae

Scomberesox saurus (Walbaum, 1792)	Atlantic saury	sauro atlántico

Belonidae

Ablennes hians (Valenciennes, 1846)	flat needlefish	agujón de golfo
Platybelone argalus (Lesueur, 1821)	keeltail needlefish	agujón aquillado
Strongylura notata (Poey, 1860)	redfin needlefish	agujón de aletas rojas
Strongylura timucu (Walbaum, 1792)	timicu	timucú
Tylosurus acus acus (Lacepède, 1803)	agujon	agujón
Tylosurus crocodilus crocodilus (Perón & Lesueur, 1821)	houndfish	agujón crocodilo

Hemiramphidae

Chriodorus atherinoides Goode & Bean, 1882	hardhead halfbeak	pejerrey
Euleptorhamphus velox Poey, 1868	flying halfbeak	escribano volador
Hemiramphus balao Lesueur, 1821	balao	escribano balao
Hemiramphus brasiliensis (Linnaeus, 1758)	ballyhoo	escribano de aletas rojas
Hyporhamphus unifasciatus (Ranzani, 1841)	silverstripe halfbeak	escribano

Exocoetidae

Cheilopogon comatus (Mitchill, 1815)	clearwing flyingfish	volador de alas claras
Cheilopogon cyanopterus (Valenciennes, 1847)	margined flyingfish	volador de borde oscuro
Cheilopogon exsiliens (Linnaeus, 1771)	bandwing flyingfish	volador bandeado
Cheilopogon furcatus furcatus (Mitchill, 1815)	spotfin flyingfish	volador de alas punteadas
Cheilopogon melanurus (Valenciennes, 1847)	Atlantic flyingfish	volador de banda estrecha
Exocoetus obtusirostris Günther, 1866	oceanic two-wing flyingfish	volador oceánico
Exocoetus volitans Linnaeus, 1758	tropical two-wing flyingfish	volador tropical
Hirundichthys affinis (Günther, 1866)	four-wing flyingfish	volador de 4 alas
Hirundichthys rondeletii (Valenciennes, 1847)	blackwing flyingfish	volador de alas negras
Oxyporhamphus micropterus similis Bruun, 1935	smallwing flyingfish	escribano volador
Parexocoetus brachypterus (Richardson, 1846)	sailfin flyingfish	volador velero
Prognichthys occidentalis Parin, 1999	bluntnose flyingfish	volador ñato

SERIES PERCOMORPHA

Order Stephanoberyciformes

Stephanoberycidae

Stephanoberyx monae Gill, 1883	pricklefish	

Melamphaidae

Melamphaes pumilus Ebeling, 1962		
Scopelogadus mizolepis mizolepis (Günther, 1878)	bigscale	

Cetomimidae

Cetostoma regani Zugmayer, 1914		

Order Beryciformes

Berycoidei

Trachichthyidae

Gephyroberyx darwinii (Johnson, 1866)	big roughy	carajuelo de fondo

Continued on next page

Appendix 2.1. continued

Hoplostethus mediterraneus mediterraneus Cuvier, 1829	silver roughy	
Hoplostethus occidentalis Woods, 1973[4]	Atlantic roughy	
Berycidae		
Beryx decadactylus Cuvier, 1829	red bream	alfonsino
Holocentridae		
Corniger spinosus Agassiz,1831	spinycheek soldierfish	candil espinoso
Holocentrus adscensionis (Osbeck, 1765)	squirrelfish	carajuelo de ascención
Holocentrus rufus (Walbaum, 1792)	longspine squirrelfish	carajuelo rufo
Myripristis jacobus Cuvier, 1829	blackbar soldierfish	candil barreado
Neoniphon marianus (Cuvier, 1829)	longjaw squirrelfish	carajuelo mariano
Ostichthys trachypoma (Günther, 1859)	bigeye soldierfish	candil de lo alto
Plectrypops retrospinis (Guichenot, 1853)	cardinal soldierfish	candil cardenal
Sargocentron bullisi (Woods, 1955)	deepwater squirrelfish	carajuelo profundo
Sargocentron coruscum (Poey, 1860)	reef squirrelfish	carajuelo de arrecife
Sargocentron poco (Woods, 1965)*	saddle squirrelfish	
Sargocentron vexillarium (Poey, 1860)	dusky squirrelfish	carajuelo oscuro
Order Zeiformes		
Parazenidae		
Parazen pacificus Kamohara, 1935	parazen	
Macrurocyttidae		
Zenion hololepis (Goode & Bean, 1896)		
Zeidae		
Cyttopsis rosea (Lowe, 1843)	red dory	
Grammicolepididae		
Grammicolepis brachiusculus Poey, 1873	thorny tinselfish	oropel
Xenolepidichthys dalgleishi Gilchrist, 1922	spotted tinselfish	
Caproidae		
Antigonia capros Lowe, 1843	deepbody boarfish	elevado
Antigonia combatia Berry & Rathjen, 1958	shortspine boarfish	pez pecarí
Order Syngnathiformes		
Aulostomoidei		
Aulostomidae		
Aulostomus maculatus Valenciennes, 1837	trumpetfish	trompa
Fistulariidae		
Fistularia petimba Lacepède, 1803	red cornetfish	trompetero
Fistularia tabacaria Linnaeus, 1758	bluespotted cornetfish	pez corneta
Centriscidae		
Macroramphosus gracilis (Lowe, 1839)	slender snipefish	tirador alargado
Syngnathoidei		
Syngnathidae		
Syngnathinae		
Acentronura dendritica (Barbour, 1905)	pipehorse	caballito enano
Bryx dunckeri (Metzelaar, 1919)	pugnose pipefish	trompetero nariz-corto
Cosmocampus brachycephalus (Poey, 1868)	crested pipefish	trompetero ñato
Cosmocampus elucens (Poey, 1868)	shortfin pipefish	trompetero brillante
Micrognathus crinitus (Jenyns, 1842)	banded pipefish	trompetero bandeado
Microphis brachyurus (Bleeker, 1853)*	opossum pipefish	trompetero colinegro
Pseudophallus mindii (Meek & Hildebrand, 1923)	freshwater pipefish	
Syngnathus caribbaeus Dawson, 1979	Caribbean pipefish	
Syngnathus floridae (Jordan & Gilbert, 1882)	dusky pipefish	
Syngnathus pelagicus Linnaeus, 1758	sargassum pipefish	trompetero oceánico
Syngnathus poeyi (Jordan & Evermann, 1896)*		trompetero grisáceo
Hippocampinae		
Hippocampus erectus Perry, 1810	lined seahorse	caballito erecto
Hippocampus reidi Ginsburg, 1933[9]	longsnout seahorse	caballito narizón
Hippocampus zosterae Jordan & Gilbert, 1882	dwarf seahorse	caballito oliváceo

Continued on next page

Appendix 2.1. continued

Order Synbranchiformes

Synbranchidae

Ophisternon aenigmaticum Rosen & Greenwood, 1976* swamp eel

Synbranchus marmoratus Bloch, 1795 marmorated swamp eel maporro

Order Scorpaeniformes

Scorpaenoidei

Scorpaenidae

Scorpaeninae

Neomerinthe beanorum (Evermann & Marsh, 1900)

Pontinus castor Poey, 1860 longsnout scorpionfish rascacio de lo alto

Pontinus nematophthalmus (Günther, 1860) spinythroat scorpionfish

Scorpaena albifimbria Evermann & Marsh, 1900 coral scorpionfish rascacio coralino

Scorpaena bergii Evermann & Marsh, 1900 goosehead scorpionfish rascacio ganso

Scorpaena brasiliensis Cuvier, 1829 barbfish rascacio pardo

Scorpaena calcarata Goode & Bean, 1882 smoothhead scorpionfish rascacio pelón

Scorpaena grandicornis Cuvier, 1829 plumed scorpionfish rascacio cornudo

Scorpaena inermis Cuvier, 1829 mushroom scorpionfish rascacio bongo

Scorpaena plumieri Bloch, 1789 spotted scorpionfish rascacio multicolor

Scorpaenodes caribbaeus Meek & Hildebrand, 1928 reef scorpionfish

Scorpaenodes tredecimspinosus (Metzelaar, 1919) deepreef scorpionfish rascacio espinoso

Setarchinae

Setarches guentheri Johnson, 1862 deepwater scorpionfish raspacio cabezirecto

Triglidae

Triglinae

Bellator egretta (Goode & Bean, 1896) streamer searobin espátula bandera

Prionotus punctatus (Bloch, 1793) bluewing searobin espátula gallina

Prionotus rubio Jordan, 1886 blackwing searobin espátula aletinegra

Peristediinae

Peristedion antillarum Teague, 1961

Peristedion brevirostre Günther, 1860 flathead searobin

Peristedion gracile Goode & Bean, 1896 slender searobin espátula esbelta

Peristedion imberbe Poey, 1861 espátula lampiña

Peristedion longispatha Goode & Bean, 1886 espátula larga

Peristedion truncatum Günther, 1880

Order Dactylopteriformes

Dactylopteridae

Dactylopterus volitans (Linnaeus, 1758) flying gurnard pez murciélago

Order Perciformes

Percoidei

Centropomidae

Centropomus ensiferus Poey, 1860 swordspine snook róbalo espinoso

Centropomus parallelus Poey, 1860 fat snook robalito

Centropomus pectinatus Poey, 1860 tarpon snook róbalo prieto

Centropomus undecimalis (Bloch, 1792) common snook róbalo común

Acropomatidae

Synagrops bellus (Goode & Bean, 1896) blackmouth bass

Verilus sordidus Poey, 1860 ocean bass berregüello

Symphysanodontidae

Symphysanodon berryi Anderson, 1970[4] slope bass

Symphysanodon octoactinus Anderson, 1970[4] colirubia rosa

Serranidae

Serraninae

Bullisichthys caribbaeus Rivas, 1971[4]

Centropristis fuscula Poey, 1861 twospot sea bass serrano oliváceo

Continued on next page

Appendix 2.1. continued

Diplectrum radiale (Quoy & Gaimard, 1824)	aquavina	aguavina
Diplectrum formosum (Linnaeus, 1766)	sand perch	serrano
Hypoplectrus aberrans Poey, 1868	yellowbelly hamlet	vaca de vientre amarillo
Hypoplectrus chlorurus (Cuvier, 1828)	yellowtail hamlet	vaca de cola amarilla
Hypoplectrus gemma Goode & Bean, 1882[4]	blue hamlet	
Hypoplectrus gummigutta (Poey, 1851)	golden hamlet	vaca dorada
Hypoplectrus guttavarium (Poey, 1852)	shy hamlet	vaca bicolor
Hypoplectrus indigo (Poey, 1851)	indigo hamlet	vaca añil
Hypoplectrus nigricans (Poey, 1852)	black hamlet	vaca negra
Hypoplectrus puella (Cuvier, 1828)	barred hamlet	vaca barreada
Hypoplectrus unicolor (Walbaum, 1792)	butter hamlet	vaca blanca
Schultzea beta (Hildebrand, 1940)[10]	school bass	serrano sin dientes
Serranus annularis (Günther, 1880)	orangeback bass	serrano de dorso naranja
Serranus baldwini (Evermann & Marsh, 1899)	lantern bass	serrano linterna
Serranus chionaraia Robins & Starck, 1961[10]	snow bass	serrano nevado
Serranus flaviventris (Cuvier, 1829)	twinspot bass	serrano de arena
Serranus luciopercanus Poey, 1852	crosshatch bass	serrano de ley
Serranus phoebe Poey, 1851	tattler	diana
Serranus subligarius (Cope, 1870)	belted sandfish	serrano leopardo
Serranus tabacarius (Cuvier, 1829)	tobaccofish	jácome
Serranus tigrinus (Bloch, 1790)	harlequin bass	serrano tigre
Serranus tortugarum Longley, 1935	chalk bass	serrano tortuga
Epinephelinae		
Alphestes afer (Bloch, 1793)	mutton hamlet	guaseta
Bathyanthias cubensis (Schultz, 1958)		
Cephalopholis cruentata (Lacepède, 1802)	graysby	enjambre
Cephalopholis fulva (Linnaeus, 1758)	coney	guatívere
Dermatolepis inermis (Valenciennes, 1833)	marbled grouper	cherna jaspeada
Epinephelus adscensionis (Osbeck, 1765)	rock hind	cabra mora
Epinephelus flavolimbatus Poey, 1865	yellowedge grouper	mero de aletas amarillas
Epinephelus guttatus (Linnaeus, 1758)	red hind	cabrilla
Epinephelus itajara (Lichtenstein, 1822)	jewfish	guasa
Epinephelus morio (Valenciennes, 1828)	red grouper	cherna americana
Epinephelus mystacinus (Poey, 1852)	misty grouper	cherno de lo alto
Epinephelus nigritus (Holbrook, 1855)	warsaw grouper	cherno prieto
Epinephelus niveatus (Valenciennes, 1828)	snowy grouper	cherna manchada
Epinephelus striatus (Bloch, 1792)	Nassau grouper	cherna criolla
Gonioplectrus hispanus (Cuvier, 1828)	Spanish flag	biajaiba de lo alto
Jeboehlkia gladifer Robins, 1967[4]	bladefin bass	
Liopropoma aberrans (Poey, 1860)	eyestripe bass	guardia chino
Liopropoma mowbrayi Woods & Kanazawa, 1951	cave bass	guardia rojo
Liopropoma rubre Poey, 1861	peppermint bass	guardia suizo
Mycteroperca acutirostris (Valenciennes, 1828)	comb grouper	bonací rojo
Mycteroperca bonaci (Poey, 1860)	black grouper	aguají
Mycteroperca interstitialis (Poey, 1860)	yellowmouth grouper	abadejo
Mycteroperca microlepis (Goode & Bean, 1879)	gag	
Mycteroperca tigris (Valenciennes, 1833)	tiger grouper	bonací gato
Mycteroperca venenosa (Linnaeus, 1758)	yellowfin grouper	arigua
Paranthias furcifer (Valenciennes, 1828)	creole-fish	rabirrubia de lo alto
Rypticus bistrispinus (Mitchill, 1818)	freckled soapfish	jaboncillo pecoso
Rypticus randalli Courtenay, 1967		jaboncillo
Rypticus saponaceus (Bloch & Schneider, 1801)	greater soapfish	jaboncillo máximo
Rypticus subbifrenatus Gill, 1861	spotted soapfish	jaboncillo punteado
Anthiinae		
Plectranthias garrupellus Robins & Starck, 1961[4]	apricot bass	
Pronotogrammus martinicensis (Guichenot, 1868)[4]	roughtongue bass	
Grammatidae		
Gramma linki Starck & Colin, 1978	yellowlined basslet	linki

Continued on next page

Appendix 2.1. continued

Gramma loreto Poey, 1868	royal gramma	loreto
Gramma melacara Böhlke & Randall, 1963	blackcap basslet	gramma violeta
Lipogramma evides Robins & Colin, 1979[4]	banded basslet	
Lipogramma robinsi Gilmore, 1997[4]	yellowbar basslet	
Priacanthidae		
Heteropriacanthus cruentatus (Lacepède, 1801)	glasseye snapper	catalufa espinosa
Priacanthus arenatus Cuvier, 1829	bigeye	catalufa toro
Pristigenys alta (Gill, 1862)	short bigeye	catalufa de lo alto
Incertae sedis		
Amiichthys diapterus Poey in Jordan, 1887*		cardenal de aletas amarillas
Apogonidae		
Apogon affinis (Poey, 1875)	bigtooth cardinalfish	cardenal dentudo
Apogon aurolineatus (Mowbray, 1927)[7]	bridle cardinalfish	
Apogon binotatus (Poey, 1867)	barred cardinalfish	cardenal barreado
Apogon lachneri Böhlke, 1959	whitestar cardinalfish	
Apogon maculatus (Poey, 1860)	flamefish	cardenal manchado
Apogon phenax Böhlke & Randall, 1968	mimic cardinalfish	cardenal mimético
Apogon planifrons Longley & Hildebrand, 1940[9]	pale cardinalfish	cardenal pálido
Apogon pseudomaculatus Longley, 1932	twospot cardinalfish	cardenal de dos puntos
Apogon quadrisquamatus Longley, 1934	sawcheek cardinalfish	cardenal espinoso
Apogon robinsi Böhlke & Randall, 1968	roughlip cardinalfish	cardenal
Apogon townsendi (Breder, 1927)	belted cardinalfish	cardenal con cinto
Astrapogon alutus (Jordan & Gilbert, 1882)	bronze cardinalfish	cardenal bronceado
Astrapogon puncticulatus (Poey, 1867)	blackfin cardinalfish	cardenal punteado
Astrapogon stellatus (Cope, 1867)	conchfish	cardenal del cobo
Phaeoptyx conklini (Silvester, 1915)	freckled cardinalfish	cardenal pecoso
Phaeoptyx pigmentaria (Poey, 1860)	dusky cardinalfish	cardenal pigmentado
Phaeoptyx xenus (Böhlke & Randall, 1968)	sponge cardinalfish	cardenal violáeo
Epigonidae		
Epigonus macrops (Brauer, 1906)		
Epigonus occidentalis Goode & Bean, 1896		
Sphyraenops bairdianus Poey, 1861	barracuda cardinalfish	cardenal barracuda
Malacanthidae		
Malacanthinae		
Malacanthus plumieri (Bloch, 1786)	sand tilefish	matejuelo blanco
Latilinae		
Caulolatilus chrysops (Valenciennes, 1833)	goldface tilefish	tumba dorada
Caulolatilus cyanops Poey, 1866	blackline tilefish	tumba
Caulolatilus intermedius Howell Rivero, 1936	anchor tilefish	tumba parda
Pomatomidae		
Pomatomus saltatrix (Linnaeus, 1766)	bluefish	anjova
Scombropidae		
Scombrops oculatus (Poey, 1860)	Atlantic scombrops	escolar chino
Rachycentridae		
Rachycentron canadum (Linnaeus, 1766)	cobia	cobia
Echeneidae		
Echeneis naucrates Linnaeus, 1758	sharksucker	guaicán
Echeneis neucratoides Zouiev, 1786	whitefin sharksucker	
Phtheirichthys lineatus (Menzies, 1791)	slender suckerfish	pez pega lineado
Remora brachyptera (Lowe, 1839)	spearfish remora	pez pega robusto
Remora osteochir (Cuvier, 1829)	marlinsucker	pez pega disco
Remora remora (Linnaeus, 1758)	remora	remora
Remorina albescens (Temminck & Schlegel, 1845)	white suckerfish	
Carangidae		
Alectis ciliaris (Bloch, 1787)	African pompano	flechudo
Caranx bartholomaei Cuvier, 1833	yellow jack	cibí amarillo
Caranx crysos (Mitchill, 1815)	blue runner	cojinúa
Caranx hippos (Linnaeus, 1766)	crevalle jack	jiguagua

Continued on next page

Appendix 2.1. continued

Caranx latus Agassiz, 1831	horse-eye jack	gallego
Caranx lugubris Poey, 1860	black jack	tiñosa
Caranx ruber (Bloch, 1793)	bar jack	cibí carbonero
Chloroscombrus chrysurus (Linnaeus, 1766)	Atlantic bumper	casabe
Decapterus macarellus (Cuvier, 1833)	mackerel scad	antonino caballita
Decapterus punctatus (Cuvier, 1829)	round scad	antonino punteado
Decapterus tabl Berry, 1968*	redtail scad	
Elagatis bipinnulata (Quoy & Gaimard, 1825)	rainbow runner	salmón cubano
Hemicaranx amblyrhynchus (Cuvier, 1833)	bluntnose jack	chicharra
Naucrates ductor (Linnaeus, 1758)	pilotfish	piloto
Oligoplites saurus (Bloch & Schneider, 1801)	leatherjack	zapatero
Selar crumenophthalmus (Bloch, 1793)	bigeye scad	chicharro
Selene setapinnis (Mitchill 1815)	Atlantic moonfish	cara de caballo
Selene vomer (Linnaeus, 1758)	lookdown	jorobado
Seriola dumerili (Risso, 1810)	greater amberjack	coronado de ley
Seriola fasciata (Bloch, 1793)	lesser amberjack	medregal
Seriola rivoliana Valenciennes, 1833	almaco jack	coronado
Seriola zonata (Mitchill, 1815)	banded rudderfish	coronado de bandas
Trachinotus carolinus (Linnaeus, 1766)	Florida pompano	palorneta común
Trachinotus falcatus (Linnaeus, 1758)	permit	pámpano
Trachinotus goodei Jordan & Evermann, 1896	palometa	palometa
Uraspis secunda (Poey, 1860)	cottonmouth jack	segundo
Coryphaenidae		
Coryphaena equiselis Linnaeus, 1758	pompano dolphinfish	dorado enano
Coryphaena hippurus Linnaeus, 1758	dolphinfish	dorado
Bramidae		
Brama brama (Bonnaterre,1788)	Atlantic pomfret	brama
Collybus drachme Snyder, 1904		
Eumegistus brevorti (Poey, 1860)		brama de lo alto
Taractichthys longipinnis (Lowe, 1843)	bigscale pomfret	brama aletuda
Lutjanidae		
Apsilus dentatus Guichenot, 1853	black snapper	arnillo
Etelis oculatus (Valenciennes, 1828)	queen snapper	cachucho
Lutjanus analis (Cuvier, 1828)	mutton snapper	pargo criollo
Lutjanus apodus (Walbaum, 1792)	schoolmaster	cají
Lutjanus buccanella (Cuvier, 1828)	blackfin snapper	pargo sesí
Lutjanus campechanus (Poey, 1860)	red snapper	pargo colorado
Lutjanus cyanopterus (Cuvier, 1828)	cubera snapper	cubera
Lutjanus griseus (Linnaeus, 1758)	gray snapper	caballerete
Lutjanus jocu (Bloch & Schneider, 1801)	dog snapper	jocú
Lutjanus mahogoni (Cuvier, 1828)	mahogany snapper	ojanco
Lutjanus purpureus Poey, 1866*	Caribbean red snapper	
Lutjanus synagris (Linneaus, 1758)	lane snapper	biajaiba
Lutjanus vivanus (Cuvier, 1828)	silk snapper	pargo del alto
Ocyurus chrysurus (Bloch, 1791)	yellowtail snapper	rabirrubia
Pristipomoides aquilonaris (Goode & Bean, 1896)[4]	wenchman	
Pristipomoides macrophthalmus (Müller & Troschel, 1848)	cardinal snapper	voraz
Rhomboplites aurorubens (Cuvier, 1829)	vermilion snapper	cotorro
Lobotidae		
Lobotes surinamensis (Bloch, 1790)	tripletail	biajaca de mar
Gerreidae		
Diapterus auratus Ranzani, 1842	Irish pompano	patao común
Diapterus rhombeus (Cuvier, 1829)	Caitipa mojarra	patao
Eucinostomus argenteus Baird & Girard, 1855	spotfin mojarra	mojarra plateada
Eucinostomus gula (Quoy & Gaimard, 1824)	silver jenny	mojarra de ley
Eucinostomus harengulus Goode & Bean, 1879	tidewater mojarra	

Continued on next page

Appendix 2.1. continued

Eucinostomus havana (Nichols, 1912)	bigeye mojarra	mojarrita manchada
Eucinostomus jonesii (Günther, 1879)	slender mojarra	mojarrita esbelta
Eucinostomus lefroyi (Goode, 1874)	mottled mojarra	mojarrita
Eucinostomus melanopterus (Bleeker, 1863)[4]	flagfin mojarra	mojarra bandera
Eugerres brasilianus (Cuvier, 1830)	Brazilian mojarra	patao brasileño
Eugerres plumieri (Cuvier,1830)	striped mojarra	patao rayado
Gerres cinereus (Walbaum, 1792)	yellowfin mojarra	mojarra blanca
Haemulidae		
Anisotremus surinamensis (Bloch, 1791)	black margate	pompón
Anisotremus virginicus (Linnaeus, 1758)	porkfish	catalineta
Haemulon album Cuvier, 1830	margate	jallao
Haemulon aurolineatum Cuvier, 1830	tomtate	jeniguano bocón
Haemulon bonariense Cuvier, 1830	black grunt	ronco prieto
Haemulon carbonarium Poey, 1860	caesar grunt	ronco carbonero
Haemulon chrysargyreum Günther, 1859	smallmouth grunt	jeníguano amarillo
Haemulon flavolineatum (Desmarest, 1823)	French grunt	ronco condenado
Haemulon macrostomum Günther, 1859	Spanish grunt	jeníguano español
Haemulon melanurum (Linnaeus, 1758)	cottonwick	ronco de lomo prieto
Haemulon parra (Desmarest, 1823)	sailors choice	ronco blanco
Haemulon plumieri (Lacepède, 1801)	white grunt	ronco arará
Haemulon sciurus (Shaw, 1803)	bluestriped grunt	ronco amarillo
Haemulon striatum (Linnaeus, 1758)	striped grunt	jeníguano rayado
Orthopristis chrysoptera (Linnaeus, 1766)	pigfish	burro
Pomadasys corvinaeformis (Steindachner, 1868)	roughneck grunt	ticopa gris
Pomadasys crocro (Cuvier, 1830)	burro grunt	ticopa
Inermiidae		
Inermia vittata Poey, 1860	boga	boga
Sparidae		
Archosargus rhomboidalis (Linnaeus, 1758)	sea bream	salema
Calamus bajonado (Bloch & Schneider, 1801)	jolthead porgy	bajonao violáceo
Calamus calamus (Valenciennes, 1830)	saucereye porgy	pez de pluma
Calamus penna (Valenciennes, 1830)	sheepshead porgy	pez de pluma manchado
Calamus pennatula Guichenot, 1868	pluma	bajonao plateado
Calamus proridens Jordan & Gilbert, 1884	littlehead porgy	pez de pluma rayado
Diplodus argenteus caudimacula (Poey, 1860)	silver porgy	sargo
Diplodus holbrooki (Bean, 1878)	spottail pinfish	
Lagodon rhomboides (Linnaeus, 1766)	pinfish	chopa espina
Sciaenidae		
Bairdiella batabana (Poey, 1860)	blue croaker	corvina barriga blanca
Bairdiella ronchus (Cuvier, 1830)	ground croaker	corvina espinosa
Bairdiella sanctaeluciae (Jordan, 1890)	striped croaker	corvina Santa Lucía
Bairdiella subequalis (Poey, 1875)⋆		
Equetus acuminatus (Bloch & Schneider, 1801)	high-hat	vaqueta rayada
Equetus lanceolatus (Linnaeus, 1758)	jackknife fish	vaqueta de cinta
Equetus punctatus (Bloch & Schneider, 1801)	spotted drum	vaqueta punteada
Larimus breviceps Cuvier, 1830	shorthead drum	corvino cabezón
Micropogonias furnieri (Desmarest, 1823)	whitemouth croaker	verrugato
Micropogonias undulatus (Linnaeus, 1766)	Atlantic croaker	roncadina
Odontoscion dentex (Cuvier, 1830)	reef croaker	corvina dentuda
Pogonias cromis (Linnaeus, 1766)	black drum	corvina negra
Umbrina coroides Cuvier, 1830	sand drum	roncador
Mullidae		
Mulloidichthys martinicus (Cuvier, 1829)	yellow goatfish	salmonete amarillo
Mullus auratus Jordan & Gilbert, 1882	red goatfish	salmonete rojo
Pseudupeneus maculatus (Bloch, 1793)	spotted goatfish	salmonete colorado
Upeneus parvus Poey, 1852	dwarf goatfish	salmonete rayado
Pempheridae		
Pempheris poeyi Bean, 1885	shortfin sweeper	pemferis colinegro

Continued on next page

Appendix 2.1. continued

Pempheris schomburgkii Müller & Troschel, 1848	glassy sweeper	pemferis bandeado
Bathyclupeidae		
Bathyclupea schroederi Dick, 1962		
Kyphosidae		
Kyphosus incisor (Cuvier, 1831)	yellow chub	chopa amarilla
Kyphosus sectatrix (Linnaeus, 1758)	Bermuda chub	chopa blanca
Ephippidae		
Chaetodipterus faber (Broussonet, 1782)	Atlantic spadefish	paguala
Chaetodontidae		
Chaetodon aculeatus (Poey, 1860)	longsnout butterflyfish	parche narizón
Chaetodon capistratus Linnaeus, 1758	foureye butterflyfish	parche ocelado
Chaetodon guyanensis Durand, 1960[4]	French butterflyfish	
Chaetodon ocellatus Bloch, 1787	spotfin butterflyfish	parche ocelado amarillo
Chaetodon sedentarius Poey, 1860	reef butterflyfish	parche mariposa
Chaetodon striatus Linnaeus, 1758	banded butterflyfish	parche rayado
Pomacanthidae		
Centropyge argi Woods & Kanazawa, 1951	cherubfish	angelote pigmeo
Holacanthus bermudensis Goode, 1876	blue angelfish	isabelita azul
Holacanthus ciliaris (Linnaeus, 1758)	queen angelfish	isabelita reina
Holacanthus tricolor (Bloch, 1795)	rock beauty	vaqueta de dos colores
Pomacanthus arcuatus (Linnaeus, 1758)	gray angelfish	chivirica gris
Pomacanthus paru (Bloch, 1787)	French angelfish	chivirica francesa
Cirrithidae		
Amblycirrhitus pinos (Mowbray, 1927)	redspotted hawkfish	rayadito
Opistognathidae		
Lonchopisthus micrognathus (Poey, 1860)	swordtail jawfish	guardián lanceolado
Opistognathus aurifrons (Jordan & Thompson, 1905)	yellowhead jawfish	guardián cabeziamarillo
Opistognathus lonchurus Jordan & Gilbert, 1882	moustache jawfish	guardián verdoso
Opistognathus macrognathus Poey, 1860	banded jawfish	guardián bocón
Opistognathus maxillosus Poey, 1860	mottled jawfish	guardián jaspeado
Opistognathus nothus Smith-Vaniz, 1997	yellowmouth jawfish	
Opistognathus whitehursti (Longley, 1927)	dusky jawfish	guardián escamudo
Mugiloidei		
Mugilidae		
Agonostomus monticola (Bancroft, 1834)	mountain mullet	dajao
Joturus pichardi Poey, 1860	bobo mullet	joturo
Mugil cephalus Linnaeus, 1758	striped mullet	lisa
Mugil curema Valenciennes, 1836	white mullet	lisa blanca
Mugil curvidens Valenciennes, 1836	dwarf mullet	lisa enana
Mugil gaimardianus Desmarest, 1831	redeye mullet	lisa conejo
Mugil hospes Jordan & Culver, 1895	hospe mullet	
Mugil incilis Hancock, 1830	Parassi mullet	liseta
Mugil liza Valenciennes, 1836	liza	lebrancho
Mugil longicauda Guitart Manday & Alvarez-Lajonchere, 1976		rabúa
Mugil trichodon Poey, 1875	fantail mullet	lisa de abanico
Polynemoidei		
Polynemidae		
Pentanemus quinquarius (Linnaeus, 1758)*	royal threadfin	
Polydactylus virginicus (Linnaeus, 1758)	barbu	barbudo
Labroidei		
Labridae		
Bodianus pulchellus (Poey, 1860)	spotfin hogfish	pez perro de cola amarilla
Bodianus rufus (Linneaus, 1758)	Spanish hogfish	pez perro español
Clepticus parrae (Bloch & Schneider, 1801)	creole wrasse	rabirrubia genízara
Decodon puellaris (Poey, 1860)	red hogfish	doncella de lo alto
Doratonotus megalepis Günther, 1862	dwarf wrasse	doncella enana
Halichoeres bivittatus (Bloch, 1791)	slippery dick	doncella rayada

Continued on next page

Appendix 2.1. continued

Halichoeres caudalis (Poey, 1860)	painted wrasse	doncella pintada
Halichoeres cyanocephalus (Bloch, 1791)	yellowcheek wrasse	doncella cabeciplateada
Halichoeres garnoti (Valenciennes, 1839)	yellowhead wrasse	doncella cabeciamarilla
Halichoeres maculipinna (Müller & Troschel, 1848)	clown wrasse	doncella payaso
Halichoeres pictus (Poey, 1860)	rainbow wrasse	doncella arcoiris
Halichoeres poeyi (Steindachner, 1867)	blackear wrasse	doncella ojinegra
Halichoeres radiatus (Linnaeus, 1758)	puddingwife	doncella pudín
Lachnolaimus maximus (Walbaum, 1792)	hogfish	pez perro
Thalassoma bifasciatum (Bloch, 1791)	bluehead	cara de cotorra
Xyrichtys martinicensis Valenciennes, 1840	rosy razorfish	doncella llorona
Xyrichtys novacula (Linnaeus, 1758)	pearly razorfish	doncella verde
Xyrichtys splendens Castelnau, 1855	green razorfish	doncella de lunar
Scaridae		
Scarinae		
Cryptotomus roseus Cope, 1871	bluelip parrotfish	loro bembiazul
Nicholsina usta usta (Valenciennes, 1840)	emerald parrotfish	loro esmeralda
Scarus coelestinus Valenciennes, 1840	midnight parrotfish	loro de medianoche
Scarus coeruleus (Bloch, 1786)	blue parrotfish	loro azul
Scarus guacamaia Cuvier, 1829	rainbow parrotfish	loro guacamayo
Scarus iserti (Bloch, 1789)	striped parrotfish	loro listado
Scarus taeniopterus Desmarest, 1831	princess parrotfish	loro princesa
Scarus vetula Bloch & Schneider, 1801	queen parrotfish	loro reina
Sparisomatinae		
Sparisoma atomarium (Poey, 1861)	greenblotch parrotfish	loro de lunar verde
Sparisoma aurofrenatum (Valenciennes, 1840)	redband parrotfish	vieja lora
Sparisoma chrysopterum (Bloch & Schneider, 1801)	redtail parrotfish	loro colirrojo
Sparisoma radians (Valenciennes, 1840)	bucktooth parrotfish	loro dientuso
Sparisoma rubripinne (Valenciennes, 1840)	redfin parrotfish	loro aletirojo
Sparisoma viride (Bonnaterre, 1788)	stoplight parrotfish	loro
Cichlidae		
Cichlasoma ramsdeni Fowler, 1938[e]	joturo	jotura
Cichlasoma tetracanthus (Valenciennes, 1831)[e]	biajaca	biajaca
Pomacentridae		
Abudefduf saxatilis (Linnaeus, 1758)	sergeant major	píntano
Abudefduf taurus (Müller & Troschel, 1848)[11]	night sergeant	píntano toro
Chromis cyanea (Poey, 1860)	blue chromis	cromis azul
Chromis enchrysura Jordan & Gilbert, 1882	yellowtail reeffish	
Chromis insolata (Cuvier, 1830)	sunshinefish	cromis sol
Chromis multilineata (Guichenot, 1853)	brown chromis	cromis prieto
Chromis scotti Emery, 1968	purple reeffish	cromis púrpura
Microspathodon chrysurus (Cuvier, 1830)	yellowtail damselfish	chopita de cola amarilla
Stegastes adustus (Troschel, 1865)	dusky damselfish	chopita prieta
Stegastes diencaeus (Jordan & Rutter, 1897)	longfin damselfish	chopita miel
Stegastes leucostictus (Müller & Troschel, 1848)	beaugregory	chopita de cola amarilla
Stegastes otophorus (Poey, 1860)	damselfish	chopita manchada
Stegastes partitus (Poey, 1868)	bicolor damselfish	chopita bicolor
Stegastes planifrons (Cuvier, 1830)	threespot damselfish	chopita amarilla
Stegastes variabilis (Castelnau, 1855)	cocoa damselfish	chopita cacao
Trachinoidei		
Uranoscopidae		
Kathetostoma cubana Barbour, 1941	marbled stargazer	pez curioso
Percophidae		
Bembrops anatirostris Ginsburg, 1955	duckbill flathead	
Bembrops gobioides (Goode, 1880)	goby flathead	
Bembrops macromma Ginsburg, 1955		
Bembrops magnisquamis Ginsburg, 1955		

Continued on next page

Appendix 2.1. continued

Bembrops raneyi Thompson & Suttkus, 1998
Chrionema squamentum (Ginsburg, 1955)[4]

Blennioidei

Tripterygiidae

Enneanectes altivelis Rosenblatt, 1960	lofty triplefin	sapito barreado
Enneanectes atrorus Rosenblatt, 1960	blackedge triplefin	sapito orleado
Enneanectes boehlkei Rosenblatt, 1960	roughhead triplefin	sapito rugoso
Enneanectes jordani (Evermann & Marsh, 1899)	mimic triplefin	sapito tres aletas

Labrisomidae

Labrisomus bucciferus Poey, 1868	puffcheek blenny	sapito fumador
Labrisomus filamentosus Springer, 1960	quillfin blenny	sapito filamentoso
Labrisomus gobio (Valenciennes, 1836)	palehead blenny	sapito ojudo
Labrisomus guppyi (Norman, 1922)	mimic blenny	sapito prieto
Labrisomus haitiensis Beebe & Tee-Van, 1928	longfin blenny	sapito de roca
Labrisomus nigricinctus Howell Rivero, 1936	spotcheek blenny	sapito lunado
Labrisomus nuchipinnis (Quoy & Gaimard, 1824)	hairy blenny	sapito cabezón
Malacoctenus aurolineatus Smith, 1957	goldline blenny	sapito lincado
Malacoctenus boehlkei Springer, 1959	diamond blenny	
Malacoctenus delalandei (Valenciennes, 1836)	delalande blenny	sapito de lalandi
Malacoctenus erdmani Smith, 1957	imitator blenny	sapito imitador
Malacoctenus gilli (Steindachner, 1867)	dusky blenny	sapito pardo
Malacoctenus macropus (Poey, 1868)	rosy blenny	sapito rosado
Malacoctenus triangulatus Springer, 1959	saddle blenny	sapito de inontura
Malacoctenus versicolor (Poey, 1876)	barfin blenny	sapito multicolor
Paraclinus barbatus Springer, 1955	goatee blenny	
Paraclinus cingulatus (Evermann & Marsh, 1899)	coral blenny	sapito coral
Paraclinus fasciatus (Steindachner, 1876)	banded blenny	sapito ocelado
Paraclinus grandicomis (Rosén, 1911)	horned blenny	sapito tarrudo
Paraclinus marmoratus (Steindachner, 1876)	marbled blenny	sapito punteado
Paraclinus nigripinnis (Steindachner, 1867)	blackfin blenny	sapito aletinegro
Starksia atlantica Longley, 1934	smootheye blenny	sapito pelón
Starksia fasciata (Longley, 1934)	blackbar blenny	sapito bandeado
Starksia lepicoelia Böhlke & Springer, 1961	blackcheek blenny	sapito carinegro
Starksia nanodes Böhlke & Springer, 1961	dwarf blenny	
Starksia ocellata (Steindachner, 1876)	checkered blenny	sapito tablero

Chaenopsidae

Acanthemblemaria asper (Longley, 1927)	roughhead blenny	
Acanthemblemaria chaplini Böhlke, 1957	papillose blenny	sapito papiloso
Acanthemblemaria maria Böhlke, 1961	secretary blenny	sapito erizo
Acanthemblemaria spinosa Metzelaar, 1919	spinyhead blenny	sapito espinoso
Chaenopsis limbaughi Robins & Randall, 1965	yellowface pikeblenny	
Chaenopsis ocellata Gill, 1865	bluethroat pikeblenny	sapito afilado
Coralliozetus cardonae Evermann & Marsh, 1899*	twinhorn blenny	
Emblemaria pandionis Evermann & Marsh, 1900	sailfin blenny	sapito dragón
Emblemariopsis occidentalis Stephens, 1970[4]	redspine blenny	sapito aletón
Emblemariopsis signifera (Ginsburg, 1942)	flagfin blenny	sapito aletón
Hemiemblemaria simulus Longley & Hildebrand, 1940	wrasse blenny	sapito doncella
Lucayablennius zingaro (Böhlke, 1957)	arrow blenny	
Stathmonotus gymnodermis Springer, 1955	naked blenny	sapito esperanza
Stathmonotus hemphilli Bean, 1885*	blackbelly blenny	sapito morenita
Stathmonotus stahli (Evermann & Marsh, 1899)	eelgrass blenny	sapito anguila

Dactyloscopidae

Dactyloscopus crossotus Starks, 1913	bigeye stargazer	
Dactyloscopus poeyi Gill, 1861	shortchin stargazer	mirón ojicorto
Dactyloscopus tridigitatus Gill, 1859	sand stargazer	mirón ojilargo
Gillellus greyae Kanazawa, 1952	arrow stargazer	mirón flecha
Gillellus uranidea Böhlke, 1968	warteye stargazer	
Leurochilus acon Böhlke, 1968	smoothlip stargazer	

Continued on next page

Appendix 2.1. continued

Platygillellus rubrocinctus (Longley, 1934)	saddle stargazer	mirón barreado
Blenniidae		
Entomacrodus nigricans Gill, 1859	pearl blenny	blenio perlado
Hypleurochilus aequipinnis (Günther, 1861)[9]	oyster blenny	blenio ostra
Hypleurochilus geminatus (Wood, 1825)	crested blenny	blenio crestado
Hypsoblennius exstochilus Böhlke, 1959	longhorn blenny	blenio cornudo
Lupinoblennius vinctus (Poey, 1867)[e]	mangrove blenny	blenio de mangle
Ophioblennius atlanticus macclurei (Silvester, 1915)	redlip blenny	blenio bembirrojo
Parablennius marmoreus (Poey, 1876)	seaweed blenny	blenio marmoreo
Scartella cristata (Linnaeus, 1758)	molly miller	blenio peineta
Gobiesocoidei		
Gobiesocidae		
Acyrtops artius (Briggs, 1955)	papillate clingfish	pequita gusarapo
Acyrtops beryllinus (Hildebrand & Ginsburg, 1926)	emerald clingfish	pequita esmerilada
Acyrtus rubiginosus (Poey, 1868)	red clingfish	pequita roja
Arcos macrophthalmus (Günther, 1861)	padded clingfish	
Gobiesox lucayanus Briggs, 1963	Bahama skilletfish	
Gobiesox nudus (Linnaeus, 1758)	riverine clingfish	
Gobiesox punctulatus (Poey, 1876)	stippled clingfish	pequita punteada
Gobiesox strumosus Cope, 1870	skilletfish	cazoleta
Tomicodon fasciatus (Peters, 1860)	barred clingfish	pequita barreada
Callionymoidei		
Callionymidae		
Diplogrammus pauciradiatus (Gill, 1865)	spotted dragonet	dragoncillo moteado
Paradiplogrammus bairdi (Jordan, 1888)	lancer dragonet	dragoncillo coralino
Synchiropus agassizii (Goode & Bean, 1888)	spotfin dragonet	
Gobioidei		
Eleotridae		
Dormitator maculatus (Bloch, 1792)	fat sleeper	guavina mapo
Eleotris pisonis (Gmelin, 1789)	spinycheek sleeper	guavina espinosa
Erotelis smaragdus (Valenciennes, 1837)	emerald sleeper	esmeralda negra
Gobiomorus dormitor Lacepède, 1800	bigmouth sleeper	guavina de ley
Guavina guavina (Valenciennes, 1837)	guavina	guavina cabezona
Gobiidae		
Awaous tajasica (Lichtenstein 1822)	river goby	sirajo
Bathygobius curacao (Metzelaar, 1919)	notchtongue goby	gobio jaspeado
Bathygobius mystacium Ginsburg, 1947	island frillfin	gobio bandeado
Bathygobius soporator (Valenciennes, 1837)	frillfin goby	gobio mapo
Coryphopterus alloides Böhlke & Robins, 1960	barfin goby	
Coryphopterus dicrus Böhlke & Robins, 1960	colon goby	gobio colón
Coryphopterus eidolon Böhlke & Robins, 1960	pallid goby	gobio pálido
Coryphopterus glaucofraenum Gill, 1863	bridled goby	gobio con brida
Coryphopterus hyalinus Böhlke & Robins, 1962	glass goby	gobio de cristal
Coryphopterus lipernes Böhlke & Robins, 1962	peppermint goby	gobio linterna
Coryphopterus personatus (Jordan & Thompson, 1905)	masked goby	gobio enmascarado
Ctenogobius boleosoma (Jordan & Gilbert, 1882)	darter goby	esmeralda flechera
Ctenogobius fasciatus Gill, 1858*	blackbar goby	
Ctenogobius munizi (Vergara, 1978)*		
Ctenogobius saepepallens (Gilbert & Randall, 1968)	dash goby	gobio guión
Ctenogobius smaragdus (Valenciennes, 1837)	emerald goby	esmeralda cabezona
Ctenogobius stigmalophius (Mead & Böhlke, 1958)[10]	spotfin goby	gobio manchado
Ctenogobius stigmaticus (Poey, 1860)	marked goby	esmeralda
Ctenogobius stigmaturus (Goode & Bean, 1882)	spottail goby	
Evorthodus lyricus (Girard, 1858)	lyre goby	gobio lira
Garmannia saucra (Robins, 1960)	leopard goby	gobio severo
Ginsburgellus novemlineatus (Fowler, 1950)	ninelined goby	
Gnatholepis thompsoni Jordan, 1904	goldspot goby	gobio puntidorado

Continued on next page

Appendix 2.1. continued

Gobioides broussoneti Lacepède, 1800	violet goby	esmeralda de río
Gobionellus oceanicus (Pallas,1770)	highfin goby	esmeralda de mar
Gobiosoma dilepis (Robins & Böhlke, 1964)	orangeside goby	gobio naranja
Gobiosoma evelynae Böhlke & Robins, 1968	sharpnose goby	gobio hocicudo
Gobiosoma gemmatum (Ginsburg, 1939)	frecklefin goby	gobio de aleta pecosa
Gobiosoma genie Böhlke & Robins, 1968	cleaner goby	gobio limpiador
Gobiosoma horsti Metzelaar, 1922	yellowline goby	gobio de banda amarilla
Gobiosoma louisae Böhlke & Robins,1968[4]	spotlight goby	
Gobiosoma macrodon Beebe & Tee-Van, 1928	tiger goby	
Gobiosoma multifasciatum Steindachner, 1876	greenbanded goby	gobio rayado
Gobiosoma pallens (Ginsburg, 1939)	semiscaled goby	gobio semidesnudo
Gobiosoma spes (Ginsburg, 1939)	gobio espes	
Lophogobius cyprinoides (Pallas, 1770)	crested goby	gobio encrestado
Lythrypnus elasson Böhlke & Robins, 1960	dwarf goby	gobio enano
Lythrypnus heterochroma Ginsburg, 1939	diphasic goby	gobio barreado
Lythrypnus nesiotes Böhlke & Robins, 1960	island goby	gobio isla
Lythrypnus spilus Böhlke & Robins, 1960	bluegold goby	gobio marcado
Microgobius microlepis Longley & Hildebrand, 1940	banner goby	gobio bandera
Microgobius signatus Poey, 1876		gobio señal
Nes longus (Nichols, 1914)[11]	orangespotted goby	gobio de manchas naranja
Priolepis hipoliti (Metzelaar, 1922)	rusty goby	gobio oxidado
Psilotris alepis Ginsburg, 1953	scaleless goby	gobio de poceta
Psilotris batrachodes Böhlke, 1963	toadfish goby	gobio de sapo
Risor ruber (Rosén, 1911)	tusked goby	gobio de boca chica
Sicydium buscki Evermann & Clark, 1906 [12]	Busck's stone-biting goby	
Sicydium plumieri (Bloch, 1786)	Plumier's stone-biting goby	
Sicydium punctatum Perugia, 1896 [12]	spotted algae-eating goby	
Varicus bucca Robins & Böhlke, 1961		
Microdesmidae		
Cerdale floridana Longley, 1934	pugjaw wormfish	
Microdesmus longipinnis (Weymouth, 1910)	pink wormfish	
Ptereleotridae		
Ptereleotris helenae (Randall, 1968)	hovering goby	gobio azu
Acanthuroidei		
Acanthuridae		
Acanthurus bahianus Castelnau, 1855	ocean surgeon	barbero
Acanthurus chirurgus (Bloch, 1787)	doctorfish	barbero rayado
Acanthurus coeruleus Bloch & Schneider, 1801	blue tang	barbero azul
Scombroidei		
Sphyraenidae		
Sphyraena barracuda (Walbaum, 1792)	great barracuda	picúa
Sphyraena borealis De Kay, 1842	northern sennet	picudilla
Sphyraena guachancho Cuvier, 1829	guaguanche	guaguancho
Gempylidae		
Diplospinus multistriatus Maul, 1948	striped escolar	sable estirado
Epinnula magistralis Poey, 1854	domine	dómine
Epinnula orientalis Gilchrist & von Bonde, 1924	sackfish	dómine chico
Gempylus serpens Cuvier, 1829	snake mackerel	dómine añil
Lepidocybium flavobrunneum (Smith, 1843)	escolar	petróleo
Nealotus tripes Johnson, 1865	black snake mackerel	dómine negro
Nesiarchus nasutus Johnson, 1862	black gemfish	dómine narizón
Promethichthys prometheus (Cuvier, 1832)	Roudi escolar	conejo
Ruvettus pretiosus Cocco, 1833	oilfish	escolar
Trichiuridae		
Aphanopodinae		
Benthodesmus elongatus (Clarke, 1879)	frostfish	sable delgado
Benthodesmus tenuis (Günther, 1877)[4]	ribbon scabbardfish	

Continued on next page

Appendix 2.1. continued

Lepidopodinae
Evoxymetopon taeniatus Gill, 1863 — channel scabbardfish — tirante

Trichiurinae
Trichiurus lepturus Linnaeus, 1758 — Atlantic cutlassfish — sable

Xiphiidae
Xiphias gladius Linnaeus, 1758 — swordfish — emperador

Istiophoridae
Istiophorus platypterus (Shaw & Nodder, 1792) — sailfish — aguja de abanico
Makaira nigricans Lacepède, 1802 — blue marlin — castero
Tetrapturus albidus Poey, 1860 — white marlin — aguja blanca
Tetrapturus pfluegeri Robins & de Sylva, 1963 — longbill spearfish — aguja del Pacífico

Scombridae
Acanthocybium solandri (Cuvier, 1832) — wahoo — peto
Auxis rochei rochei (Risso,1810) — bullet mackerel — melva aleticorto
Auxis thazard thazard (Lacepède, 1800) — frigate mackerel — melva aletilargo
Euthynnus alletteratus (Rafinesque, 1810) — little tunny — comevíveres
Katsuwonus pelamis (Linnaeus, 1758) — skipjack tuna — bonito listado
Sarda sarda (Bloch, 1793) — Atlantic bonito — bonito
Scomber colias Gmelin, 1789 — Atlantic chub mackerel — macarela estornino
Scomberomorus cavalla (Cuvier, 1829) — king mackerel — sierra
Scomberomorus maculatus (Mitchill, 1815) — Spanish mackerel — serrucho
Scomberomorus regalis (Bloch, 1793) — cero — pintada
Thunnus alalunga (Bonnaterre, 1788) — albacore — albacora
Thunnus albacares (Bonnaterre, 1788) — yellowfin tuna — atún de aleta amarilla
Thunnus atlanticus (Lesson, 1831) — blackfin tuna — falsa albacora
Thunnus obesus (Lowe, 1839) — bigeye tuna — atún ojo grande
Thunnus thynnus (Linnaeus, 1758) — bluefin tuna — atún aleta azul

Stromateoidei
Ariommatidae
Ariomma bondi Fowler, 1930 — silver-rag — pastorcillo lucia
Ariomma regulus (Poey, 1868) — spotted driftfish — pastorcillo aquillado

Nomeidae
Nomeus gronovii (Gmelin, 1789) — man-of-war fish — pastorcillo
Psenes cyanophrys Valenciennes, 1833 — freckled driftfish — pastorcillo amarillento

Stromateidae
Peprilus paru (Linnaeus, 1758) — — palometa moneda
Peprilus triacanthus (Peck, 1804)* — butterfish — palometa estrecha

Order Pleuronectiformes
Paralichthyidae
Citharichthys arenaceus Evermann & Marsh, 1900 — sand whiff — lenguado arenero
Citharichthys cornutus (Günther, 1880) — horned whiff — lenguado cornudo
Citharichthys dinoceros Goode & Bean, 1886[6]
Citharichthys gymnorhinus Gutherz & Blackman, 1970[6] — anglefin whiff
Citharichthys spilopterus Günther, 1862* — bay whiff — lenguado pardo
Etropus crossotus Jordan & Gilbert, 1882 — fringed flounder — lenguado orlado
Syacium gunteri Ginsburg, 1933 — shoal flounder
Syacium micrurum Ranzani, 1842 — channel flounder — lenguado anillado
Syacium papillosum (Linnaeus, 1758) — dusky flounder — lenguado moreno

Bothidae
Bothus lunatus (Linnaeus, 1758) — peacock flounder — lenguado lunado
Bothus maculiferus (Poey, 1860) — maculated flounder — lenguado manchado
Bothus ocellatus (Agassiz, 1831) — eyed flounder — lenguado ocelado
Bothus robinsi Topp & Hoff, 1972 — twospot flounder

Continued on next page

Appendix 2.1. continued

Bothus spinosus (Poey, 1868)*		
Engyophrys senta Ginsburg, 1933	spiny flounder	
Poecilopsettidae		
Poecilopsetta beanii (Goode 1881)		
Achiridae		
Achirus lineatus (Linnaeus, 1758)	lined sole	acedía rayada
Gymnachirus nudus Kaup, 1858		acedía nudosa
Trinectes inscriptus (Gosse, 1851)	scrawled sole	acedía reticulada
Cynoglossidae		
Symphurus caribbeanus Munroe, 1991		
Symphurus diomedeanus (Goode & Bean, 1885)	spottedfin tonguefish	lengua de vaca
Symphurus ommaspilus Böhlke, 1961[7]	ocellated tonguefish	
Symphurus piger (Goode & Bean, 1886)	deepwater tonguefish	lengua de vaca perezosa
Symphurus plagiusa (Linnaeus, 1766)	blackcheek tonguefish	lengua de vaca gris
Symphurus plagusia (Bloch & Schneider, 1801)	duskycheek tonguefish	lengua de vaca parda
Symphurus urospilus Ginsburg, 1951	spottail tonguefish	lengua de vaca colipunteada
Order Tetraodontiformes		
Balistoidei		
Triacanthodidae		
Hollardia hollardi Poey, 1861	reticulate spikefish	rombo verde
Hollardia meadi Tyler, 1966	spotted spikefish	rombo pardo
Johnsonina eriomma Myers, 1934		
Balistidae		
Balistes capriscus Gmelin, 1789	gray triggerfish	cochino gris
Balistes vetula Linnaeus, 1758	queen triggerfish	cochino
Canthidermis maculata (Bloch, 1786)	rough triggerfish	sobaco manchado
Canthidermis sufflamen (Mitchill, 1815)	ocean triggerfish	lija
Melichthys niger (Bloch, 1786)	black durgon	negrito
Xanthichthys ringens (Linnaeus, 1758)	sargassum triggerfish	cocuyo
Monacanthidae		
Aluterus monoceros (Linnaeus, 1758)	unicorn filefish	lija barbuda
Aluterus schoepfi (Walbaum, 1792)	orange filefish	lija anaranjada
Aluterus scriptus (Osbeck, 1765)	scrawled filefish	lija trompa
Cantherhines macrocerus (Hollard, 1853)	whitespotted filefish	lija de lunares blancos
Cantherhines pullus (Ranzani, 1842)	orangespotted filefish	lija colorada
Monacanthus ciliatus (Mitchill, 1818)	fringed filefish	sobaco común
Monacanthus tuckeri Bean, 1906	slender filefish	lija reticulada
Stephanolepis hispidus (Linnaeus, 1766)	planehead filefish	lija áspera
Stephanolepis setifer (Bennett, 1831)	pygmy filefish	lija ciliada
Ostraciidae		
Acanthostracion polygonius Poey, 1876	honeycomb cowfish	torito hexagonal
Acanthostracion quadricornis (Linnaeus, 1758)	scrawled cowfish	torito común
Lactophrys bicaudalis (Linnaeus, 1758)	spotted trunkfish	chapín de lunares negros
Lactophrys trigonus (Linnaeus, 1758)	trunkfish	chapín de lunares
Lactophrys triqueter (Linnaeus, 1758)	smooth trunkfish	chapín común
Tetraodontoidei		
Tetraodontidae		
Canthigaster rostrata (Bloch, 1786)	sharpnose puffer	tamboril narizón
Lagocephalus laevigatus (Linnaeus, 1766)	smooth puffer	tamboril gigante
Lagocephalus lagocephalus (Linnaeus, 1758)[9]	oceanic puffer	tamboril oceanico
Sphoeroides dorsalis Longley, 1934	marbled puffer	
Sphoeroides nephelus (Goode & Bean, 1882)	southern puffer	tamboril sureño
Sphoeroides spengleri (Bloch, 1785)	bandtail puffer	tamboril manchado
Sphoeroides testudineus (Linnaeus, 1758)	checkered puffer	tamboríl rayado
Diodontidae		
Chilomycterus antennatus (Cuvier, 1816)	bridled burrfish	guanábana manchada
Chilomycterus antillarum Jordan & Rutter, 1897	web burrfish	guanábana antillana

Continued on next page

Appendix 2.1. continued

Chilomycterus atinga (Linnaeus, 1758)	spotted burrfish	guanábana
Chilomycterus schoepfii (Walbaum, 1792)	striped burrfish	guanába rayada
Diodon holocanthus Linnaeus, 1758	balloonfish	pez erizo
Diodon hystrix Linnaeus, 1758	porcupinefish	puerco espín
Molidae		
Masturus lanceolata (Liénard, 1840)	sharptail mola	pez luna coliagudo
Mola mola (Linnaeus, 1758)	ocean sunfish	pez luna
Ranzania laevis (Pennant, 1776)	slender mola	pez luna colitruncado

References for recent records: [1]L. Espinosa, personal communication; [2]Pol et al., 1991; [3]D. Guitart, personal communication; [4]Claro et al. 2000; [5]Tariche, in press b; [6]Valdés et al. 1999; [7]Tariche 1998; [8]Tariche, in press a; [9]Claro et al., in press a; [10]Tallet, in press a; [11]Tallet and Fernández 1999; [12]Watson 2000.

e: endemic.

3

Behavior of Marine Fishes of the Cuban Shelf

EMILIO VALDÉS-MUÑOZ AND ANDREI D. MOCHEK

3.1 Introduction

The high diversity of fish species characteristic of tropical coastal waters fosters high behavioral diversity. Behavioral patterns associated with successful feeding, defense, and reproduction are major features of the individual, school, and assemblage and are fundamental components of the ethological structure of fish communities. Coral reef fishes have been classified into a variety of categories according to behavior (e.g., permanent residents and water column transients; Smith and Tyler 1973a). Because within- and between-species interactions are often influenced by habitats or food, we grouped species into broad behavioral patterns according to the use of spatial resources (see Table 3.1), which might be limiting factors in some fish communities (Smith and Tyler 1973a, 1973b; Jones 1991). These broad categories can then be evaluated in terms of the periods of greatest activity (diurnal, nocturnal, crepuscular; Table 3.1). This system provides a useful template for categorizing the behavior of the fishes we have studied. We recognize that additional factors also influence reef fish assemblages (Sale 1991).

During daylight, many coastal fish species in Cuban waters gather in groups over relatively small areas of uneven relief, typically reefs. Fish diversity is lower in large nonestuarine sandy areas, seagrass beds, and softbottom areas. For example, around Punta del Este (Isla de la Juventud), 96 species were found on coral reefs, 57 in mangroves, 55 on seagrass beds, and 25 in sandy areas (Valdés-Muñoz 1982; Fig. 3.1). Our behavioral studies typically focused on the habitats with high species diversity.

Our study methods were based on field assessments of some of the primary elements of fish behavior (Mochek and Valdés-Muñoz 1983; Valdés-Muñoz and Mochek 1994). Several parameters were measured, including vertical positioning of fishes, degree of mobility, presence of group formation, social and aggressive behaviors, and defensive behavior. The assessments were conducted on at least 25 individuals of each species. Additional quantitative indices, such as total observation time and distance covered by individuals (in meters), allowed us to compare the behavior of species among a variety of habitats (Mochek and Valdés-Muñoz 1983; Valdés-Muñoz and Mochek 1994).

3.2 Diurnal Fishes

3.2.1 Inshore Pelagic Fishes

Few reef fishes are associated with the upper water column without using benthic resources. Aside from the planktonic stages (eggs and larvae), this diurnal ichthyofauna consists typically of needlefishes (Belonidae), jacks (Carangidae), and barracudas (*Sphyraena* spp.). Great mobility is one of the main features of these fishes: species either appear to have no permanent territory or their home range extends over great distances (e.g., jacks). Rapid movements occur when they chase prey or are threatened. For example, needlefishes in retreat perform a pseudoflight, a sustained fluttering of the caudal fin's lower lobe over the water surface prompted by an initial impelling force. Needlefishes and barracudas are less dynamic than oceanic–pelagic species such as tuna and some jacks; swimming alternates with periods of relative immobility, which allows them to examine both the upper water column and the area above the reef without alarming potential prey.

Intraspecific cooperation and gregarious behavior is frequently observed in schooling pelagic fishes, in contrast to the sedentary fishes of demersal habitats with irregular relief. Our use of the term *school* follows Breder (1959). Intra- and interspecific cooperation makes pelagic predators successful

Table 3.1.   Behavioral categories of selected coastal fishes in Cuba, based on spatial–temporal activity patterns and feeding

Spatial Category	Activity Period		
	Diurnal	Nocturnal	Crepuscular
Inshore pelagics	Belonidae (T) Carangidae (T) Sphyraenidae (T)	Clupeidae (T)	—
Epibenthic pomacentrids/apogonids	Pomacentridae (R)	Apogonidae (R)	—
Suprabenthic fishes	Scaridae (R) Labridae (R) Acanthuridae (R) Chaetodontidae (R)	Lutjanidae (R) Haemulidae (R) Holocentridae (R)	—
Benthic zone	Pomacentridae (Te)	—	Serranidae (R) Muraenidae (R)

T: transients; R: residents; Te: territorial.

Species diversity by habitat type

	Sand	Mangroves	Grassbeds	Reefs
Number of families	14	19	20	29
Number of species	25	57	55	96

Species similarity by habitat type

	Sand	Mangroves	Grassbeds	Reefs
Reef	8	56	13	–
Grassbeds	40	11	–	
Mangroves	10	–		
Sand	–			

Fig. 3.1. Species diversity and similarity of the fish fauna of different habitats at Punta del Este, Isla de la Juventud. Species similarity values are based on the Sorensen index (from Valdés-Muñoz 1982).

foragers. Moreover, in the relatively homogenous water column, structural habitats are typically absent, and the mobility of gregarious species combined with disorienting maneuvers can create an effective defensive behavior (Radakov 1972). For this reason, use of shelters is not essential for pelagic fishes. In jacks, schooling has several benefits: in a school it is easier to discover and disorient prey, and schooling provides a critical defensive role for each individual.

Interspecific interactions in pelagic species can be reflected by the symbiotic relationship between large predators and some smaller fishes, such as that between great barracuda (*Sphyraena barracuda*) and small jacks (mainly the bar jack, *Caranx ruber*). The cooperation of both species is analogous to the well-known symbiotic relationship between sharks and pilot fishes. The great barracuda–bar jack association is well coordinated; the barracuda establishes the speed and direction, and the jacks follow when the barracuda chases prey. Both benefit: jacks and barracudas are protected from other predators and jacks are guided toward prey.

Some water column dwellers (e.g., barracudas and needlefishes) are common above seagrass habitats. They generally do not display the maneuvers of typical schooling fish and are solitary in the water column over the seagrass beds. Small prey disperse quickly and individual hunting is important, in contrast to schooling as a means of capturing food. The solitary hunter is not conspicuous and can meet food requirements by capturing solitary prey. During hunting, barracuda use an ambush tactic, suddenly attacking their prey from a considerable distance (Hobson 1974).

Euhaline mangrove areas usually have less than 1 m water depth, so the number of fishes moving through the water column beneath roots and branches is potentially limited. Among inshore pelagic fishes, some solitary species move throughout this zone, and a more numerous group of sedentary predator species typically swims close to the mangrove roots. Among the latter, barracuda and needle-fishes are important; they can remain for hours in this narrow zone, ambushing prey. Individuals more than 1 m long can be found here. During the night, all representatives of this group appear to be inactive. We detected little movement in either barracudas or needlefishes close to mangrove roots or over seagrass beds (Mochek and Valdés-Muñoz 1983). They were usually extremely passive and did not react to our flashlights, only moving away when chased.

3.2.2   Epibenthic Pomacentrids
These small fishes are often resident and move over the reef in the water column, feeding on plankton. Davis and Birdsong (1973) considered this trophic specialization to be the most important feature for separating reef fishes that feed in the

water column into different trophic groups. The primary species of this group are the damselfishes (Pomacentridae), particularly *Abudefduf,* and *Chromis*. The sergeant major (*Abudefduf saxatilis*) is common in Cuba on irregular bottom types, such as patch reefs and ledges, where it finds shelter. Sergeant majors spend most of their time in the water column eating planktonic crustaceans (Randall 1967; Böhlke and Chaplin 1993) and other foods. They become more active as water currents increase and when plankton flows faster over the reef. While eating, they venture in groups beyond the reef, sometimes reaching the water surface.

Water column reef fishes are characterized by relatively organized groups (Davis and Birdsong 1973). While feeding, groups of sergeant majors separate and display chaotic movements, manifesting weak imitation behaviors (i.e., one or more individuals behaving in a similar manner), according to Manteifel (1980). Defensive reactions appear in localized sites within the group or extend outward like a wave, involving all individuals. For example, during the aggressive approach of a diver, sergeant majors exhibited the disorienting movement known as the θ-maneuver: the group splits into two, surrounding the predator and then merging again behind it (Radakov 1972). Unlike pelagic fishes that display the entire movement, sergeant majors tended to quickly approach the reef, showing only the initial splitting phase. Synchronization of fish movements ceased abruptly when the individuals reached the bottom. Most of the fish scattered in different directions or quickly hid in bottom crevices.

Sergeant majors are abundant in or near mangrove roots and on reefs. Groups can be found in all sections of these habitats, and movements to adjacent seagrass beds can also occur. Our diurnal observations (Fig. 3.2) suggest that in mangroves, sergeant majors almost always stay in the water column feeding on plankton, and consistently move horizontally for distances of 1.5 m (average). They are gregarious: 95% of the time they were seen forming groups with a high rate of imitation behavior (96%) and fairly low aggressiveness among individuals.

Blue chromis (*Chromis cyanea*) and brown chromis (*C. multilineata*) are abundant in deeper habitats. They show behavior similar to that of the sergeant major. During the day groups were observed in the water column where they fed constantly on plankton (Hobson 1991). When chromis were threatened, they formed a compact school and swam together to the shelters in the immediate reef, remaining hidden in the crevices until the predator moved away (Hartline et al. 1972). At dusk the groups descended to the reef to individual shelters that they occupied during the night (Collette and Talbot 1972; Hobson 1991).

3.2.3 Suprabenthic Fishes

Resident fishes living over reefs are primary representatives of the reef fish community in both number and biomass. This diverse group consists of several speciose families.

Among the most common are parrotfishes (Scaridae), surgeonfishes (Acanthuridae), wrasses (Labridae), and butterflyfishes (Chaetodontidae). Despite their differences in morphology, coloration, size, feeding habits, and defense mechanisms, fishes of this group share an important feature: diurnal foraging above the bottom.

SCARIDAE.   Parrotfishes feed mainly on epiphytic algae (Randall 1961, 1965; Hobson 1974; Ogden and Lobel 1978); their nocturnal shelters include coral crevices and refuges (Winn and Bardach 1960). In Cuba, large parrotfishes were found in small groups or alone, from the reef crest vicinity of the fore reef to the first fore reef slope. The crunching noise parrotfishes make while feeding on corals with their powerful jaws can be heard underwater from a distance of several dozen meters. They feed during the day and stop only when threatened. No coordinated reaction as a group was observed, even during attack by a predator. Most likely the gathering of large parrotfishes is aimed at facilitating feeding and defense, but not as a mechanism for attracting one to another, which is typical of gregarious behavior. Thus this type of grouping might not be considered schooling, but aggregating (Nikolsky 1963). Gregariousness is important for parrotfishes and is associated with the competition for feeding grounds (Robertson et al. 1976).

The general pattern of behavior of the larger parrotfish species is similar to that of the smaller parrotfishes, such as the spotlight parrotfish (*Sparisoma viride*). Unlike larger parrotfishes, however, the smaller species prefer shallower waters and are less responsive to the observer. The most characteristic features are high motor activity and ample horizontal movement at the reef's surface (Fig. 3.3). Fast movements alternate with relatively quiet periods, particularly while grazing. Defensive reaction in spotlight parrotfishes declines as soon as the individual leaves the dangerous area. This species does not typically use shelters; even a long chase is not enough to make them hide in reef crevices, but only evokes the fish's acceleration and attempts to escape. Spotlight parrotfishes were seen alone or rarely in groups. When threatened, however, they became more gregarious than larger parrotfishes and displayed more imitation behaviors.

Other small parrotfishes, such as the striped parrotfish (*Scarus iserti* = *croicensis*), were typically seen in groups (Fig. 3.2). They gathered in impermanent groups for most of the day and exhibited characteristic imitation behaviors. Group composition changed continually as individuals entered and left the group, and its size ranged from two individuals to dozens. Solitary individuals were common over the reefs. Research in Panama indicates that their hierarchical association includes some elements of territoriality (Buckman and Ogden 1973). Varied forms of social organization were observed in striped parrotfishes, from large unstructured groups to hierarchical–territorial associations of few individuals or solitary individuals. We observed the fish moving constantly and exploring wide reef areas while grazing.

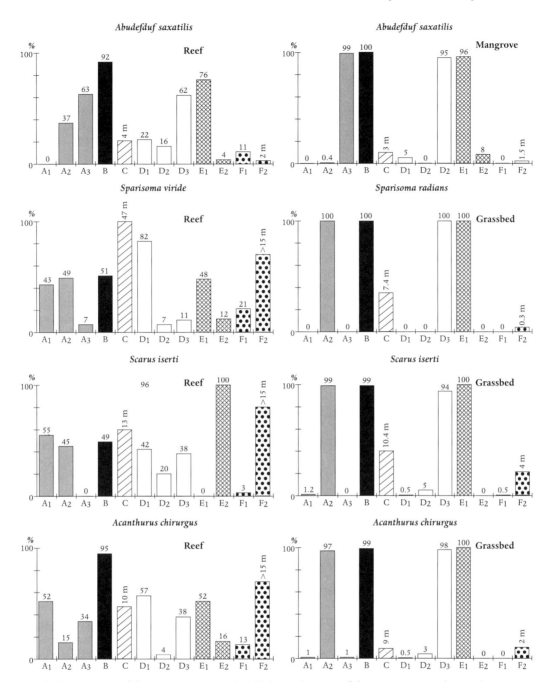

Fig. 3.2. Behavioral characteristics of the sergeant major (*Abudefduf saxatilis*), parrotfishes (*Sparisoma viride, S. radians, Scarus iserti* [= *croicensis*]), and doctorfish (*Acanthurus chirurgus*) in several habitats. A: Relationship with the substrate; A1: time on the bottom or among mangrove roots; A2: time close to the bottom (< 2 cm); A3: time far from the bottom (> 20 cm). B: Mobility. C: Mean distance covered (m) during 5-minute surveys. D: Group formation; D1: solitary; D2: in pairs; D3: in groups of 3 or more individuals. E: Social responses; E1: imitation; E2: aggressiveness. F: Defensive behavior; F1: permanence in the shelter; F2: distance to closest refuge (m). The numbers above the columns represent total time spent in that activity as a percentage of the time surveyed, or distance in meters for C and F2 (modified from Mochek and Valdés-Muñoz 1985).

Feeding by some individuals prompted others to imitate this behavior. The halting of feeding by some individuals was followed by the whole group. When a striped parrotfish was threatened, as a fast and mostly gregarious species, it escaped quickly using the stereotypical disorienting maneuver, swimming around bottom obstacles, rushing away from the chaser.

Some members of this family inhabit not only reefs but also seagrass beds, where they gather in mixed groups (e.g., striped parrotfishes and bucktooth parrotfishes [*Spar-*

isoma radians]) during the day. In seagrass beds they showed some behavioral elements of nomadic fishes: a high degree of motor activity, the formation of large schools, and marked imitation behavior during feeding and defense (Fig. 3.2). Thus behaviors displayed by parrotfishes in reef and seagrass beds differ substantially, even within a single species, because of differences in bottom relief.

ACANTHURIDAE.   Surgeonfishes are active, mainly demersal fishes, although the family as a whole shows great ecological variability. Three species of surgeonfishes inhabit Cuban waters: ocean surgeon (*Acanthurus bahianus*), doctorfish (*A. chirurgus*), and blue tang (*A. coeruleus*). Surgeonfishes occupy varied habitats; they can be found on all types of coral reef and at different depths, from shallow mangrove waters to several dozen meters deep. They prefer benthic and suprabenthic strata, are mobile, and generally gather in groups, exhibiting the typical social reaction. They make little use of shelters. Their main food target is macrophyto-benthos. Our data showed that while grazing during the day they spent most time on the bottom (Fig. 3.2 and 3.3).

Surgeonfishes, along with other members of this group, can be classified as species with broad ranges of territory and food resource utilization (Shilov 1977). They are rather nomadic, in contrast to typical territorial fishes whose isolated individuals or groups dominate a specific territory containing vital resources. On their feeding grounds, the three surgeonfish species are almost constantly in motion, performing broad migrations across the reef (Figs. 3.2 and 3.3). They also tended to form schools, often as mixed species. We recorded mutual imitation of the individuals within the schools; when threatened, the whole group escaped from the enemy, showing elements of the defensive θ-maneuver.

Despite the marked gregarious behavior of surgeon-fishes within the group, they are aggressive with one another, particularly with individuals of the same species. Most of the attacks against group-mates were recorded in blue tang and might reflect a hierarchical organization. Surgeonfishes could represent an intermediate step between two extreme ecological forms: pelagic fishes that are often gregarious, with no aggression among individuals (Radakov 1972), and bottom dwellers.

Two main species are found in seagrass beds: ocean surgeon and doctorfish. In this habitat, surgeonfishes and parrotfishes are characterized by a number of features typical of nomadic species: schooling, imitation behaviors, and an absence of territorial aggressiveness (Figs. 3.2 and 3.3). Their behavior in seagrass beds differs from that in coral reefs, since in seagrass beds they move constantly across the suprabenthic stratum. In seagrass beds, they not only move faster, but hide among seagrass blades when threatened.

LABRIDAE.   Wrasses are one of the most common families inhabiting Cuban coral reefs. In hogfish (*Lachnolaimus*

maximus) the intensity of body coloration and the distribution of stripes and spots can vary according to the surrounding landscape. When the fish is chased by a predator, body coloration changes to broken stripes, which allows it to blend among background habitats and evade capture. Thus, for such a relatively scarce species with large size and nomadic behavior, mimicry is an important element of defense. In small wrasses of other species, body coloration might also be used for camouflage. The abundance of small wrasses on the reef suggests that social contact might influence behavioral patterns. Thus the markings or conspicuous forms of a member of the group determines the likelihood of being chased, as well as the response when in danger, during foraging, or that evoked from a mate. Apparently a number of important aspects of imitation, hierarchical organization, and sexual and symbiotic reactions are performed using visual signals.

Wrasses are active only during the day, when foraging. At dusk they hide in reef crevices and shelters, bury themselves in the sand or in seagrass beds, or shelter among mangrove roots (Collette and Talbot 1972; Hobson 1974). Many small wrasses maintain a symbiotic relationship with other fishes. Cleaning symbioses are widely known, and in small wrasses, play an important behavioral role, as evidenced by ritual postures and movements. Small wrasses remove parasites not only from harmless fish, but also from many predators (Federn 1966; Losey 1972; Slobodkin and Fishelson 1974; Lowe-McConnell 1979). Ectoparasite ingestion by small wrasses has particular biological significance; in some fishes parasites are part of their diet, but ectoparasite ingestion also benefits fishes by controlling parasitism on them (Federn 1966). Small wrasses not only gather together with other fish species for cleaning, they also gather when those other fish are feeding.

The bluehead (*Thalassoma bifasciatum*) is a mobile fish found in many areas of the reef. The fast and constant movements, recorded in 99% of the time surveyed (Fig. 3.3), appear chaotic. They swim toward the bottom and then rise abruptly. This behavior could be explained in part by the bluehead's wide food spectrum, which includes planktonic crustaceans, benthic invertebrates, fish eggs and larvae, and ectoparasites (Randall 1967). The varied behavior of their prey prompts them to switch feeding tactics to obtain a constant supply of food. Consequently, the distance above the bottom changes frequently, but they usually spend a similar amount of time close to the bottom as they do in the midwater column (Fig. 3.3).

In 84% of the surveys, bluehead wrasses showed a strong tendency to gather in groups. We often observed groups consisting of dozens of bluehead individuals eating the benthic eggs of sergeant majors and yellowtail damsel-fishes (*Microspathodon chrysurus*). Imitation behavior is well developed in the bluehead and contributes not only to finding high concentrations of food organisms, but also to feeding on benthic egg nests defended by adult damselfishes.

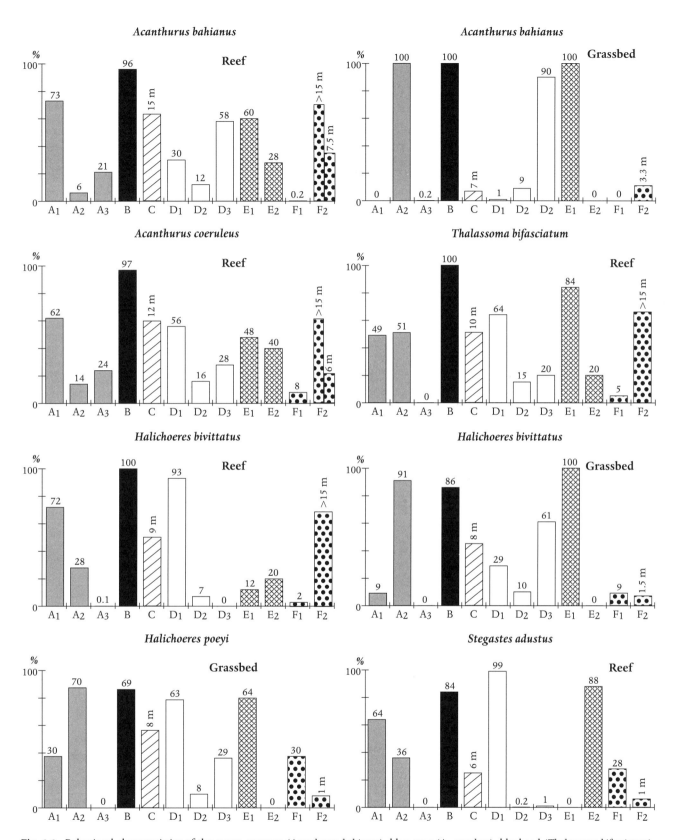

Fig. 3.3. Behavioral characteristics of the ocean surgeon (*Acanthurus bahianus*), blue tang (*A. coeruleus*), bluehead (*Thalassoma bifasciatum*), slippery dick (*Halichoeres bivittatus*), blackear wrasse (*H. poeyi*), and dusky damselfish (*Stegastes adustus* [= *dorsopunicans*]) in reefs and seagrass beds. Abbreviations as in Fig. 3.2.

In approximately 20% of the cases observed, isolated bluehead individuals showed aggressive reactions against other individuals of the same species (Fig. 3.3). Most individuals made little use of shelter; in a few cases they were observed examining reef crevices, searching for food. On the other hand, their defensive behavior was based on rapid movements, helped by bottom irregularities. When we chased groups of individuals, they reacted with coordinated group maneuvers.

Slippery dick (*Halichoeres bivittatus*) are also common on the reef, but are not as abundant as blueheads. Slippery dicks were seen actively searching for prey and continuously moving, mostly along the bottom (72% of the time) where they find and capture prey, but they were also in the suprabenthic stratum (28% of the time). Their incursions into the water column were extremely rare (< 0.1%, see Fig. 3.3). Slippery dick were typically solitary and exhibited aggressive reactions more often than imitation of neighbors. These two types of social reactions seemed to be relatively unimportant in species behavior. When in danger, they fled from predators by swimming away through the irregular bottom relief. We did not see them use shelters during the daytime surveys.

Some small wrasses, particularly blackear wrasse (*Halichoeres poeyi*) and slippery dick, also inhabit seagrass beds. The blackear wrasse is a diurnal species strongly associated with this habitat (Fig. 3.3). They were typically seen alone rather than in coordinated aggregations. It does not seem to be a mobile species, as the individuals were frequently seen descending to the bottom and remained among macrophytes (30% of the time). Such behavior, together with the green body coloring that blends with seagrasses, makes blackear wrasse difficult prey to catch. In contrast, slippery dick individuals showed a more dynamic behavior. They were recorded moving frequently (86% of the time) in the stratum immediately over seagrasses. Slippery dicks exhibited a markedly gregarious behavior that included imitation behaviors, whereas solitary individuals were seen only 29% of the time. They used seagrasses as a shelter only when in sudden danger. As soon as the threat disappeared they left the shelter. We conclude that there is a gradient from a dynamic gregarious behavior in more active species (e.g., slippery dick) to a more sedentary, sheltered behavior in the species living among macrophytes (blackear wrasse).

CHAETODONTIDAE. Butterflyfishes have been described as diurnal species (Starck and Davis 1966; Collette and Talbot 1972); nevertheless, Hobson (1974) found one species with nocturnal habits (*Chaetodon lunula,* not recorded in Cuba), and several other species that fed at dusk. Individuals of foureye butterflyfish (*Chaetodon capistratus*), banded butterflyfish (*C. striatus*) and spotfin butterflyfish (*C. ocellatus*) were most commonly recorded feeding actively near the bottom. During that time, they held no territory but moved over the reef, swimming a distance of 7–12 m in 5 minutes.

The small amplitude of movements and permanence over the bottom of foureye butterflyfishes can be explained by their euryphagy and intense feeding activities (Birkeland and Neudecker 1981). For long periods they were seen gathering in pairs showing imitation behaviors in different situations, such as when threatened, while feeding, or when moving over the reef. Among the three chaetodontids studied, only foureye butterflyfishes showed slight elements of aggressiveness. The three *Chaetodon* species did not use shelters, but escaped from predator attack by moving quickly. The disorientation created by zigzagging among the irregularities of bottom relief curtailed predator effectiveness. Fig. 3.4 shows the activity of foureye butterflyfishes in mangroves and along the interface with adjacent seagrass beds. They commonly gathered close to the mangrove roots, moving a distance of 2.7 m (average of several 5-minute surveys) with constant, sometimes chaotic, movements.

The social behavior of the foureye butterflyfish was variable: it was observed alone, in pairs, and in groups of three or more individuals. According to the interaction exhibited with neighboring individuals, 56% of the cases showed imitation behavior and 4% displayed aggressive reactions, although most of the individuals seemed to be indifferent to conspecifics. When threatened by predators, foureye butterflyfish fled instantly, rarely using disorienting maneuvers, and rarely hiding in shelters. This behavior is in contrast to that of territorial demersal fishes (e.g., pomacentrids).

3.2.4   Territorial Benthic Fishes

Territorial benthic species are abundant in tropical coastal fish communities. The concept of territoriality is applied here only to isolated fishes occupying permanently specific spaces that they defend from intrusion. Similar behavior is observed in many species from different regions during various life-history stages, particularly during the breeding season. Nevertheless, only in tropical and subtropical regions are there species that defend individual territories throughout their lifetime. Many of them belong to the family Pomacentridae, the damselfishes.

In most cases, territorial damselfishes have strong and contrasting coloration, are small, and are not fast swimmers, all of which makes them easy prey while outside their shelters (Clarke 1977). The defensive behavior of such fishes is their ability to retreat to shelters. Territorial damselfishes catch food (preferably detritus, benthic and planktonic animals, and phytobenthos) within their territory; therefore, the defended territory (which can be a shelter, the feeding area, or the spawning ground) is crucial for survival. Territorial behavior provides considerable advantages in competition for critical environmental resources (i.e., food, space, and shelter) and explains why their aggressive behavior addresses potential competitors (Lowe 1971; Myrberg and Thresher 1978; Ogden and Lobel 1978). Because territorial behavior is effective when competing for food, the defended

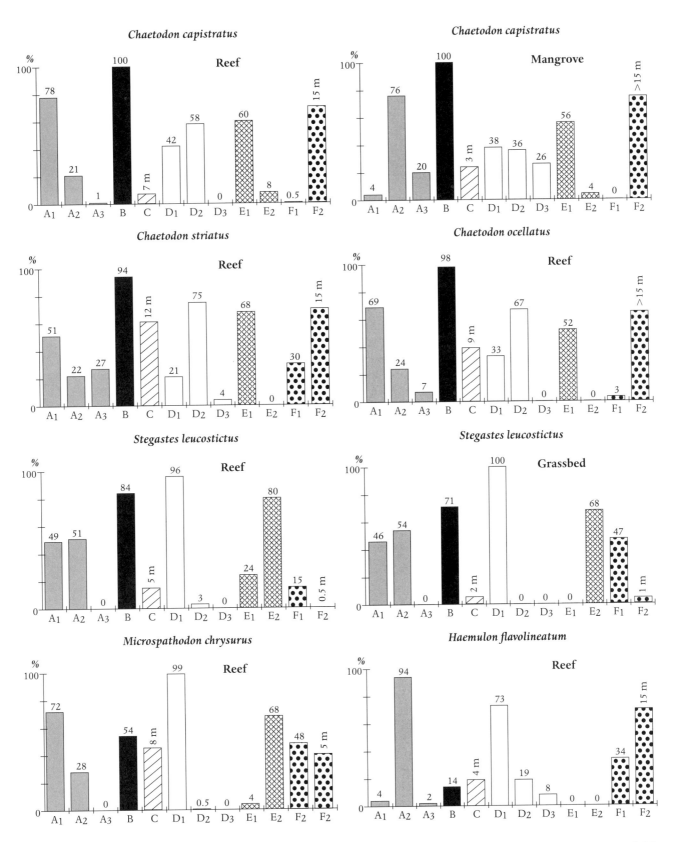

Fig. 3.4. Behavioral characteristics of the foureye butterflyfish (*Chaetodon capistratus*), banded butterflyfish (*C. striatus*), spotfin butterflyfish (*C. ocellatus*), beaugregory (*Stegastes leucostictus*), yellowtail damselfish (*Microspathodon chrysurus*), and French grunt (*Haemulon flavolineatum*) in several habitats. Abbreviations as in Fig. 3.2.

area usually exceeds the minimum needed to provide vital resources (Montgomery 1981). Only schooling fishes can effectively compete for food with territorial residents (Barlow 1974; Robertson et al. 1976; Ogden and Lobel 1978). Sessile benthic organisms, as well as suprabenthic organisms living in the water column, can benefit from the territorial defensive behaviors of fishes. For example, schools of mysid shrimp concentrate over damselfish shelters (Emery 1968).

Territorial fishes are numerous in Cuban coral reefs; among them, dusky damselfish (*Stegastes adustus*), beaugregory (*S. leucostictus*), and yellowtail damselfish (*Microspathodon chrysurus*) are common. The individual territories of dusky damselfishes occupy wide zones within the reef. This small (about 15 cm long), dark brown (mostly black in adults) fish moves quickly across its territory (Fig. 3.3) searching for food and aggressively attacking other fishes. The diet of this species consists of algae, detritus, and benthic animals (Randall 1967); thus part of the time (64%) they were observed feeding on the bottom, and part of the time (36%) they defended themselves from other fishes, above the substrate.

Territorial aggression in dusky damselfishes is high; this behavior was displayed against territorial neighbors or to defend territory in 88% of the surveys. During daytime surveys, this species used shelters for a short period, but when threatened, they quickly took shelter. The boundaries of the feeding ground defended by a single individual were within a 70–80 cm radius of the shelter, although the territory averaged 2.25 m² (Mochek 1977). Typically, dusky damselfishes performed most of the attacks (Fig. 3.5), leaving the shelter with fast and brusque movements in search of the intruder. When this reaction was insufficient, the fish seized and bit the intruder. Almost all dusky damselfish adults were found alone. On only a few occasions, for example, at great food concentrations, were territorial individuals observed gathering in an area.

The beaugregory showed a high degree of territorial behavior, even in young individuals. All sizes displayed the complete set of territorial reactions. Adult individuals, whose coloration (brown) differs from that of the young, spent most of the time solitary (Fig. 3.4), in fixed spots, exhibiting territorial aggression toward neighbors. Small groups were observed in a few cases, for example, feeding on an egg nest. Imitation behaviors were seen 24% of the time. The beaugregory feeds only on bottom objects (phytobenthos, zoobenthos, demersal eggs, etc.); individuals were observed feeding on these items 49% of the time surveyed. The rest of the time was dedicated to defending territory and attacking in the suprabenthic stratum; they were never found in the water column. When threatened, they sheltered in bottom refuges as their only defense against predators. They stayed in the shelters only briefly, although they remained relatively near them.

The beaugregory is also common in mangroves, where they find food and shelter. Fig. 3.4 provides data on

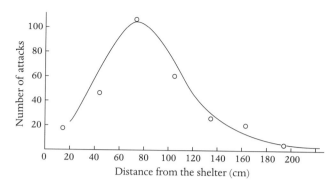

Fig. 3.5. Intensity of the attacks of the dusky damselfish on an aggressor relative to the distance of the aggressor from the shelter.

behavior of this species in mangroves. Despite the vertical zonation of refuges in mangroves, the beaugregory uses benthic and suprabenthic strata, as in reef areas. They were observed 46% of the time among mangrove roots and 54% of the time in the adjacent area. They performed relatively few movements. Their horizontal movements were also limited (2 m, average), and were performed mainly within the defended territory. The individualistic behavior of the beaugregory was evident: they were never observed in groups or in pairs. Contacts with conspecifics were of a strictly antagonistic nature, directed at the defense of their territory. Almost half the time beaugregory were hidden in shelters. This type of behavior might be critical for the survival of these small fishes, which likely bear high predation pressure in the mangrove habitat. Exit from shelters is prompted only to protect their own territory from intruders.

The yellowtail damselfish is one of the largest pomacentrid fishes (up to 15 cm long). Territoriality is typical of this species and is displayed in adults as well as in juveniles, but to a lesser extent than in dusky damselfish and beaugregory. During our surveys the territorial aggressive reactions were manifested less often (Fig. 3.4). Most of the attacks were directed toward individuals of the same species, whereas they showed indifference toward other species. The yellowtail damselfish lives in a fixed territory and we nearly always found them solitary. Its feeding ground is larger than that of any other damselfish. While moving they rose off the bottom 28% of the time, although they generally swam over the substrate, feeding mainly on benthic algae (Starck and Davis 1966). At any sign of danger they hid in shelters where they were found 48% of the time. In addition, their shelters were almost enclosed, which provides greater protection for individuals. Typically, all territorial species take shelter in reef crevices or among mangrove roots at nightfall and remain there until dawn.

3.3   Nocturnal Fishes

Behavioral patterns of species with nocturnal–crepuscular habits can be grouped into three main categories: inshore pelagic, epibenthic apogonids, and suprabenthic fishes.

3.3.1   Inshore Pelagic Fishes

Various sardine species (Clupeidae) are the most common pelagic nocturnal fishes inhabiting Cuban shallow-water areas. These include the redear sardine (*Harengula humeralis*), scaled sardine (*H. jaguana*), false pilchard (*H. clupeola*), and dwarf herring (*Jenkinsia lamprotaenia*). Schools are composed of more than one species.

During the day, sardine species gather in the water column, forming large schools in which individuals move almost constantly with little motion of the shoal itself. Because of their strong gregarious conduct, imitation behaviors are constantly occurring. The defensive reaction is also typical. On several occasions we observed that, before a predator attacked, sardines moved in different directions, leaving a space at the point where the attack occurred. When the attack is performed by a number of individuals, this prompts such disorder that sardines "fly off the water" in all directions (Silva Lee 1975b). All species of this group are planktivores. It is common to find large concentrations of these fish feeding on organisms (crustaceans and others) attracted to boat lights or submarine objects (Starck and Davis 1966; Smith and Tyler 1972).

Dwarf herring typically aggregate in the same areas as *Harengula* spp., although there is certain separation between the schools, probably because the dwarf herring are smaller. Dwarf herring schools show all the characteristics of equipotential gregariousness: a highly developed form of fish aggregation characterized by strong imitation, mutual attraction, and lack of a leader. Dwarf herring schools are commonly found within mangrove root canopies (Rooker and Dennis 1991; Dennis 1992a; Claro and García-Arteaga 1993). These schools are compact and frequently become spherical. Before any attack they show the typical θ-maneuver: the group splits into two and quickly merges into a single school behind the predator.

3.3.2   Epibenthic Apogonids

Cardinalfishes (Apogonidae) are among the most common of nocturnal epibenthic planktivores. These fishes remain inactive, either hiding in reef shelters and crevices during the day, forming groups, or remaining solitary. Some cardinalfishes use other organisms for protection: the conchfish (*Astrapogon stellatus*) exhibits commensalism with the queen conch (*Strombus gigas*), the sponge cardinalfish (*Phaeoptyx xenus*) uses the oscules of *Aplysina* sponges for shelter during the day, and the dusky cardinalfish (*P. pigmentaria*) can shelter under long spine sea urchins (*Diadema antillarum*) (Starck and Davis 1966; Smith and Tyler 1972).

Cardinalfish activity increases at nightfall, when diurnal species move to reef shelters. They feed in the water column, replacing the diurnal activity of damselfishes. For example, the bigtooth cardinalfish (*Apogon affinis*) might occupy the position of *Chromis* and show a typically gregarious behavior (Collette and Talbot 1972). At dawn, all fishes of this group return to their reef shelters.

3.3.3   Suprabenthic Fishes

The most abundant members of the suprabenthic group are snappers (Lutjanidae), grunts (Haemulidae), and squirrelfishes (Holocentridae). Grunt and snapper aggregations are the largest among the species associated with the bottom. They are most active at night, when they move off the reefs and forage in neighboring areas (Starck and Davis 1966; Helfman et al. 1982). Although they can be classified as inactive during the day, they react promptly when frightened or are offered food.

HAEMULIDAE AND LUTJANIDAE.   Because of the great variety of ecological habits in grunts and snappers, school formation and mechanisms for disorienting predators are quite different among these species and perhaps within some species in differing conditions. Schooling behaviors of juvenile grunts, coupled with the disruptive lateral stripes of early juveniles in all western Atlantic species, may disorient visual predators (Lindeman 1986). An inverse relationship can exist between schooling and sheltering, the two prototypes of fish defensive behavior. However, grunts and snappers can perform schooling behaviors while occupying shelters (Mochek and Silva Lee 1975). For example, juveniles (Helfman et al. 1982) and newly settled stages (Lindeman and Snyder 1999) of at least six grunt species can form multispecies schools while associating with individual structures. Diverse gregarious behaviors have been observed in different *Haemulon* species among a variety of life stages. Older individuals of some grunts were seen displaying certain elements of territoriality, whereas some showed highly gregarious behavior (Mochek and Silva Lee 1975).

We were able to gather considerable data on two species: the French grunt (*Haemulon flavolineatum*) and the bluestriped grunt (*H. sciurus*). During the day, the French grunt spent 94% of the time close to the bottom, moved little, and moved an average of 4 m during 5-minute surveys (Fig. 3.4). Schools were not dense (73% of the time, individuals were not in contact with neighboring fishes); however, grouping was seen in a number of cases (pairing in 19% of the surveys and three or more individuals in 8%). Group formation was particularly frequent in the smallest fishes. No signs of aggressiveness against neighbors were detected in the French grunt. They frequently used shelters. However, when threatened, flight to shelters was seen only in some French grunts; most of the individuals moved away from the area of attack. Fig. 3.6 presents characteristics of

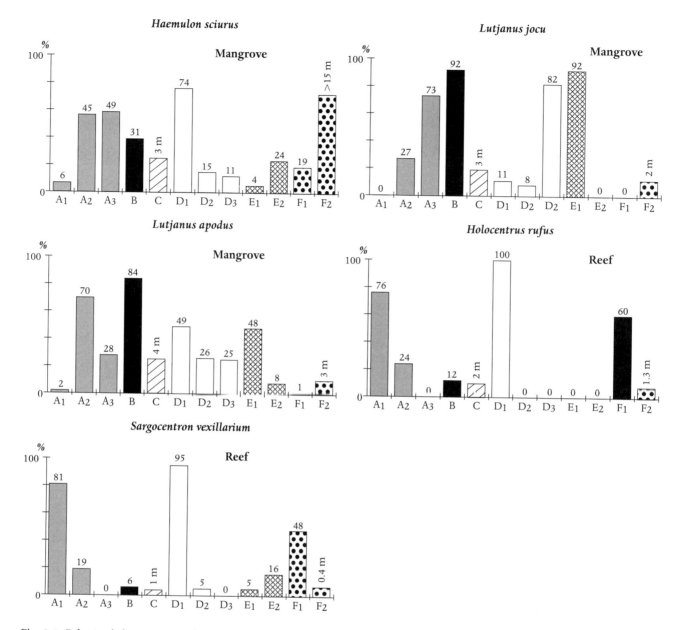

Fig. 3.6. Behavioral characteristics of the bluestriped grunt (*Haemulon sciurus*), dog snapper (*Lutjanus jocu*), schoolmaster (*L. apodus*), and longspine squirrelfish (*Holocentrus rufus*) and dusky squirrelfish (*Sargocentron vexillarium*) in reefs and mangroves. Abbreviations as in Fig. 3.2.

bluestriped grunt behavior in mangroves. This species displayed little activity during the day and stayed close to the bottom or in the water column beneath mangrove branches. They were typically solitary, with a marked territorial-defensive behavior. Bluestriped grunts spent little time in shelters; they did not hide among mangrove roots, but swam away before an attack. Thus, these fishes showed characteristics of both solitary species and migratory groups.

Like grunts, snappers of Cuban waters have a wide spectrum of social conduct. Yellowtail snapper (*Ocyurus chrysurus*) is gregarious, lane snapper (*Lutjanus synagris*) is often gregarious, and mutton snapper (*L. analis*) is often solitary. Other species, such as gray snapper (*L. griseus*), dog snapper (*L. jocu*), and schoolmaster (*L. apodus*), have gregarious-suprabenthic habits and use shelters when in danger. During the day, it is common to see the whole range of behavioral types on one reef. Mueller et al. (1994) recorded considerable size-specific and diurnally variable interactions within a mutton snapper population in the Exuma Cays, Bahamas, and concluded that a dominance hierarchy was present.

In our surveys, groups of yellowtail snappers were typically found in the water over reefs. Closer to the reefs and in mangroves, large numbers of gray snappers were found aggregating with other snapper species. Dog snappers

(Fig. 3.6), like gray snappers, typically live in groups. They are both active and display strong imitation behaviors. They generally use bottom shelters when chased by predators. Schoolmasters (Fig. 3.6), in comparison with dog and gray snappers, used the suprabenthic stratum more frequently and were solitary half of the time. Imitation behaviors were observed 48% of the time and aggressive behavior against conspecific individuals was weak. Sometimes these fishes were seen sheltering at the bottom, typically during the attack of predators. Our data suggest that mutton and cubera (*Lutjanus cyanopterus*) snappers are often solitary (particularly the largest individuals), and when threatened, quickly take shelter for long periods.

HOLOCENTRIDAE.   Relative to grunts and snappers, squirrel-fishes display little activity. In Cuba, there are nine species in five genera, all with similar behavioral characteristics. Fig. 3.6 provides data on the behavior of two species: dusky squirrelfish (*Sargocentron vexillarium*) and longspine squirrel-fish (*Holocentrus rufus*). Most of the day these fishes stay inactive on the bottom. They do not gather in groups and in dusky squirrelfish antagonistic reactions with conspecific individuals were observed. In general, individuals of both species defend themselves by moving quickly to the closest refuge (typically reef shelters and crevices). During the day they were in the reef shelters, hidden from observers almost half of the time.

3.4   Crepuscular Demersal Fishes

This group includes mostly piscivorous species of the Serranidae, Scorpaenidae, Muraenidae, and other families. They use an ambush tactic to capture prey, while hiding in bottom irregularities. The large reef predators are cautious and remain hidden most of the day. For that reason we were unable to obtain quantitative behavioral data. Their move-ments are restricted to moving near the bottom from one shelter to another. Only when threatened do they react with fast movement, then hide for long periods in the closest shelters. During the nonreproductive period, these predators are often solitary and rarely show social reactions (particu-larly antagonistic ones).

Masking coloration allows predators to capture prey at nightfall and dawn. At dusk, evidently, predators have a greater likelihood of hiding and approaching prey; however, presence and activity of potential prey is lower during the night. Hunting activity of reef predators is more intense and successful during those periods of the day when it is more difficult for prey to detect danger (Collette and Talbot 1972; Hobson 1974); thus, numerous piscivorous fishes have crepuscular habits. Fish activity is usually not confined to a particular period of the day (Helfman 1978); for example, Nassau groupers do not have a marked diurnal feeding pattern (Silva Lee 1974b). This might also be true of other grouper species (Serranidae) that do not interrupt feeding activity during the day.

3.5   Behavioral Mechanisms of Community Differentiation

Tropical shallow-water ichthyofaunas, particularly those of coral reefs, are species-rich and exhibit complex biotic relationships. Some workers correlate the high diversity of reef fish fauna with the degree of ecological specialization of co-occurring species (Clarke 1971; Smith and Tyler 1973a, 1973b; Fishelson et al. 1974; Clarke 1977; Gladfelter and Gladfelter 1978). Other workers offer a variety of alternative explanations involving density-independent factors (Sale 1980, 1991; Doherty and Williams 1988). On reefs, the daily variation of fish behavior is so marked that communities can be differentiated into diurnal, nocturnal, and crepuscular components. However, on several spatial and behavioral scales the degree of specialization can be difficult to determine empirically or is outweighed by abundant evidence of behavioral plasticity. In the next sections, we examine some aspects of ethological specializa-tion and opportunism among species with potentially similar niches.

To examine behavioral differentiation in the use of ecological niches, we examined quantitative indicators of behavior of diurnal fishes with similar ecological patterns. We measured variables such as bottom association and formation of groups with three or more individuals. In only a few cases, particularly in comparing butterflyfishes (foureye and spotfin) and striped parrotfish with other species, we found no significant differences for the variable "permanence in the shelter" (Figs. 3.2 and 3.4). During the day, the simultaneous activity of a large number of fishes within reef habitats, which are often quite small, might foster inter-species competition for resources.

In reef environments, where space and shelter might be limiting factors, differences in fish behavior–particularly between species with similar ecological features—might allow the community to compensate for limited resources (Mochek and Valdés-Muñoz 1983). However, behavior can also be highly variable within species (Manteifel 1980). Some degree of behavioral plasticity could explain the opportunis-tic settlement patterns within several grunt and snapper species. For example, newly settled stages of sailor's choice (*Haemulon parra*), white grunt (*H. plumieri*), lane snapper, and yellowtail snapper can occur at the same sizes (10 to 20 mm standard length) in habitats with different structural characteristics, such as hardbottom areas and seagrass beds, whereas settlers of other species (e.g., porkfish, *Anisotremus virginicus,* and gray snapper) are less opportunistic (Lindeman et al. 1998).

Considerable evidence suggests that daily patterns of fish activity can be independent of location, size, morphol-ogy, and species composition on the studied reefs

(Valdés-Muñoz and Mochek 1994). Behavior at dawn and dusk is associated with the type of food the fishes consume and the diurnal activity they display. During the day, the dominant active fishes are those that consume benthic sessile or slowly mobile organisms, such as algae, mollusks, some crustaceans, and zooplankton. These fishes belong mainly to Acanthuridae (surgeonfishes), Scaridae (parrotfishes), Labridae (wrasses), and Pomacentridae (damselfishes). Most of the piscivorous fishes, and those feeding on large active benthic animals, spend the day on the reef, relatively inactive.

At dusk, radical changes in the behavior of most reef fishes take place. Diurnal fishes remain inactive in shelters until dawn or use other defensive mechanisms, such as hard body covers, spines, and toxins that make them less accessible to predators. Small nocturnal fishes, after leaving their shelters, display great mobility, and capturing them becomes difficult. Thus, nocturnal planktivores, such as cardinalfishes (Apogonidae), remain in shelters during the day and at night replace diurnal planktivores such as *Chromis* (Pomacentridae). They have developed a dynamic defensive behavior.

Unlike most diurnal fishes that display restricted movements, many nocturnal fishes are highly mobile. This is particularly notable in snappers and grunts. For fishes that consume mobile crustaceans, nocturnal hunting over seagrass beds is not only less risky, but also more productive because many invertebrates leave their nocturnal shelters during this period. This is also an important source of nutrient transfer between habitats (Parrish 1989).

Data on the daily distribution and zonation of fishes in seagrass beds are limited, and the information provided here is based on indirect data and tentative assumptions. Shortly before sunset (10–15 minutes), the distribution and behavior patterns are the same as during the rest of the day: numerous individuals (solitary or in groups) of slippery dick and blackear wrasse move among the seagrass blades; parrotfishes are less conspicuous during this period because of their coloration. In addition, because of the small rocky formations scattered among the seagrass beds, large schools of French grunt and yellowtail snapper are observed. Immediately after sundown, wrasses and parrotfishes begin to vanish among the seagrass blades. In the 10 minutes after sundown, their activity declines drastically, and toward the end of this period they are completely hidden. They remain inactive among the macrophytes all night. Not until dawn do wrasses and parrotfishes leave the seagrass. In the afternoon, and even after sunset, small grunts and snappers remained active and do not shelter. Typically, groups become less dense, and solitary individuals are seen. Even in complete darkness, grunts do not stop moving. Using flashlights, we could observe them over the seagrass beds.

Diurnal fishes, represented mostly by small consumers of phytobenthos and zoobenthos, are replaced at dusk by species that consume larger benthic invertebrates (mainly crustaceans). The large consumers of benthic invertebrates, and even the piscivores, are not usual inhabitants of the seagrasses, but move toward this habitat from adjacent areas. Other neighboring species, such as sardines (Clupeidae) and silversides (Atherinidae), feed on zooplankton over the seagrass beds. Only the spiny puffers (Diodontidae) and other less abundant species—consumers of benthic organisms of limited movements—are permanent dwellers of this biotope, where they forage during the nocturnal–crepuscular periods. The daily pattern of large predators can depend on the homogeneity of bottom relief that is characteristic of seagrass beds lacking large shelters.

Both reefs and mangroves are coastal biotopes of irregular relief. Unlike reefs where shelters are arranged horizontally, mangrove roots form a wall where shelters are located vertically. Thus some fishes that move vertically on the reef for feeding, such as sergeantfishes, are distributed horizontally in mangroves in four well-delineated zones: the water column, the border of the mangrove, the corridor under the mangrove branches between the roots and the seagrass, and the mangrove wall itself. Fish faunal composition and behavior are different in each of the four zones, so that species can be classified according to use of space.

In mangroves, as in seagrass beds and reefs, fishes redistribute themselves during the night as their behavior changes. Our survey provides the only data on this daily pattern; the complexity of the phenomenon demands further investigation. Only a few minutes after sunset, diurnal fishes, the inhabitants of the mangrove corridors, move toward nocturnal shelters in the seagrass bed. These fishes typically move a few meters into the seagrass bed, where the vegetation is tall and dense. Within 10 minutes after sunset, striped parrotfishes disappear among the blades, followed by slippery dicks and foureye butterflyfishes, among others. Surgeonfishes (Acanthuridae) are among the last species sheltering in the seagrass.

In euhaline, low-turbidity areas, damselfishes find nocturnal shelter among mangrove roots; they disappear relatively late, 25 minutes after sunset. At the moment when the mangrove is virtually dark, beaugregory, which are abundant during the day, cease defense of their territories and hide in the intricate root structures. In sergeant majors, this shift in territorial behavior corresponds with a behavioral transition: individuals abandon earlier school formations in the water column to take shelter in mangrove structures. The entire sheltering process of diurnal fishes usually occurs within 30 minutes after sunset.

At the end of sunset, nocturnal fishes are dominant in these mangrove areas. Commonly occurring families and trophic patterns include planktivorous cardinalfish; squirrelfishes, spiny puffers, and batfishes (Ogcocephalidae) among the secondary consumers; and piscivores such as snooks (Centropomidae). A little later, snappers are observed, but not entirely inside the mangrove. At night, in the corridors under the mangrove branches, solitary gray, dog, and cubera snappers can be found. They remain over the bottom, under the drop roots, in a quiet state and react only to the threat of

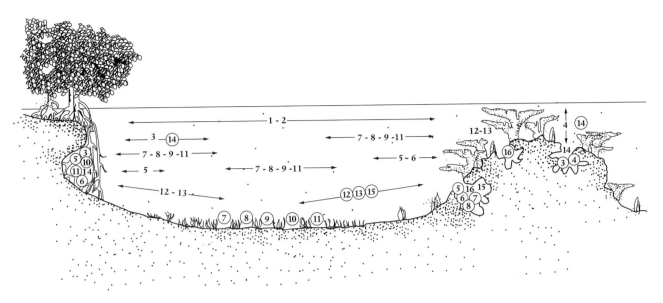

Fig. 3.7. Diurnal–nocturnal activity of fishes in mangroves, seagrass beds, and reefs. The circled numbers indicate presence during the night; uncircled numbers indicate presence during the day. The species represented are (1) great barracuda (*Sphyraena barracuda*), (2) jacks (*Caranx* spp.), (3) sergeant major (*Abudefduf saxatilis*), (4) *Chromis* spp., (5) *Stegastes* spp., (6) yellowtail damselfish (*Microspathodon chrysurus*), (7) *Chaetodon* spp., (8) bluehead (*Thalassoma bifasciatum*), (9) slippery dick (*Halichoeres bivittatus*), (10) *Sparisoma* spp., (11) surgeonfishes (*Acanthurus* spp.), (12) grunts (*Haemulon* spp.), (13) snappers (*Lutjanus* spp.), (14) *Apogon* spp., (15) squirrelfishes (*Holocentrus* and *Sargocentron* spp.), and (16) groupers (*Epinephelus* spp.).

capture. This nocturnal pattern of behavior in mangroves can also apply to some grunt species, such as white and French grunt. Like snappers, these grunt species can distribute along bottom irregularities and not show any foraging activity. Barracuda, an active diurnal predator, were occasionally seen beneath the mangrove branches.

Fig. 3.7 represents the spatial–temporal activity patterns of the Cuban coastal ichthyofauna. Daily patterns of fish activities and their distributions across the main biotopes during the day and night periods for the more common families are depicted. Many coastal fishes inhabit different habitats; thus their adaptations are not isolated, but represent a complex system of migratory patterns at different trophic levels. Each habitat plays a different role in the assemblage: the reef can be a trophic defensive zone, the extensive seagrass beds play a critical role in feeding, and mangroves provide additional shelter for the defense of older juveniles. Not only coastal fishes use the resources of shelf areas, open-water species also use these resources; the presence of these fishes is also of trophic significance (see Chapter 5).

3.6 Summary

Fishes of the Cuban shelf can be grouped according to their activity pattern as diurnal, nocturnal, or crepuscular and according to both their substrate associations and their feeding patterns. Diurnal or nocturnal fish species can also show vertical distribution patterns across different water column strata. Quantitative analyses of several behavioral attributes indicate that some species either have no permanent territory or have home ranges that extend for several kilometers (as in jacks and barracudas). These species are also characterized by well-developed schooling behaviors. Some species forage in the water column near the bottom, with a relatively organized school structure. Examples include the pomacentrid genera *Abudefduf* and *Chromis*.

Suprabenthic fishes exhibit a broad spectrum of social conduct. Interspecific behaviors vary according to species, day or night activity patterns, individual size, and other factors. Some suprabenthic fishes maintain continuous activity during the day in near-bottom strata; they typically do not use shelters when threatened, but leave the area. Some species are solitary, whereas others (parrotfishes, wrasses, and surgeonfishes) form mixed groups that show clear imitation behaviors when they are moving in response to predators. Other suprabenthic species, among which grunts and snappers dominate, are most active during the night, when they move away from diurnal resting areas to feeding sites. Some of these species perform defensive group maneuvers and sometimes make use of shelters. Other species, such as the squirrelfishes (Holocentridae), are often solitary and remain hidden most of the time.

Territorial benthic fishes are solitary and occupy individual spaces that they defend aggressively from intruders. Damselfishes (*Stegastes* spp.) are typical members of this group. Fishes of nocturnal habits often occupy areas used by diurnal species. Transient pelagic species school in the water column during the day, forming groups with little large-scale displacement. Imitation behaviors are frequent and little aggression is observed among individuals. At dusk, schools scatter and feed in areas where their preferred organisms are concentrated. Clupeids are the most representative species of this group.

Nocturnal foragers are characterized by inactivity during the day. They hide in shelters until dusk, then begin searching for food across the water column, replacing diurnal species with the same behavior. Crepuscular benthic fishes, mostly piscivorous species such as groupers (Serranidae), moray eels (Muraenidae), and scorpionfishes (Scorpaenidae), remain largely hidden during the day. They use ambush tactics and attack prey at dusk and dawn during transitions in prey activity. Comparisons of primary behavioral variables yielded notable interspecific differences and some intraspecific differences among habitats (reef, mangroves, and seagrass beds). Habitat-specific variations in behavior could be an important organizational feature within fish communities because of potentially limiting shelter and food resources.

4

Reproductive Patterns of Fishes of the Cuban Shelf

ALIDA GARCÍA-CAGIDE, RODOLFO CLARO, AND BORIS V. KOSHELEV

4.1 Introduction

The complex cellular, organismal, and population processes associated with reproduction exercise a profound influence over the dynamics of reef fish species. Detailed knowledge of reproduction requires research on the interplay among the diverse physiological and ecological factors associated with spawning. In this chapter we have focused on integrated laboratory and field studies of gametogenesis, sexual maturation, annual reproductive cycles, and spawning events. Information on these patterns and how they vary among species and geographic regions is of vital significance to informed fisheries management.

Reduced temperature fluctuations in the tropics make fish reproductive processes different from those of temperate and polar populations. Additional seasonal climatic fluctuations that might influence reproduction in tropical fishes include precipitation cycles and, in some areas, changes in wind patterns and associated upwellings (Lowe-McConnell 1979). However, knowledge is limited on the direct environmental control of reproductive cycles in tropical marine fishes.

The reproductive biology of tropical fishes has been examined in a variety of families, particularly those with commercial value. Intraspecific variations in spawning among regions (e.g., Grimes 1987; Robertson 1991), as well as variations among similar species (as in groupers, Sadovy 1996; Coleman et al. 1999), complicates the broad identification of both spawning characteristics and management strategies. This chapter summarizes information on reproductive physiology and spawning of a variety of species that inhabit tropical waters (e.g., Oven and Saliejova 1970; Munro et al. 1973; Erdman 1977; García-Cagide et al. 1983; Munro 1983a; Thresher 1984; García-Cagide and Claro 1985; García-Cagide et al. 1985; Reshetnikov 1985; Grimes 1987; Shapiro

1987; Colin and Clavijo 1988; Colin 1992; Shapiro et al. 1993a, 1993b; Sadovy et al. 1994; Carter and Perrine 1994; Sadovy and Colin 1995; Domeier et al. 1996; García-Cagide and García 1996; Sadovy 1996; Domeier and Colin 1997; Luckhurst 1998; Hood and Johnson 1999; Sadovy and Eklund 1999; García-Cagide et al. 1999a, 1999b).

4.2 Sexuality

Most of Cuba's marine fishes are gonochoristic. However, six of the 20 families summarized here include hermaphroditic species, many of which play a critical role in reef communities because of their abundance and ecological importance. Many species are also sexually dimorphic, a phenomenon that can be associated with sex change.

4.2.1 Hermaphroditism

Hermaphroditism is common in tropical and subtropical fish families such as the Serranidae, Labridae, Scaridae, Sparidae, Pomacanthidae (Table 4.1), Grammistidae, and Muraenidae, as well as others (Reinboth 1962; Robertson 1972; Shapiro 1979, 1981, 1985; Thresher 1984; Koenig et al. 1996). Three basic types of hermaphroditism are found in families of the Cuban shelf: (1) synchronous, in which the ovarian and testicular tissue are simultaneously functional; (2) protogynous, in which the gonads function first as ovaries and then transform into testes; and (3) protandric, in which the gonads function first as testes and then transform into ovaries. Protogynous hermaphroditism is the most common type among tropical marine fishes.

Synchronous hermaphroditism has been reported for various serranids (Longley and Hildebrand 1941; D'Ancona

Table 4.1.   Hermaphroditic species occurring on the Cuban shelf

Family Subfamily Species	Type of Hermaphroditism	Source
Serranidae		
Serraninae		
Diplectrum formosum	Synchronous	Obando and Léon 1989
Hypoplectrus chlorurus	Synchronous	Barlow 1975
Hypoplectrus nigricans	Synchronous	FAO 1978
Hypoplectrus puella	Synchronous	Barlow 1975
Hypoplectrus unicolor	Synchronous	Barlow 1975
Serranus phoebe	Synchronous	Smith 1959
Serranus tabacarius	Synchronous	Smith 1959
Serranus tigrinus	Synchronous	Smith 1959
Epinephelinae		
Alphestes afer	Protogynous	Smith 1959
Cephalopholis cruentata	Protogynous	Smith 1959
Cephalopholis fulva	Protogynous	Smith 1959
Epinephelus guttatus	Protogynous monandric	Smith 1959; Sadovy et al. 1992, 1994
Epinephelus itajara	Protogynous	Smith 1959
Epinephelus itajara	No conclusive evidence of protogyny	Bullock et al. 1992; Sadovy and Eklund 1999
Epinephelus morio	Protogynous	Moe 1969
Epinephelus niveatus	Protogynous	Wyanski et al. 2000
Epinephelus striatus	Protogynous monandric	Smith 1959; García-Cagide and Espinosa 1991
Epinephelus striatus	Gonochoric (some tendency to protogynous diandric)	Sadovy and Colin 1995; Sadovy and Eklund 1999
Mycteroperca bonaci	Protogynous monandric	Smith 1959; García-Cagide and García 1996; Crabtree and Bullock 1998
Mycteroperca interstitialis	Protogynous	Smith 1959
Mycteroperca microlepis	Protogynous monandric	McErlean 1963; Koenig et al. 1996; Collins et al. 1987, 1998
Mycteroperca tigris	Protogynous monandric	Smith 1959; García-Cagide et al. 1999a
Mycteroperca venenosa	Protogynous monandric	Smith 1959; García-Cagide and García 1996
Rypticus saponaceus	Protogynous	Smith 1959
Sparidae		
Calamus bajonado	Protogynous	Dubovistky 1974; Druzhinin 1976
Calamus pennatula	Protogynous	Druzhinin 1976
Calamus proridens	Protogynous	Dubovistky 1977
Pomacanthidae		
Holacanthus tricolor	Protogynous monandric	Hourigan and Kelly 1985
Labridae		
Bodianus rufus	Protogynous monandric	Warner and Robertson 1978
Clepticus parrae	Protogynous monandric	Warner and Robertson 1978
Halichoeres bivittatus	Protogynous diandric	Roede 1972; Warner and Robertson 1978
Halichoeres garnoti	Protogynous monandric	Roede 1972; Warner and Robertson 1978
Halichoeres maculipinna	Protogynous diandric	Roede 1972; Warner and Robertson 1978
Halichoeres pictus	Protogynous diandric	Warner and Robertson 1978
Halichoeres poeyi	Protogynous diandric	Roede 1972; Warner and Robertson 1978
Halichoeres radiatus	Protogynous monandric	Warner and Robertson 1978
Lachnolaimus maximus	Protogynous monandric	Colin 1982; Claro et al. 1989
Thalassoma bifasciatum	Protogynous diandric	Roede 1972; Warner and Robertson 1978
Xyrichtys martinicensis	Protogynous	Roede 1972
Xyrichtys splendens	Protogynous	Roede 1972
Scaridae		
Scarinae		
Cryptotomus roseus	Protogynous monandric	Robertson and Warner 1978
Scarus coelestinus	Protogynous	Robertson and Warner 1978

Continued on next page

Table 4.1. continued

Scarus coeruleus	Protogynous	Robertson and Warner 1978
Scarus guacamaia	Protogynous	Robertson and Warner 1978
Scarus iserti	Protogynous diandric	Robertson and Warner 1978
Scarus taeniopterus	Protogynous	Robertson and Warner 1978
Scarus vetula	Protogynous diandric	Robertson and Warner 1978
Sparisomatinae		
Sparisoma atomarium	Protogynous monandric	Robertson and Warner 1978
Sparisoma aurofrenatum	Protogynous monandric	Robertson and Warner 1978
Sparisoma chrysopterum	Protogynous monandric	Robertson and Warner 1978
Sparisoma radians	Protogynous monandric	Robertson and Warner 1978
Sparisoma rubripinne	Protogynous monandric	Robertson and Warner 1978
Sparisoma viride	Protogynous monandric	Robertson and Warner 1978

1950; Smith 1959, 1961, 1965; Bortone 1971, 1977; Barlow 1975; Obando and León 1989; Koenig et al. 1996). Physiological or genetic barriers may impede self-fertilization of synchronous hermaphrodites under natural conditions (Smith 1959), ensuring the genetic integrity of species. Under natural conditions, the spawning of such species occurs in pairs, or in some cases, in groups. Clark (1959) demonstrated the possibility of self-fertilization in the belted sandfish (*Serranus subligarius*), and Tucker and Woodward (1995) demonstrated repetitive cross-fertilization and reversal of sex roles in this species. Johnson (1932) noted the possibility of self-fertilization in a teratological hermaphrodite, the sparid *Stenotomus chrysops*.

Among the species of the serranid subfamily Serraninae, eight are known to function as synchronous hermaphrodites (Table 4.1). Barlow (1975) recorded the spawning in Puerto Rican waters of three species exhibiting synchronous hermaphroditism: the yellowtail hamlet (*Hypoplectrus chlorurus*), butter hamlet (*H. unicolor*), and barred hamlet (*H. puella*) (see also Domeier 1994). Four serranine species inhabiting Bermudan waters—butter hamlet, tattler (*Serranus phoebe*), tobaccofish (*S. tabacarius*), and harlequin bass (*S. tigrinus*)—are considered synchronous hermaphrodites (Smith 1959), as well as sand perch (*Diplectrum formosum*) (Bortone 1971). Individuals bearing simultaneously ripe oocytes and spermatozoids have been recorded (Bortone 1971).

Most species of the subfamily Epinephelinae (genera *Epinephelus, Mycteroperca,* and *Paranthias*) appear to be protogynous, except the creole-fish (*Paranthias furcifer*), which might be gonochoristic, although little information is available. In all of these species, the entire gonad is a mixture of ovarian and testicular tissue, but the tissues do not function simultaneously (Smith 1965).

Fish length at time of sex change in protogynous serranids is variable. In red grouper (*Epinephelus morio*), sex-change was reported to occur between 45 and 65 cm (standard length, SL) in Florida (Moe 1969) and at 42 cm (total length, TL) in the Yucatán (Brulé et al. 1991). Because sexual maturation in females occurs at 30 cm TL (more typically in fish >53 cm TL) in the Yucatán (Brulé et al 1991),

and at 42.5 cm TL (in all individuals >50 cm TL) in Florida (Moe 1969), part of the population might change sex before the individuals spawn as females. However, in most species, sex change occurs after individuals have spawned at least once as a female.

We have found male specimens of tiger grouper (*Mycteroperca tigris*) at 37 cm TL, but most Cuban specimens change sex above 50 cm TL (García-Cagide et al. 1999a). Smith (1959) states that sexual transition in the tiger grouper is fast and total, and occurs at approximately 40 cm SL, whereas in the red hind (*Epinephelus guttatus*) the change is gradual and is completed between 27.5 and 40 cm SL (mean, 38 cm; Thompson and Munro 1978). Thompson and Munro state that the coney (*Cephalopholis fulva*) switches sex at 27 cm SL on average. McErlean and Smith (1964) indicate that the gag (*Mycteroperca microlepis*) matures as a female at 5–6 years old and changes to a male at 10–11 years old. Nassau grouper in Bermuda start to change sex between 30 and 35 cm SL (Bardach 1958). However, in Cuba, García-Cagide and Espinosa (1991) found transitional individuals from 30 cm TL and many females 42–46 cm TL (mature and immature) without any trace of testicular tissue in their gonads (Plate 4.1, a and b).

Evidence that jewfish (*Epinephelus itajara*) are protogynous hermaphrodites is not definite (Bullock et al. 1992; Sadovy and Eklund 1999). Sadovy and Colin (1995) suggest that sexual development in the Nassau grouper (*Epinephelus striatus*) is primarily gonochoristic with the potential for sex change. Their data suggest that most individual males and females develop directly from a juvenile bisexual phase. Other evidence also suggests that a large proportion of Nassau grouper (Bardach 1958; Claro et al. 1990c) and red grouper (Moe 1969; Brulé et al. 1991) females do not switch sex (they are primary females). In both species, the sex ratio in larger individuals is almost equal. In red grouper this ratio is reached at approximately 15 years of age. At the population level, the transitional age classes are 5–15 years, but this is more common after the 13-year age class and some individuals can reach 30 years of age before transition (Moe 1969).

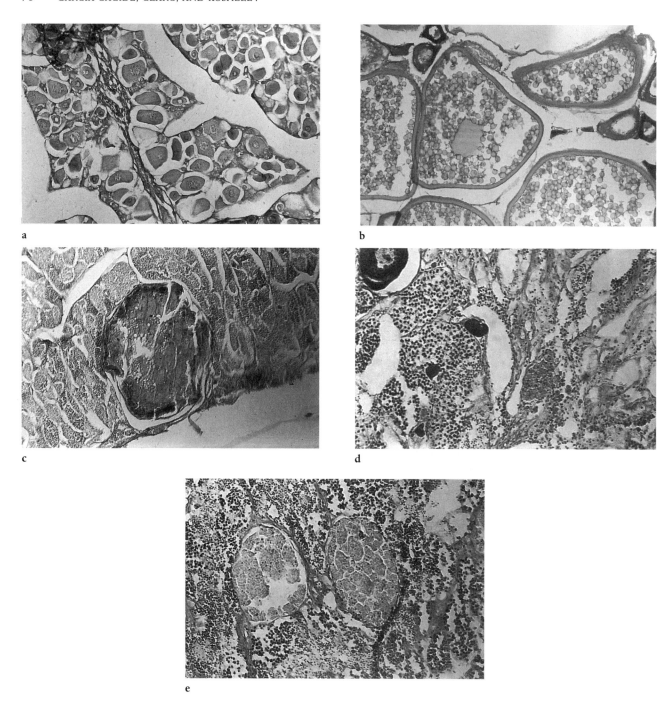

Plate 4.1. Photomicrographs of gonad histological sections: (a) Nassau grouper (*Epinephelus striatus*), immature female (52.0 cm TL, GSI = 0.5), with nonvitellogenic oocytes; (b) Nassau grouper mature female (63.0 cm TL, GSI = 1.44) with vitellogenic and nonvitellogenic oocytes (40 ×); (c) Hogfish (*Lachnolaimus maximus*) testes (30 to 38 cm FL), with a mature oocyte in an early phase of resorption, and masculine cells in different development stages; (d) Nonvitellogenic oocytes, an atretic body, and crypts with spermatids and spermatozoa (50 ×); and (e) Two yellow or atretic bodies at higher magnification (125 ×).

A pattern similar to that seen in jewfish and Nassau grouper (Sadovy and Eklund 1999) has been described by Roede (1972, 1975) for diandric hermaphrodites of the Labridae. Roede also suggested this pattern might represent an intermediate step in the evolution of hermaphroditism from gonochorism. Almost all species of Labridae and Scaridae inhabiting Cuba are protogynous hermaphrodites (Table 4.1). Approximately half of the labrids on the Cuban shelf are diandric (i.e., not all individuals switch sex) and variable proportions of primary or secondary males have been found within different populations. The remaining species are known to be monandric (all individuals switch sex). Roede (1972) pointed out that during the period of sex change, gonad activity is reduced in Labridae. Therefore, functional hermaphroditism in these species cannot be detected during a certain period, in contrast with what has been described for Serranidae and Sparidae.

Claro et al. (1989) found nonvitellogenous oocytes, ripe oocytes in resorption, or atretic bodies in testes of small individuals (28–38 cm FL) of hogfish (*Lachnolaimus maximus*) (Plates 4.1 c, d, and e). Atretic bodies are considered to represent the most advanced phase of the resorption of ovarian tissue, so we concur with Colin (1982) that this is a monandric species, that is, no primary males exist in the population. In hogfish, sex change at small sizes can be the result of differential growth in some individuals, most likely caused by genetic factors. Sex change in larger sizes (which occurs in most individuals) is probably associated with specific behavioral patterns. In another labrid (*Labroides dimidiatus*), Robertson (1972) reported that aggressive behavior suppressed the tendency of females in the harem to change sex. As soon as the primary male dies, the dominant female changes. The existence of a high proportion of females larger than 58 cm suggests that the timing of sexual transition might also be determined by a similar behavioral pattern.

At least 13 of the 14 scarid species inhabiting Cuban waters are protogynous hermaphrodites. The species belonging to the subfamily Scarinae (genera *Scarus* and *Cryptotomus*) may be diandric or monandric, whereas all species of the subfamily Sparisomatinae (genus *Sparisoma*) are monandric. In the sparisomatines, sex change before the first sexual maturation is common. Therefore, these individuals always reproduce first as males, analogous to scarid primary males (Robertson and Warner 1978). Robertson and Warner suggested that this resulted from development of a secondary gonochorism and the apparent loss of primary males during evolution. This sexual pattern was also observed among the Indo-Pacific Scaridae (Choat and Robertson 1975).

Several authors (D'Ancona 1950; Ginsburg 1952; Hoar 1957; Atz 1964, 1965; Druzhinin 1976) have pointed out that species with protogynous, synchronous, and protandric hermaphroditism can be found among the Sparidae. Unlike the gonads of groupers, testicular and ovarian tissues of sparids are well separated by connective tissue. Among the species of the wider Caribbean, protogyny has been observed in the pluma (*Calamus pennatula*), knobbed porgy (*C. nodosus*), jolthead porgy (*C. bajonado*), and littlehead porgy (*C. proridens*) (Dubovitsky 1973, 1974, 1977; Druzhinin 1976). In the last three species, Dubovitsky (1973) found an initial hermaphroditic phase (nonfunctional) in which the fishes are defined first as females and then change their sex. There is no reference to protandric fish species in the Western Atlantic, but several species have been found off coastal Africa (Druzhinin 1976).

Few criteria have been established for determining which factors prompt sex change in hermaphroditic fishes. Behavioral factors might be involved, as in the Pacific species *Labroides dimidiatus* (Robertson 1972) and other species of labrids, as well as in serranids and pomacanthids (Aldenhoven 1984; Moyer and Nakazone 1978; Warner 1978; Shapiro 1979). Other aspects, such as population density (Aldenhoven 1984; Moyer and Zaiser 1984), might also be associated. Reinboth (1988) considered it unlikely that social factors determine sex change in the protogynous *Epinephelus* species; these species tend to be solitary and only aggregate for spawning. On the other hand, Olsen and LaPlace (1979) stated that sex change is a population regulatory mechanism. Moe (1969) considered that external factors rather than genetic factors induce sex change in red grouper. We found a high proportion of transitional individuals (36%) in Nassau grouper populations in Cuba (Claro et al. 1990c). This appears to be an effect of altered sex ratios due to overfishing, particularly for larger individuals, most of which are males. Evidence for such shifts exists for several grouper species (Koenig et al. 1996, 2000; McGovern et al. 1998; Harris and Collins 2000).

4.2.2 Sexual Dimorphism

Pronounced sexual dimorphism is relatively uncommon among teleost fishes in general, but it occurs among some reef fishes. This phenomenon is manifested mainly in color differentiation (dichromatism) or changes in body shape during growth, both of which are often associated with protogynous hermaphroditism. This type of dimorphism is widely represented among labrid and scarid species.

Size- and sex-related color changes of the most common labrid species inhabiting Caribbean waters have been thoroughly described by Roede (1972) and Warner and Robertson (1978). Roede (1972), however, questioned the relationship between color phase and sex based on the following arguments: (1) in some species (bluehead, *Thalassoma bifasciatum*, and slippery dick, *Halichoeres bivittatus*), there are males with ripe testes in the adult primary phase; (2) in other species (yellowhead wrasse, *H. garnoti*), some females have functional ovaries in the intermediate and terminal phases; (3) color patterns can develop from the adult primary phase to one of the terminal phases in individuals that are still females; (4) sex change from female to male can start in individuals that are still in the adult primary phase coloration; and (5) intermediate colors coincide approximately with a decline in gonad activity (however, mature females and males with this transition color have been found).

Roede (1972) examined diandric species characterized by the occurrence of primary males that are not hermaphrodites (see Warner and Robertson 1978). These fishes pass through color stages typical of monandric species, and the

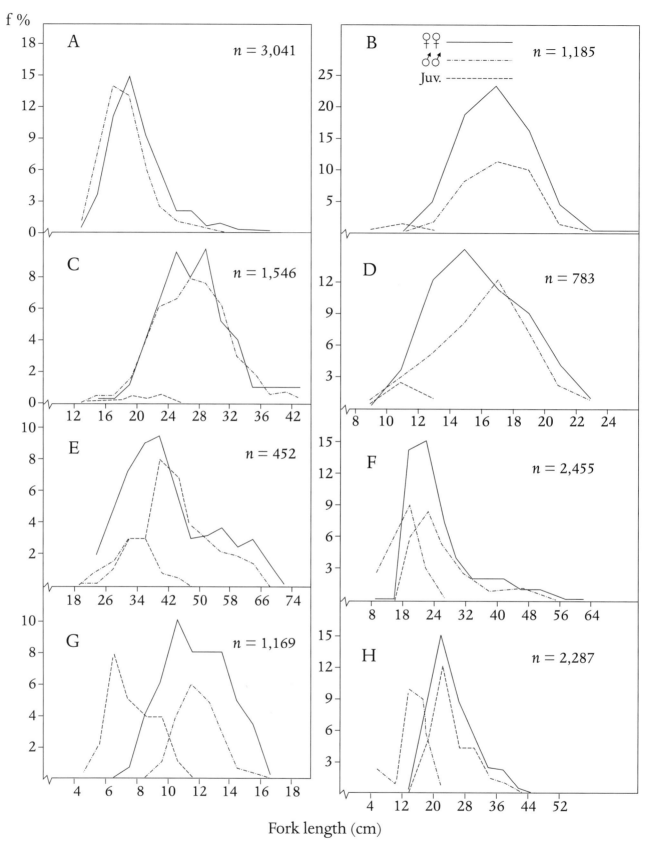

Fig. 4.1. Size and frequency distributions of juveniles, and adult females and males of Golfo de Batabanó fish populations. (A) lane snapper, *Lutjanus synagris;* (B) bluestriped grunt, *Haemulon sciurus;* (C) gray snapper, *Lutjanus griseus;* (D) white grunt, *Haemulon plumieri;* (E) mutton snapper, *Lutjanus analis;* (F) margate, *Haemulon album;* (G) redear sardine, *Harengula humeralis;* (H) bar jack, *Caranx ruber.*

terminal phase matches its actual sex. The examples provided by Roede (1972) are not typical and often show variable color patterns. Despite the importance of these attributes, a strong relationship between sex and body coloration is observed in most labrids. There are color differences among hogfish individuals and notable morphological differences between the sexes (Claro et al. 1989).

Both labrids and scarids exhibit a pattern in which smaller individuals with brownish coloration are females and larger individuals with bright coloration are males; however, functional testes have been found in both large and small individuals. Among scarids, the pattern of sexual dichromatism is more uniform than in labrid species, but is perhaps more complex. Hourigan and Kelly (1985) also reported sexual dichromatism in rock beauty (*Holacanthus tricolor*), a pomacanthid.

Distinct growth rates between sexes might be considered a sexual dimorphism in some species. In protogynous and protandric hermaphroditic species, males will be larger in the former and smaller in the latter. However, size overlap of the sexes is common, particularly during the transition of size classes. In gonochoristic species, differences in size composition between the sexes are also observed (Fig. 4.1). In snappers, grunts, jacks, mullets, and mojarras, differences might not be due to a higher growth rate, but to greater longevity in females (see Chapter 6).

4.2.3 Sex Ratio

Sex ratio is a population characteristic that plays an important role in reproduction; however, it is usually difficult to get an overall measure of sex ratio because life histories vary considerably. The size distributions of males and females in *Epinephelus* and *Mycteroperca* can show a bimodal distribution, typical of protogynous hermaphroditic species (Sadovy and Shapiro 1987), in which the smaller fishes are females, and the proportion of males increases in the larger length classes, with a marked overlapping of the two size distributions. In underexploited oceanic banks of Jamaica, Thompson and Munro (1978) found a male to female ratio of 1:0.85 in yellowfin grouper (*Mycteroperca venenosa*), 1:0.72 in Nassau grouper, and 1:2.8 in red hind. However, in the Port Royal fishery, a 1:5.6 ratio for Nassau grouper and 1:6.0 for graysby (*Cephalopholis cruentata*) was recorded. Collins et al. (1998) reported a 1:49 ratio for gag grouper from northeastern Mexico.

Evidently, larger fish, which are mostly male in highly exploited protogynous populations, are the most heavily fished, making these populations particularly vulnerable to overfishing (Harris and Collins 2000). Moe (1969) stated that "continuing exploitation may diminish the masculine component to below a critical level, thereby causing a fast decline of the number of individuals in the population." In Cuba, sexually transitional individuals account for more than 25% in almost all length classes, even after the fisheries declines of the species mentioned above, and the decline in numbers of larger individuals in the stocks.

For Nassau grouper in Bermuda, Bardach (1958) recorded more females than males in smaller length classes; the ratio gradually approached 1:1 in larger individuals (see Appendix 4.1 in García-Cagide et al. 1994). Moe (1969) described a 1:1 ratio for red grouper, although parity was obtained only in older fishes (15 years old). In both cases, such a sex ratio in exploited populations was attained because sex does not change in a large part of the population.

In six labrid species, Roede (1972) showed that the number of females in the population was 2 to 4 times higher than that of males. In hogfish, we found one male per five females (Claro et al. 1989), whereas in Puerto Rico and Florida, sex ratios of 1:3 and 1:10 have been reported (Davis 1976; Colin 1982). These differences might be associated with local fisheries pressure.

Size-dependent patterns of sex ratios in snappers, jacks, and grunts suggest that females are more abundant in almost all length classes and reach a larger size than do males (Claro 1981c, 1982, 1983c, 1983d; García-Cagide 1985, 1986a, 1986b, 1987, 1988). Female dominance in the population might be explained by a greater survivorship, but also by differences in habitat preferences of both sexes, such as in the gray snapper (*Lutjanus griseus*). Starck (1970) stated that male gray snappers are typically more abundant in reef zones off the coast, whereas females prefer nearshore areas. Grimes (1987) suggested that the sex ratio in some snappers might be caused by distinct growth and mortality rates in both sexes. Billings and Munro (1974) reported that a higher proportion of females in some haemulid populations in Jamaica might have been caused by their greater survival or by females entering fish traps more readily than males. Nevertheless, in our seine surveys, females were also more abundant (García-Cagide et al. 1994).

4.3 Gonad Development and Spawning Patterns

4.3.1 Sexual Differentiation and Attainment of Sexual Maturity

In Cuban fishes, the transition from juvenile to preadult occurs quickly and at an early age. This change is characterized by the first evidence of sexual differentiation. Oocytes change to the protoplasmic growth phase, and testes start the initial phase of spermatogenesis. In some fishes, this transformation is accompanied by changes in body coloration and the shift of both habitat and feeding habits. In Cuban snapper populations, individuals start feeding on fishes (which have higher caloric content than invertebrates) and, as a result, tend to migrate gradually from shallow nearshore zones to broader shelf areas. Similar changes have been observed in grunts and jacks (see Section 5.3).

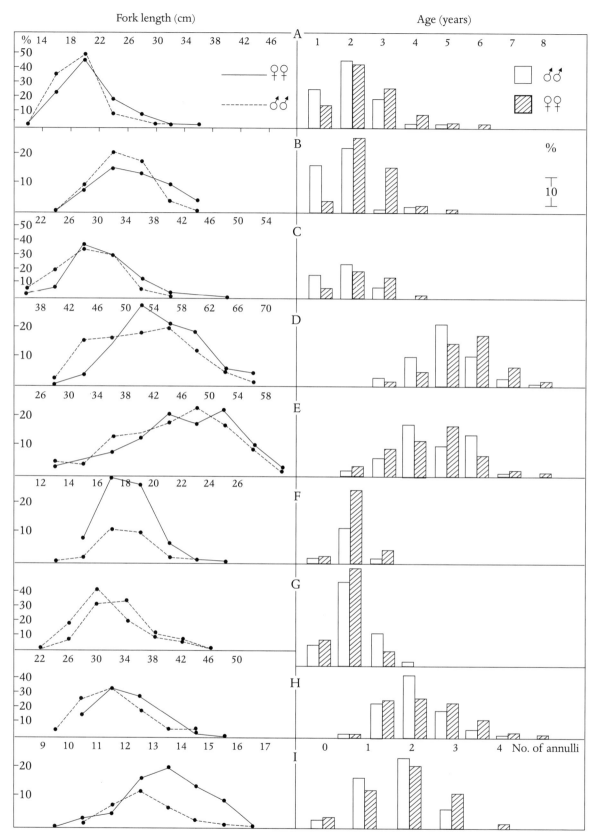

Fig. 4.2. Size and age distribution of adult males and females of Golfo de Batabanó fish populations. (A) lane snapper, *Lutjanus synagris;* (B) gray snapper, *Lutjanus griseus;* (C) yellowtail snapper, *Ocyurus chrysurus;* (D) mutton snapper, *Lutjanus analis;* (E) margate, *Haemulon album;* (F) bluestriped grunt, *Haemulon sciurus;* (G) white grunt, *Haemulon plumieri;* (H) bar jack, *Caranx ruber,* (I) redear sardine, *Harengula humeralis.* y-axis: percent frequency.

Table 4.2. Description of the gonad stages used in this book

Gonad Stage	Main Features
I	**Immature:** Young individuals. Difficult to determine sex visually. Oogonia and oocytes in different degrees of development up to the one-layer follicle. Testes with spermatogonia. This stage occurs only once during a lifetime.
II	**Early Maturation or Resting:** At the first spawning (II) and at the beginning of each annual cycle (II_n). Small, pink, semitransparent ovaries. Oocytes are not visually distinguishable and are in protoplasmatic growth stages or earlier. Thin, flat, white-grayish testes. Spermatogonia in the first spermatogenesis phase.
III	**Developing:** At the beginning of annual sexual cycle. Enlarged gonads. Yellow or light orange ovaries, fairly vascularized externally. Oocytes in early stages of trophoplasmatic growth, somewhat united to one another. Can be distinguished visually. Very white testes, sometimes pinkish. All phases of spermatogenesis are seen. Gonadosomatic index (GSI) increasing, particularly in females.
IV	**Mature:** Gonads very developed, high GSI. Ovary coloration more intense than in Stage III. White testes, enlarged. Oocytes in intense trophoplasmatic growth period, until final size. Mature spermatozoa fill the seminiferous channels. A compact mass of eggs can be expressed from the female abdomen; dense semen flow in males. In individuals with intermittent spawning, an intermediate gonadal stage (VI-III or VI-IV) occurs: gonads are less turgent and GSI is lower.
V	**Ovulation and Sperm Release:** Gonads close to or in spawning condition. A short stage. Hydrated oocytes transparent; diameter much larger than Stage IV. Gametes flow with light pressure on the abdomen. GSI high because of ovary hydration.
VI	**Spent:** Flaccid ovaries, purple red. Rapid process of resorption of empty follicles, and remnant oocytes occur. Testes white, reddish, slender. Seminiferous channels empty, with some remnant sperm. GSI low.

Sexual differentiation in the lane snapper (*Lutjanus synagris*) was visible to the naked eye at a size of 12–13 cm FL (before attaining the first year of life). Most individuals attain sexual maturity and spawn for the first time shortly thereafter, at 13–16 cm FL and age class I (Claro 1982). However, this pattern varied in the population throughout a 10-year period: Rodríguez Pino (1962) reported that males matured at 18 cm and females at more than 20 cm FL (Appendix 4.1). This difference is probably due to the intensive fishery exploitation that lane snapper stocks underwent during the period between the two studies (1960 and 1972). Overfishing likely provoked a marked decline in mean size (Claro 1981a) and consequently an earlier maturation of individuals. From 1970 to 1973, schools of lane snapper spawners consisted of 20–24% 1-year-old individuals, 50–60% 2-year-old individuals, 20–30% 3-year-old individuals, and only 5% older individuals (Fig. 4.2 and Appendix 4.1).

In gray and yellowtail (*Ocyurus chrysurus*) snappers, sexual differentiation also occurs slightly earlier than sexual maturity (Claro 1983c, 1983d); in yellowtail snappers, it occurs at an early age, 1–2 years old (Appendix 4.1). This pattern is similar in other small to medium-sized species, such as the bluestriped (*Haemulon sciurus*) and white (*H. plumieri*) grunts, (García-Cagide and Claro 1983; García-Cagide 1986b, 1987). However, in Jamaica most individuals of these species attain sexual maturity at a larger size (Billings and Munro 1974) than in Cuban populations. In contrast, the bar jack (*Caranx ruber*) attains sexual maturity later in life in Cuba than in Jamaica, demonstrating that the population parameters are specific to the habitat parameters of each population. Food availability probably plays an important role in this pattern. In Venezuelan waters (Guerra and Bashirullah 1975), the gray snapper attains the adult

phase at a larger size and greater age than in Cuba (Claro 1983c) and Florida (Starck 1970). In other species in Cuban waters that attain large sizes, such as mutton snapper (*Lutjanus analis*), margate (*Haemulon album*), and yellow jack (*Caranx bartholomaei*), most individuals reach sexual maturity at a relatively larger size (Claro 1981c; García-Cagide 1986a; Sierra et al. 1986). In these species, however, sexual differentiation occurs relatively early; thus fish remain with undeveloped gonads (Stage II) for a period of two or more years.

Time of sexual differentiation and maturity can be altered by external factors, such as intense fisheries exploitation (e.g., lane snapper). Grimes (1987) states that snapper species attain maturity at 40–50% of their maximum length and affirms that insular populations of these species mature at a larger size than those inhabiting continental coasts.

Fishes of Cuban waters attain sexual maturity at 40–70% of their maximum length (40–60% in males, 50–70% in females), which corresponds to 1–2 years of age in small to medium-sized species (e.g., bluestriped grunt; gray, lane, and yellowtail snappers; redear sardine [*Harengula humeralis*]) (Fig. 4.2). In larger species (bar and yellow jacks, margate, mutton snapper, and others), this transition occurs at 3–5 years of age, although this group sexually differentiates much earlier.

4.3.2 Gametogenesis

Patterns of sexual cell development (gametogenesis), particularly oogenesis, have been detailed in many temperate freshwater fishes (see review by Koshelev 1984), as well as in many marine species. Numerous papers contain information on this important aspect of tropical marine fish reproduction (Andreu 1956; Schaefer and Orange 1956; Andreu and dos Santos Pinto 1957; Bara 1960; Bowers and Holliday 1961;

Abraham 1963; De Sylva 1963; Gorbunova 1965; Smith 1965; Abraham et al. 1966; Naumov 1968; Fontana 1969; Moe 1969; Overko 1969, 1971; Oven and Saliejova 1970, 1971; Roede 1972; Beaumarriage 1973; Mester et al. 1974; Pien and Liao 1975; Oven 1976; Ros and Pérez 1978; García 1979; Nagelkerken 1979; Claro 1982; Ramos 1983; García-Cagide and Claro 1983; Rodríguez 1984; García-Cagide 1985, 1986a, 1986b, 1987, 1988; García-Cagide and Espinosa 1991; Shapiro et al. 1993a; Sadovy et al. 1994; Sadovy and Colin 1995; García-Cagide and García 1996; Sadovy 1996; Collins et al. 1998; Brulé et al. 1999).

Many of these investigations are descriptive and few couple the physiological and ecological aspects of gonad development. In our studies we have attempted to relate patterns of gametogenesis with ecological patterns of reproduction. Because it is general and easy to apply, we adopted the maturation scale proposed by Koshelev (1984) with some modifications according to the particularities of the gonads of the fishes inhabiting Cuban coastal waters (Table 4.2). This gonad scale is compatible with other scales that have been used for many species. Koshelev (1978, 1984) defined six periods, each of which is characterized by specific processes within the gonad and the organism.

To identify cellular patterns of gametogenesis, standards for the different phases of oogenesis and spermatogenesis are also necessary. Considering the criteria of some authors (e.g., Naumov 1956; Yamamoto and Yamazaki 1961; Abraham 1963; Sakun and Butskaia 1963; Smith 1965; Moe 1969; Oven 1976; Wallace and Selman 1981; and Wyanski and Pashuk 1990), we developed a scale based on the microscopic features of different phases of oogenesis (Table 4.3; Plate 4.2) that can apply to many tropical marine fishes.

Four fundamental linkages between gametogenesis and field spawning patterns were found in the marine fishes we have studied in Cuba. These spawning patterns largely result from diverse and complicated variations in oogenesis and can be described using four spawning categories:

TYPE A. Discontinuous asynchronous development of oocytes, with each individual spawning all oocytes in only one month. All oocytes that will be spawned within a reproductive cycle split from those that remain in the protoplasmic phase. Different batches of oocytes mature asynchronously so that the ovaries carry different sizes and stages of vitellogenic oocytes. Spawning occurs by batches as each group completes maturation (Type A in Fig. 4.3 and Fig. 4.4). Examples include tiger grouper, black grouper (Mycteroperca bonaci), yellowfin grouper, yellowtail snapper, and dog snapper (Lutjanus jocu).

TYPE B. Continuous asynchronous development of oocytes, with each individual spawning over several months of the year. This pattern differs from Type A in that during the entire reproductive cycle, batches of proto-plasmic oocytes continue to pass to the trophoplasmic phase. This pattern allows the female to produce a large number of egg batches (Type B in Fig. 4.3 and Fig. 4.4). Examples include many species of Haemulon as well as the redear sardine.

TYPE C. Synchronous development of oocytes and total spawning (one spawning event only per individual). Growth, maturation, and spawning of all oocytes occur simultaneously (Type C in Fig. 4.3 and Fig. 4.4). Examples include the liza (Mugil liza) and white mullet (Mugil curema).

TYPE D. Synchronous development of oocytes with each individual spawning all oocytes in one month. Vitellogenesis occurs as in Type C, but ovulation occurs intermittently so that eggs are released in several batches (Type D in Fig. 4.3 and Fig. 4.4). Examples include Nassau grouper and gray, mutton, and lane snappers.

Although the ecological characteristics of species exhibiting Type A and Type D spawning patterns in Cuba appear similar (several batches released over a 5- to 10-day period during only one month), the underlying physiological patterns of oogenesis differ. This explains why available evidence suggests that Type A species might be able to release more batches (7 or 8 batches) over their spawning period than Type D species (4 or 5 batches).

General characteristics of oogenesis are further represented in Fig. 4.4; synchronous types of vitellogenesis are segregated from asynchronous types using the terminology of von Götting (1961) that was further developed by Oven (1974, 1976) and Koshelev (1981). The initial formation of ovarian tissue only occurs once in the fish's lifetime (gonad Stage I). Oogonia develop over a long period. These cells are located in the germinal epithelium at the base of the ovarian folds and are found at the beginning and end of the reproductive period; part of them become oocytes during each reproductive cycle. As oogonia protoplasm increases, fishes enter gonad Stage II. At this time, the sex can be identified through a simple macroscopic observation of the gonads. In this period, oogonia and oocytes in Phases C_1 and C_2 can be found in the ovaries (Table 4.3).

In fishes with synchronous vitellogenesis, all yolk-bearing oocytes pass simultaneously through all phases of the trophoplasmic process until ovulation and spawning is complete (Type C in Fig. 4.4). In another type of synchronous oogenesis, the final period of oocyte maturation (ovulation, from Phase E to F) occurs with oocytes being released in batches (Type D in Fig. 4.4). In contrast, Oven (1974) classified this as discontinuous asynchronous with intermittent spawning.

In the asynchronous trophoplasmic growth phase of development, the groups of oocytes develop in separate groups and spawning always occurs in batches. Most intermittent spawners exhibit continuous vitellogenesis

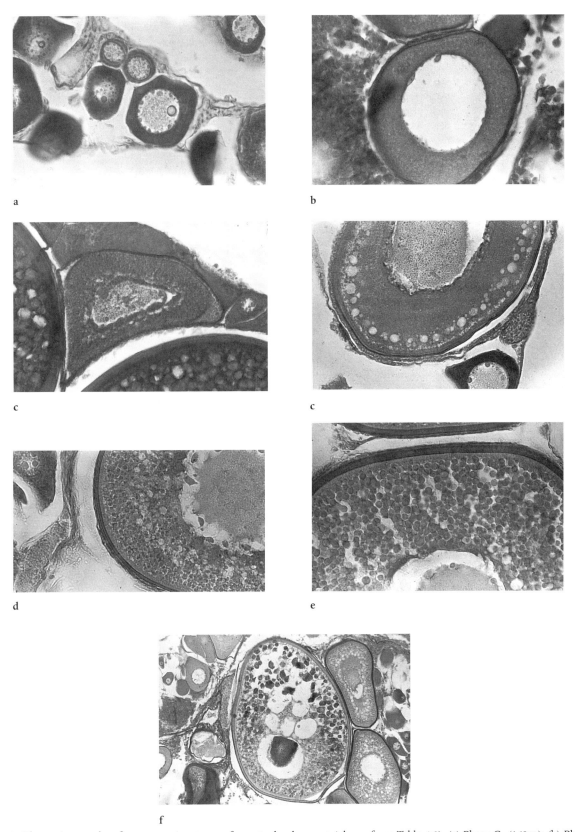

Plate 4.2. Photomicrographs of representative stages of oocyte development (phases from Table 4.3). (a) Phase C_1 (160×); (b) Phase C_2 (160×); (c) Phase D_1, two different forms of early vitellogenesis are shown (160×); (d) Phase D_2 (160×); (e) Phase E, well-defined oocyte membrane layers are shown (160×); (f) Phase F, during which the nucleus migrates to the animal pole and oil droplets and then yolk granules merge (40×).

Table 4.3.   Microscopic characteristics of the different phases of fish oocyte development (see Plate 4.2)

Oocyte Phase	Characteristics
C	One-layer follicle. Previtellogenic oocyte. Protoplasmatic growth.
C_1	Irregular-shape, sometimes polyhedric oocytes. Diameter, 10 to 80 μm. Cytoplasm strongly basophilic. Central nucleus round and vesicular; typically occupying two-thirds of the cell. Nucleoli irregularly distributed throughout the nucleus.
C_2	Round oocytes. Diameter 60 to 150 μm. Cytoplasm starting to show a granular aspect; basophily diminishing. Single-layered follicular epithelium surrounding the oocyte. Round nucleus, occupying half of the cell. Nucleoli in variable numbers, spherical, and of different sizes; located at the nucleus periphery.
D	Onset of vitellogenesis. Two-layered follicle. Trophoplasmatic growth.
D_1	Primary stage of vitellogenesis. Diameter, 100 to 200 μm. Nucleoli still in the nucleus periphery. The first vacuoles appear, forming an uncolored ring surrounding the nucleus, or close to the oocyte membrane (depending on the species). Cytoplasm basophily diminishes and is stained light blue with a Mallory solution. The oocyte membrane structure begins to be defined.
D_2	Secondary stage of vitellogenesis. Diameter, 140 to 350 μm. The oocyte membrane is completely defined during this stage; formed by the theca, the follicular epithelium, and the inner membrane (composed of the external and internal *zona radiata*). As the oocyte matures, the number and size of the yolk vesicles increase. Numerous round yolk granules, which stain orange with a Mallory solution, are observed. Depending on the way D_1 transpired, cytoplasm filled with yolk granules and vesicles, from the center to the periphery of the cell.
E	Completion of yolk accumulation or tertiary stage of vitellogenesis.
	The yolk granules and oil droplets increase in size and number. Oocyte diameter (250 to 600 μm) increases notably. Nucleus still in central position, occupying about a quarter of the cell volume. Oocyte membrane with different thicknesses, according to the species characteristics.
F	Mature oocyte, migration of the nucleus.
	Final stage of oocyte maturation. Diameter, 500 to 850 μm. Oocyte reaches maximum diameter before starting ovulation. Oil droplets and yolk granules merge altogether. The nucleus migrates to the animal pole; nuclear membrane disappears, and in some cases, the nucleus is not distinguishable.

(von Götting 1961; Oven 1974, 1976; Koshelev 1981). Histological sections of their gonads show oocytes in all maturation phases. Groups of reserve oocytes pass continuously to the vitellogenesis phase so that the separation between groups of oocytes at different stages in their size distribution curve is small or nonexistent.

Among Cuban marine fishes we found two rather generalized types of continuous asynchronism. In some species (Figs. 4.3B and 4.4), gonads in Stage V of maturity contain oocytes in Phases C, D, and F, and hydrated ovules, but not oocytes in Phase E. It appears that after an oocyte batch is released, the ovary returns to Stage VI-III, after which there is a certain period of recovery before maturation of the next batch (passage of Phase D to E, Table 4.3). In species with Type B spawning (Figs. 4.3 and 4.4), we observed ovaries in Stage V with ovules and oocytes in all phases of development, including Phase E. This suggests that after release of an oocyte batch, gonads pass to an intermediate stage closer to spawning, called Stage VI-IV. These intermediate stages (VI-III and VI-IV) of oocyte maturation are short, lasting only a few hours. The passage from Stage V to VI-IV$_n$ is repeated n times during the spawning season, equivalent to the number of oocyte batches the female produces.

As soon as all eggs are released, the ovaries pass to a spent stage (VI) in which the following is observed: oocytes are in early developmental phases, empty follicles are in late stages of resorption, some ripe oocytes remain (those that

were not spawned), and some atretic bodies remain. The atretic bodies, described thoroughly by Smith (1965) and Moe (1969), are the final phase of the oocyte resorption process and constitute the last evidence of spawning (Koshelev 1984).

Histological sections of ovaries in Stage IV showed oocytes compactly arranged in the lamellae and no empty follicles or atretic oocytes (Plate 4.3a). Partially spent ovaries (Stage VI-III or VI-IV) contain empty follicles at the beginning (Plate 4.3b) and final phase (Plate 4.3, c and d) of the resorption process. In ovaries during Stage VI, both unspawned ovules (Plate 4.3g) and empty follicles were seen at the final resorption phase, although in many cases the latter could not be seen at all. In such cases, some disorganization of the ovary lamellae, marked blood vessels, some necrotic zones, and atretic bodies are indicators of the occurrence of recent spawning (Plate 4.3h). After the reproductive cycle ends, there is a recovery period in individuals (Stage II$_n$) during which new gametes are formed along with the final resorption of the unreleased oocytes.

4.3.3   Ecological Patterns of Spawning

Much of our work has emphasized the interrelationships between gametogenesis and spawning processes in representative species. As mentioned, spawning patterns are influenced by the degree of asynchrony of oocyte develop-

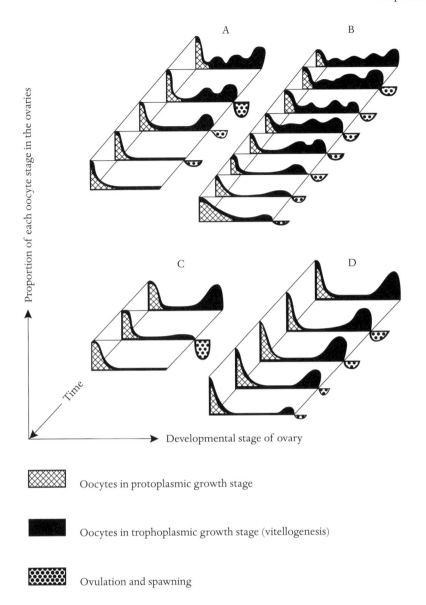

Proportion of each oocyte stage in the ovaries

Time

Developmental stage of ovary

Oocytes in protoplasmic growth stage

Oocytes in trophoplasmic growth stage (vitellogenesis)

Ovulation and spawning

Fig. 4.3. Schematic representation of primary spawning patterns of fishes, based on field and laboratory studies of gametogenesis (modified from Koshelev 1981). Type A: discontinuous asynchronous oogenesis with each individual spawning all batches in one month; Type B: Continuous asynchronous oogenesis with each individual spawning over several months of the year; Type C: synchronous oogenesis and total spawning (only one spawning event per individual); Type D: synchronous oogenesis with each individual spawning several batches in one month.

ment within the ovary. However, the variety of intermittent-type patterns (spawning Types A, B, and D in Fig. 4.3) complicates the determination of the spawning process and fecundity. We have stratified some of this information in Appendix 4.2, which summarizes information on vitellogenesis, spawning type, and total duration of the reproductive season for many species that occur in Cuba. Appendix 4.3 details monthly spawning information from most studies of marine fishes in Cuba.

Among tropical species, relatively few species display synchronous gametogenesis (Appendix 4.2). Because vitellogenesis does not seem to require a high energy expenditure from females it can occur at any time of the year, specific to the species. Here, vitellogenesis is short and occurs before spawning (1 to 3 weeks in most cases). Among Cuban fishes with synchronous gametogenesis, two quite different groups

have been reported: mullets (Type C; Alvarez-Lajonchere 1979, 1980a; García and Bustamante 1981) and some snapper and grouper species (Type D). In mullets, ovulation is synchronous, spawning is total, and all oocytes are released within a few hours.

Synchronous vitellogenesis and total spawning (Type C) are most common in fishes of higher latitudes. Their populations endure severe climatic conditions, so the season suitable for spawning is short. Total spawning allows production of a large number of eggs in a brief period. Many species inhabiting these areas can spawn only every 2 to 3 years (Oven 1976; Koshelev 1984). In such fishes, oocyte vitellogenesis starts after a long feeding period. Koshelev (1984) and Shatunovsky (1980) reported this type of reproduction in both freshwater (e.g., *Perca fluviatilis*) and marine fishes (e.g., Atlantic herring, *Clupea harengus harengus;* Atlantic cod,

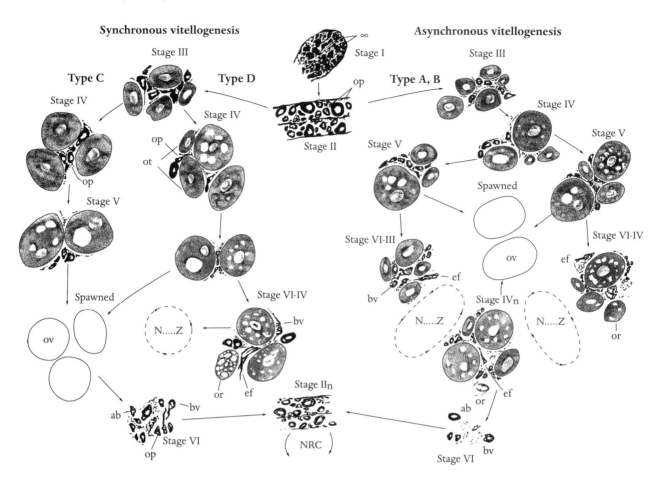

Fig. 4.4. Primary features of synchronous and asynchronous vitellogenesis in the species studied. ab: atretic bodies; bv: blood vessels; ef: empty follicles; oo: oogonia; op: oocytes in protoplasmic growth stage; or: oocytes in resorption; ot: oocytes in trophoplasmic growth stage; ov: ovules; NRC: next reproductive cycle. From left to right across the top, spawning Types C, D, A, and B (described in Section 4.3.2).

Gadus morhua; and others). This type of gametogenesis was also described for *Platichthys flesus* by Oven (1967), and *Theragra chalcogramma* by Gorbunova (1954) and Zverkova (1969); it seems to be similar in *Thunnus obesus,* according to Sharov's (1973) data.

Some snapper and grouper species show synchronous gametogenesis, but spawning occurs in batches and all eggs are released over a few days. This Type D pattern of intermittent spawning might allow greater fecundity for species whose oocytes develop synchronically. This pattern is well-documented in lane snapper. Spawners exhibited notable variation in the daily gonadosomatic index (GSI) and flaccid gonads suggested a recent spawning (Claro 1982). Histological examination of ovaries revealed that all vitellogenic oocytes were in Phase E, although some empty follicles from spawned eggs and unspawned oocytes were observed undergoing resorption (Plate 4.3). Lane snapper usually remained in the spawning area for a period no longer than 7 to 10 days, after that individuals in gonad Stage VI were present in the samples (see Section 4.6.1). Spawning induced in captive lane

snapper by the injection of human chorionic gonadotropin (Millares et al. 1979a) yielded eggs with a 24-hour interval between batches. Histological sections of mature ovaries in gray and mutton snappers and schoolmaster (*Lutjanus apodus*) (Plate 4.3) and the occurrence of newly spawned individuals in catches immediately after their first spawning peak suggest similar Type D spawning patterns. The oocyte size distribution in the ovaries of these fishes (e.g., Fig. 4.5) shows a continuity in the oocyte development process, from those in reserve (< 100 μm) to the mature ones (> 300 μm). A separate peak is only seen with ovulation (see García-Cagide et al. 1994). Using hormone injections, Soletechnik et al. (1988) were able to initiate a second reproductive phase one to three months after the first spawning.

The data of Moe (1969) and Nagelkerken (1979) show similar patterns of gametogenesis in red grouper and graysby. Histological analysis of gonads (Claro et al. 1990c; García-Cagide and Espinosa 1991) and spawning observations (Smith 1972; Olsen and LaPlace 1979; Carter et al. 1994; Sadovy and Colin 1995) suggest that gametogenesis occurs similarly in

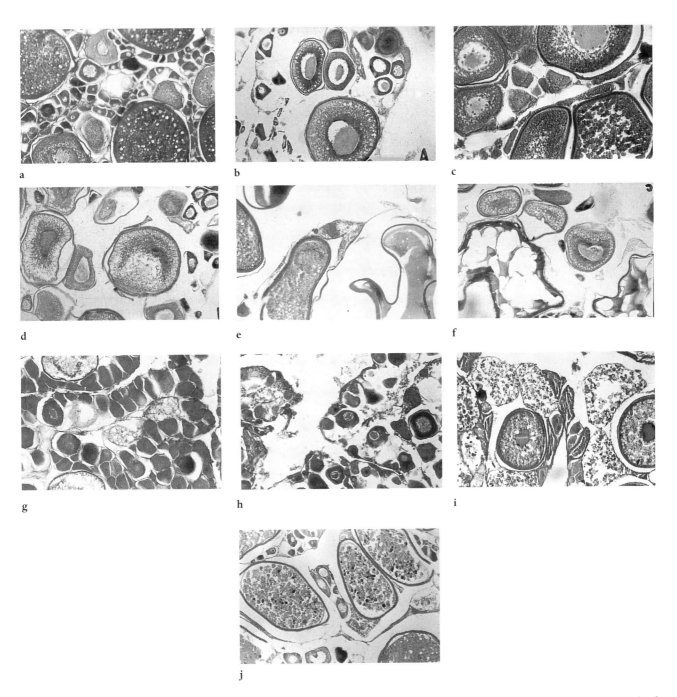

Plate 4.3. Histological sections of ovaries of species with different spawning patterns. Images a to h: Continuous asynchronous growth of oocytes and batch spawning (Type B) during different development stages in (a) margate, *Haemulon album,* Stage IV; (b) redear sardine, *Harengula humeralis,* Stage VI-III; (c) bar jack, *Caranx ruber,* Stage VI-IV; (d) redear sardine, *Harengula humeralis,* Stage VI-IV; (e) bluestriped grunt, *Haemulon sciurus,* Stage V; (f) redear sardine, *Harengula humeralis,* Stage V (hydrated ovules appear deformed because of the histological technique); (g) margate, *Haemulon album,* Stage VI; and (h) white grunt, *Haemulon plumieri,* Stage VI. Images i and j: Synchronous growth of oocytes and batch spawning (Type D) during development in (i) lane snapper, *Lutjanus synagris* (specimen caught at spawning site), with empty follicles from a spawned batch, oocytes in resorption from a pending batch, and a batch of oocytes in Phase E (Table 4.3); (j) schoolmaster, *Lutjanus apodus,* with oocytes in Phase E and two empty follicles (in different phases of resorption) as indicators of recent spawning.

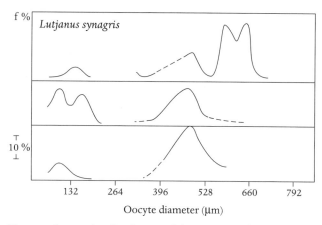

Fig. 4.5. Oocyte size distribution of three ovaries of lane snapper (*Lutjanus synagris*), characterized by synchronic vitellogenesis and intermittent spawning. Dashed lines: no data.

Nassau grouper. After hormone injections, Nassau grouper produced one or two batches (Kelly et al. 1994) and females could spawn two or three times at intervals of 28–75 days (Head et al. 1996). Importantly, certain environmental conditions might trigger alterations of gametogenesis within species, and asynchronous vitellogenesis has been observed in lane snapper (García-Cagide and Claro 1985).

The duration of the reproductive period in fishes with synchronous vitellogenesis (Type C or D) is variable, although it is generally shorter than in fishes with asynchronous gametogenesis and intermittent spawning (Appendix 4.2). Mullet are a typical example. Liza focus their spawning activity in November–December. Ripe individuals of white mullet can be found all year, but mostly in June–July and November–December (Alvarez-Lajonchere 1976, 1979, 1980a). White mullet can have two spawning periods (winter and summer). In Cuba, the Nassau grouper breeds mainly in December and January; in the Bahamas, this species usually spawns in January (Smith 1972). In the Bahamas, aggregations occur in June–July when the average temperature is near 25.5°C (Colin 1992). Red grouper females with developing oocytes can be found from December to August, but are more abundant in April–May (Moe 1969). Mature lane snappers appear throughout half the year, with a marked peak in May. Mutton and gray snappers also concentrate their breeding activity within 1–2 months (May–June for mutton snapper, July–August for gray snapper) and their reproductive season is shorter than that of lane snapper.

Of the 55 Cuban species studied, 75% have asynchronous vitellogenesis and intermittent spawning (Appendix 4.2) within Type A or B spawning patterns. A Type A pattern was observed in several snapper species (dog, yellowtail, blackfin [*Lutjanus buccanella*], and silk [*Lutjanus vivanus*]) and in groupers (black, tiger, and yellowfin) in which vitellogenesis is typically asynchronous (Pozo et al. 1984; Pozo and Espinosa 1983; García-Cagide and García 1996; García-

Cagide et al. 1999a, 1999b). Discontinuous gametogenesis also occurs in the bar jack (García-Cagide 1985) and yellow jack (Sierra et al. 1986). Mature bar jacks were caught from March to August in the Golfo de Batabanó, but two marked peaks of spawning were found in April and July.

Although available information does not allow us to completely characterize reproductive patterns among the Cuban ichthyofauna, our data showed that grunts (*Haemulon* spp.) and redear sardine are good representatives of the Type B spawning pattern: continuous asynchronous development of oocytes with each individual spawning over several months of the year. In white grunt, bluestriped grunt, and margate, oogenesis is continuous and spawning is recurrent. The reproductive period of these species is longer than in species with synchronous development of oocytes. In margate, mature and spent individuals were observed almost monthly. Stage VI (spent individuals) was verified through histological analysis of gonads; atretic bodies and remnant oocytes of a recent spawning in different stages of resorption were observed. This shows that although spawning is continuous and recurrent, all batches are released in a relatively short period of time.

White and bluestriped grunts show the same pattern of gametogenesis and spawning as margate. A histological section of an ovary in gonad Stage VI-IV shows oocytes in all stages of vitellogenesis and an empty follicle in a primary stage of resorption, indicating the recent release of one egg batch. Another batch of oocytes was already mature, in the stage of intensive vitellogenesis. Plate 4.3e shows a histological section of a gonad from a female in Stage V; together with hydrated ovules, oocytes in Phases E and D are shown. The release of the different batches occurs almost continuously. In bluestriped grunts in Stage VI, empty follicles were rarely found, probably because they are quickly absorbed. However, the remnant unspawned oocytes are reabsorbed more slowly, so they might serve as indicators of the end of the spawning cycle.

Continuous gametogenesis was also recorded in the redear sardine (García-Cagide 1988). Unspawned mature individuals were caught from November to March, and mature but partially spawned fish were caught from December to August (Plate 4.3, b and d; Stages VI-III and IV$_n$); individuals in Stage VI also appear in the samples. This means that the redear sardine also spawns all its batches in a short period. Nevertheless, the histological sections of various ovaries in Stage V (Plate 4.3f) showed certain discontinuity among the sizes of oocytes in early vitellogenesis (D) and ripe (F) stages. This pattern suggests an interruption in the development of vitellogenic oocytes, probably associated with the need to attain a certain level of physiological condition before the oocyte's final maturation. Andreu and dos Santos Pinto (1957) reported an analogous interruption in the vitellogenesis of *Sardina pilchardus*. This period of physiological recuperation might be short (a few days or hours), according to the occurrence of entirely spent individuals shortly after the start of a new reproductive cycle.

4.3.4 The Resorption Process

An important aspect of fish gametogenesis involves the structures that remain in the ovaries after spawning, visible in the histological sections of ovaries. These residual structures are the result of the oocyte resorption process, triggered either by normal physiological processes or by unfavorable spawning conditions. Koshelev (1981) states that under normal conditions the following can happen: (1) resorption of part of the vitellogenous oocytes caused by constraints on individual fecundity; (2) resorption of unspawned oocytes in different vitellogenesis phases (residual oocytes); and (3) resorption of empty follicles after ovulation. In the second type, a massive resorption of mature oocytes can occur because environmental conditions are unfavorable for reproduction (e.g., regular routes to breeding grounds are closed, quality of the spawning habitat is inadequate, or individuals of the other sex are absent). We have not encountered the first type in the species we have studied and it seems to be uncommon in tropical species. The second and third types are found in the Cuban fish fauna, as well as in temperate and polar species, but they occur faster in tropical fishes.

Plate 4.4 shows different stages of the resorption process for empty follicles and unspawned remnant oocytes. As resorption progresses, the cells of the follicular membrane are reabsorbed until they almost disappear. The cells are found embedded in the lamellae lining the ovary, often close to mature oocytes or blood vessels. Resorption starts at the cell membrane; the *zona radiata* and the nuclear membrane are quickly reabsorbed, and the nuclear and cytoplasmic inclusions blend. Afterward, the granules and yolk vesicles are reabsorbed and only the lipid inclusions remain. In the last phase of this process, atretic bodies remain in the ovary and are situated toward the borders of the ovarian septa close to the blood vessels.

We found massive resorption of ripe oocytes in species of economic significance such as lane snapper, bar jack, and grunts. In lane snapper, this phenomenon might be related to the use of set nets by fishermen (Claro 1983c). Ripe fish caught by this method sometimes escape after hours of stress, causing atresia of their ovules. Likewise, mature bar jacks captured in these set nets, as well as the white and bluestriped grunts surveyed in fish traps that were immersed for substantial periods, showed similar alterations in their near-spawning ovaries. In species with synchronous oogenesis and total spawning, resorption processes within the ovary were complete long before the start of the next reproductive cycle. In species with asynchronous oogenesis, resorption processes take place simultaneously with the next cycle of gametogenesis and oocyte maturation. Resorption processes in tropical fishes, as a natural phenomenon or

provoked by fishing, do not seem to affect the individual's next reproductive cycle. Resorption occurs quickly in the tropics and during the most advanced phases of oocyte development.

4.3.5 Gonadosomatic Index

The weight of a fish's gonad relative to its body weight (whole or gutted) is known as the gonadosomatic index (GSI). This index, as well as being a measure of gonad development, provides information on energy expenditure during the reproductive process. The magnitude of such energy expenditure for gonad maturation depends mainly on the individual fecundity and the amount of reserves necessary for normal egg development (see Chapter 7).

The GSI of ripe, temperate fishes ranges from 10 to 25%, whereas in polar fishes it is usually higher than 30%, and in some species can reach 50% (García-Cagide et al. 1983). In Cuban fishes (Table 4.4), the index for female spawners ranges from 1 to 20%, although it is generally less than 6%. The highest GSI values are found in species that are total spawners such as mullets (liza and white mullet). However, since many tropical fishes exhibit intermittent spawning, their GSI provides information for only a certain period of the reproductive cycle, rather than for the entire year. Therefore, energy expenditure for the entire annual reproductive cycle is higher than for one spawning period alone. In some tropical species, energy expenditure can be roughly equivalent to the expenditure of temperate fishes. In tropical total spawners, there is no evidence of more than one reproductive period within the year, but further maturation and spawning after a period of recovery may be possible. This phenomenon requires further study.

Fig. 4.6 presents the monthly variation of GSI of some fish populations of the Golfo de Batabanó for individuals in different stages of gonad development. Individuals with immature gonads are present in almost all months. Their GSI is generally lower than 0.5%, but the index increases quickly before the spawning period. The highest GSI values usually coincide with the spawning peak. Thus, although mutton and lane snappers breed from March to August–September, the greatest proportion of individuals spawn in May–June, when the adult population shows the highest GSI values.

A similar pattern was observed in gray and yellowtail snappers, as well as in bar jack and redear sardine. Nevertheless, in some cases, annual variation of GSI does not clearly depict reproductive peaks, because sampling might not properly cover all portions of the population. This is the case with the margate. One annual spawning peak is in March, according to the high proportion of mature and spent individuals found, and the second peak occurs in July. However, the highest mean GSI value was obtained in September, according to the occurrence of individuals found with gonads in an advanced stage of maturation.

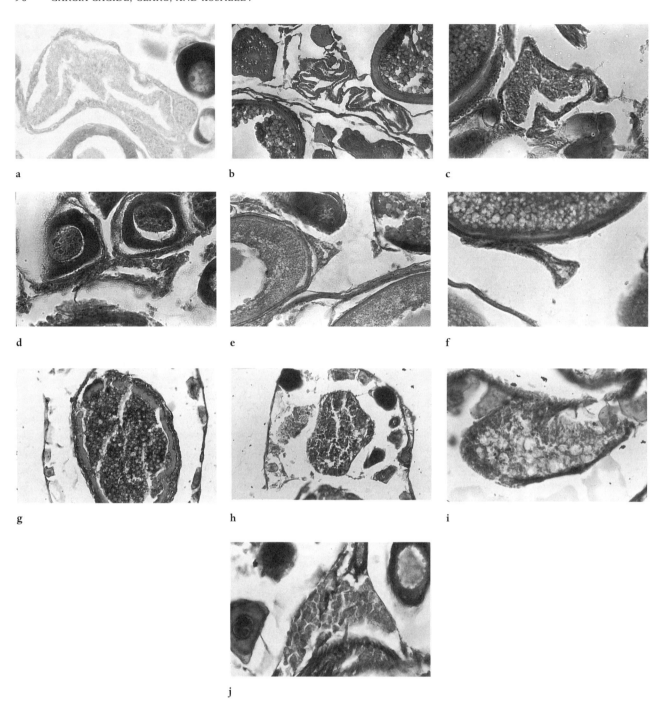

Plate 4.4. Representative sections of ovaries in which resorption of the empty follicle and the mature oocyte are observed. Empty follicle: (a) and (b) in early phases of resorption; (c) and (d) in intermediate phases; (e) and (f) in final phases. Mature oocyte: (g) in the early phase of resorption; (h) in intermediate; and (i) in the final phase, and (j) yellow or atretic bodies.

Likewise, mature bluestriped grunt individuals exhibited low GSI values during winter months, although this is the main spawning period of the population inhabiting the Golfo de Batabanó. This happened because most of the fish surveyed had already spawned some batches, as verified by gonad histological analysis (gonad Stage VI-IV). To better define peak spawning periods, seasonal variations of GSI must be examined in conjunction with other aspects such as the monthly proportion of individuals in different stages. In snappers, mature individuals caught after the spawning peak had lower GSIs than in previous months (among individuals of the same maturation stage). This demonstrates either a lower fecundity or a smaller amount of nutritional reserve in the oocytes of such individuals (see Chapter 7).

Table 4.4. Gonadosomatic index (GSI) of ripe females in Stages III-IV, IV, VI-IV, and V (*with hydrated oocytes) of different species of the Cuban ichthyofauna

Family / Species	GSI Average	GSI Range	Source
Clupeidae			
Harengula clupeola	3.2	1.1–6.1	García-Cagide et al. 1994
Harengula humeralis	5.0	2–10.0 (13–20*)	García-Cagide 1988
Harengula jaguana	5.5	4.1–9.0*	Martínez and Houde 1975
Jenkinsia lamprotaenia	3.7	1.0–12.0	García-Cagide et al. 1994
Serranidae			
Cephalopholis cruentata	3.2	1.8–5.6	Nagelkerken 1979
Cephalopholis fulva	5.0	—	Smith 1965
Epinephelus striatus	4.0	1.4–5.9 1.5–15.0	Claro et al. 1990c; Tucker et al. 1993
Mycteroperca bonaci	2.21	0.85–4.9	García-Cagide and García 1996
Mycteroperca microlepis		0.4–2.2	Hood and Schlieder 1992
Mycteroperca tigris	3.5	1.1–7.3	García-Cagide et al. 1999a
Mycteroperca venenosa	1.29	1.2–1.5	García-Cagide and García 1996
Carangidae			
Caranx bartholomaei	2.5	—	Sierra et al. 1986
Caranx ruber	3.0	1.2–7.0	García-Cagide 1985
Lutjanidae			
Lutjanus analis	4.0	2.0–9.0	Claro 1981c
Lutjanus apodus	2.9	1.7–3.7	García-Cagide et al. 1994
Lutjanus griseus	4.0	2.0–7.0	Claro 1983c
Lutjanus griseus	4.4	0.6–10.8	González et al. 1979
Lutjanus jocu	3.13	1.2–5.0	García-Cagide et al. 1999b
Lutjanus synagris	7.1	2.0–16.0*	Claro 1982
Ocyurus chrysurus	2.5	1.0–5.0	Claro 1983d
Gerreidae			
Eugerres brasilianus	4.0	1.0–5.3	Báez et al. 1983
Eugerres plumieri	1.8	0.5–8.5	Millares et al. 1979b
Haemulidae			
Haemulon album	4.0	1.5–8.0	García-Cagide 1986a
Haemulon aurolineatum	6.55	2.9–10.1	García-Cagide et al. 1994
Haemulon carbonarium	3.07	—	García-Cagide et al. 1994
Haemulon parra	3.59	—	García-Cagide et al. 1994
Haemulon plumieri	4.0	2.0–8.0	García-Cagide 1987
Haemulon sciurus	3.5	1.0–9.7	García-Cagide 1986b
Sparidae			
Calamus calamus	—	3.2–5.3	Liubimova and Capote 1971
Sciaenidae			
Bairdiella ronchus	5.0	1.5–8.3	García 1979
Micropogonias furnieri	8.0	—	García 1979
Labridae			
Lachnolaimus maximus	2.73	1.8–3.1	Claro et al. 1989
Mugilidae			
Mugil curema	7	1.0–18*	Alvarez-Lajonchere 1976
Mugil curema	—	12–14* (14–30 ovulation)	García and Bustamante 1981
Mugil hospes	—	1.0–12*	Alvarez-Lajonchere 1981b
Mugil liza	14	2–20*	Alvarez-Lajonchere 1979
Mugil trichodon	4.0	1.0–16*	Alvarez-Lajonchere 1980c
Trichiuridae			
Trichiurus lepturus	2.9	0.1–6.7	Ros and Pérez 1978
Istiophoridae			
Istiophorus platypterus	—	1–9 (8–13*)	Jolley 1977

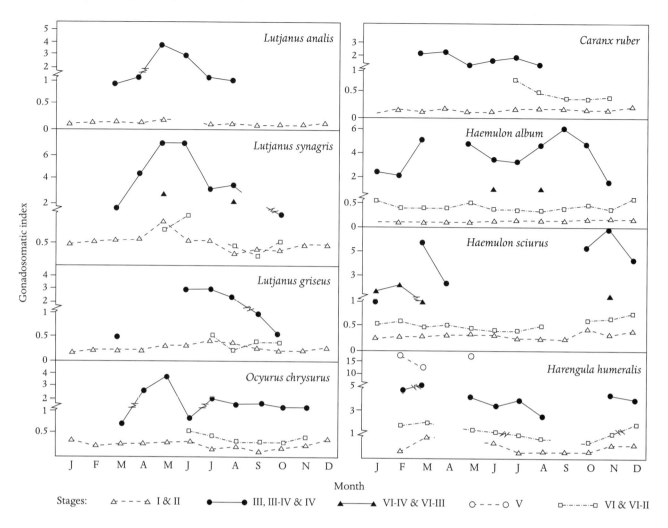

Fig. 4.6. Monthly variation in the gonadosomatic index of different sexual stages in several species from the eastern Golfo de Batabanó. Gonad stages are described in Table 4.2. Mixed stages (e.g., III-IV, VI-III) are discussed in Section 4.3.2.

4.4 Annual Reproductive Cycle

Species-specific annual cycles of sexual development and spawning are correlated with habitat. The reproductive period of each fish population is determined by complex interactions of various factors, some of which might occur simultaneously. Seasonal changes of photoperiod (generally coinciding with temperature variation) and lunar cycles are among the most important factors if the organism has attained appropriate physiological conditions (see Section 7.3). In temperate and polar fishes, the spawning period is short. During the spawning season, fish must not only grow and restore fat reserves for the winter, but must also ensure the development of gonads either before or after the winter. In such fishes, summer is the season of intensive feeding. Nutritive substances (mostly yolk) accumulate in gametes in the fall. Individuals spend the winter with mature gonads and typically spawn during the spring, depending on environmental conditions. Each individual completes spawning in a

few hours and the whole population spawns over a period generally of not more than 1 or 2 months.

Tropical marine fishes do not require large fat reserves for the winter because food supplies fluctuate little seasonally (see Section 5.3.1). Thus, energy requirements for sexual development can be met gradually over a long period of time. Gonad maturation can occur in tropical fishes during any season, and the reproductive period is generally more prolonged than in temperate fishes with similar gametogenesis. Nevertheless, there is great intraspecies variability, both in duration and timing of the reproductive period.

In most of the species studied in Cuba, the individuals are in Stage II or II_n of maturation during most of the year. Oocytes mature rapidly; typically, vitellogenesis (the transition from Stage II to IV) takes place in a few days, and the final maturation of oocytes happens in a few hours just before spawning. Thus, individuals with oocytes in Phase F (Table 4.3) are uncommon.

In spawning aggregations, we often found individuals in early Stage III with variable GSI values. For example, mutton snapper GSI values ranged from 0.5 to 8.6 in females, and from 0.4 to 6.0 in males. Similar values were recorded among lane, gray, and yellowtail snappers, as well as in bar jack, margate, and others. The individual spawning process is short, even in those species with intermittent spawning and an extended spawning period. For example, Nassau grouper spawning typically occurs within 10 minutes of sunset, and the hydration of vitellogenic oocytes occurs shortly before spawning (Colin 1992). Thus, depending on the type of gametogenesis, release time of all ripe eggs within one spawning event is characteristic for each species. In Cuba, mullet spawning occurs in a few hours (García and Bustamante 1981). In various snappers (Lutjanidae) it extends over 7 to 10 days (Claro 1981c, 1982, 1983c, 1983d). In various grouper (Serranidae) populations, it takes about 7 days (Randall and Brock 1960; Smith 1972; Claro et al. 1990c). In grunts (Haemulidae) and herring (*Harengula humeralis*) it seems to take approximately a month or slightly longer in Cuba.

In most species, we found entirely spent (Stage VI) individuals during the first month of the spawning period, as well as spent individuals with no sign of recent spawning in the middle of the reproductive season. This suggests that the duration of the reproductive season for each species is the result of differing size/age groups. This phenomenon has been summarized for various tropical and subtropical fish species (e.g., Oven 1976, 1985).

In mutton snapper, Claro (1981c) suggested that female spawners were larger in May (55.5 ± 1.5 cm FL) than in June (52.7 ± 1.9 cm). In addition, experienced fishermen in Cuba confirm that mutton snapper spawning in May and June are notably larger (weight, 2–5 kg) than those breeding in July and August (weight, 1.5–3 kg; see Claro 1981c). In the bar and the yellow jack, larger individuals also spawned earlier in the season. For example, in a sample of 86 bar jacks and 79 yellow jacks caught together in March south of Cayo Flamenco, all bar jack individuals larger than 35 cm were ripe, whereas those 28–30 cm (most of the group) were immature. Among the yellow jacks, all of the individuals (38–40 cm) were immature, with the exception of two individuals (> 50 cm) spearfished in the same area that were found with mature gonads.

Similar patterns were reported by De Sylva (1963) for the great barracuda (*Sphyraena barracuda*), by Dubovistky (1974) for the littlehead porgy (*Calamus proridens*), and by Claro (1982, 1983a) for lane and gray snappers. Occurrence of more than one spawning peak during the reproductive period could be related to the differential gametogenesis patterns of fish of different size and age groups.

Fig. 4.7 summarizes basic patterns of the reproductive cycle in various species of the Cuban shelf. The curves were drawn using data on gametogenesis type and the seasonal variation in the proportion of fishes at different stages of

maturity and GSI. The pattern shown by lane snapper is similar to that of gray and mutton snappers. These three snapper species have synchronous vitellogenesis and an intermittent Type D spawning pattern, although they differ from each other in length of reproductive cycle. Lane snapper can appear in Stage III of maturation in early spring (March), but only at the end of April can they be found largely mature and in fair numbers. In the southwestern part of Cuba, mass spawning takes place typically in May. In the northwest region, spawning occurs one month earlier, whereas in the northeast and southeast regions it occurs one month later. These differences could be due to the seasonal pattern of oceanographic conditions and adjustment of the reproductive strategy of each species to the environmental conditions (Section 4.5).

The annual reproductive cycle of the mutton snapper is similar to but slightly shorter than that of the lane snapper. Individuals in Stage II of maturation appear in the southwest region, around the shelf drop-off, starting from April; the spawning peak occurs in May, becoming less intensive in June. In the northwest region, however, mutton snapper spawn massively a month later. Catch data also show that the former type of spawning is applicable to the population of the south coast and the latter type occurs on the north coast. In Cabo Corrientes (southwest coast) and Cabo San Antonio (westernmost end of the main island), the spawning peaks seem to take place in June and fishes that aggregate here come from the north and south coasts. Mature mutton snappers in Stages III and IV of maturation can be found until August, which suggests that individuals can mature and spawn for the first time within one reproductive season from early spring to August. On the other hand, the reproductive cycle of the gray snapper is similar to that of the mutton snapper, but it occurs later (and during the warmest time) in the summer, from June to September, with a peak in July–August.

In the lane snapper, the spawning peak is associated with lunar phases. In the southwest Golfo de Batabanó, spawning occurs typically 5 to 7 days before the fifth full moon of the year, which coincides approximately with late April or early May (usually May). This immense spawning aggregation may be the largest of all commercial reef fishes in Cuba and is the target of an important fishery (Section 8.6.1). We have found mature lane snappers (Stages III and IV) until September (but in decreasing percentages after May). However, Olaechea and Quintana (1975) reported mature lane snappers until November in the western part of the Golfo de Batabanó. During the spawning run, partially spawned fishes (in Stage VI-III or VI-IV) can be found and, several days later, spent individuals (in Stage VI) that disperse after spawning.

In some locations new lane snapper spawners arrive at the spawning area every month, extending the reproductive period by several months (Claro 1982). A similar process has been observed in the mutton, gray, and yellowtail snapper.

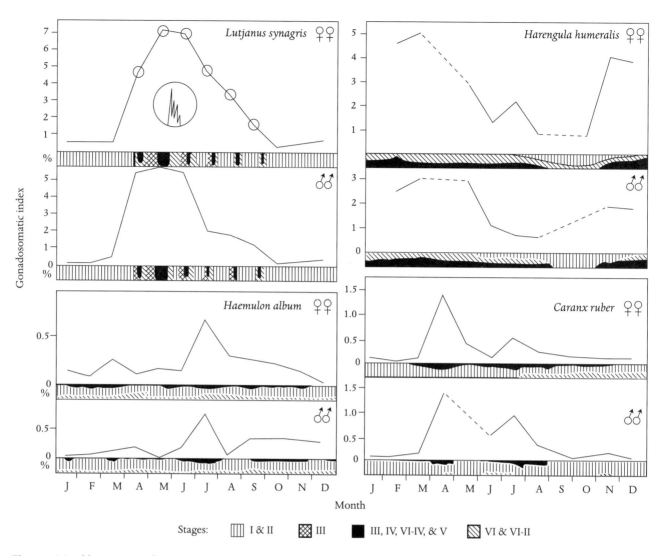

Fig. 4.7. Monthly variation of average gonadosomatic index in spawners (curve) and the proportion of individuals in different stages of maturation (below) for various Cuban coastal fishes. The insert for female lane snapper (*Lutjanus synagris*) represents the fine-scale response of GSI to the concentrated release of four spawning batches within each monthly spawning event.

Data on red grouper in Bermuda (Moe 1969) and Nassau grouper (Smith 1972; Sadovy and Eklund 1999) suggest that these groupers have a reproductive cycle like that of snappers, but with a winter-spring breeding season. In northeastern Cuba, Nassau groupers once formed large aggregations in December and January, mainly in the Banco de Jagua, south of Cienfuegos (south-central coast of Cuba).

The deep-water snappers—silk and blackfin snappers—are also intermittent spawners, but have asynchronous vitellogenesis (Pozo and Espinosa 1983; Pozo et al. 1984) (Type A spawners) and a longer reproductive cycle than the snappers mentioned above. In the oceanic banks of Jamaica, Munro et al. (1973) reported mature individuals of both species from January to November. Thus, Munro et al. suggest that those species spawn year-round, but more intensively in September and October (both species), March (silk snapper), and April (blackfin snapper).

Jacks (bar and yellow jack), also Type A spawners, have a moderately long reproductive period. We have found mature individuals of bar jack starting from March, and until August, with two marked peaks in spawner abundance in April and July (Fig. 4.7). The yellowtail jack apparently has three periods of maximum breeding intensity: April, July, and November–December. However, these data are limited and should be confirmed.

Grunts (Haemulidae), Type B spawners, have a more extended reproductive cycle than jacks. We have found margate with mature gonads almost year-round, although in larger numbers in March and July (Fig. 4.7). In addition, spent individuals were found throughout the year. However, in April and July–August, a little after the spawning peaks, the number of spent individuals was higher. This shows that despite spawning that might be intermittent or continuous, each individual spawns all ovules in a short time. The period of maximum abundance of margate spawners from the eastern Golfo de Batabanó differed from that reported by Billings and Munro (1974) in southern Jamaica. These authors found spawning mainly in January–April, with a second lower peak in September–October. They also confirmed that spawning is not synchronous in all locations and that mature fishes can be found year-round. We found mature bluestriped and white (*H. plumieri*) grunt almost year-round, but spawning seems to be more intensive in winter. The highest GSI values were obtained for these species in November and January, respectively.

Dwarf herring (*Jenkinsia lamprotaenia*) aggregate to spawn monthly during every full moon of the year (Coblentz 1995). We found mature and spent individuals in all months. During studies of the redear sardine, we found mature individuals (Stages IV and VI-III) from November to August, but more abundantly in February–May, which is similar to the seasonal pattern found by Mester et al. (1974) on the La Habana coast. Starting from December, spent individuals (gonad stage VI) appear in the surveys; this might suggest that, as in grunts, the redear sardine releases all egg batches in about one month.

The information presented here shows that spawning periods of tropical fish populations are typically long, but the reproductive period of each individual is relatively short, even in species that release numerous egg batches. The pattern of fat accumulation in some species (see Chapter 7) suggests that the process of reserve storage for maturation and spawning can happen more than once a year. Reserves might also be stored in small, mostly territorial reef fishes such as pomacentrids, labrids, and scarids that spawn throughout the year (Roede 1972; Robertson and Warner 1978; Warner and Robertson 1978).

Reproductive cycles might be related to habitat usage patterns. Species that inhabit coastal lagoons or nearshore areas (and perform long spawning migrations to the shelf edge), such as the lane, mutton, and gray snappers, and mullets, typically have a shorter (but more concentrated) reproductive period than fish species that live closer to the fringing or barrier reefs along the shelf edge (yellowtail snapper, jacks, grunts, and others) where environmental conditions are more stable.

Fig. 4.8 gives estimates of the number of spawning species per family per month. A considerable number of species spawn year-round. There is a marked peak in March–April, and a notable decrease in October–November. Evidently, small geographic differences in environmental conditions can create a difference in spawning periods. Thus, most studied fishes in Jamaica breed during the coldest period of the year (January–April; Munro et al. 1973). Munro et al. estimated the amount of planktonic fish eggs based on the biomass of the major species and the monthly rate of mature individuals and fecundity. They found that February–March was the period of peak egg

production, and that in February–March, egg production was 12 times the production of July–December. Fig. 4.8 shows that on the Cuban shelf 50% of the species examined reproduce between March and August; most of those species are most abundant in the shallow-water areas. However, in both locations (Jamaica and Cuba) spawning occurs when the temperature is 26–27°C. Another notable species is gray snapper, which has its maximum spawning intensity in Florida (Starck 1970), Cuba (Báez et al. 1982b; Claro 1983c), and Venezuela (Guerra and Bashirullah 1975) in July–August, when both photoperiod and water temperature reach a maximum. Similar associations between reproduction, temperature, and lunar phase were also noted in natural and captive Nassau grouper (Tucker et al. 1993, 1996; Sadovy and Eklund 1999).

Despite the occurrence of spawning at any time of the year, there is usually a clear sequence of spawning periods, particularly for the most abundant species and for those with similar ovulation patterns. Spawning peaks are in May, June, July–August in lane, mutton, and gray snapper, respectively, whereas yellowtail snapper spawning peaks in April and less intensively in September. Importantly, the peak spawning in all of these species does not overlap temporally (Fig. 4.9).

The margate appears to spawn more intensively when the bluestriped grunt and white grunt do not. Similar sequences in spawning can be found in other families such as mullets, large groupers, and some jacks. Even for some species whose spawning periods coincide, there are differences in the moon phases. Lane snapper and mutton snapper both spawn in May and June, but generally lane snappers spawn between the quarter and full moon and mutton snappers spawn after the full moon.

4.5 Fecundity

Fish fecundity is closely related to environmental factors and individual physiological conditions. Relative fecundity (number of ovules in ripe ovaries per unit of fish body weight) can reflect, better than any other indicator, the physiological status of the organism (Nikolsky 1974a). All factors combined determine the reproductive rate of the species or population. The concept of fecundity as species-characteristic (Zotin 1961) is inadequate. Fecundity can differ across populations within the same species, or within the population, depending on habitat conditions and human activities. Independently, absolute and relative fecundity do not characterize the reproductive capacity of the fish population over a certain period of time because these parameters depend on the age of sexual maturity and the spawning frequency throughout the individual's lifetime (Nikolsky 1963).

Estimating the fecundity of tropical fishes is complex and can be inaccurate in many cases because detailing intermittent spawning requires comprehensive sampling efforts. Typically,

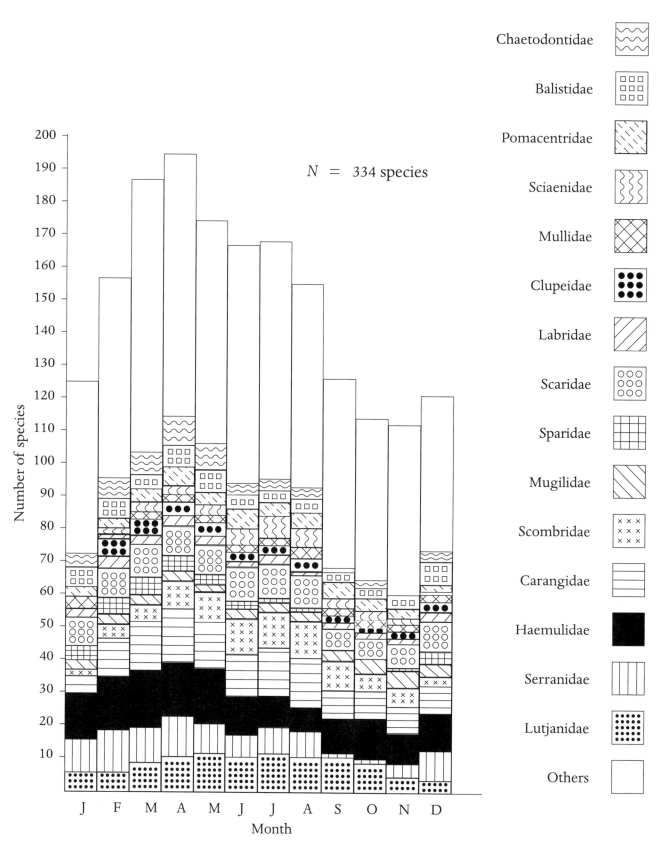

Fig. 4.8. Spawning periods of marine fishes inhabiting the Cuban shelf.

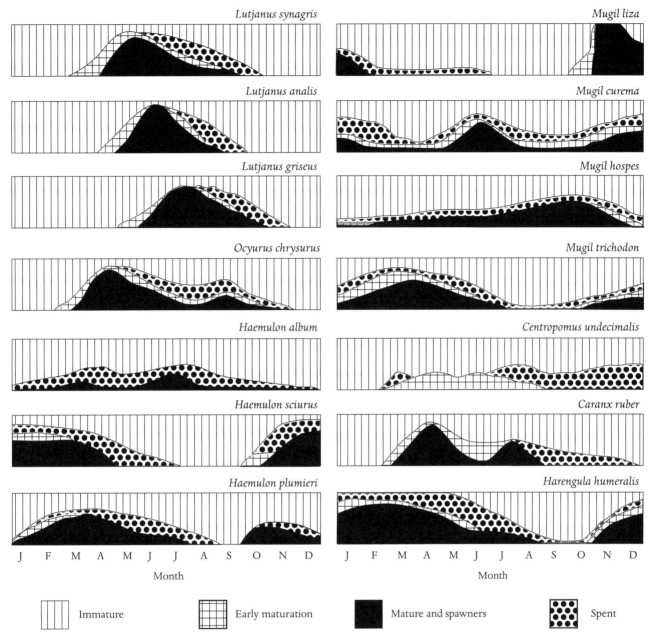

Fig. 4.9. Temporal patterns of spawning in selected species of the Cuban shelf. The *y*-axis represents the proportion of individuals in the indicated stage (modified from García-Cagide and Claro 1985).

the calculation of fecundity involves counting only the yolk-bearing oocytes of the ovaries. This is valid for total spawners and for some species such as snappers and groupers that have discontinuous ovulation but synchronous gametogenesis (Type D spawning). However, in species with continuous vitellogenesis, the oocytes still in the protoplasmic phase (the reserve oocytes) will mature and be spawned within the current reproductive period (Oven 1976). In this case, and in those involving intermittent spawners, the investigator can underestimate the actual individual fecundity for one entire reproductive cycle. Appendix 4.4 summarizes individual and relative fecundity of some fishes inhabiting the Cuban shelf.

To obtain a more accurate value of fecundity, histological sections of the ovaries should be examined to confirm that the fish has not yet spawned. The number of egg batches to be released can also be calculated using the "batch rate" (Oven 1976). This indicator provides information on the percentage of mature oocytes (in Phase F and hydrated; that is, the number of oocytes in one batch), and the total number of vitellogenous oocytes within the ovaries in Stage IV with no marks of recent spawning (assuming that all batches have a similar number of oocytes).

The batch rate was calculated for several individuals of redear sardine and hogfish in gonad Stages IV-V and V

Table 4.5. Fecundity of fishes in gonad Stage IV of several species with asynchronous vitellogenesis and multibatch spawning (Type B)

Species	n	Mean Length (cm)	Mean Weight (g)	Individual Fecundity Mean (Range)	Mean GSI (%)	Relative Fecundity[a] (Mean)	No. Batches Spawned/year Mean (Range)	Oocytes in One Batch (Mean)
Harengula humeralis	16	14.4	60.3	103,791 (41–208 × 10³)	8.06	1,722	20 (15–30)	7,000
Caranx ruber	3	37.5	938.3	926,416 (8–10 × 10⁵)	4.28	1,008	—	—
Caranx bartholomaei	2	73.9	6,760	(71–77 × 10⁵)	2.60	1,118	—	—
Haemulon album	5	54.6	3,463	1,447,067 (8–22 × 10⁵)	5.15	415	—	—
Haemulon plumieri	5	24.1	293	163,455 (64–312 × 10³)	3.06	537	—	—
Haemulon sciurus	10	21.8	3238	153,717 (47–250 × 10³)	4.65	678	—	—
Haemulon aurolineatum	5	13.1	51.9	54,800 (29–81 × 10³)	7.45	1,075	—	—
*Lachnolaimus maximus**	11	39.1	1,222	324,364 (10–83 × 10⁴)	2.43	257	6 (4–9)	39,000
Jenkinsia lamprotaenia	92	3.6	0.458	—	7.5	907	—	378

GSI: gonadosomatic index.
[a] Number of oocytes per gram of the individual fish.
* Specimens with recent spawning marks in the ovaries.

(García-Cagide et al. 1994). The number of Phase F oocytes and hydrated ovules (which constitute the spawning batch) of four redear sardine individuals ranged from 4,300 to 11,000 (mean, 7,000), whereas the number of vitellogenous oocytes in Stage IV individuals ranged from 103,600 to 208,900 (mean, 144,000). These data allowed us to estimate about 20 batches, each batch involving 5% of the total mature eggs. Taking into account that each redear sardine female can continuously generate oocytes (from the ovary reserve), the total number of batches could be even higher, and the spawning period could last more than 20 days, if one oocyte batch is produced every day. For the hogfish we calculated about 39,000 oocytes per batch (Table 4.5). Nevertheless, all individuals in Stage IV had marks of recent spawning; therefore, the batch rate may have been underestimated (García-Cagide et al. 1994). This species could spawn more than 10 batches in each reproductive cycle, according to the number of eggs of the individuals with a higher GSI.

Hunter and Goldberg (1980) developed a method of evaluating fecundity based on counting the number of hydrated oocytes in the ovaries. This method accounts for only partial fecundity (i.e., the number of oocytes per batch) of fishes with multibatch or recurrent spawning, and has been used in several *Engraulis* species (Hunter and Macewicz 1980; Alheit et al. 1983; Hunter et al. 1984): the Peruvian sardine (*Sardinops sagax;* Alegre and Alheit 1986; Lo et al. 1986), and the chub mackerel (*Scomber japonicus;* Peña et al. 1986). In some cases, the spawning periodicity was estimated by counting the empty follicles and mature oocytes in different stages of resorption in histological sections of ovaries. Using this method, we estimated the batch fecundity of 92 dwarf herrings. The small size of the ovaries allowed us to count all the oocytes in the spawning batch (those larger than 400 μm); these averaged 378 oocytes (standard deviation, 168) and relative fecundity was estimated at 907 eggs per gram of body weight (Table 4.5).

Appendix 4.4 summarizes fecundity data for some fishes found in Cuba. Despite differences in methodology, we noticed the following patterns: (1) pelagic oceanic species, such as tuna, have lower relative fecundity than species inhabiting shelf waters; (2) some fishes that usually inhabit offshore reefs, such as butterflyfishes and angelfishes, exhibit a slightly lower fecundity than grunts and groupers, which are also found on inshore reefs; and (3) snappers, clupeids, and mullets have even higher fecundity values. This gradient might be associated with predation rates on early stages during development and recruitment.

4.6   Reproductive Strategy

The reproductive strategy of most temperate and polar fishes involves spawning in the right place and time for the larvae to get the greatest amount of food (Qasim 1955; Nikolsky 1963; Lasker 1970; Cushing 1972). However, this might not be a critical factor in tropical fishes where plankton density is generally low and lacks a marked seasonality. Intraspecific variability in social systems, the role of adult fishes in the timing of spawning, and the spatial and temporal patterns necessary for optimal recruitment (Doherty 1991; Robertson 1991; Shapiro 1991; Warner 1991; Sadovy 1996) account for many observed reproductive patterns of tropical reef fishes.

Johannes (1978) suggests that reef fishes time and locate spawning so that eggs can avoid the great predation pressure of reef zones. Based on this premise, Johannes argues that predation is the determining adaptive element in the reproductive strategy of tropical fishes. Although these generalizations might apply to many shallow-water species, it seems contradictory that many open-sea species place their offspring in inner shelf waters. Other authors have a different view. For example, Barlow (1981) states that egg and larvae dispersal is a critical component of reef fish reproductive behavior, whereas antipredation is a secondary factor. He affirms that dispersal reduces mortality, which depends on larval density and competition, and that dispersal temporally and spatially spreads mortality risk across the new generation. However, Barlow (1981) states that dispersal also takes the progeny away from the environment to which the parents are adapted.

Shapiro et al. (1988) questioned these hypotheses, analyzing the great diversity of environmental factors that determine large-scale variability of circulation patterns and their effects on the transport of fish eggs and larvae. These authors list many species whose reproductive behaviors do not fit these hypotheses. In many cases, information on the relationships of oceanographic processes and reproductive strategies is inadequate.

Johannes (1978) analyzed the reproduction of relatively large species with high fecundity and wide distribution and mobility, most of which have fisheries significance. Barlow (1981) emphasized small reef species whose limited fecundity and mobility make them more spatially restricted.

There is no reason to expect only one reproductive strategy for all tropical fishes, given their great ecological variability, and no strategy can be considered applicable to all members of a species or population. García-Cagide et al. (1994) described the reproductive strategies of some common reef fishes, which are summarized below in two general groups.

4.6.1   Spawning at the Shelf Edge

The formation of spawning aggregations at a particular site and time is a common reproductive strategy for some reef fishes (Shapiro et al. 1993a; Carter and Perrine 1994; Aguilar-Perera and Aguilar-Dávila 1996; Koenig et al. 1996; Domeier and Colin 1997; Bolden 2000; Eklund et al. 2000). Many species of fisheries significance (snappers, grunts, groupers, jacks, mullets, and others) perform spawning migrations from the inner shelf waters to areas close to the shelf drop-off

or offshore banks. Based largely on information from commercial fishermen, Lindeman et al. (2000) identified potential spawning aggregation sites for eight snapper species at 22 locations in the lower Florida Keys. Using comparable information in Cuba, we have identified a variety of potential aggregation sites for lane and mutton snapper. The following spawning sites are among the most well-known and exploited by fishermen: Corona de San Carlos (north of Pinar del Río), Cabo Corrientes, Cabo San Antonio, Peninsula de Hicacos, south of Cayo Avalos, and Puntalón de Cayo Guano. In most places, the shape of the shelf is similar to that shown in Fig. 4.10, which depicts the migratory process of mutton snapper spawners on part of the northwest coast.

Mutton snappers aggregate mainly on certain reef promontories at the shelf edge. During May and June, a large number of ripe individuals gather close to the shelf edge on the rocky coralline bottom in depths of 20 to 40 m. Currents here are strong, generally parallel to the coast, and flow according to the tidal cycle. Spawning occurs during the tidal shift when water movement is minimal. During spawning, snappers feed sparingly and cannot be caught by baited fishing gear. Tidal and wind-driven currents may disperse eggs and larvae offshore. Prevailing winds and tidal movements might favor further recruitment of juveniles along a large strip of the shelf.

Corona de San Carlos is the spawning ground of different species throughout the year: mutton snapper in May and June; gray and cubera snappers in July and August; Nassau grouper in December and January; red hind in February and March; and other less abundant groupers in winter and spring. In Cabo Corrientes (Fig. 1.3), a known spawning ground for mutton snapper and other species, the current systems may also favor the return of larvae to the shelf in some instances. Some larvae may be transported to the northwest coast; experienced fishermen suggest that adults from the Archipiélago Los Canarreos may migrate to Cabo Corrientes to spawn.

The Golfo de Cazones (Fig. 4.11) is known as a spawning area for some fishes inhabiting the Golfo de Batabanó, especially lane snapper, which is the dominant commercial species in this region. The water mass flows westerly from the Golfo de Cazones at an approximate velocity of 4.5 nautical miles per day (see Section 1.4). This may help direct fish eastward to aggregate in the narrow shelf edge running from Cayo Diego Pérez to the edge of the Golfo de Cazones. These large aggregations typically gather at dusk during the days before the full moon in May and move toward the shelf border where spawning takes place. After releasing each batch of eggs, the fish move back to the shelf where large predators (sharks, barracudas, groupers) are scarce. These nocturnal visits to the spawning ground by each individual can be repeated for a few days until all the ripe eggs are released.

Ichthyoplankton tows conducted during the lane snapper spawning peak in May (Claro 1982) showed that egg densities were higher in the shelf slope waters (128 eggs/minute) than in the inner part of the shelf (55) and the oceanic waters (4) (see García-Cagide et al. 1994). However, larval density was higher in the ocean (6.5 larvae/minute) than on the drop-off (3.8) and shelf (0.9) waters (García-Cagide et al. 1994). Eggs spawned at the shelf edge are transported away, ideally to the ocean, although those spawned during full tide or unfavorable wind conditions can be advected onto the shelf. Embryonic and larval development might occur in oceanic waters close to the shelf drop-off so that postlarvae can move back from the ocean to shallow waters (Fig. 4.11).

In our plankton surveys (Claro 1982), we tentatively identified late planktonic stages of three snapper species: one mutton snapper of 20 mm FL, two lane snappers of 13 and 15 mm, and one dog snapper of 15 mm. These fishes were probably from eggs that were fertilized in the previous lunar cycle spawning, so they should be about one month old. Our underwater visual censuses and demersal collections (with a seine designed for small individuals) in shallow areas near the Golfo de Cazones, yielded only snappers larger than 3 cm.

The bathymetry of the Golfo de Cazones might contribute to the formation of an anticyclonic gyre with a net movement to the north that retains larvae for at least several weeks. This might explain why juveniles of various species were always more abundant in the shallow waters at the north of the gulf than in the area closer to the spawning grounds (Fig. 4.11). Other recirculation systems off southern Cuba might also be present that contribute to larval retention (Lindeman et al. 2001).

A reproductive strategy similar to that seen in lane snapper also seems to occur for mutton (Claro 1981c), gray (Claro 1983c), and yellowtail (Claro 1983d) snappers, as well as bar jack (García-Cagide 1985) and margate (García-Cagide 1986a) in the eastern Golfo de Batabanó, although spawning aggregations are less notable in all of these species. In the southeastern part of the shelf, these species perform similar spawning migrations to the east and then aggregate near the shelf slope between Cabo Cruz and Cayo Cabeza del Este.

South of Bajo La Vela at the southwest boundary of the Golfo de Ana María is a promontory traditionally known by fishermen as a spawning area for mutton, cubera, and dog snappers and groupers, and other species. In July 1986 we conducted a dive in that area during a spawning aggregation of hundreds of cubera (*Lutjanus cyanopterus*) and dog snappers. On the shelf border drop-off, on a reef at 20–30 m, mixed-species schools (all larger than 50 cm FL) swam in a circle. These schools were seen in the same place for 9 to 10 days. The few individuals speared in the morning were in gonad Stage IV, not yet ready for spawning. Ovulation and spawning might occur at night. Unfortunately, little information is available about this aggregation. Data in Victoria del Río and Penié (1998) on regional circulation systems suggest that spawning products might recruit to the southwestern Cuban shelf, or, because of the many eddy systems recently found off southern Cuba (Fig. 1.7), there may also be some local recruitment on the southeastern shelf.

Cuban fishermen know other snapper spawning sites whose oceanographic characteristics may be similar–Cayos San Felipe, Los Indios, Cantiles, and Avalos, all in the southwest region. However, many other reef fishes do not aggregate to spawn. A number of eggs and larvae can also be advected inshore by strong currents on the shelf, especially through areas with no emergent reefs, or through channels among islets and keys.

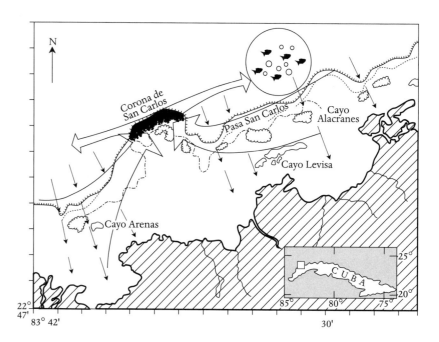

Fig. 4.10. Schematic of mutton snapper (*Lutjanus analis*) migrations to form spawning aggregations at Corona San Carlos, northwestern Cuba (from Claro 1981). Wide arrows: spawning migrations; medium arrows: dispersal of eggs and larvae; fine arrows: recruitment to shelf.

4.6.2 Spawning on the Inner Shelf

Some small species with pelagic eggs spawn in the inner waters of the shelf. Among them, clupeids, engraulids, atherinids, gobiids, labrisomids, and gerreids are the most common. They spawn far from the reefs, in areas where planktivores are few; for example, eggs and larvae are not frequently found in the stomachs of sardines (*Harengula* spp.), the most abundant planktivorous fish group on the Cuban shelf (see Section 5.3.1). These small spawners are usually under high predation pressure and their large numbers may be a selective advantage. For small, nonschooling species, such as gobiids and labrisomids, spawning on the outer shelf, where planktivores are more common, would increase their vulnerability to predators.

For these species a critical factor might be the vulnerability of their young to environmental conditions (particularly temperature, salinity, and pollution) in near-shore areas. Both high predation pressure on adults and unfavorable conditions for offspring must be compensated for with a higher reproductive rate. For the redear sardine, that pressure results in a higher fecundity and a larger number of egg batches released over a longer period of time.

4.6.3 Other Spawning Strategies

Many territorial and generally small reef species do not perform spawning migrations. Among these, the most common are species of Scaridae, Labridae (Warner and Robertson 1978; Appeldoorn et al. 1994), Mullidae (Winn and Bardach 1960; Randall and Randall 1963), Acanthuridae (Randall 1961), Pomacanthidae, Chaetodontidae, and small serranids (Barlow 1975). More details on these and other families are provided in Thresher (1984).

Some species have demersal eggs; examples are discussed in Thresher (1984) and Robertson (1991). Many species must protect their nests against numerous predators (Myrberg et al. 1967; Pressley 1980; Thresher 1984). Defense can be passive, such as the parental care of eggs inside the mouth (e.g., Ariidae, Apogonidae, and Opisthognathidae), pouch (Syngnathidae), lateral body surface (Antennariidae; see Pietsch and Grobecker 1980), or in coral, shells, and sponge crevices (Pomacentridae). The eggs of such species are a similar size or even smaller (0.5–0.6 mm) than those of pelagic fishes (Myrberg et al. 1967; Pressley 1980), despite their longer embryonic development (5–6 days, in yellowtail damselfish, *Microspathodon chrysurus*; Pressley 1980). Nests of

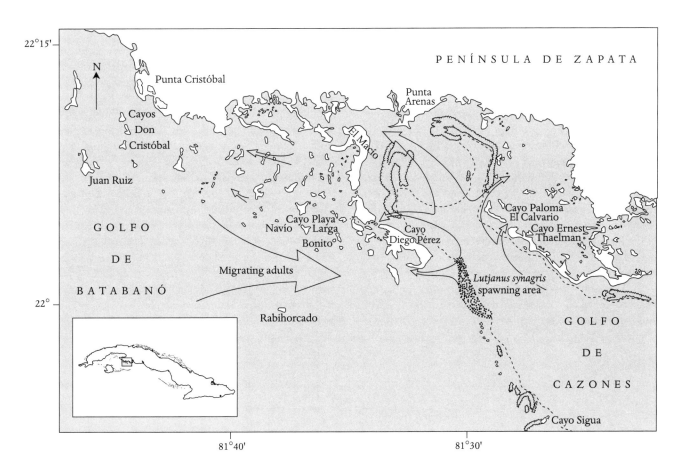

Fig. 4.11. Potential larval recruitment pathways to shallow-water areas of the eastern Golfo de Batabanó. The stippled area east-southeast of Cayo Diego Pérez shows a primary spawning area of lane snapper (*Lutjanus synagris*) and other coastal fishes (modified from García-Arteaga et al. 1990).

these species contained 19,000 eggs (mean, 244–256/cm²), but few data are available on the number of nests produced by each female during the reproductive cycle (Pressley 1980).

4.7   Summary

Although most Cuban marine fish species are gonochoristic, several important families are hermaphroditic, and protogyny has been recorded in 39 species. Sexual dimorphism is largely chromatic and associated with protogynous transitions in the Scaridae and Labridae. Sex ratios among hermaphroditic species can be related to habitat conditions, growth, fisheries exploitation, and reproductive seasonality. In heavily exploited species, overfishing of large males might affect sex ratios and reproductive output with negative effects that cascade through successive generations. Among many gonochoristic species, sex ratios are variable. However, in some species (e.g., snappers), sex ratios are nearer 1:1. Habitat-related differences in the distributions of sexes have often been recorded, but the ratios tend to equalize during the reproductive period.

Vitellogenesis is usually of short duration, starting 1 to 3 weeks before spawning. Many species show asynchronous gametogenesis and can release multiple batches, some within one month (Type A spawners) and some over several months (Type B), although the process of vitellogenesis may vary. Some species with synchronous vitellogenesis (*Mugil* spp.) have a discrete, or total, spawning event (Type C), but other species (some lutjanids and serranids) show synchronous vitellogenesis and batch spawning (Type D), with the release of 3 to 5 batches during certain phases of the lunar cycle.

The duration of the reproductive period for a species is the result of the spawning of different groups of individuals at different times. At the individual level, Type C spawners release their eggs in a few hours and Type D spawners, over 7 to 10 days. Type A and Type B spawners can release eggs for more than 10 days, in some instances, for more than a month (Type B). Each year the larger individuals (or those attaining a certain level of nutritional reserves) form the primary spawning stock.

Although spawners of most coastal fishes can be found year-round, many spawning peaks occur from April to August, with a second, smaller peak in the winter (December–February). This pattern can be specific to the environmental conditions prevalent in the area. Despite the large amount of spawning at any time of the year, discernible patterns of reproductive timing seem to occur (at least for the most abundant species with similar reproductive strategies). Even within a given period, different species spawn in different phases of the lunar cycle. Water temperature (or associated factors) greatly influences some species: spawning periods might occur at a certain monthly environmental temperature, no matter the geographic location of the population (within the wider Caribbean).

According to GSI estimates (typically lower than 10% in spawners), tropical fishes have lower energetic requirements for generative metabolism. Within species, the maximum GSI generally coincides with the spawning peak and declines during the breeding season. Fishes that spawn late in the season appear to have lower fecundity and/or batch spawning. Sexual differentiation and the start of the protoplasmic growth period in oocytes of most species studied in Cuba generally occurs at an early age. In small species with a short life span, sexual maturity is attained at 1 to 2 years of age (in some cases even earlier). In large species, sexual maturity often occurs at 3 to 4 years of age. Sexual maturity is often attained at a size that is 50–70% of the maximum length for females and 40–60% for males.

The estimation of fecundity in tropical fishes is complex in many cases because spawning is often intermittent. Typically, calculations of fecundity account for only a certain part of the total number of oocytes that will be released throughout an annual reproductive cycle. Relative fecundity (number of oocytes per gram of body weight) is high in Cuban coastal fishes (about 700 oocytes/g). It might also be higher for species inhabiting unstable eutrophic ecosystems (e.g., some estuaries) than for those dwelling in other environments (reefs and open ocean).

In the field, the marine fishes of Cuba display a variety of reproductive strategies. In many species, spawning occurs near promontories on the shelf edge. Juvenile settlement in shallow nursery areas occurs when fishes reach a size that helps them avoid predators. An extended spawning period contributes significantly to a steady influx of new recruits throughout the year. Nevertheless, many abundant species focus their individual reproductive activity over a short period and some also concentrate reproduction spatially using spawning aggregations.

Appendix 4.1. Sizes at sexual differentiation (SD) and maturation (fork length unless otherwise indicated) of some marine fishes inhabiting Cuban waters. Values that are common to both sexes are in the center of M and F columns.

Family / Species	SD (cm)	Min. Size (cm) F	Min. Size (cm) M	Mean Size (cm) F	Mean Size (cm) M	Age (years) F	Age (years) M	Mean Size / L_{max} F	Mean Size / L_{max} M	Region	Source
Elopidae											
Megalops atlanticus	—	13	10	15–17	11–13	—	—	0.70*	0.60*	Brazil	Ferreira de Menezes and dos Santos Pinto 1966
Albulidae											
Elops saurus	20	—	—	—	—	2	—	—	—	Cuba	Carles 1967
Albula vulpes	—	22–36	—	—	—	1	—	—	—	S Florida	Bruger 1974
Clupeidae											
Harengula clupeola	<7	7	—	—	9–10	—	—	0.60	—	SW Cuba	This chapter
Harengula humeralis	6	9–10	8–9	11	10	—	1–2	0.64	0.62	SW Cuba	García-Cagide 1988
Opisthonema oglinum	8 TL	13 TL	—	—	—	—	—	0.64	—	Gulf of Mexico	García-Abad et al. 1998
Jenkinsia lamprotaenia	1.6	2.4	—	2.8	3.0	—	2 mo.	0.64	0.68	SW Cuba	This chapter
Engraulidae											
Anchoa parva	—	—	—	4.8 TL	4.5 TL	—	—	—	—	Colombian Caribbean	Caselles-Osario and Acero 1996
Anchovia clupeoides	—	—	—	16.2 TL	15.1 TL	—	—	—	—	Colombian Caribbean	Caselles-Osario and Acero 1996
Cetengraulis edentulus	—	12 TL	—	14	—	—	—	0.74	—	E Venezuela	Simpson and Griffths 1973
Holocentridae											
Holocentrus adscensionis	—	—	—	—	—	14–15	—	0.54	—	Jamaica	Wyatt 1983
Holocentrus rufus	—	—	—	—	—	13–14	—	0.62	—	Jamaica	Wyatt 1983
Centropomidae											
Centropomus undecimalis	22	42–45	30–33	—	—	—	—	—	—	SE Cuba	Alvarez-Lajonchere et al. 1982
Centropomus undecimalis	—	44	36 TL	—	—	—	—	—	—	Venezuela	Carvajal 1975
Serranidae											
Cephalopholis cruentata	—	—	—	16	—	—	—	0.47	—	Jamaica	Thompson and Munro 1978
Cephalopholis cruentata	—	14	18	16	24	3–4	5–6	0.52	0.78*	Curaçao	Nagelkerken 1979
Cephalopholis fulva	—	—	—	16	—	—	—	0.47	—	Jamaica	Thompson and Munro 1978
Epinephelus guttatus	—	21	26	25	—	—	—	0.48	—	Jamaica	Thompson and Munro 1978
Epinephelus guttatus	—	15 SL	17	20–25	25–27	—	—	—	—	West Indies	Colin et al. 1987
Epinephelus itajara	—	120–135	110–115	—	—	6–7	4–6	—	—	E Gulf of Mexico	Bullock et al. 1992
Epinephelus morio	—	42–45	39–42	>40	>52	>6	>10	—	—	Campeche	Brulé et al. 1991, 1999
Epinephelus morio	12.5	27 SL	42	>40	>52	>6	>10	0.66*	—	E Gulf of Mexico	Moe 1969

Continued on next page

Appendix 4.1. continued

Species									Location	Reference
Epinephelus morio	35	—	40–45	—	—	—	—	—	Campeche	Valdés-Alonso and Padrón 1980
Epinephelus niveatus	47	—	54	—	—	5	—	—	North and South Carolina	Wyanski et al. 2000
Epinephelus striatus	—	—	48	—	—	—	0.53	—	Jamaica	Thompson and Munro 1978
Epinephelus striatus	36	—	—	—	—	—	—	—	Cuba	Claro et al. 1990c
Epinephelus striatus	42 SL	—	—	—	—	—	—	—	Cayman I.	Collins et al. 1987
Epinephelus striatus	50	—	—	—	—	—	—	—	Cayman I.	Tucker et al. 1993
Epinephelus striatus	—	—	58 SL	58	5	5	—	—	Virgin I.	Olsen and LaPlace 1979
Mycteroperca bonaci	57	—	85	100	—	5	—	—	SW Cuba	García-Cagide and García 1996
Mycteroperca bonaci	50.8	—	82.6	—	6–33	5	0.63	—	Florida	Crabtree and Bullock 1998
Mycteroperca microlepis	—	—	60	—	5	5	—	—	SE US	Collins et al. 1987
Mycteroperca microlepis	48	—	69.5	—	4	—	—	—	E Gulf of Mexico	Hood and Schlieder 1992
Mycteroperca microlepis	57	—	—	—	—	—	—	—	NE Gulf of Mexico	Collins et al. 1998
Mycteroperca tigris	34	37	46	55	9–10	6–7	0.74	0.74	SW Cuba	García-Cagide et al. 1999a
Mycteroperca venenosa	51	—	51	—	—	—	0.59	—	Jamaica	Thompson and Munro 1978
Mycteroperca venenosa	58	77	70–80	>85	—	—	—	—	SW Cuba	García-Cagide and García 1996
Carangidae										
Caranx bartholomaei	45	—	23	—	—	—	0.53	—	Jamaica	Thompson and Munro 1983
Caranx bartholomaei	32	30	—	—	16	—	—	—	SW Cuba	Sierra et al. 1986
Caranx crysos	66	—	28	26	—	—	0.38	0.36	Jamaica	Thompson and Munro 1983
Caranx hippos	34	—	55	—	—	—	—	—	Jamaica	Thompson and Munro 1983
Caranx latus	38	37	—	42	—	—	0.53	0.60	Jamaica	Thompson and Munro 1983
Caranx lugubris	23	—	—	42	—	—	0.46	—	Jamaica	Thompson and Munro 1983
Caranx ruber	26	—	42	—	16	—	0.46	0.46	Jamaica	Thompson and Munro 1983
Caranx ruber	25	22	—	20–30	12–13	3	0.68	0.68	SW Cuba	García-Cagide 1985
Trachinotus carolinus	—	—	23	—	—	—	—	—	Florida	Moe et al. 1968
Coryphaenidae										
Coryphaena equiselis	22 SL	—	—	—	—	—	—	—		Scherbachev 1973
Coryphaena hippurus	40 SL	—	—	—	—	—	—	—	Florida	Beardsley 1967
Lutjanidae										
Apsilus dentatus	18–20	—	—	44	—	—	0.71	0.76	Jamaica	Thompson and Munro 1983
Lutjanus analis	41	38	52	50	5–6	—	0.65*	—	SW & NW Cuba	Claro 1981c
Lutjanus apodus	25	—	—	—	—	—	0.44	—	Jamaica	Thompson and Munro 1983
Lutjanus buccanella	21–25	—	26–30	31	—	—	0.63*	0.66*	Cuba	Pozo et al. 1984
Lutjanus buccanella	26	—	—	—	—	—	0.48	—	Jamaica	Thompson and Munro 1983
Lutjanus griseus	23	—	22	—	2	—	0.48*	0.46*	Florida	Starck 1970
Lutjanus griseus	18	28	—	26	2	—	0.53*	0.61*	SW Cuba	Claro 1983a
Lutjanus griseus	14	—	45 TL	—	—	—	—	—	Venezuela	Guerra and Bashirullah 1975
Lutjanus jocu	32	—	—	—	—	—	0.45	—	Jamaica	Thompson and Munro 1983

Continued on next page

Appendix 4.1. continued

Taxon										Location	Reference
Lutjanus jocu	30	43	48	—	51	5–6	—	0.71	0.66	SW Cuba	García-Cagide et al. 1999b
Lutjanus synagris	—	18	—	—	—	—	—	—	—	Jamaica	Thompson and Munro 1983
Lutjanus synagris	15	20	—	—	—	—	—	—	—	Cuba	Rodríguez Pino 1962
Lutjanus synagris	12–13	—	19	—	18	2	—	0.53*	0.50*	SW Cuba	Claro 1982
Lutjanus vivanus	—	—	52	57	—	5	—	0.72	0.84	Jamaica	Thompson and Munro 1983
Lutjanus vivanus	—	31–35	46–50	48–53	—	5	—	0.70*	—	SE Cuba	Pozo and Espinosa 1983
Lutjanus campechanus	—	—	30–32	—	—	2	—	0.44*	0.47*	W Florida	Futch and Bruger 1976
Ocyurus chrysurus	9	—	13–14	—	—	2	—	—	—	Cuba	Piedra Castañeda 1965
Ocyurus chrysurus	16	17	25	24	—	2	—	0.49*	0.47*	SW & NW Cuba	Claro 1983d
Pristipomoides macrophthalmus	—	18	—	—	—	—	—	0.51	—	Jamaica	Thompson and Munro 1983
Rhomboplites aurorubens	—	31	—	—	—	—	—	—	—	Jamaica	Thompson and Munro 1983
Gerreidae											
Eugerres brasilianus	—	14	12	15	13	1	—	—	—	Cuba	Millares et al. 1979
Eugerres plumieri	—	18	18	22	—	—	—	—	—	Colombia	Rubio 1975
Eugerres plumieri	—	18	—	—	—	—	—	—	—	Cuba	Millares et al. 1979b
Gerres cinereus	—	16	19	20	17	1–2	—	—	—	Cuba	Báez and Álvarez-Lajonchere 1983
Haemulidae											
Haemulon album	—	30–32	26–28	36–38	34–36	3–4	—	0.58	0.56	SW Cuba	García-Cagide 1986a
Haemulon album	—	—	24	18	—	—	—	0.37	—	Jamaica	Billings and Munro 1974
Haemulon aurolineatum	—	—	13	13	—	—	—	0.57	—	Jamaica	Billings and Munro 1974
Haemulon flavolineatum	—	—	16	16	—	—	—	0.64	—	Jamaica	Billings and Munro 1974
Haemulon melanurum	—	—	19	19	—	—	—	0.63	—	Jamaica	Billings and Munro 1974
Haemulon plumieri	—	14	22	20	—	—	—	—	0.52	Jamaica	Billings and Munro 1974
Haemulon plumieri	9	11	12	16	18	2	2	0.66	0.75	SW Cuba	García-Cagide 1987
Haemulon plumieri	—	—	21	16	—	—	2	0.74*	0.50*	Puerto Rico	Román-Cordero 1991
Haemulon sciurus	10	14–16	17	—	—	2	2	0.50	0.61	SW Cuba	García-Cagide 1986b
Haemulon sciurus	—	—	20	—	—	—	—	0.50	—	Jamaica	Billings and Munro 1974
Sparidae											
Calamus bajonado	—	30	—	—	—	—	—	—	—	Cuba	Liubimova and Capote 1971
Calamus bajonado	—	17	—	—	—	—	3	—	—	Campeche	Dubovitsky 1973
Mullidae											
Mulloidichthys martinicus	—	—	16	—	—	—	—	0.64	—	Jamaica	Munro 1976
Pseudupeneus maculatus	—	—	18	—	—	—	—	0.67	—	Jamaica	Munro 1976
Pomacanthidae											
Holacanthus ciliaris	—	—	13	—	—	—	—	0.33	—	Jamaica	Aiken 1983a
Holacanthus tricolor	—	—	10	—	—	—	—	0.29	—	Jamaica	Aiken 1983a
Pomacanthus arcuatus	—	—	13	—	—	—	—	0.22	—	Jamaica	Aiken 1983a
Scaridae											
Scarus coeruleus	30.5	—	—	—	—	—	—	—	—	Jamaica	Reeson 1983a
Sparisoma aurofrenatum	15	—	—	—	—	—	—	—	—	Jamaica	Reeson 1983a
Sparisoma rubripinne	16	27	—	—	—	—	—	—	—	Jamaica	Reeson 1983a

Continued on next page

Appendix 4.1. continued

Species									Locality	Source
Sparisoma viride	18	—	—	—	—	—	—	—	Jamaica	Reeson 1983a
Mugilidae										
Mugil curema	14	20–23	23	25	2	2	0.57*	0.67*	Cuba	Alvarez-Lajonchere 1976
Mugil hospes	16	15	19	17	2	2	0.67*	0.70*	Cuba	Alvarez-Lajonchere 1981b
Mugil liza	22–24	40–41	31–33	—	—	—	—	—	Cuba	Alvarez-Lajonchere 1979
Mugil trichodon	12	13	14	16	2	2	0.57*	0.68*	Cuba	Alvarez-Lajonchere 1980c
Sphyraenidae										
Sphyraena barracuda	58	46	66	50	—	—	0.47*	0.35*	Florida	De Sylva 1963
Acanthuridae										
Acanthurus bahianus	9	11	15–16	—	—	—	—	0.68*	Jamaica	Reeson 1983b
Acanthurus chirurgus	17	14–15	—	—	—	—	—	—	Jamaica	Carles 1971
Acanthurus coeruleus	13	11	—	—	—	—	—	—	Jamaica	Reeson 1983b
Trichiuridae										
Trichiurus lepturus	44	71	75	—	—	—	—	—	S Cuba	Ros and Pérez 1978
Scombridae										
Katsuwonus pelamis	33	—	—	—	—	3	—	—	Cuba	Carles 1971
Scomberomorus maculatus	25	28	35	—	—	—	—	—	S Florida	M. E. León and Guardiola pers. comm.
Scomberomorus maculatus	40	35	50	43	—	3	—	—	Florida	Powell 1975
Scomberomorus maculatus	33	—	—	—	2	—	—	0.47*	México	Mendoza 1968
Scomberomorus regalis	43	—	—	—	—	—	0.61*	—	SE Cuba	M. E. León and Guardiola pers. comm.
Scomberomorus cavalla	50	—	—	—	—	—	0.41*	—	SE Cuba	Carles 1971
Thunnus atlanticus	50	60	—	—	—	—	0.75	—	NE Brazil	Monte 1964
Thunnus atlanticus	35	—	—	—	—	3	—	—	Cuba	Carles 1971
Istiophoridae										
Istiophorus platypterus	44	115	121–146	—	3–4	3	0.71*	—	S Florida	Jolley 1977
Balistidae										
Balistes vetula	16	17	23	26	—	—	0.56	—	Jamaica	Aiken 1983b
Ostracidae										
Acanthostracion quadricornis	17 TL	16 TL	22 TL	20 TL	—	—	—	—	Venezuela	Ruiz et al. 1999

L_max: Maximum observed length; F: female; M: male.
* Values calculated in the present study from data in the literature.

Appendix 4.2. Types of vitellogenesis, spawning characteristics, and duration (for populations, not individuals) in some fish species inhabiting the Cuban shelf.

Family Species	Types of Vitellogenesis	Spawning Characteristics	Spawning Duration (months)	Source
Albulidae				
Albula vulpes	Asynchronous	Batch	12	Bruger 1974
Clupeidae				
Harengula clupeola	Continuous asynchronous	Multibatch	—	Mester et al. 1974
Harengula humeralis	Continuous asynchronous	Multibatch	10–12	Mester et al. 1974; García-Cagide, 1988
Jenkinsia lamprotaenia	Continuous asynchronous	Multibatch	12	This chapter
Belonidae				
Tylosurus acus acus	Asynchronous	Batch	—	This chapter
Serranidae				
Cephalopholis cruentata	Synchronous	Batch	—	Nagelkerken 1979*
Epinephelus morio	Synchronous	Batch	3–4	Moe 1969*
Epinephelus striatus	Synchronous	Batch	3–6	Claro et al. 1990c
Mycteroperca bonaci	Asynchronous	Batch	7	García-Cagide and García 1996
Mycteroperca interstitialis	Asynchronous	Batch	—	This chapter
Mycteroperca tigris	Asynchronous	Batch	6	García-Cagide et al. 1999a
Mycteroperca microlepis	Asynchronous	Batch	8–27	Collins et al. 1997
Mycteroperca venenosa	Asynchronous	Batch	6	García-Cagide and García 1996
Rachycentridae				
Rachycentron canadum	Asynchronous	Batch	—	This chapter
Carangidae				
Caranx bartholomaei	Discontinuous asynchronous	Batch	5–6	Sierra et al. 1986
Caranx ruber	Discontinuous asynchronous	Batch	5–6	García-Cagide 1985
Trachinotus falcatus	Asynchronous	Batch	—	This chapter
Coryphaenidae				
Coryphaena equiselis	Asynchronous	Batch	—	Scherbachev 1973
Coryphaena hippurus	Asynchronous	Batch	—	Beardsley 1967
Lutjanidae				
Lutjanus analis	Synchronous	Batch	4–5	This chapter
Lutjanus apodus	Synchronous	Batch	—	This chapter
Lutjanus buccanella	Asynchronous	Batch	6–7	Pozo et al. 1984
Lutjanus cyanopterus	Synchronous	Batch	—	This chapter
Lutjanus griseus	Synchronous	Batch	4–5	Claro 1983c
Lutjanus jocu	Asynchronous	Batch	10–12	García-Cagide et al. 1999b
Lutjanus synagris	Synchronous	Batch	4–5	Claro 1982
Lutjanus vivanus	Asynchronous	Batch		Pozo and Espinosa 1983
Ocyurus chrysurus	Asynchronous	Batch	7–8	Claro 1983c
Gerreidae				
Eugerres plumieri	Asynchronous	Batch	12	Millares et al. 1979b
Haemulidae				
Haemulon album	Continuous asynchronous	Multibatch	12	García-Cagide 1986a
Haemulon aurolineatum	Continuous asynchronous	Multibatch	—	This chapter
Haemulon carbonarium	Continuous asynchronous	Multibatch	—	This chapter
Haemulon parra	Continuous asynchronous	Multibatch	—	This chapter
Haemulon plumieri	Continuous asynchronous	Multibatch	8–9	García-Cagide 1987
Haemulon sciurus	Continuous asynchronous	Multibatch	8	García-Cagide 1986b
Sparidae				
Calamus bajonado	Asynchronous	Batch	4	Hernández-Corujo 1975
Calamus proridens	Asynchronous	Batch	—	Dubovistky 1977
Sciaenidae				
Bairdiella ronchus	Asynchronous	Batch	12	García 1979

Continued on next page

Appendix 4.2. continued

Micropogonias furnieri	Asynchronous	Batch	12	García 1979
Mullidae				
Pseudupeneus maculatus	Asynchronous	Batch	—	This chapter
Labridae				
Halichoeres bivittatus	Asynchronous	Batch	12	Roede 1972
Halichoeres garnoti	Asynchronous	Batch	12	Roede 1972
Halichoeres maculipinna	Asynchronous	Batch	12	Roede 1972
Halichoeres poeyi	Asynchronous	Batch	12	Roede 1972
Lachnolaimus maximus	Continuous asynchronous	Multibatch	4–5	Claro et al. 1989
Thalassoma bifasciatum	Asynchronous	Batch	12	Roede 1972
Scaridae				
Scarus coeruleus	Asynchronous	Batch	—	This chapter
Scarus coelestinus	Asynchronous	Batch	—	This chapter
Scarus guacamaia	Asynchronous	Batch	—	This chapter
Sparisoma chrysopterum	Asynchronous	Batch	—	This chapter
Sparisoma rubripinne	Asynchronous	Batch	—	This chapter
Sparisoma viride	Asynchronous	Batch	—	This chapter
Mugilidae				
Mugil curema	Synchronous	Total	12	Alvarez-Lajonchere 1980c; García and Bustamante 1981
Mugil liza	Synchronous	Total	2–3	Alvarez-Lajonchere 1979
Mugil trichodon	Asynchronous	Batch	12	Alvarez-Lajonchere 1980c
Sphyraenidae				
Sphyraena barracuda	Asynchronous	Batch	10	DeSylva 1963, This chapter
Trichiuridae				
Trichiurus lepturus	Asynchronous	Batch	6	Ros and Pérez 1978
Scombridae				
Katsuwonus pelamis	Asynchronous	Batch	12	Cayre and Farrugio 1986
Scomberomorus cavalla	Asynchronous	Batch		Beaumariage 1973*
Scomberomorus maculatus	Asynchronous	Batch	2–3	Mota Alvez and Tomé 1968; Powell 1975
Thunnus alalunga	Asynchronous	Batch	—	Otsu and Uchida 1959
Thunnus atlanticus	Asynchronous	Batch	—	Monte 1964*
Istiophoridae				
Istiophorus platypterus	Asynchronous	Batch	8	Jolley 1977
Ostraciidae				
Acanthostracion quadricornis	Asynchronous	Batch	12	Ruiz et al. 1999
Lactophrys triqueter	Asynchronous	Batch	—	This chapter

* Our interpretation of data and photographs in the original source.

Appendix 4.3. Spawning seasonality of coastal fish species from studies in Cuba, based on the presence of individuals in the following gonad stages (Table 4.2): Stage III (3); Stage IV (4); Stages VI or VI-III (6), presence of eggs and/or larvae (E); spawning individuals recorded (S); spawning peak (SP).

Family / Species	J	F	M	A	M	J	J	A	S	O	N	D	Source
Clupeidae													
Harengula clupeola	—	—	—	4	4	4	4	—	—	4	—	4	García-Cagide et al. 1994
Harengula humeralis	S	S	S	SP	SP	S	S	S	—	—	S	S	García-Cagide 1988
Jenkinsia lamprotaenia	S	S	S	S	S	S	S	S	S	S	S	S	García-Cagide et al. 1994
Belonidae													
Strongylura notata	—	—	—	4	—	—	—	—	—	—	—	—	García-Cagide et al. 1994
Tylosurus acus acus	—	—	4	—	—	—	—	—	—	—	—	—	García-Cagide et al. 1994
Tylosurus crocodilus crocodilus	—	—	—	—	—	—	4	—	—	—	—	—	García-Cagide et al. 1994
Centropomidae													
Centropomus undecimalis	—	S	S	—	—	—	—	—	—	—	—	—	Naranjo 1956
Centropomus undecimalis	SP	—	—	—	—	—	—	SP	SP	S	SP	SP	Alvarez-Lajonchere et al. 1982
Serranidae													
Epinephelus adscensionis	S	S	S	—	—	—	—	—	—	—	—	—	Naranjo 1956
Epinephelus morio	—	S	S	S	—	—	—	—	—	—	—	—	Naranjo 1956
Epinephelus nigritus	—	—	—	S	S	—	—	—	—	—	—	—	Naranjo 1956
Epinephelus striatus	—	S	S	—	—	—	—	—	—	—	—	—	Naranjo 1956
Epinephelus striatus	SP	SP	—	—	—	—	—	—	—	—	S	SP	San Martín 1965
Epinephelus striatus	SP	S	S	—	—	—	—	—	6	—	S	SP	Claro et al. 1990c
Mycteroperca bonaci	S	—	S	S	S	—	—	—	—	—	—	—	Naranjo 1956
Mycteroperca bonaci	S	SP	S	S	S	—	—	—	—	—	SP	S	García-Cagide and García 1996
Mycteroperca interstitialis	4	—	S	S	S	—	—	—	—	—	4	—	García-Cagide et al. 1994
Mycteroperca tigris	—	—	S	4	—	—	—	—	—	—	—	—	Naranjo 1956
Mycteroperca tigris	4	4	SP	S	—	—	—	—	—	—	—	4	García-Cagide et al. 1999a
Mycteroperca venenosa	—	—	S	—	—	—	—	—	—	—	—	—	Naranjo 1956
Mycteroperca venenosa	SP	S	S	SP	SP	S	—	—	—	—	—	—	García-Cagide and García 1996
Rachycentridae													
Rachycentron canadum	4	—	—	4	—	—	4	—	—	—	—	—	García-Cagide et al. 1994
Carangidae													
Caranx bartholomaei	—	—	—	—	—	S	S	S	—	—	—	—	Naranjo 1956
Caranx bartholomaei	4	—	—	4	—	—	—	4	—	—	—	—	Sierra et al. 1986
Caranx crysos	—	—	—	—	S	S	—	—	—	—	—	—	Naranjo 1956
Caranx hippos	—	—	—	—	S	—	—	—	—	—	—	—	Naranjo 1956
Caranx latus	—	—	—	—	—	S	S	S	S	—	—	—	Naranjo 1956
Caranx latus	—	—	—	—	—	—	—	S	S	—	—	4	García-Cagide et al. 1994
Caranx ruber	—	—	—	S	—	S	S	S	S	—	—	—	Naranjo 1956
Caranx ruber	—	—	—	S	S	S	S	SP	S	—	—	—	García-Cagide 1985
Selar crumenophthalmus	—	—	S	—	—	—	—	4	4	—	—	—	Naranjo 1956
Trachinotus falcatus	—	—	4	—	—	—	—	4	—	—	—	—	García-Cagide et al. 1994

Continued on next page

Appendix 4.3. continued

											Reference
Trachinotus goodei	—	—	—	—	S	S	—	—	—	—	Naranjo 1956
Coryphaenidae											
Coryphaena hippurus	—	—	—	—	S	S	S	S	S	—	Naranjo 1956
Lutjanidae											
Lutjanus analis	—	—	—	—	S	S	—	—	—	—	Naranjo 1956
Lutjanus analis	S	—	S	S	SP	S	S	S	—	S	Claro 1981c
Lutjanus apodus	S	—	S	S	S	S	S	—	—	S	Naranjo 1956
Lutjanus apodus	4	—	3	—	3	4	—	—	—	—	García-Cagide et al. 1994
Lutjanus buccanella	—	—	S	—	S	S	S	S	—	S	Naranjo 1956
Lutjanus buccanella	—	—	S	S	S	S	S	S	S	—	Pozo et al. 1984
Lutjanus cyanopterus	—	—	—	—	S	4	4	S	S	—	Naranjo 1956
Lutjanus cyanopterus	—	—	S	4	4	S	S	S	4	—	García-Cagide et al. 1994
Lutjanus griseus	S	—	S	S	S	S	S	S	S	—	Naranjo 1956
Lutjanus griseus	—	—	S	SP	S	SP	S	S	—	S	Claro 1983c
Lutjanus jocu	S	—	S	S	4	—	—	4	4-6	—	Naranjo 1956
Lutjanus jocu	4	4	4	4	4	4	4	4	4	4	Thompson and Munro 1983
Lutjanus mahogoni	—	—	S	S	S	S	—	—	—	S	García-Cagide et al. 1999b
Lutjanus synagris	S	—	S	S	S	S	S	—	—	S	Naranjo 1956
Lutjanus synagris	S	—	S	S	S	S	S	—	—	S	Naranjo 1956
Lutjanus synagris	S	—	SP	S	S	SP	S	—	—	S	Claro 1982
Lutjanus vivanus	S	—	S	S	S	S	S	S	—	S	Naranjo 1956
Lutjanus vivanus	SP	S	S	SP	S	S	S	SP	—	S	Pozo and Espinosa 1983
Ocyurus chrysurus	—	—	S	S	S	—	S	S	—	—	Naranjo 1956
Ocyurus chrysurus	S	SP	SP	S	S	SP	SP	S	—	S	Claro 1983d
Gerreidae											
Eugerres brasilianus	S	S	S	S	S	S	S	SP	SP	SP	Báez et al. 1983
Eugerres plumieri	—	S	—	S	S	—	S	S	S	S	Naranjo 1956
Eugerres plumieri	S	S	S	S	S	S	S	S	S	S	Millares et al. 1979b
Gerres cinereus	—	S	S	S	S	—	S	S	—	—	Naranjo 1956
Gerres cinereus	4	—	—	S	—	SP	S	—	S	S	Puga and Wong 1978
Gerres cinereus	S	S	—	S	—	S	S	S	S	S	Báez and Alvarez-Lajonchere 1983
Haemulidae											
Anisotremus surinamensis	—	—	S	S	S	S	S	—	—	—	Naranjo 1956
Haemulon album	—	S	S	S	S	S	S	—	—	—	Naranjo 1956
Haemulon album	SP	SP	S	S	S	SP	S	S	S	S	García-Cagide 1986a
Haemulon aurolineatum	—	—	S	S	S	S	S	—	—	—	Naranjo 1956
Haemulon aurolineatum	—	4	4	4	4	—	4	—	4	—	García-Cagide et al. 1994
Haemulon carbonarium	—	—	S	S	S	S	S	S	—	S	Naranjo 1956
Haemulon carbonarium	—	—	—	—	—	—	4	4	4	—	García-Cagide et al. 1994
Haemulon macrostomum	—	—	S	S	S	S	S	—	—	—	Naranjo 1956
Haemulon parra	4	—	4	S	S	—	—	—	—	—	García-Cagide et al. 1994
Haemulon plumieri	S	S	S	SP	SP	S	SP	S	SP	S	García-Cagide 1987

Continued on next page

Appendix 4.3. continued

Species										Reference
Haemulon sciurus	SP	S	S	—	—	—	—	S	S	García-Cagide 1986b
Sparidae										
Archosargus rhomboidalis	—	—	4	—	—	—	4	—	4	García-Cagide et al. 1994
Calamus bajonado	S	S	S	S	S	S	—	—	—	Naranjo 1956
Calamus bajonado	—	SP	—	—	—	—	S	—	3	García-Cagide et al. 1994
Calamus bajonado	S	SP	SP	SP	S	S	S	S	S	Olaechea and Sauskan 1974
Sciaenidae										
Bairdiella ronchus	S	S	S	S	S	S	S	S	S	García 1979
Micropogonias furnieri	SP	SP	SP	SP	SP	S	S	S	S	García 1979
Mullidae										
Pseudupeneus maculatus	—	—	—	—	S	S	S	—	—	Naranjo 1956
Pseudupeneus maculatus	4	4	4	4	4	—	—	4	4	García-Cagide et al. 1994
Pempheridae										
Pempheris schomburgkii	—	—	—	—	—	4	—	—	—	García-Cagide et al. 1994
Labridae										
Doratonotus megalepis	—	—	—	4	4	—	—	—	—	Claro et al. 1989
Lachnolaimus maximus	S	—	—	—	—	—	S	S	S	Claro et al. 1989
Lachnolaimus maximus	—	—	—	—	—	S	S	—	—	Naranjo 1956
Scaridae										
Scarus coelestinus	—	6	6	4	6	6	—	4	—	García-Cagide et al. 1994
Scarus guacamaia	—	4	4	—	4	—	—	—	4	García-Cagide et al. 1994
Sparisoma chrysopterum	4	—	—	4	—	—	4	—	—	García-Cagide et al. 1994
Sparisoma radians	—	—	4	—	4	4	—	—	—	García-Cagide et al. 1994
Sparisoma viride	4	4	—	4	—	—	—	—	—	García-Cagide et al. 1994
Mugilidae										
Mugil curema	SP	S	S	S	S	SP	SP	S	SP	Alvarez-Lajonchere 1976, 1980a
Mugil liza	—	S	—	—	—	—	S	S	S	Alvarez-Lajonchere 1979
Mugil hospes	S	S	S	S	S	SP	S	SP	S	Alvarez-Lajonchere 1981b
Mugil trichodon	SP	S	SP	S	SP	S	S	S	SP	Alvarez-Lajonchere 1980c
Sphyraenidae										
Sphyraena barracuda	—	—	—	—	—	4	4	—	—	García-Cagide et al. 1994
Sphyraena guachancho	—	—	—	S	S	S	S	—	—	Naranjo 1956
Trichiuridae										
Trichiurus lepturus	SP	S	SP	—	—	—	—	S	S	Ros and Pérez 1978
Scombridae										
Katsuwonus pelamis	—	E	E	E	E	E	E	E	—	Montolio and Juárez 1976
Katsuwonus pelamis	S	S	SP	SP	SP	SP	SP	S	S	C. Carles, pers. comun.
Scomberomorus maculatus	—	—	—	—	4	—	—	—	—	García-Cagide et al. 1994
Scomberomorus regalis	4	4	4	4	4	4	4	4	4	García-Cagide et al. 1994
Xiphiidae										
Xiphias gladius	—	S	S	S	S	S	S	S	S	Guitart Manday 1964
Ostraciidae										
Lactophrys trigonus	—	4	4	4	4	4	—	—	—	García-Cagide et al. 1994

Appendix 4.4. Individual fecundity (total number of oocytes in the ovaries) and relative fecundity (number of oocytes per gram of fish body weight) of some marine fishes inhabiting Cuban waters.

Family Species	n	Individual Fecundity	Relative Fecundity	Source
Anguillidae				
Anguilla rostrata	—	$5-10 \times 10^{6}$	—	Centro de Investigaciones Pesqueras 1978
Clupeidae				
Harengula humeralis	24	$21-208 \times 10^{3}$	2,000	García-Cagide 1988
Harengula jaguana	22	$5-52 \times 10^{3}$	548	Martínez and Houde 1975
Jenkinsia lamprotaenia	92	$141-1,156\dagger$	500–3,300	This chapter
Engraulidae				
Anchoa parva	29	$4-22 \times 10^{2}$	—	Caselles-Osario and Acero 1996
Anchovia clupeoides	38	$35-280 \times 10^{2}$	—	Caselles-Osario and Acero 1996
Centropomidae				
Centropomus undecimalis	1	144×10^{4}	—	Volpe 1959
Centropomus undecimalis	2	$3-4 \times 10^{6}$	377–614	Carvajal 1975
Serranidae				
Alphestes afer	4	$15-22 \times 10^{4}$	573–746	Thompson and Munro 1978
Cephalopholis cruentata	1	26×10^{4}	639	Thompson and Munro 1978
Cephalopholis cruentata	12	29×10^{4}	1,100*	Nagelkerken 1979
Cephalopholis fulva	4	$67-280 \times 10^{3}$	291–1,086	Thompson and Munro 1978
Epinephelus guttatus	6	$96-526 \times 10^{3}$	305–740	Thompson and Munro 1978
Epinephelus guttatus	—	20×10^{4}	—	Olsen and LaPlace 1979
Epinephelus guttatus	—	23×10^{4}	—	Smith 1961
Epinephelus guttatus	—	$90-336 \times 10^{4}$	—	Burnett-Herkes 1975
Epinephelus morio	14	14×10^{5}	200*	Moe 1969
Epinephelus morio	39	$85-4,035 \times 10^{3}$	—	Valdés-Alonso and Fuentes-Castellanos 1987
Epinephelus striatus	—	78×10^{4}	—	Smith 1961
Epinephelus striatus	42	48×10^{5}	—	Olsen and LaPlace 1979
Epinephelus striatus	—	$35-650 \times 10^{4}$	—	Carter et al. 1994
Epinephelus striatus	—	785×10^{3}	—	Bardach 1958
Mycteroperca bonaci	—	5×10^{5}	—	Smith 1961
Mycteroperca microlepis	39	$10-865 \times 10^{3}\dagger$	—	Collins et al. 1998
Mycteroperca venenosa	—	14×10^{5}	—	Smith 1961
Carangidae				
Caranx bartholomaei	2	72×10^{5}	1,100	Sierra et al. 1986
Caranx crysos	25	$41-1,546 \times 10^{3}$	712*	Goodwin and Finucane 1985
Caranx ruber	7	$19-11 \times 10^{4}$	750	This chapter
Caranx ruber	3	14×10^{4}	—	Thompson and Munro 1974
Trachinotus carolinus	1	42×10^{4}	—	Moe et al. 1968
Coryphaenidae				
Coryphaena equiselis	3	$6-18 \times 10^{4}$	—	Scherbachev 1973
Coryphaena hippurus	—	$24-30 \times 10^{5}$	—	Beardsley 1967

Continued on next page

Appendix 4.4. continued

Coryphaena hippurus	2	17×10^5	—	Scherbachev 1973
Lutjanidae				
Lutjanus analis	16	$7–40 \times 10^5$	600*	Claro 1981c
Lutjanus analis	—	13×10^5	—	Rojas 1970
Lutjanus buccanella	—	67×10^4	—	Pozo et al. 1984
Lutjanus griseus	1	6×10^5	—	Starck 1970
Lutjanus griseus	30	6×10^5	1,500*	González et al. 1979
Lutjanus griseus	16	4×10^6	1,718	Guerra and Bashirullah 1975
Lutjanus jocu	13	$128–846 \times 10^4$	453–1,449	García-Cagide et al. 1999
Lutjanus synagris	60	$47–926 \times 10^3$	754	Damas et al. 1979
Lutjanus synagris	6	$347–995 \times 10^3$	—	Rodríguez Pino 1962
Lutjannus synagris	42	$9–110 \times 10^4$	1,900*	Claro 1982
Ocyurus chrysurus	4	$10–147 \times 10^4$	1,100*	Piedra Castañeda 1965
Rhomboplites aurorubens	—	$5–23 \times 10^3$	—	Hood and Johnson 1999
Gerreidae				
Eugerres plumieri	5	$18–70 \times 10^3$	341	Millares et al. 1979b
Eugerres plumieri	14	1,083	—	Angell 1976
Haemulidae				
Haemulon album	1	602×10^3	198	Billings and Munro 1974
Haemulon album	5	$8–22 \times 10^5$	415	This chapter
Haemulon aurolineatum	13	3×10^4	435	Billings and Munro 1974
Haemulon aurolineatum	6	$8–22 \times 10^3$	920	This chapter
Haemulon aurolineatum	42	82×10^3	—	Vasconcelos and Sobreira 1976
Haemulon bonariense	4	22×10^3	190	Billings and Munro 1974
Haemulon chrysargyreum	1	32×10^3	432	Billings and Munro 1974
Haemulon plumieri	34	235	—	Billings and Munro 1974
Haemulon plumieri	—	$31–206 \times 10^3$	—	Román Cordero 1991
Haemulon plumieri	6	$27–312 \times 10^3$	430	This chapter
Haemulon sciurus	3	32×10^3	113	Billings and Munro 1974
Haemulon sciurus	23	$23–250 \times 10^3$	406	This chapter
Haemulon flavolineatum	9	31×10^3	284	Billings and Munro 1974
Sciaenidae				
Micropogonias furnieri	54	$44–776 \times 10^3$	—	Vazzoler 1969
Chaetodontidae				
Chaetodon capistratus	5	$3–13 \times 10^3$	181–478	Aiken 1983a
Chaetodon ocellatus	4	$12–64 \times 10^3$	114–464	Aiken 1983a
Chaetodon sedentarius	3	$7–23 \times 10^3$	433–848	Aiken 1983a
Chaetodon striatus	4	$11–25 \times 10^3$	220–600	Aiken 1983a
Pomacanthidae				
Holocanthus tricolor	2	$10–12 \times 10^3$	79–161	Aiken 1983a
Pomacanthus arcuatus	3	$16–126 \times 10^3$	50–123	Aiken 1983a
Pomacanthus paru	2	$31–37 \times 10^3$	46–51	Aiken 1983a

Continued on next page

Appendix 4.4. continued

	n			
Mugilidae				
Mugil curema	15	700–1,000	41–57 × 10⁴	García and Bustamante 1981
Mugil curema		750*	100	Alvarez-Lajonchere 1976
Mugil liza	80	—	1,600*	Alvarez-Lajonchere 1979
Mugil trichodon	50	—	850*	Alvarez-Lajonchere 1980c
Mugil hospes	60	·	1,200*	Alvarez-Lajonchere 1981b
Sphyraenidae				
Sphyraena barracuda	2	—	56–67 × 10⁴	De Sylva 1963
Trichiuridae				
Trichiurus lepturus	45	56–263*	18–193 × 10³	Ros and Pérez 1978
Scombridae				
Katsuwonus pelamis	—	70	40 × 10⁴ (Caribbean)	Goldberg and Au 1983
Katsuwonus pelamis	—	62	86–1,341 × 10³ (W Caribbean)	Montolio and Juárez 1976
Katsuwonus pelamis	—	—	1–10 × 10⁵ (Atlantic)	Cayre and Ferrugio 1986
Scomberomorus maculatus	2	125*	3–15 × 10⁵	Klima 1959
Balistidae				
Balistes vetula	3	73	49–83 × 10³	Aiken 1983b
Canthidermis sufflamen	4	217	219–62 × 10³	Aiken 1983b
Ostraciidae				
Acanthostracion quadricornis	30	194–902	30–256 × 10³	Ruiz et al. 1999

n: Number of individuals.

* Values calculated in the present study using data from the original reference.

† Number of eggs in one batch.

5

Trophic Biology of the Marine Fishes of Cuba

LUIS M. SIERRA, RODOLFO CLARO, AND OLGA A. POPOVA

5.1 Introduction

Feeding is one of the most complex interactions between fishes and their environment. It involves ontogenetically and geographically variable energy exchange among organisms and ecosystems, and fuels tissue repair, growth, respiration, motion, and reproduction. Adaptation of individuals to specific prey items depends not only on inherited morphological characteristics and environmental factors, but also on behavioral mechanisms that allow fishes to capture food while avoiding predators (Manteifel 1961). Trophic interactions among species that exploit similar energy resources might influence community composition of fishes of both temperate (Nikolsky 1974a) and tropical areas (Hixon 1991; Polunin 1996; Beets 1997; Hixon and Carr 1997). This trophic complexity is also fostered by substantial variations in feeding across differing ontogenetic (Wootton 1990; Rooker 1995; Sierra 1996) and spatial scales (Wootton 1990; Turingan et al. 1995). Geographic differences in feeding habits might reflect opportunistic responses to locally variable qualities and quantities of food resources (Starck 1970). Conclusive explanations of causal mechanisms of observed trophic patterns will require (1) descriptive information further linking gut contents with relative prey abundances (Jones et al. 1991), and (2) experimental manipulation of both prey and predators (e.g., Steele 1996; Danilowicz and Sale 1999).

5.2 Major Trophic Groups

Appendix 5.1 summarizes the dietary composition of 365 species inhabiting the Cuban shelf following different authors. These data are difficult to compare because of the wide variety of methods employed in obtaining them. Many sources do not provide quantitative information or are based on small samples. Nonetheless, we attempted to arrange the food items in order of importance. If more than one source was available for a certain species, we selected the data that seemed most representative of the species' diet. In each case, up to five taxa, the most important and representative, are shown.

As summarized in Table 5.1, 138 species (38%) consume fishes as their main food (fishes make up more than 50% of the food spectrum); in 190 species (52%), fishes were present in the stomach contents. Benthic crustaceans are eaten by 225 species and were the principal component in 28%. Fishes and crustaceans are particularly important in the diets of snappers, jacks, and grunts. Of the other species summarized in Table 5.1 and Appendix 5.1, 41 species consume plankton (accounting for more than 70% of the food in 31 of these species), 31 species are herbivores, and 5 are detritivores (only the mullets [Mugilidae] are strictly detritivores). Planktivores include Clupeidae, Engraulidae, and other less abundant families. None of these fishes is a strict phytoplanktivore. The Atlantic anchoveta (*Cetengraulis edentulus*), however, has been reported to consume large numbers of diatoms (Cervigón 1994). Unlike temperate and polar species in which planktivores account for a significant proportion of the fish biomass, nearly all shallow-water planktivorous fishes in Cuba are small (adult length < 20 cm fork length [FL]) and not abundant. Coastal planktivorous fishes constitute less than 10% of the total Cuban finfish catches and are represented by only three species of the family Clupeidae: the redear sardine (*Harengula humeralis*), false pilchard (*Harengula clupeola*), and Atlantic thread herring (*Opisthonema oglinum*).

Many species can feed on zooplankton during early demersal life stages, including snappers (Lutjanidae; Sierra 1996), jacks (Carangidae), and grunts (Haemulidae). Some

Table 5.1. Primary food items of 365 fish species studied in Cuba

Food Item	Species That Consume Item in This Category		Species That Use Item as a Principal Food [a]	
	No.	%	No.	%
Zooplankton	58	15.9	41	11.3
Fishes or their larvae	190	52.2	138	37.9
Benthic crustaceans	225	61.6	102	28.0
Crabs	104	28.5	27	7.4
Shrimp	97	26.6	28	7.6
Stomatopods	19	5.2	1	0.3
Macrurids	5	1.3	2	0.5
Mollusks	108	29.6	17	4.7
Pelecypods	20	5.5	3	0.8
Gastropods	34	9.3	6	1.6
Cephalopods	54	14.8	5	1.4
Echinoids	30	8.2	6	1.6
Annelids	67	18.4	12	3.3
Sponges	10	2.7	6	1.6
Other invertebrates	67	18.4	6	1.6
Algae/seagrasses	54	14.8	31	8.5
Detritus	17	4.6	5	1.4

Source: Appendix 5.1.

[a] Constitutes more than 50% of the food spectrum within a species.

species, such as yellowtail snapper (*Ocyurus chrysurus*), are able to consume zooplankton throughout their life cycle (Claro 1983d). Many species inhabiting coral reefs, such as damselfishes (Pomacentridae), wrasses (Labridae), filefishes (Monacanthidae), triggerfishes (Balistidae), bigeyes (Priacanthidae), cardinalfishes (Apogonidae), and squirrelfishes (Holocentridae), consume plankton in addition to other food sources.

Of the species summarized in Table 5.1, 31 consumed plants as a primary food (see Appendix 5.1). These fishes belong to strictly herbivorous families such as Scaridae (14 species), Acanthuridae (3), Blenniidae (5), Kyphosidae (2), or other trophically diverse families. Macrophytes constitute more than 20% of the diet for omnivores such as *Stegastes adustus, S. leucostictus, S. planifrons, S. variabilis, Alutera scriptus, Cantherhines pullus, Monacanthus ciliatus,* and *Coryphopterus* spp. (Randall 1967). Algae growing on top of corals is eaten most typically by parrotfishes (Scaridae). This behavior provides parrotfishes with a notable advantage over other herbivores when plants are lacking (Randall 1967). Spermatophytes, mainly *Thalassia testudinum,* are an important food for parrotfishes (Scaridae), surgeonfishes (Acanthuridae), filefishes, triggerfishes (Balistidae), and to a lesser extent, trunkfishes (Ostraciidae). Parrotfishes and surgeonfishes feed on the plants that surround reefs, forming a halo, the radius of which depends on the number of fishes inhabiting this biotope (Randall 1965). The damselfishes (Pomacentridae) consume a great variety of green, brown, and red algae. Floating algae (*Sargassum natans, S. fluitans*) and

seagrass (*Thalassia testudinum*) fragments are ingested principally by chubs (Kyphosidae) and halfbeaks (*Hemiramphus* spp.).

As in planktivores, in most herbivorous fishes the stomach contents include a large variety of taxa (Appendix 5.1). For example, Randall (1967) recorded 10–20 species of algae and spermatophytes in nearly all examined parrotfish stomach contents. In the bucktooth parrotfish (*Sparisoma radians*), however, he found 88% seagrass and only two species of algae. Our data on the redtail parrotfish (*Sparisoma chrysopterum*) suggest that in Cuba this species might occupy the feeding niche that the bucktooth parrotfish holds in other Caribbean islands (Randall 1967). Vegetation made up more than 90% of scarid diets, although other organisms, such as sponges (see Dunlap and Pawlik 1998), gorgonians, echinoderms, and crustaceans, are also found in their stomach contents.

Trophic opportunism is recorded in various families, including some species of Pomacentridae, Ostraciidae, Balistidae, and Diodontidae. For example, the queen triggerfish (*Balistes vetula*) consumes echinoids, crabs, pelecypods, ophiuroids, polychaetes, anomurans, gastropods, asteroids, algae, sipunculids, shrimps, tunicates, fishes, corals, chitons, macrurans, stomatopods, amphipods, and anthozoans (Appendix 5.1; Randall 1967). Such a high degree of euryphagy might be attributed to the large number of species in the reef fauna.

Trophic opportunists might narrow their diet, depending on the availability of food. Such is the case of some species occupying artificial habitats. For example, a narrow food spectrum was found in lane and gray snappers on artificial refuges located at Cayo Tablones, Golfo de Batabanó: only one prey species, the scarecrow toadfish (*Opsanus phobetron*), dominated (Valdés-Muñoz and Silva Lee 1977). This diet was much narrower than that recorded by Claro (1981b, 1983c) in the reef–seagrass bed complexes of nearby areas of the Golfo de Batabanó. This local stenophagy is not characteristic of the overall feeding habits of the species.

At least 10–12 species inhabiting Cuban waters are specialized to consume the ectoparasites of other species, although parasites do not constitute an important food object for any of them. These species include porkfish (*Anisotremus virginicus;* juvenile), sharksucker (*Echeneis naucrates;* juvenile), remora (*Remora remora;* juvenile), bluehead (*Thalassoma bifasciatum;* juvenile and adult, but not the final terminal male phase), and juveniles of *Holocanthus* species and yellowtail damselfish (*Microspathodon chrysurus*) (Randall 1967).

5.3 Trophic Relationships of the Cuban Marine Fish Fauna

5.3.1 Food Composition

Sierra et al. (1994) summarized the existing information on feeding habits of fishes in the wider Caribbean and particularly in Cuba. In this section, we discuss features of some important species that represent considerable fish biomass on the Cuban shelf. Appendix 5.2 summarizes the overall diet of various Cuban fishes.

Snappers (Lutjanidae) are among the most important (in biomass) fish species of Cuban shelf waters. Snappers are predators with greatly varied feeding habits. They are carnivorous, mainly piscivorous, although they consume a wide variety of benthic organisms, principally crustaceans. For example, more than 80 taxa were recorded in the stomachs of lane snappers (*Lutjanus synagris*) in the Golfo de Batabanó (Claro 1981b). Grunts (Haemulidae), usually more abundant than snappers, eat a wide variety of invertebrates.

The lane snapper feeds mainly on the benthic organisms and small fishes that inhabit the seagrass bottom. Feeding habits vary locally, and even within a locality in different years (Rodríguez Pino 1962; Buesa 1970; Claro 1981b). Crustaceans were the most frequent organisms found in the stomach contents, which included a wide variety of species, mainly crabs. In contrast, among lane snappers dwelling in artificial refuges on seagrass bottoms far from reefs, crustaceans made up only 8% of the entire food content volume, whereas fishes constituted 80%; one fish species, the scarecrow toadfish, was 56% of the total (Appendix 5.2) (Valdés-Muñoz and Silva Lee 1977).

The mutton snapper preys on many of the same species as the lane snapper (Claro 1981c; Mueller et al. 1994), but there are some differences in prey size that might limit competition between the two species. Because the mutton snapper is larger, the prey is larger. In both mutton and lane snapper, crabs are the most frequent prey, although in terms of biomass, fishes are more important to mutton snapper.

For gray snappers (*Lutjanus griseus*), fishes were the dominant prey in the Golfo de Batabanó population, mainly Scaridae and Batrachoididae, both by occurrence (61% of individuals examined had fishes in the stomach) and by weight (74%). Crustaceans (principally Portunidae and Penaeidae), although frequent food items (58%), constituted only 25% by weight of the food spectrum in the region. The role of mollusks, polychaetes, and other invertebrates is less vital for gray than for lane and mutton snappers. However, as with lane snappers, gray snappers dwelling in artificial refuges far from coral reefs exhibited a food spectrum narrower than those individuals inhabiting the coral reef areas. Gray snappers of coastal lagoons feed mainly on crustaceans (González-Sansón 1979). Evidently this is associated with the great abundance of swimming crabs and shrimps in such ecosystems. Although small fishes, juvenile mullets, and mojarras are also abundant in the coastal lagoons of the south-central part of Cuba, they were rarely found in the gray snapper stomach contents (Claro 1983c).

The main food of the yellowtail snapper is fishes and, to a lesser extent, crustaceans and other invertebrates. This predation behavior coexists with zooplanktivory in this species, which demonstrates its great adaptive plasticity to different food conditions. The large number of gill rakers in the first branchial arch allows this species to trap plankton (Davis and Birdsong 1973; Hobson 1991). The dog snapper (*Lutjanus jocu*) is an opportunistic piscivorous species whose diet depends mainly on the availability of local ichthyofauna. Fishes were the primary food, by occurrence and weight, with parrotfishes (Scaridae) and grunts (Haemulidae) dominating in the southwest, northwest, and northeast regions of the shelf (Claro et al. 1999; Appendix 5.2). In addition to fishes, some cephalopods (octopus and squid) and crustaceans (swimming crab and lobster) were identified in dog snapper stomach contents (Claro et al. 1999).

Groupers are bottom-dwelling, solitary predators that take a wide variety and size range of fishes and invertebrates (Randall 1967; Parrish 1987; Sadovy and Eklund 1999). *Epinephelus* and *Mycteroperca,* the largest genera, are often considered piscivores, although they also consume crustaceans, mollusks, and other invertebrates (Randall 1967; Silva Lee 1974b; Claro et al. 1990c; Carter et al. 1994; Sullivan and de Garine-Wichatitsky 1994; García-Arteaga et al. 1999). They usually feed at dawn and dusk (Starck and Davis 1966; Randall 1967), but recently digested food has been found in stomachs during the day (Silva Lee 1974b; Claro et al. 1990c), indicating daylight feeding as well.

Data collected on the Nassau grouper (*Epinephelus striatus*) in the northeast and southwest regions of the Cuban shelf showed a diet composed mainly of fishes (>80% of food weight), whereas in the northwest region, fishes constituted only 43% of the stomach contents (Claro et al. 1990c). Among fishes found in its diet, grunts (Haemulidae) dominated the diets of individuals caught in the southwest shelf, and parrotfish (Scaridae) dominated the diets of specimens from the northwest and northeast regions (Appendix 5.2). The high consumption of grunts in the southwest shelf area might be associated with their greater availability in the region. Interestingly, studies during 1973 found a mostly crustacean diet for Nassau grouper in southwest Cuba (Silva Lee 1974b). The difference in these studies suggests that changes in the Nassau groupers' feeding patterns could have been caused by an increased density of grunts (see Section 2.4) in the area (Claro 1991).

The tiger grouper (*Mycteroperca tigris*) consumes mainly haemulids, scarids, and acanthurids and showed similar dietary variations in northwest and southwest Cuba (Valdés-Muñoz 1980; García-Arteaga et al. 1999). Such variation in feeding among both regions and years enhances the trophic diversity and functional relationships of reef fishes. Jewfish (*Epinephelus itajara*) usually feed on slow-moving fishes and invertebrates (Bullock and Smith 1991). Crustaceans, particularly lobsters and crabs, are important parts of their diet although fishes and sometimes turtles are consumed (Longley and Hildebrand 1941; Randall 1967). Juveniles consume shrimp, mollusks, and sea catfish (Bullock and Smith 1991).

Jacks (Carangidae), which are active predators that spend much of their time searching for food, have been divided into three groups according to their food source: piscivores (*Caranx* and *Seriola*), planktivores (*Decapterus* and *Selar*), and mollusk-feeders (*Trachinotus*) (Randall 1967). These are not, however, absolute categories; many species are trophic opportunists that shift feeding habits throughout their life history.

Bar jacks (*Caranx ruber*) of the Golfo de Batabanó feed mainly on fishes (82% of food weight; Sierra and Popova 1982). Scaridae and Labridae were dominant (occurrence and weight) in their diet; species of the families Balistidae, Blenniidae, Synodontidae, and Muraenidae were also recorded. Most jacks have a relatively small mouth and a reduced and muscular stomach, so slender fishes such as small parrotfishes and wrasses are likely prey. Off the northwest coast of Cuba, Puerto Rico, and the Virgin Islands (Randall 1967), food composition for the bar jack is broader than in the shelf area of southwest Cuba. The shelf in northeastern Cuba and Puerto Rico is narrower and largely covered by reefs, which have a more diverse fauna than seagrass meadows, the largest habitat type (by area) of the inner shelf of southwest Cuba.

Yellow jacks (*Caranx bartholomaei*) eat demersal fishes almost exclusively, mostly during the day. In the Golfo de Batabanó, Sierra et al. (1986) recorded 97% fishes (by weight and occurrence) and only 3% crustaceans and mollusks (combined) in yellow jack stomachs. Among the prey of yellow jacks, members of the families Labridae, Scaridae,

Haemulidae, Sparidae, Gerreidae, and others were found in the stomach contents; the genera *Halichoeres* (Labridae), *Sparisoma*, and *Scarus* (Scaridae) were dominant. These latter species were also recorded in the bar jack's diet. Scarid and labrid species constituted 80% of the food spectrum and occurred in 84% of the examined individuals; grunts (Haemulidae) constituted 11% by weight and 4% by occurrence. Crustaceans appeared in the stomachs of 12% of the individuals examined, but they represented less than 2% of the diet weight, and squid (*Loligo*) constituted less than 1% (Appendix 5.2).

Grunts are carnivorous but rarely consume fishes. Their diet consists mainly of small benthic organisms (Randall 1967; Dennis 1992b). Newly-settled stages and early juveniles feed principally on zooplankton (Longley and Hildebrand 1941; Helfman et al. 1982). Older grunts grind food with plate-shaped pharyngeal teeth; consequently the food in their intestines is always a heterogeneous mass that makes it difficult to identify and quantify food items.

The margate (*Haemulon album*) is a typical invertebrate consumer that forages in sandy and seagrass areas. Crustaceans (mostly decapod) were the main food target in the Golfo de Batabanó (Sierra 1983), followed by mollusks (mainly bivalves, by occurrence), echinoderms (particularly holothurians), polychaetes, and fishes, to a lesser extent. Sipunculid, foraminifera, tunicates, and others also appeared, but in low proportions. Stomach contents of the margate in other regions were similar to these findings, but the food items were in different proportions (Cummings et al. 1966; Randall 1967). These differences are most likely associated with local differences in food availability.

White and bluestriped grunts (*Haemulon plumieri* and *H. sciurus*) feed on benthic organisms (mainly crustaceans, mollusks, polychaetes, echinoderms) in varied proportions depending on the region (Beebe and Tee-Van 1928; Longley and Hildebrand 1941; Breder 1948; Reid 1954; Randall 1967) (Appendix 5.1). Populations of both species inhabiting the eastern part of the Golfo de Batabanó ate mostly crustaceans (52% and 55% of the food volume, respectively; Valdés-Muñoz and Silva Lee 1977; Appendix 5.2). These proportions may be underestimated because the prey in the stomachs were highly degraded (40% of the food material could not be identified). An index of diet overlap (Menge 1972) was estimated at 97.8% for the two species. White and bluestriped grunts caught in artificial refuges close to Cayo Tablones (Golfo de Batabanó) were found to feed principally (28% and 17% of the food volume, respectively) on polychaetes, although much of the stomach contents could not be identified (59% and 75%, respectively). Evidently, as for lane and gray snappers, for grunts aggregating on artificial refuges the diet depends on the local food supply.

Clupeidae are nocturnal, shallow-water zooplanktivores. During the day they gather in dense schools close to key mangroves and at night they scatter for feeding (Starck and Davis 1966; Silva Lee 1974a; Randall 1967). The diet of the redear sardine of the Golfo de Batabanó was 54% planktonic organisms (total weight) and plankton appeared in 87% of the stomachs examined. In addition, a large amount of caridean shrimp and polychaetes, taken in the plankton during spawning, were also found. Overall, the diet composition of the redear sardine depends on the zooplankton supply. Surveys conducted on the north coast near La Habana (Sierra and Díaz-Zaballa 1984) showed a great proportion of eunicid polychaetes (79% of the food weight); a massive reproductive aggregation of these polychaetes in the water column was recorded during one of the surveys. The redear sardine can act as an active predator and has been observed attacking dwarf herring (D. Guitart Manday, personal communication).

The false pilchard and the redear sardine, which form mixed schools, have similar behavior but differ remarkably in diet. Within one school surveyed, planktonic copepods constituted 45% of the total organisms in the stomach contents of the false pilchard; whereas in the redear sardine this item constituted only 17%. On the other hand, the copepods eaten by the false pilchard were mostly *Calanoida*; the redear sardine consumed mainly *Oithona*. In other surveys in the same region (Sierra and Díaz-Zaballa 1984), polychaetes were the main food object for the redear sardine (78.4% of the food weight), but false pilchard individuals from the same school had a small amount of polychaetes in their stomachs (1.3%). The diet of the false pilchard consisted of mostly (98.7%) smaller planktonic organisms, mainly copepods and decapod larvae (Appendix 5.2). These differences could be attributed to several factors: (1) the larger size attained by redear sardines than by false pilchards; (2) morphological differences between the two species, particularly the number of gill rakers and the distance between them, and the size of mouth and eyes; or (3) physiological divergence in the daily feeding pattern (associated with distinct visual characteristics).

The dwarf herring (*Jenkinsia lamprotaenia*), one of the smallest species of the family Clupeidae, feeds exclusively on plankton. In the southwest section of the shelf, this species was found to eat mainly shrimp larvae (60%), mostly penaeid postlarvae (34%). Copepods were rare in the food content of dwarf herrings in Cuba (they made up only 8% of the ingested food). Small amounts of polychaetes recorded in the alimentary tract (< 1%) might have been captured during the vertical migrations to the surface. The data collected support Durbin's (1979) criterion that planktivorous fishes are food-selective because the food consumed does not always correspond with the abundance of planktonic organisms.

Benthic invertebrates with exoskeletons are the main food for labrids, except for the creole wrasse (*Clepticus parrae*), which is a pelagic fish that feeds on plankton. Wrasses, which usually feed during the day, use their canine teeth to pull up gastropod mollusks and other organisms attached to the bottom. Their highly developed pharyngeal teeth allow them to grind hard parts and ingest whole prey.

Small wrasses (*Halichoeres* spp., *Xyrichtys* spp. and the bluehead) consume mainly small crustaceans, mollusks, ophiuroids, and polychaetes. Some wrasses are known to be parasite cleaners, but this does not seem to play an important role in their nutrition. Unlike other labrids, the green razorfish (*Xyrichtys splendens*) principally consumes planktonic organisms, along with other small benthic invertebrates (amphipods, gastropods, pelecypods, crustaceans; Randall 1967).

The hogfish (*Lachnolaimus maximus*), the largest species of the labrid family, feeds principally on mollusks, crustaceans, and echinoderms, although it can eat other invertebrates and small fishes. In surveys in the Golfo de Batabanó, mollusks were present in 85% of the examined stomachs, pelecypods in 41%, gastropods in 29%, crustaceans in 36% (Brachyura, 10%, and Anomura, 9%), echinoids in 5%, pieces of corals in 4%, and algae and turtle grass in 6% (these two latter seem to be ingested incidentally) (Claro et al. 1989). Morphological adaptations and feeding habits of the Spanish hogfish (*Bodianus rufus*) are analogous to the hogfish, but the Spanish hogfish may consume a higher proportion of crustaceans, ophiuroids, and echinoids (Randall 1967). Relationships between labrid morphology and feeding ecology are examined in Wainwright (1988).

Yellow and spotted goatfish (*Mulloidichthys martinicus* and *Pseudupeneus maculatus*) were reported feeding on polychaetes, bivalves, crabs, and shrimp in the U.S. Virgin Islands and Puerto Rico (Randall 1967). In the Golfo de Batabanó, yellow goatfishes eat polychaetes, crustaceans, and zooplankton, whereas spotted goatfishes consumed crustaceans, mollusks, and fishes.

The existing information on feeding should not be considered definitive. Some data from Cuba show great intraspecific differences among different regions. Large changes might also occur through time relative to environmental conditions. For example, the diets of several important species (redear sardine, yellow jack, bar jack, and margate) varied over several years (Table 5.2). Changes in weather conditions (e.g., rainfall) might have influenced the availability of food items and their proportions in fish diets (see Section 5.4.3 for more details).

5.3.2 Seasonal Variations in Feeding Habits

Feeding habits of temperate fishes vary markedly with the sequential abundance of a few food species throughout the year. In the tropics, the relatively uniform climatic conditions lead to a more stable food supply for fishes, at least in areas not heavily influenced by large river discharges. In reef areas, a less marked seasonality of fish nutrition should be expected, but fluctuations in the proportion of food items has been found in some species. These annual variations might be related mostly to local changes in prey availability (Randall 1967). Although some monthly shifting in food composition for several coastal species of Cuba has been suggested (Sierra

Table 5.2. Interannual variations in the diets of four common fish species of the Cuban shelf

Species Food Item	Percentage of the Stomach Contents			
	1978	1979	1980	1981
Harengula humeralis				
Juvenile fishes	1.0	8.0	4.0	—
Zooplankton	6.0	54.0	62.0	—
Shrimp	32.0	3.0	18.0	—
Polychaetes	23.0	27.0	14.0	—
Siphonophores	—	2.0	—	—
Unidentified	38.0	6.0	2.0	—
n	178	253	382	—
Caranx bartholomaei				
Wrasses	—	30.0	51.0	36.0
Parrotfishes	—	47.0	37.0	56.0
Other fishes	—	19.0	8.0	4.0
Crustaceans	—	4.0	4.0	2.0
Other organisms	—	—	—	2.0
n	—	113	145	94
Caranx ruber				
Wrasses	31.0	18.0	37.0	—
Parrotfishes	30.0	45.0	20.0	—
Other Fishes	30.0	23.0	16.0	—
Zooplankton	2.0	7.0	24.0	—
Shrimp	5.0	4.0	2.0	—
Squid	2.0	3.0	1.0	—
n	403	855	582	—
Haemulon album				
Fish	6.0	14.0	14.0	—
Crustaceans	51.0	45.0	57.0	—
Polychaetes	13.0	23.0	16.0	—
Mollusks	37.0	46.0	35.0	—
Holothuroids	7.0	33.0	26.0	—
n	151	334	480	—

et al. 1994), we could not define a seasonal pattern. In some species, such as the lane snapper, however, a certain potential to increase consumption of crustaceans in the summer was seen; this also was the peak season of feeding intensity for this species (Section 5.4.5). It is likely that increased energy requirements with the increased temperature cause fishes to consume larger amounts of food and, possibly, to catch less accessible prey. Such an increase, however, might also be associated with a season-related increase in prey availability. Bar jacks and yellow jacks also consumed fishes during the summer, particularly *Haemulon* and *Eucinostomus,* although these prey might be less appropriate for jacks because jacks have a relatively small mouth considering the deep body of grunts and mojarras.

Coastal lagoon fauna are often subjected to more acute environmental changes than are reef faunas. Large nutrient inputs and salinity declines during the rainy season provoke notable variations in bioproductivity, which modifies the food supply. During the rainy season, feeding activity of the Brazilian mojarra (*Eugerres brasilianus*) in the Tunas de Zaza coastal lagoons (southeast Cuba) focused mainly on sediment-dwelling organisms such as nematodes, cirripedian larvae, copepods, foraminifera, and ostracods (González-Sansón and Rodríguez-Viñas 1983); during the dry season, however, they principally consumed organisms living among filamentous algae (amphipods and other small crustacean, vegetation, and organic material). During the wet season, the ground croaker (*Bairdiella ronchus*) eats primarily amphipods, but in the dry season it eats mostly porcellanid crabs and callianassid prawns (García and Nieto 1978).

Unfortunately, little information is available on the seasonality of food consumption by tropical fishes. Seasonality is more likely to be detected in lower-level taxonomic groups (species and genera). Seasonal shifts in feeding might contribute to a sequential exploitation of food resources throughout the year. To identify such patterns, intense monthly sampling during an entire year and better taxonomic identification of food items are needed.

5.3.3 Ontogenetic Variations in Feeding Habits

Fish food composition usually varies with fish growth as the individual moves from one life history stage to another. These changes in feeding habits constitute important adaptations that might allow individuals to optimize food resources (Shoriguin 1952; Stoner and Livingston 1984, Mullaney and Gale 1995; Wainwright 1995). A change in life stage is often accompanied by a shift in habitat, and thus also in food supply. An individual predator's increased size enhances its ability to consume larger prey and thus enhances its potential prey selection. Switching from one prey type to another is usually a critical life history phase—particularly during the early life stages when the organism undergoes a series of complex adaptations. These changes can therefore provoke alterations in fish growth rate (Chapter 6). Ontogenetic variations in feeding have been studied in various coastal fishes of Cuba (Silva Lee 1974b; Claro 1981b, 1981c, 1983c, 1983d; Sierra et al. 1986; Sierra and Popova 1982; Sierra 1996) and are shown in Figs. 5.1 and 5.2 for some species.

Studies in Cuba suggest that young snappers are largely planktivorous until they reach 6–8 cm FL. In the stomach contents of lane snapper up to 6.5 cm FL, only planktonic crustaceans (isopods, amphipods, copepods, and decapod nauplii) were found (Claro 1981b). Juvenile yellowtail snappers of the same size (6.9–8 cm) have similar feeding habits, but they also feed on fish larvae (Claro 1983d). Studies on gray snappers from south Florida (Starck 1970; Hettler 1989), schoolmasters (*Lutjanus apodus*) from southwest Puerto Rico (Rooker 1995), and lane and gray snappers from the Mississippi Sound (Franks and VanderKooy 2000) found that demersal juveniles can feed on a variety of benthic crustaceans and fishes.

In the Cuban studies, most snapper species were found to feed exclusively on benthic organisms or small fishes by approximately 8 cm length (Sierra 1996). Zooplankton continue to be an important food for the yellowtail snapper during much of its life cycle, although irregularly. The yellowtail snapper, like the gray snapper, begins to consume

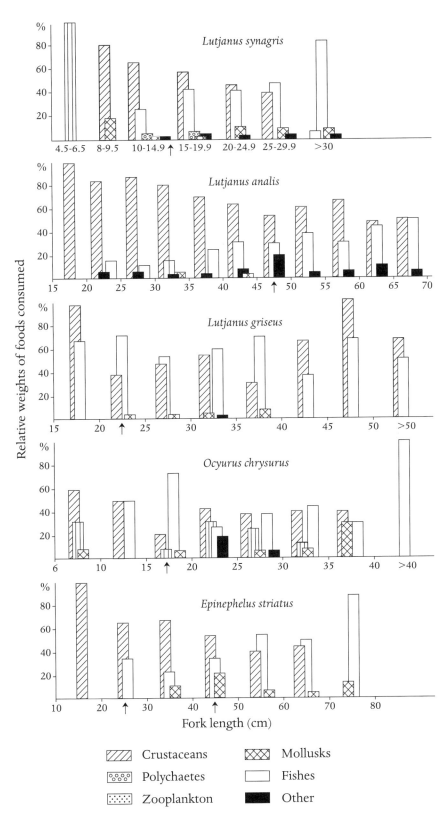

Fig. 5.1. Ontogenetic variation in food consumption in lane (*Lutjanus synagris*), mutton (*L. analis*), gray (*L. griseus*), and yellowtail (*Ocyurus chrysurus*) snappers, and Nassau grouper (*Epinephelus striatus*) (modified from Claro 1981c, 1982, 1983c, 1983d, and Claro et al. 1990c). Arrows pointing upward indicate the size of sexual maturation.

fishes earlier (at 10 cm FL), which can be attributed to a more hydrodynamic body shape than other snappers. Lane snappers from 12 cm FL eat small fishes, but in lower proportions than do yellowtail snappers (Fig. 5.1). The mutton snapper, one of the largest species of the family, does not begin to consume fishes until it is relatively larger than the other snappers. This limitation probably contributes to its relatively slow growth rate during the first year of life (see Chapter 6). Individuals of the four snapper species show an increment in fish consumption with growth, particularly before or during attainment of sexual maturity. Crustaceans are an important food target throughout the life cycle, particularly for lane and mutton snappers. Small crabs and shrimp are gradually replaced in the diet by larger and more mobile prey, such as swimming crabs.

Nassau groupers (*Epinephelus striatus*) show similar changes in feeding with growth (Eggleston et al. 1998). Until individuals reach 20 cm total length (TL), crustaceans are the only food type, but from that size on, fishes become a dominant prey (Silva Lee 1974b). During late larval stages, Nassau groupers feed on zooplankton and shift to a progressively more benthic diet after they settle (Grover 1993; Sullivan and de Garine-Wichatitsky 1994; Grover et al. 1998).

Grunts feed on benthic invertebrates for all but the earlier stages of the life cycle, during which they eat zooplankton (Randall 1967). From an early stage (9–10 cm FL), the margate exhibits a varied diet of invertebrates. Smaller individuals eat mainly mollusks, polychaetes, and crustaceans. As they grow, the proportion of crustaceans, echinoderms (holothurians), and polychaetes increases; the presence of mollusks diminishes in adults. With the attainment of sexual maturity, the margate starts to consume small fishes, and the proportion of fishes increases gradually with age (Fig. 5.2). Mollusks appear in significant amounts in the stomachs of adult margate, unlike the species mentioned earlier. Unfortunately, we have no data on the size-related feeding changes of other grunt species.

Carangids also have planktonic feeding habits during the juvenile stage. Bar jacks, although piscivorous as adults, eat zooplankton until they are larger than snappers when they begin to consume fishes. Bar jacks at 15–17 cm FL still feed on zooplankton (> 50% of food weight), and they continue to consume a large proportion of zooplankton (30–50%) until they reach 23–25 cm FL (Sierra and Popova 1982). For younger individuals, copepod larvae constitute the main food object. At approximately 15–17 cm, bar jacks start to eat caridean shrimp, small squid, and fishes (dwarf herring, wrasses, and juvenile parrotfishes) (Fig. 5.2). The attainment of an entirely piscivorous habit occurs shortly before the onset of maturity, the period when energy requirements for gonad development increase.

Feeding shifts can occur at different sizes, depending on food supply conditions. For example, the switch to a piscivorous habit for bar jacks occurred at a larger size in 1980 than in 1978 and 1979. In 1980, individuals of 25 cm FL were recorded eating mostly zooplankton (Fig. 5.3). This was due to a large zooplankton bloom in the late winter and early spring of 1979/80, a result of the intense rainfall of tropical storm Frederick (September–October 1979) and subsequent increased coastal productivity (Sierra and Popova 1988; Bustamante and Schwartzberg, unpublished data).

The yellow jack, unlike the bar jack, is piscivorous from 15–20 cm FL. At this size, small fishes make up 70% of the diet (by weight); squid (25%), crustaceans (3%), and decapod larvae (2%) were also found in the stomach contents. From 25 cm FL on, the individuals feed exclusively on fishes (Sierra et al. 1986). The differences between the ontogenetic patterns of bar and yellow jacks might be attributed to their different mouth sizes. The mouth of the yellow jack is larger than that of the bar jack for the same body length. They attain the same mouth height when the yellow jack is 19–20 cm FL and the bar jack is 28–30 cm FL; from this size on, the proportion of fishes in the diet of the two species is similar. In addition, the number of gill rakers in the first branchial arch of bar jack (20–34) is higher than in yellow jack (19–21; Guitart Manday 1974–1978), which might contribute to the bar jack's continued consumption of planktonic organisms until it attains a larger size.

The redear sardine and false pilchard, despite planktivory throughout the life cycle, exhibit some ontogenetic changes in their prey (Fig. 5.2). Small planktonic organisms (decapod larvae, ostracods, copepods, and others) prevail in the diet of smaller individuals. As the sardine individuals grow, their food composition changes to larger items, such as crustacean megalops, caridean shrimp, amphipods, and polychaetes (Sierra and Díaz-Zaballa 1984; Sierra 1987). Qualitative changes of food size and composition at the size of maturity were observed in most studied species (Fig. 5.2). These transitions tend to replace lower caloric organisms (crustaceans) with higher caloric ones (fishes).

5.4 Feeding Intensity

5.4.1 Diurnal Cycle

Diurnal cycles of feeding activity are species-related and can vary ontogenetically according to the prey types that individuals consume in different stages of their life history. Feeding patterns are associated with the species' ability to forage and catch prey while avoiding predators and are also influenced by environmental conditions. When food is abundant, feeding activity can be concentrated during a certain period of the day; if food supply is poor, two peaks of activity are common, or there may be a whole-day foraging pattern. The decline of feeding activity during the night is considered a defensive reaction against nocturnal predators (Disler 1950; Manteifel 1980).

Among tropical marine fishes, herbivores or those feeding on sessile organisms are typically diurnal. That is true

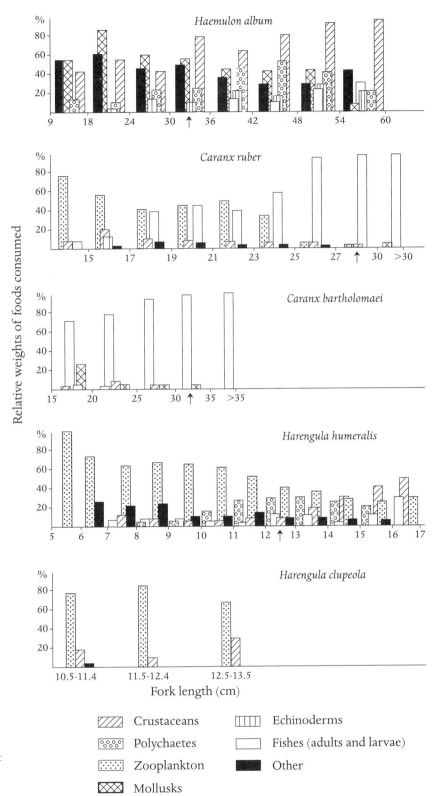

Fig. 5.2. Ontogenetic variation of food consumption in margate (*Haemulon album*), bar jack (*Caranx ruber*), yellow jack (*C. bartholomaei*), redear sardine (*Harengula humeralis*), and false pilchard (*H. clupeola*) (from Sierra and Popova 1982; Sierra 1983, 1987; Sierra and Díaz-Zaballa 1984; Sierra et al. 1986). Arrows pointing upward indicate the size of sexual maturation.

for Scaridae, Labridae, Chaetodontidae, Pomacentridae, and for most territorial species, and strongly for reef-associated fishes. For these species, light is an important factor, both for foraging and for the ability to see predators, because they are under high predation pressure due to their small size.

During a feeding study on groupers in Cuba (Reshetnikov et al. 1974), the fullest stomachs were recorded in the

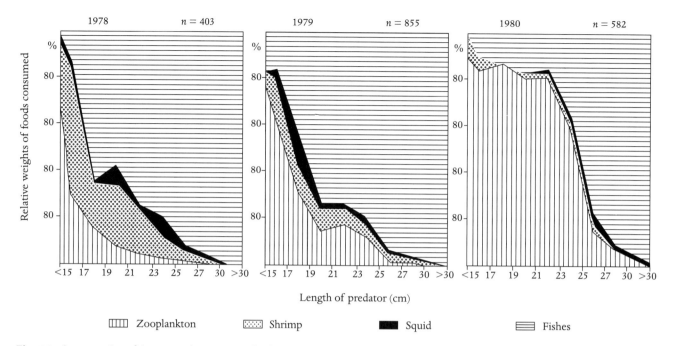

Fig. 5.3. Ontogenetic and interannual variation in food consumption of the bar jack (*Caranx ruber*), 1978–1980 (modified from Sierra and Popova 1982).

morning and at nightfall, although empty stomachs were also recorded at all times of the day. Underwater observations by Reshetnikov and colleagues showed that these predators were more active at dawn and dusk. Other studies have found Nassau groupers with empty and full stomachs at all times of the day, but the highest percentage with recently digested food was encountered early in the morning (6:00 to 8:00 am) and at dusk (Silva Lee 1974b).

Snappers seem to feed during both the day and the night, although the daily pattern differs with species. Lane snappers usually gather in groups around reefs and display low activity during the day. At nightfall they scatter toward the seagrass meadows to eat benthic crustaceans and small nocturnal fishes such as sardines. Many crustaceans and fishes (such as *Opsanus* sp.) remain buried in the bottom sediment during the day and come out at night (Valdés-Muñoz 1987). Most individuals found with fresh food in their stomachs during the day were larger and thus perhaps better equipped for defense. The large size of mutton snapper might make it less vulnerable to predators, allowing it to display an intense diurnal activity, especially in the afternoon. Experimental studies on mutton snapper in fish tanks showed a faster ingestion of food in the afternoon (2:00 to 5:00 pm), although the fishes were fed as much as they would take daily (Claro and Colás 1987).

At dusk, gray snappers disperse to neighboring seagrass beds and forage (Valdés-Muñoz 1987). The highest feeding activity of gray snappers seems to occur early in the night and at dawn. During the day they remain on reefs or among

mangrove roots where they display low feeding activity; nevertheless, fresh food content has been found in large individuals during the day. Cubera snappers (*Lutjanus cyanopterus*) and schoolmasters also feed primarily at night (Randall 1967; Rooker 1995).

Yellowtail snappers usually feed during the day, but some individuals were recorded with half-digested stomach contents early in the morning, indicating nocturnal feeding. Other studies have stated that this species feeds during both the day and the night (Longley and Hildebrand 1941). During daylight underwater observations we saw constant foraging activity in both adults and juveniles. The dog snapper also feeds during both the day and the night. Many dog snappers caught in deep reefs during the day had fresh food in their stomachs.

The highest feeding activity of carangid species occurs during the day when jacks feed mainly on fishes (Starck and Davis 1966). Schools of jack were observed also feeding on zooplankton concentrations at night. In bar jacks and yellow jacks, we found a nearly constant daily pattern of feeding activity throughout the year. The daily variation of the index of gastric repletion (IGR, the weight of prey in the stomach relative to total weight of the predator) showed two peaks during the day: one in the morning and the other in the early afternoon (Fig. 5.4). In bar jack, the morning peak occurred at 8:00 to 9:00 am; in winter it was at 10:00 to 11:00 am. The second (and highest) peak of the day was observed at 12:00 to 2:00 pm in the summer, and at 2:00 to 3:00 pm in the winter (Sierra and Popova 1988). This pattern might be associated with the availability of their main prey species, parrotfishes and small wrasses, whose diurnal foraging activity makes them more accessible to jacks.

Most grunts (Haemulidae) have nocturnal habits (Randall 1967). For these small-sized species, foraging in the seagrass beds or sandy areas during the day would expose them to predators. This might explain why the margate—the largest of the grunt species—feeds during the day. An analogous pattern was demonstrated for the mutton snapper among coastal snappers. The daily fluctuation of IGR for margate had two peaks: one early in the morning and another, larger, peak in the afternoon (Fig. 5.4). Semidigested food in individuals caught in the morning suggested some feeding activity during the night. Studies have shown that small margate (less than 30 cm FL) remain on the reefs

during the day but leave at night for foraging, whereas adults do the opposite (Cummings et al. 1966; Starck and Davis 1966). The hogfish feeds nearly exclusively during the day, mainly in the afternoon (Fig. 5.4).

Redear sardine and false pilchard move at dusk to less sheltered open shelf areas, apparently in small groups. At dawn the schools move onshore and all individuals were found with full stomachs. At this time, they exhibited an IGR of 9.1% for the redear sardine and 5.1% for the false pilchard (Sierra and Díaz-Zaballa 1984). The IGR dropped rapidly during the day, reaching its minimum at 5:00 pm in winter (Fig. 5.4). In one of two surveys we conducted to describe this cycle, we observed that the individuals in a single school of redear sardines at 2:00 pm had a higher (5.1%) IGR than at

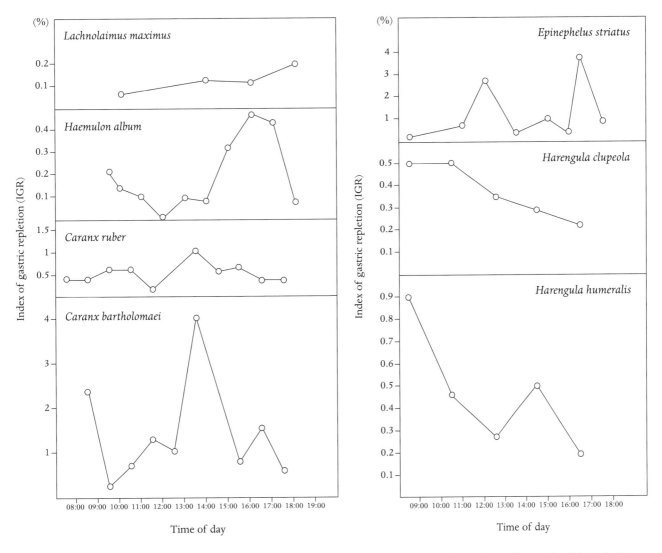

Fig. 5.4. Diurnal variation in the index of gastric repletion (IGR) in various Cuban fish species, from Claro et al. 1989 (hogfish, *Lachnolaimus maximus*), Sierra 1983 (margate, *Haemulon album*), Sierra and Popova 1982 (bar jack, *Caranx ruber*), Sierra et al. 1986 (yellow jack, *C. bartholomaei*), Silva-Lee 1974b (Nassau grouper, *Epinephelus striatus*), and Sierra and Díaz-Zaballa 1984 (false pilchard, *Harengula clupeola*, and redear herring, *H. humeralis*).

noon (2.7%), which suggests an afternoon feeding. This finding agrees with other studies (Silva Lee 1974a) that found that redear sardine might eat during the day. False pilchard individuals in the same school with the redear sardines, however, did not have food in their stomachs, which suggests a different food selectivity for the two species.

5.4.2 Digestion Rate

Digestion rate, as part of the process of food assimilation, can be considered an indirect indicator of metabolism intensity. Digestion rate is known to be higher for active foragers with a high metabolic expenditure (Reshetnikov et al. 1975). Experimental studies in both marine and freshwater fishes have demonstrated a strong relationship between water temperature and digestion rate (Arnold and Fortunatova 1937; Karpevich and Bokova 1937; Fabian et al. 1963; Molnar and Tolg 1967; Smith 1967; Tyler 1970; Edwards et al. 1971; Fortunatova and Popova 1973; Reshetnikov et al. 1974; Brett and Groves 1979; Sierra and Claro 1979). Within a species, the size and age of an individual and the size and density of its prey's scaled skin might influence the rate of food digestion (Manteifel et al. 1965). Biochemical and physiological factors also intervene in the process. The physiological

condition of the fish, as well as the water temperature, might affect the activity of digestive fluids, which is higher in more active fishes (Fabian et al. 1963).

Experiments on the feeding habits of snappers (Reshetnikov et al. 1974; Sierra and Claro 1979), bar jack, and margate (Sierra and Popova 1988) are summarized in Fig. 5.5. Dog snapper, schoolmaster, and gray snapper (Reshetnikov et al. 1974) and lane and gray snappers (Sierra and Claro 1979) fed sardines showed a digestive period lasting 45 to 50 hours in winter and 22 to 26 hours in summer; intermediate spring and fall values were also obtained. The highest digestion rate was recorded one month after spawning peaks for the lane (22 hours, June) and gray snapper (25 hours, October). For both species, water temperatures were relatively high. The fishes were also recovering from spawning and in their main growth period; both of these processes might require a higher food intake. In nearly all experiments (except for the one conducted in October), the food digestion period was 1–3 hours longer in gray snappers than in lane snappers, possibly due to the lane snapper's higher swimming activity.

Bar jack, in both winter and summer, entirely digested food in half the time taken by snappers. Bar jack is an active,

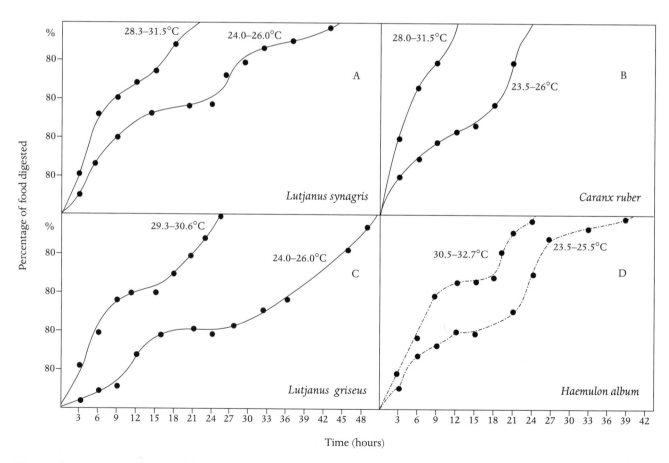

Fig. 5.5. Digestion rates of various fishes in winter and summer while feeding on fishes (A, B, and C) or crabs (D) (modified from Reshetnikov et al. 1974).

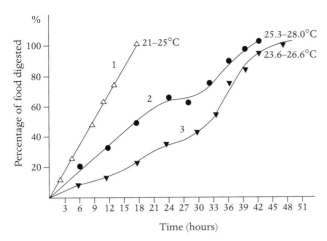

Fig. 5.6. Digestion rate of lane snapper (*Lutjanus synagris*) fed 1: dwarf herring (*Jenkinsia lamprotaenia*); 2: redear sardine (*Harengula humeralis*); or 3: crabs (modified from Sierra and Claro 1979).

piscivorous forager that swims constantly, whereas snappers shelter in refuges or remain close to them. Bar jacks also have a smaller stomach than snappers and groupers, so they must eat smaller prey. Small food organisms must be ingested quickly, so the fish can eat a larger number of prey to meet nutritional needs.

The digestion process also depends on the characteristics of the prey. In experiments in which lane snappers were fed crabs, digestion took twice as long as when they were fed dwarf herring (Fig. 5.6; Sierra and Claro 1979). The thick calcareous exoskeleton of the crabs could account for this difference. Dwarf herring are small and have no scales, so digestion of them might be faster than for scaled redear sardines, as well as in crabs.

Small seasonal differences in water temperature (5–6°C) can also generate significant changes in physiological processes. Even small temperature changes of 1 or 2°C can have a notable effect. Decreased digestive activity occurs in the early morning, in the part of the diurnal cycle when temperatures are at their minimum (e.g., the inflections in Fig. 5.5).

5.4.3 Food Ration

For temperate freshwater fish species, data are abundant on the daily, monthly, and annual food rations and their spatial and seasonal variations, but little is known about the food rations of marine fishes. Of the studies conducted on marine species, most have concerned tropical and subtropical species (Edwards et al. 1971; Gorelova 1974; Graber 1974; Lipskaia 1975; Ryzhov and Formoso 1975; Clarke 1978, 1980; Wakeman et al. 1979; Gruber 1982; Stilwell and Kohler 1982). These studies provide only isolated information on fish food rations because samples were collected in different seasons or during experimental conditions.

In our studies, we estimated the daily, monthly, and annual rations of several Cuban coastal fish species, in both natural and experimental conditions, using the formula proposed by Fortunatova (1961):

$$DR = \sum_{i=1}^{n} \frac{S_i/n_i}{V} \times \frac{n}{N}$$

where DR is daily ration, the amount of food consumed in one day, expressed as a percentage of the fish's weight; N is the total number of fishes sampled; n is the number of fishes with gut content; V is the time of completion of digestion (in days or fractions of days); n_i is the number of fishes with gut content in each subsample; and S_i is the amount of food ingested by all examined individuals in each subsample, expressed as a percentage of the reconstructed food weight in relation to fish weight. Comparative tables were constructed to estimate from the food content the original weight of the prey (when they were ingested). The tables provide a relationship between a certain part of the prey and its whole length/weight (Popova and Sierra 1983). To help identify the prey, we selected bones (cleithrum, lower jaw) that are frequently found in the stomach contents.

Studies on bar jack and yellow jack and other species from southwest Cuba in 1979–1980 provide data on feeding patterns among different size groups and temporal scales (Figs. 5.7 to 5.9). The daily ration for juvenile (12–18 cm) bar jacks was higher in 1980 (mean 4.1% of the fish's weight) than in 1979 (3.1%) because of the intensive consumption of zooplankton by fishes in 1980. Abundant and long-lasting runoffs after heavy rainfall in September 1979 provoked a plankton bloom in coastal areas that lasted more than six months. In individuals of 18–30 cm that consume both zooplankton and fishes, the average daily ration in 1980 was slightly higher (3.2%) than in 1979 (2.8%). Bar jacks larger than 30 cm (piscivorous adults) had daily rations of 0.4–6.5%, depending on the season. In this part of the population, daily rations were similar in 1979 and 1980 (Fig. 5.9). In yellow jacks, daily and annual food ration (mostly fishes) were high in both adults (mean, 3.9%) and subadults (4.2%; Sierra et al. 1986). Daily rations for bar jacks (18–30 cm length) were smaller than for yellow jacks of similar size (20–30 cm). This might be because yellow jacks appear to feed on fishes earlier than bar jacks (Sierra 1986).

The estimated daily ration of Nassau groupers in natural conditions was 1.7–5.7% of the fish's weight (Reshetnikov et al. 1975). The daily ration of mutton snappers in tanks, fed as many fishes as they would take, was 2.2–4.5% of fish weight (Fig. 5.10), for an annual ration of 1,167% of fish weight (Claro and Colás 1987). Because food ration might differ with fish size, Claro and Colás (1987) placed two individuals (58 and 83 g weight) in separate tanks; their food ration in March was 4.3% (daily average). Larger experimental fishes (mean weight, 299 g) ingested only 2.8% of their daily weight in the same time period. Analogous results were obtained in March from two schoolmaster individuals (see Claro and Colás 1987; Sierra et al. 1994, for details). The daily ration of 31 mutton snappers under

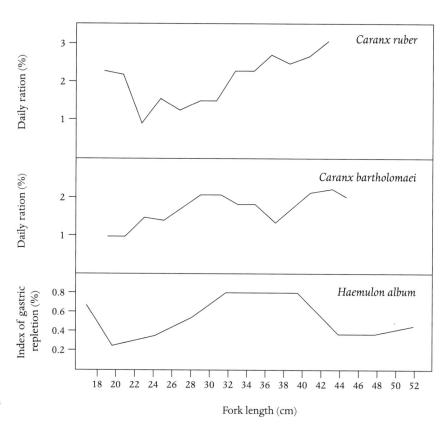

Fig. 5.7. Ontogenetic variation in feeding intensity in three species (modified from Sierra and Popova 1982; Sierra 1983; Sierra et al. 1986).

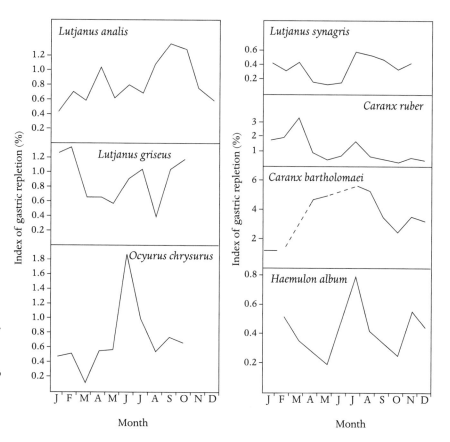

Fig. 5.8. Seasonal variations of feeding intensity in mutton snapper (*Lutjanus analis*), gray snapper (*L. griseus*), lane snapper (*L. synagris*), yellowtail snapper (*Ocyurus chrysurus*), bar jack (*Caranx ruber*), yellow jack (*C. bartholomaei*), and margate (*Haemulon album*) from Cuba. The dotted line indicates no data available. (From Claro 1981c, 1982, 1983c, 1983d; Sierra and Popova 1982; Sierra 1983; Sierra et al. 1986.)

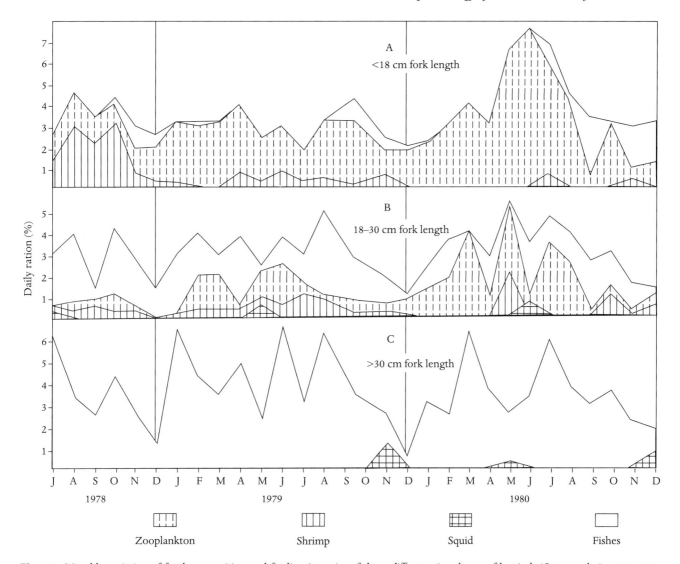

Fig. 5.9. Monthly variation of food composition and feeding intensity of three different size classes of bar jack (*Caranx ruber*), 1978–1980 (modified from Sierra and Popova 1982).

natural conditions in the winter (mean, 2.3%) was similar to the ration calculated for fish in tanks. In 53 lane snapper individuals we obtained a daily ration of 1.4%, although 75% of the individuals had empty stomachs, which reduced the calculated value. A 1.5% daily ration has been reported for red snappers under experimental conditions (Wakeman et al. 1979).

Daily and annual food rations for the coastal fishes of Cuba (Reshetnikov et al. 1975; Sierra 1986; see Table 5.6 in Sierra et al. 1994) showed similar values for mutton snappers, Nassau groupers, and bar jacks. However, data obtained for yellow jacks were somewhat higher,. The data collected on small-sized fishes, such as Spanish hogfish, beaugregory (*Stegastes leucostictus*), and tobaccofish (*Serranus tabacarius*) (4–10 cm length) in the Cuban National Aquarium (under experimental conditions) provided values of 11–30% of fish weight, depending on the type of food supplied (Pastor 1987).

5.4.4   Ontogenetic Patterns in Feeding Intensity

A change in feeding habits from smaller to larger organisms during the life cycle can be associated with a change in growth rate and metabolism. Shatunovsky (1980) stated that strictly zooplanktivorous fishes are an exception to this hypothesis. Our data showed that ontogenetic shifts toward the consumption of larger, more caloric food during particularly critical periods of the life cycle can include a decrease in daily rations as a consequence of a temporal- and size-associated increase in mobility to capture larger and more mobile prey. As a result, fishes expend more energy on searching for food, and the actual food ration is lower than would be expected (Bustamante 1988; Bustamante et al. 1988).

Data collected in Cuba for bar jacks showed a decline in food consumption at 20–22 cm length (attainment of the first year of life) (Fig. 5.7). This stage marks the beginning of the gradual change from feeding mostly on zooplankton and

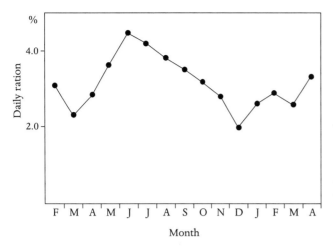

Fig. 5.10. Monthly variation of mean daily food ration (as a percentage of fish body weight) in mutton snapper (*Lutjanus analis*) under experimental conditions in aquaria (from Claro and Colás 1987).

shrimp to feeding on fishes. This temporal crisis ends when the individual attains 28–30 cm; at this length they become entirely piscivorous and the food ration increases substantially. From this size on, all individuals are adults. Such a marked ontogenetic shift in fish feeding habits and physiological condition was not clear for the yellow jack; the food ration increased gradually up to attainment of sexual maturity (30–32 cm) (Fig. 5.7).

In the margate, the IGR dropped with the decline of small crustaceans in the diet and an increase of mollusks. After the margate attains the first year of life, at 18–22 cm (García-Arteaga 1983), the average IGR per size class increases to the maximum value for adult individuals (> 30 cm length) at maturation.

5.4.5   Seasonal Variations in Feeding Intensity

The annual rhythm of fish feeding intensity is strongly associated with environmental conditions and their effect on the food supply. Just as temperature affects fish metabolism, it also affects the organism's food requirements. IGR can be considered an indirect indicator of feeding intensity, although food ration is a more appropriate indicator. IGR in four species of snapper typically showed higher values in the summer (Fig. 5.8) and a decline during the spawning peak of each species, when many individuals were found with empty stomachs. The individual IGR of the few spawners found with some food content was as high as the IGR for the summer period. The decrease in IGR at the population level might be caused by higher competition for food among spawners because they are in high concentrations during spawning migrations. This period of low feeding might span two to three weeks during migrations to and from the spawning sites. This process and the great energy expendi-

tures during ovulation and migration lead to marked alteration of physiological condition (see Chapter 7). Thus, spawning is followed by a period of intensive feeding.

Experiments on juvenile mutton snappers (1–2 years old) in tanks (Fig. 5.10) with naturally varying water temperature showed maximum daily food rations in June–August (4.0–4.5% of fish weight, monthly mean) and minimum values in December (2.2%) when the temperature declined (Claro and Colás 1987). This pattern is analogous to that found in natural conditions (Fig. 5.8) if we ignore the May and June IGR values that are influenced by spawning.

The annual variation of food ration and composition for bar jacks of three different size classes (12–18 cm FL, juveniles; 18–30 cm, subadults; and >30 cm, adults) is shown in Fig. 5.9. During 1979 and 1980, the food of 12- to 18-cm individuals was dominated by copepod larvae, particularly during January–March 1980 (2–7.4% fish weight, monthly average). The food ration increased from January to April in 1979 and 1980 (no data in 1978), and from August to October in 1978 and 1979. This species showed faster growth (García-Arteaga and Reshetnikov 1992) and intensive fat accumulation (Bustamante et al. 1981) during this latter period. The low values of food consumption in November–December might be attributable to falling water temperatures. Growth studies showed that growth rate declined during this period (García-Arteaga and Reshetnikov 1992). For fishes larger than 30 cm, the seasonal cycle was similar for the three studied years. Two peaks of feeding intensity were evident: January–March, then decreasing during the reproductive season from April to June, and a new peak in summer after spawning, followed by a marked decline from July–August to December (see Section 4.4).

5.5   Predator–Prey Size Relationships

Prey size (length, diameter, body depth, etc.) and predator mouth aperture are the most important factors involved in prey selectivity (Vasnetsov 1953; Ivlev 1955; Werner 1974; Juanes 1994). Mijeev (1983) emphasized that prey size is related to the predator's energy expenditure for catching the prey and to its energy value to the predator. The predator–prey size dependence has been reported for many fish species (Ivlev 1955). Prey size has been found to increase proportionally with predator size; the maximum and range of prey size depends on the type of food and the size of the predator's mouth cavity. Popova (1967) demonstrated that prey size was generally 10% to 60% of predator size, decreasing with the growth of the latter.

Sierra et al. (1994) summarized the predator–prey size information available for Cuban fishes (see Table 5.3). Although clear correlations between these two parameters are not always present, enough information is available to propose several logical relationships. The predator's mouth opening seems to restrict the prey's perimeter rather than its

Table 5.3.   Relationships between predator and prey sizes in some Cuban marine fishes

Family Species	Predator Length (cm)	Prey Length		Source
		Total Length (cm)	Relative Length (%) (Range)	
Belonidae				
Tylosurus acus acus	43–46	3.0	6.7 (6.5–6.9)	Sierra et al. 1994
Serranidae				
Epinephelus itajara	191	12.5–63.5	20 (6.5–33.2)	Sierra et al. 1994
Epinephelus striatus	20–65	1.0–62.0	15 (1.5–107.0)	Silva Lee 1974b
Mycteroperca bonaci	25–101	8.2–30.0	27 (16.5–33.1)	Sierra et al. 1994
Mycteroperca tigris	35–63	7.5–18.0	26 (15.9–42.8)	Sierra et al. 1994
Mycteroperca venenosa	34–44	4.2–9.0	17 (12.3–20.4)	Sierra et al. 1994
Carangidae				
Caranx bartholomaei	15–15	1.7–75.0	20 (11.3–27.2)	Sierra et al. 1986
Caranx ruber	15–43	2.3–14.5	22 (15.3–33.7)	Sierra and Popova 1982
Lutjanidae				
Lutjanus analis	24–68	1.0–20.0	25 (2.0–41.0)	Reshetnikov et al. 1974
Lutjanus apodus	22–23	1.0–13.0	15 (4.0–19.0)	Reshetnikov et al. 1974
Lutjanus griseus	25–40	2.0–19.0	18 (8.0–34.0)	Reshetnikov et al. 1974
Haemulidae				
Haemulon album	15–60	1.5–30.0	18 (18.7–52.6)	Sierra et al. 1994
Sciaenidae				
Bairdiella ronchus	16–17	3.0–12.0	21 (17.7–25.0)	García and Nieto 1978
Sphyraenidae				
Sphyraena barracuda	25–60	3.0–4.0	16 (12.0–25.0)	Reshetnikov et al. 1974
Sphyraena guachancho	23–40	3.0–12.0	23 (12.0–40.0)	Reshetnikov et al. 1974
Scombridae				
Scomberomorus regalis	41–63	4.0–28	24 (9.7–44.4)	Sierra et al. 1994

length. A large mouth with a consequently broad buccal cavity apparently has an adaptive advantage for widening the range of potential prey. Although overall prey size increases with predator size in the margate and Nassau grouper (Fig. 5.11), our data on the bar jack and yellow jack showed that, above a certain length, the range of prey size in the former species narrowed as the maximum size of the predator is reached. This might be related to the increased effectiveness of the prey's defense as it grows. In the margate, the minimum prey size increased in proportion to predator size. This relationship was not evident in groupers. Because these species are ambush predators, they might be able to consume larger prey without a corresponding increase in energy expenditure, thereby compensating for the ingestion of small prey.

5.6   Trophic Comparisons with other Habitats and Regions

5.6.1   Trophic Structure among Habitats

The trophic structure of fish communities in different habitats (patch reefs, coastal reefs, reef crests, slope reefs, and nonestuarine mangroves) in the major archipelagos of the Cuban shelf (Sierra et al. 1990; Claro and García-Arteaga 1993, 1994b) and on the slope reefs of Martinique, Guadeloupe, and Key West, Florida, were compared using the same visual census method (Claro et al. 1998) (Table 5.4). The highest biomass of herbivorous fishes was observed in the reef crests of the Golfo de Batabanó. Most of the sampled reef crests in this area are relatively far from fishing grounds. Among the slope reef habitats that were comparable, the highest biomass of herbivorous fishes was present in a recently designated marine protected area (Cayos de Doce Leguas) in Jardines de la Reina. This area also had the highest total fish density and biomass (see Section 2.3.1). In the heavily fished areas of north-central and southwest Cuba, and particularly in Martinique, Guadeloupe, and Key West, Florida, herbivorous fishes were scarce.

Various authors (Hiatt and Strasburg 1960; Randall 1963b, 1967; Acero 1980; Bohnsack et al. 1987; Polunin and Roberts 1996) have concluded that herbivores usually represent 10–20% of the fish species on western Atlantic coral reefs. We found this to be true for our study region, regardless of habitat: patch reefs, mangroves, and artificial refuges. Bakus (1966) suggested that the proportion of herbivores can reach 25%, and Harmelin-Vivien (1981) and Hobson (1974) reported less than 10% in the Indian and

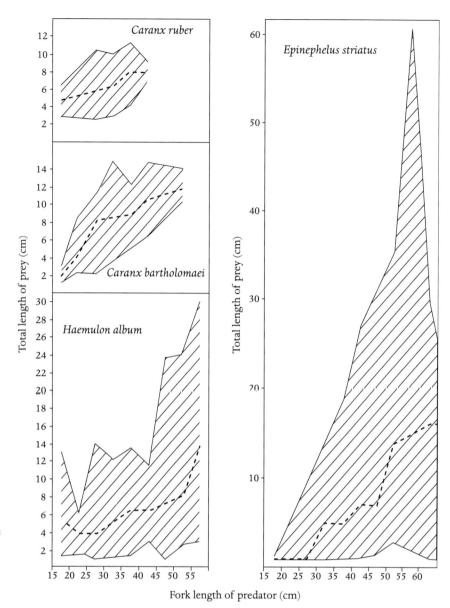

Fig. 5.11. Size relationships among predators and prey in four reef fish species found in Cuba. The dashed line indicates the mean prey size. The hatched areas encompass the maximum and minimum prey sizes.

Pacific oceans. Although our data agree with those provided by these authors, we must add that this proportion is based on the number of species, which is not necessarily a reliable representation of ecosystem structure. Many reef species reach a small maximum length or their populations are not dense, so their role in the trophic structure of the ecosystem might not be as important as that of dense populations of a few species (for example acanthurids) or large individuals (scarids for herbivores, and snappers and groupers for carnivores).

In mangroves of the three archipelagos studied in Cuba, as well as in patch reefs, the proportion of carnivorous fishes was higher (90%) than on the reef crests (Table 5.4). The proportion of carnivorous fishes in artificial refuges (in the Golfo de Batabanó) was higher than in other bottom types; this might be related to a richer food supply in the areas where the artificial refuges are placed (mainly in seagrass meadows). Randall (1967) also found that the proportion of carnivorous fishes was higher in artificial refuges than in natural refuges. Fishes from mangroves and patch reefs also feed on seagrass beds.

5.6.2   Trophic Patterns among Regions

Prey and size diversity is higher for fishes than for any other vertebrate group (Nikolsky 1963). This diversity depends on climatic conditions: many studies suggest that feeding habits in temperate fishes are more generalized (opportunistic) than in tropical species (Hartley 1948; Dobzhansky 1950; Larkin 1956; Nikolsky 1963, 1974a; Keast and Webb 1966; Pianka 1970). More recent information demonstrates great trophic variability in tropical fishes (Turingan et al. 1995). Criteria for

Table 5.4. Trophic structure (as relative fish biomass) of fish communities in different habitats and regions

		Archipiélago Jardines de la Reina (SE Cuba)			Archipiélago Los Canarreos (SW Cuba)			
		Mangroves	Reef Crests	Slope Reefs	Mangroves	Reef Crests	Patch Reefs	Slope Reefs
Omnivorous	(g/m^2)	0.03	0.51	1.08	0.04	1.99	0.59	0.75
	(%)	<0.1	0.5	0.5	0.3	1.7	0.1	0.7
Herbivorous	(g/m^2)	3.70	30.85	53.96	3.42	61.88	22.76	17.97
	(%)	2.6	30.1	24.5	2.2	53.6	3.9	16.1
Planktivorous	(g/m^2)	1.89	0.02	34.28	10.85	0.97	0.33	10.54
	(%)	1.3	<0.1	15.6	7.0	0.8	0.1	9.5
Benthophagous	(g/m^2)	40.59	62.35	50.79	63.80	27.49	398.90	14.24
	(%)	30.8	57.6	23.1	41.2	24.0	68.5	12.8
Piscivorous	(g/m^2)	71.06	3.72	26.30	54.54	9.05	131.08	38.95
	(%)	50.2	3.4	11.9	37.9	7.8	22.5	35.0
Invert + fishes	(g/m^2)	21.31	9.92	53.50	17.52	13.97	28.72	28.88
	(%)	15.0	8.2	24.4	11.4	12.1	4.9	25.9

		Archipiélago Sabana-Camagüey (North-Central Cuba)				Martinique	Guadeloupe	Florida
		Mangroves	Slope Reefs	Reef Crests	Patch Reefs	Slope Reefs	Slope Reefs	Slope Reefs
Omnivorous	(g/m^2)	0.09	0.48	0.63	0.36	4.86	1.69	1.60
	(%)	<0.1	0.3	0.5	0.3	7.7	2.7	0.7
Herbivorous	(g/m^2)	4.62	16.62	31.59	16.10	5.40	11.03	8.54
	(%)	2.7	11.9	27.0	11.6	8.6	17.4	3.9
Planktivorous	(g/m^2)	12.68	5.12	1.06	0.86	29.00	27.52	6.9
	(%)	7.5	3.7	0.9	0.6	46.1	43.0	0.3
Benthophagous	(g/m^2)	71.03	31.60	60.40	85.97	15.15	10.6	138.1
	(%)	42.1	22.6	51.6	62.5	24.1	16.9	61.9
Piscivorous	(g/m^2)	49.28	35.44	12.43	14.77	6.10	6.41	13.29
	(%)	29.7	25.3	10.6	10.4	9.7	10.1	6.0
Invert+fishes	(g/m^2)	31.16	50.60	11.02	20.44	2.42	6.36	60.71
	(%)	18.4	36.2	9.4	14.6	3.8	9.9	27.2

stenophagy are not specific and do not easily allow comparison among species. Data on tropical fishes show a great taxonomic variety of prey, even in species characterized by relatively high trophic specialization. For example, the Atlantic herring (*Clupea harengus harengus*), a temperate zooplanktivorous fish, feeds on one to three main taxa (that make up 50–100% of the diet), whereas two tropical zooplanktivores of the same family (redear sardine and false pilchard) consume a dozen different taxa that can vary with location and time (Sierra and Díaz-Zaballa 1984). Therefore, trophic specializations can vary geographically. For example, on a broad scale, parrotfishes (Scaridae), butterflyfishes (Chaetodontidae), two triggerfish (Balistidae) species, and several other reef fishes of the Marshall Islands (Pacific Ocean) are coral-polyp consumers (Hiatt and Strasburg 1960), but that feeding habit does not occur in the Antilles, even in species analogous to those living in the Pacific (Randall 1967). Randall found small amounts of coral tissue in

the stomachs of 10 fish species in Puerto Rico and the Virgin Islands, but considered coral as incidental food from scraping of the coral surface in search of algae.

A comparison of the feeding habits of fishes from different regions (Tables 5.5 and 5.6) suggests that the proportion of piscivorous fishes in the tropics is similar to that of some temperate waters (e.g., Gulf of Maine), but the number of planktivorous species is lower. Benthic feeding species constitute approximately 52% of the Cuban ichthyo-fauna; many of these also feed on fishes. Some are omnivorous and others can be planktivorous at some period of their life history. Marine phytobenthophagous fishes are also more numerous in the tropics, although many of them have a low biomass.

Despite numerous examples of trophic specialization among tropical fishes, most of the species we studied can be considered trophic opportunists with differing degrees of

Table 5.5.   Trophic composition of the ichthyofauna of different regions

	Gulf of Maine[a]		U.S. Virgin I. and Puerto Rico[b]		Marshall I., Pacific Ocean[c]		Margarita I., Venezuela[d]		Cuba[e]	
	%	n	%	n	%	n	%	n	%	n
Detritivorous	0.7	1	0.5	1	0.9	2	1.7	2	1.4	5
Omnivorous	2.0	3	5.2	11	10.5	23	5.0	6	3.8	14
Herbivorous	0.7	1	13.3	28	14.9	33	3.3	4	8.5	31
Carnivorous	96.4	144	81.0	171	73.7	163	90.0	108	86.3	315
Planktivorous	16.9	25	13.3	28	3.6	8	9.2	11	12.3	45
Benthophagous	41.2	61	37.0	78	42.5	94	44.2	53	33.1	121
Piscivorous	39.2	58	18.9	40	13.1	29	18.3	22	25.3	92
Piscivorous and benthophagous	—	—	11.8	25	14.5	32	18.3	22	15.6	57
Total no. of species	149		211		221		120		365	

Sources: [a]Bigelow and Schroeder 1953; [b]Randall 1967; [c]Hiatt and Strasburg 1960; [d]Cervigón 1994; [e]This chapter.
%: the relative proportion of species with a given feeding pattern in that region; n: number of species.

Table 5.6.   Biological characteristics of predatory fishes of different latitudes

	Latitude		
Characteristic	High	Intermediate	Tropics
Feeding period (months)	3–4.5	5.5–8.5	12
Water temperature (°C)	8–20	8–25	22–32
Digestion rate (days)	5–2	4–1	2–0.5
Annual ration (% of fish weight)	150–350	250–500	998–1,500
Growth rate in first year (cm)	2–3	5–10	10–30
Age of sexual maturation (years)	4–5	2–3	1–4
Longevity (years)	15–30	10–15	6–15

Source: For high and intermediate latitudes, data from Popova 1982.

feeding generalization. Among those inhabiting Cuban shelf waters, nearly 80% can be classified as euryphagous species by virtue of their broad feeding spectrum. Among the specialist fish species, however, relatively few species represent a notable component of the Cuban marine ecosystem. Exceptions include several species of parrotfishes (Scaridae), mullets (Mugilidae), and sardines (Clupeidae).

Euryphagy in fish is typically associated with a larger expenditure of energy for foraging and a lower expenditure for conversion of food into biomass. Adapting to the consumption of a limited number of food objects both diminishes energy expenditure (if these targets are abundant) and facilitates a high degree of food assimilation (Nikolsky 1974a). Stenophagous species feed mainly on the lowest trophic levels (plants, zooplankton, and detritus), whereas euryphagous fishes use higher trophic levels with a consequent energy loss. Tropical ichthyofaunas are characterized by considerable morphological specialization, as noted by some authors (Dobzhansky 1950; Nikolsky 1963,

1974b; Keast and Webb 1966; Cowey and Sargent 1979; Wainwright 1988; Hobson 1991), but they are also characterized by generalized feeding habits. Both of these interrelated factors play a major role in the organization of trophic relationships on reefs.

5.7   Summary

The feeding habits of the 365 fish species inhabiting the Cuban shelf for which information is available suggest that euryphagy (rather than stenophagy) is the dominant trophic strategy. This might be an adaptation to a diverse but variably abundant food supply. Although there can be extensive trophic specialization in tropical fishes, these adaptations might not be fixed and can vary widely with fluctuating local food supplies. Nearly 80% of the Cuban ichthyofauna consists of trophic generalists or opportunists. Therefore, the concept of trophic specialization in tropical waters and opportunism in temperate species suggested by some authors (Hartley 1948; Larkin

1956; Keast and Webb 1966; Hiatt 1979) might not apply to the marine species of Cuba and perhaps the wider Caribbean. However, trophic specialists can be abundant in the tropics, often as an adaptation to local food supply.

Three main groups have been identified as stenophagous: zooplanktivores, herbivores, and detritivores. Altogether, these groups account for less than 20% of the Cuban fish fauna. Zooplanktivores are relatively few (12%) in comparison with the ichthyofauna of temperate and polar regions. Strictly herbivorous fish species make up only 8.5% of the ichthyofauna, although up to 15% of the total species eat plants and/or algae. This group encompasses mainly small fishes, often those associated with reef crevices. Detritivores are restricted to mullets (Mugilidae), which live in coastal lagoons and estuaries, although other species might incidentally ingest detritus.

Zooplankton play an important trophic role in the early life-history stages of many species. The increase in complexity of food webs as fishes grow might be explained by the relatively small amount of plankton biomass in tropical waters.

About 80% of Cuban marine fish species are carnivorous predators, primarily with diets of small fishes and invertebrates. Approximately 38% of the species are typically piscivores (> 50% of the food spectrum is fish) and 52% consume fish in smaller proportions. Of the studied fish species, approximately 59% eat crustaceans, which are the main prey for 28%. Nearly all predators consume crustaceans during some part of the life history, so this group is a significant component of trophic webs. Mollusks are the main food item for 5% of species, although they are consumed by many more (34% of the total). Cephalopods are ingested principally by pelagic–oceanic fishes, whereas gastropods and pelecypods are a food item of demersal neritic fishes. Many carnivorous predators consume polychaetes (18.4%), but they are an important food item in a lesser percentage.

Feeding habits shift with age in many species, typically from a diet based on small planktonic and benthic organisms to larger organisms of wider diversity. Such changes might be partly influenced by the balance between the energy expended in foraging and the energetic contribution of the prey. The ratio between prey size and predator mouth size seems to be specific for each life-history period and dependent on energy requirements, although other behavioral features of prey and predators are also involved.

The daily cycle of food ingestion in the studied species seems to be related to the defense mechanisms of the predator and the prey. Most of the smaller fish species that use permanent shelters feed during the day and occupy shelters at night. Small and medium-sized fishes that feed away from the reef do so during the night, and are typically scattered. Larger fishes or those with efficient defense mechanisms (e.g., schooling) can feed both night and day. Prey behavior and its relative abundance combine to produce different feeding patterns, although a marked tendency for nocturnal feeding, mainly after dusk and before dawn, was observed among predators.

Temporal feeding patterns might vary according to local food supply, which is under the influence of environmental factors such as temperature and rainfall. The seasonal cycle of feeding intensity can exhibit two periods of maximum food consumption, summer and late winter or early spring. Water temperature is instrumental in this pattern. The digestion rate during the summer can be twice the winter rate. This factor, together with elevated energy expenditure, influences foraging intensity. Comparison of feeding processes in some predatory fishes of Cuba with other regions suggests that the annual food ration is often larger for tropical fishes than for temperate and polar species because tropical fishes feed nearly all year and digest food faster.

Appendix 5.1.   Feeding habits and main dietary components of most marine fish species occurring in Cuba.

Family Species	Feeding Habits[a]	Main Food Components[b]	Source
Hexanchidae			
Heptranchias perlo	P	F, C	FAO 1978
Hexanchus griseus	P	F, C	FAO 1978
Rhincodontidae			
Rhincodon typus	Pk	C, F, Mc	FAO 1978
Ginglymostomatidae			
Ginglymostoma cirratum	P	F, M	Randall 1967
Alopiidae			
Alopias superciliosus	P	F	FAO 1978
Alopias vulpinus	P	F, Mc, C	FAO 1978
Lamnidae			
Isurus oxyrinchus	P	F	Naranjo 1956
Isurus paucus	P	F	FAO 1978
Triakidae			
Allomycter dissutus	PB	C, M, F	FAO 1978
Mustelus canis insularis	PB	C, M, F	FAO 1978
Proscyllidae			
Eridacnis barbouri	P	C, M, F	FAO 1978
Carcharhinidae			
Carcharhinus acronotus	P	F	FAO 1978
Carcharhinus altimus	P	F, Mc	FAO 1978
Carcharhinus brevipinna	P	F, Mc	FAO 1978
Carcharhinus falciformis	P	F, Mc, Cc	FAO 1978
Carcharhinus leucas	P	F, C	FAO 1978
Carcharhinus limbatus	P	F, Mc	FAO 1978
Carcharhinus longimanus	P	F, Mc, Oi	FAO 1978
Carcharhinus perezii	P	F	Randall 1967
Carcharhinus plumbeus	PB	C, M, E, F	FAO 1978
Galeocerdo cuvier	PB	F, Cc, T	Randall 1967; Cervigón 1994
Negaprion brevirostris	P	F	Randall 1967
Prionace glauca	P	F, Mc	FAO 1978
Rhizoprionodon terraenovae	P	F	Randall 1967
Sphyrnidae			
Sphyrna lewini	P	F, Mc, C	FAO 1978
Sphyrna mokarran	P	F, C, M	FAO 1978
Sphyrna tiburo	P	F, C, M	FAO 1978
Sphyrna zygaena	P	F, C	Naranjo 1956
Squalidae			
Centrophorus granulosus	P	F	FAO 1978
Etmopterus hillianus	P	F	FAO 1978
Squalus cubensis	P	F	FAO 1978
Pristidae			
Pristis pectinata	PB	Oi, F	Cervigón 1994
Narcinidae			
Narcine brasiliensis	B	A	Cervigón 1994
Dasyatidae			
Dasyatis americana	B	F, Sp, Cc, A, Mp, Cm	Randall 1967
Myliobatidae			
Aetobatus narinari	B	Mg, Mp	Randall 1967
Manta birostris	Pk		FAO 1978
Mobula hypostoma	Pk		FAO 1978
Mobula mobular	Pk		FAO 1978
Elopidae			
Elops saurus	B	F, Cj, Cs	Cervigón 1994

Continued on next page

Appendix 5.1. continued

Megalops atlanticus	P	F	Randall 1967
Albulidae			
Albula vulpes	B	Cc, Cs, F, Oi, M	Bruger 1974; Crabtree et al. 1998
Anguillidae			
Anguilla rostrata	B	Mp, C, A, Fl	FAO 1978
Muraenidae			
Echidna catenata	B	Cc, Cs	Randall 1967
Gymnothorax funebris	P	F	Gudger 1929
Gymnothorax moringa	P	F	Randall 1967
Gymnothorax vicinus	PB	F, Cc, Mc	Randall 1967
Congridae			
Conger esculentus	P	F	FAO 1978
Ophichthidae			
Myrichthys breviceps	B	Cc, Ce, Cs, F, A	Randall 1967
Myrichthys oculatus	B	Cc, Ce, E	Randall 1967
Ophichthus ophis	PB	F, Mc	Randall 1967
Muraenesocidae			
Cynoponticus savana	B		FAO 1978
Clupeidae			
Harengula clupeola	Pk	C, A	García 1976; Sierra and Díaz-Zaballa 1984
Harengula humeralis	Pk	A, Cs, Co, Cs	García 1976; Sierra and Díaz-Zaballa 1984; Sierra 1987
Harengula jaguana	Pk	Z	Suárez-Caabro et al. 1961
Jenkinsia lamprotaenia	Pk	Co, Cs, Ce, Hp	Randall 1967
Opisthonema oglinum	Pk	Co, Cs, A, Cc, F	Lowe-Mc Conell 1962; Randall 1967
Sardinella aurita	Pk	Z	FAO 1978
Engraulidae			
Anchoa hepsetus	Pk	Z	FAO 1978
Anchovia clupeoides	Pk	Z	González-Sansón 1979
Cetengraulis edentulus	Pk	Z	FAO 1978
Stomiidae			
Chauliodus danae	P	F, C	Borodulina 1972
Chauliodus sloani	P	F	Borodulina 1972
Synodontidae			
Saurida brasiliensis	P	F	Cervigón 1994
Synodus foetens	P	F	Randall 1967
Synodus intermedius	P	F, Mc, Cs	Randall 1967
Synodus synodus	P	F	Randall 1967
Chlorophthalmidae			
Parasudis truculenta	P	F	Cervigón 1994
Ariidae			
Bagre marinus	B	Cc, Cs, F	Cervigón 1994; FAO 1978
Antennariidae			
Antennarius multiocellatus	P	F, Cc, Ce	Randall 1967
Antennarius striatus	P	F	Randall 1967
Ogcocephalidae			
Halieutichthys aculeatus	B		FAO 1978
Halieutichthys smithii	B		FAO 1978
Ogcocephalus nasutus	B	Ce, Mg, Va, F, Cc	Randall 1967
Ogcocephalus vespertilio	B		FAO 1978
Hemiramphidae			
Hemiramphus balao	P	F, Te, A, Z	Randall 1967
Hemiramphus brasiliensis	H	Vp, F	Randall 1967
Hemiramphus unifasciatus	O	Va, Oi	Cervigón 1994
Belonidae			
Ablennes hians	P	F	FAO 1978
Platybelone argalus	P	F, I	Randall 1967

Continued on next page

Appendix 5.1. continued

Strongylura notata	P	F, Cs	Randall 1967
Strongylura timucu	P	F	FAO 1978
Tylosurus acus acus	P	F, Z, I	Randall 1967
Tylosurus crocodilus crocodilus	P	F	FAO 1978
Atherinidae			
Atherinomorus stipes	Pk	Cs, Co, Hp	Randall 1967
Hypoatherina harringtonensis	Pk	F, Co, A, Hp	Randall 1967
Lampridae			
Lampris guttatus	PB	Mc, C, F	FAO 1978
Polymixiidae			
Polymixia lowei	B	Mc	Cervigón 1994
Holocentridae			
Holocentrus adscensionis	B	Cc, Cs, A, Mg	Randall 1967
Holocentrus rufus	B	Cc, Cs, Mg, Ec	Randall 1967
Myripristis jacobus	Pk	Cs, Cc, Ce, Cp	Randall 1967
Neoniphon marianus	B	Cs, Cc, Cot, Ce	Randall 1967
Plectrypops retrospinis	B	Cc, A	Randall 1967
Sargocenton coruscum	B	Cs, Cc, Cot	Randall 1967
Sargocentron vexillarium	B	Cc, Mg, Cs, F	Randall 1967
Aulostomidae			
Aulostomus maculatus	P	F, Cs	Randall 1967
Fistulariidae			
Fistularia petimba	P	F, Cs	FAO 1978
Fistularia tabacaria	P	F	Randall 1967
Centropomidae			
Centropomus ensiferus	PB	Mc, F	Cervigón 1994
Centropomus parallelus	PBI	F, C	FAO 1978
Centropomus pectinatus	PB	F, C	FAO 1978
Centropomus undecimalis	PB	F, Cot	Alvarez-Lajonchere et al. 1982
Serranidae			
Alphestes afer	B	Cc, F, Cs, Mc	Randall 1967
Cephalopholis cruentata	PB	F, Cm, Ce, Cc, Mg	Randall 1967
Cephalopholis fulva	PB	F, Cs, Cc, Ce	Randall 1967
Diplectrum formosum	PB	C, F	Cervigón 1994
Diplectrum radiale	PB	C, F	Cervigón 1994
Epinephelus adscensionis	B	Cc, F, Cs, Mg	Randall 1967
Epinephelus guttatus	PB	Cc, F, Ce, Cs	Randall 1967
Epinephelus itajara	B	Cr, F, Cc, T	Randall 1967; Bullock and Smith 1991; Sadovy and Eklund 1999
Epinephelus morio	PB	F, Cc, M, Cr	Randall 1967
Epinephelus niveatus	B	Cr, F, M	Dodrill et al. 1993
Epinephelus striatus	PBI	F, Cc, Ce, Mg	Randall 1967; Silva 1974b; Claro et al. 1990c; Carter et al. 1998
Hypoplectrus aberrans	B	Cs, Cc, F, Ce	Randall 1967
Hypoplectrus chlorurus	B	Cs, F, Cc	Randall 1967
Hypoplectrus nigricans	PB	F, Cs, Cc, Ce	Randall 1967
Hypoplectrus puella	B	Cs, Cc, F, Cot	Randall 1967
Hypoplectrus unicolor	B	C, F	Cervigón 1994
Mycteroperca bonaci	P	F	Randall 1967; Valdés-Muñoz 1980
Mycteroperca interstitialis	P	F	Randall 1967
Mycteroperca tigris	P	F	Randall 1967; Valdés-Muñoz 1980; García-Arteaga et al. 1999
Mycteroperca venenosa	P	F, Mc, Cs	Randall 1967; Valdés-Muñoz 1980
Paranthias furcifer	Pk	Co, Oi, Cs, Ca	Randall 1967
Rypticus saponaceus	PB	F, Cs, Cc, Ce	Randall 1967
Serranus phoebe	B	C	Cervigón 1994
Serranus tabacarius	P	F	Randall 1967
Serranus tigrinus	B	Cs, F, Cc, Ce	Randall 1967

Continued on next page

Appendix 5.1. continued

Grammatidae

Gramma loreto	Pk	Co, Cs, Cot	Böhlke and Randall 1963
Gramma melacara	Pk	C	Böhlke and Randall 1963
Priacanthidae			
Heteropriacanthus cruentatus	Pk	Fl, Cc, A, Cs	Randall 1967
Priacanthus arenatus	Pk	Fl, Cs, A, Cc	Randall 1967
Apogonidae			
Apogon maculatus	Pk	Cs, Cc, Co, A	Randall 1967
Phaeoptyx conklini	Pk	Cs, Ca, Cc, A, Co	Randall 1967
Malacanthidae			
Caulolatilus intermedius	B	C, Oi, F	FAO 1978
Malacanthus plumieri	B	Eo, Cc, Ce, F, Oi	Randall 1967
Pomatomidae			
Pomatomus saltatrix	P	F	Cervigón 1994
Rachycentridae			
Rachycentron canadum	P	F	Randall 1967
Echeneidae			
Echeneis naucrates	Pk	Fl, Ci, C	Randall 1967
Remora remora	Pk	Co, Ci, Ce, F, Ca	Randall 1967
Carangidae			
Alectis ciliaris	P	F, Mc	FAO 1978
Caranx bartholomaei	P	F, Mc, Cs	Randall 1967; Sierra et al. 1986
Caranx crysos	P	F, Cs, Cc, Ce	Randall 1967
Caranx hippos	P	F	Cervigón 1994
Caranx latus	P	F, Cs, Cot	Randall 1967
Caranx lugubris	P	F	Randall 1967
Caranx ruber	P	F	Randall 1967; Sierra and Popova 1982
Chloroscombrus chrysurus	P	C	Cervigón 1994
Decapterus macarellus	P	Mg, Co, Cs	Randall 1967
Decapterus punctatus	P	Co, Mg, Cs, Cot	Randall 1967
Elegatis bipinnulata	PB	I, F	FAO 1978
Naucrates ductor	PB	F, I	FAO 1978
Oligoplites saurus	P	F	Randall 1967
Selar crumenophthalmus	Pk	Cs, Cc, Fl, Co, Mg	Randall 1967
Selene vomer	PB	F, C	FAO 1978
Seriola dumerili	P	F, Mc	Randall 1967; Manooch and Haimovici 1983
Seriola fasciata	P	F	FAO 1978
Seriola rivoliana	P	F, Mc, Oi	FAO 1978; Manooch and Haimovici 1983
Seriola zonata	PB	Mc, F	Cervigón 1994
Trachinotus carolinus	PB	M, C, Oi, F	FAO 1978
Trachinotus falcatus	B	Mg, Ee, Mf, Cc	Randall 1967
Trachinotus goodei	P	F, Mg, Mp	Randall 1967
Coryphaenidae			
Coryphaena equiselis	PB	F, Mc	FAO 1978
Coryphaena hippurus	P	F, C, Mc	FAO 1978
Bramidae			
Brama brama	PB	F, Mc	FAO 1978
Eumegistus brevorti	PB	F, Mc	FAO 1978
Taractichthys longipinnis	PB	F, Mc	FAO 1978
Lutjanidae			
Apsilus dentatus	B	Cr, Tu	Munro 1983a
Etelis oculatus	PB	F, C, Mc	FAO 1978
Lutjanus analis	PB	F, Cc, Cm, Oi	Randall 1967; Claro 1981c; Muller et al. 1994
Lutjanus apodus	PB	F, Cc, Ce, Cs, Mc, Ca	Randall 1967; Rooker 1995
Lutjanus buccanella	PB	Ci, F, Cs, Cr, Oi	Brownell and Rainey 1971
Lutjanus campechanus	PB	Mc, Cs, F, Cc, Mg, Ce	Camber 1955; FAO 1978; Bradley and Bryan 1976
Lutjanus cyanopterus	P	F	Randall 1967
Lutjanus griseus	P	F, Cc, Cs	Starck 1970; Claro 1983c

Continued on next page

Appendix 5.1. continued

Lutjanus jocu	PB	F, Cc, Mc, Cr, Oi	Randall 1967; Claro et al. 1999
Lutjanus mahogoni	P	F, Cs, Mc, Cc	Randall 1967
Lutjanus synagris	PB	F, Cc, Cs, Oo	Randall 1967; Claro 1981b
Lutjanus vivanus	PBI	F, Cs, Cc, Oi	Brownell and Rainey 1971
Ocyurus chrysurus	PBI	F, Cc, Cs, Oi	Randall 1967; Starck 1970; Claro 1983a
Pristipomoides macrophthalmus	P	F, Fl	FAO 1978
Rhomboplites aurorubens	Pk	C, Mc, Mg, F, Fe,Tu, A, An	Grimes 1979
Gerreidae			
Diapterus auratus	B	I	FAO 1978
Diapterus rhombeus	B	C, A, V, Oi, M	González-Sansón 1979
Eucinostomus argenteus	B	Ca, A, Cc, Cs	Randall 1967
Eucinostomus gula	B	C, A, V, Oi	González-Sansón 1979
Eucinostomus havana	B	I	FAO 1978
Eugerres brasilianus	B	C, M, A	González-Sansón and Rodríguez-Viñas 1983
Eugerres plumieri	B	C, A	Naranjo 1956
Gerres cinereus	B	Cc, Mf, Mg, Cs, A, V	Randall 1967; González-Sansón and Rodríguez-Viñas 1983
Haemulidae			
Anisotremus virginicus	B	Ee, Cc, Cs, A, Mp	Randall 1967
Anisotremus surinamensis	B	Ee, Mg, Cc, Cs, Oi	Randall 1967
Haemulon album	B	Cc, Cs, Ce, Mp, Eh, F, Ee, A	Cummings et al. 1966; Randall 1967; Sierra 1983
Haemulon aurolineatum	B	Cs, A, Cc, Ca, Mp	Randall 1967
Haemulon carbonarium	B	Cc, Mg, Ee, A, Oi	Randall 1967
Haemulon chrysargyreum	B	Cs, Ce, A, Mg, Oi	Randall 1967
Haemulon flavolineatum	B	A, Cc, Eh, Cs, Mp	Randall 1967
Haemulon macrostomum	B	Ee, Cc	Randall 1967
Haemulon melanurum	B	C	FAO 1978
Haemulon parra	B	Cs, Cc, Ca, Mg, A	Randall 1967
Haemulon plumieri	B	Cc, A, Ee, Mg, Oi	Randall 1967; Valdés-Muñoz and Silva 1977
Haemulon sciurus	B	Cc, Mp, Cs, Ee, Oi	Randall 1967; Valdés-Muñoz and Silva 1977
Haemulon striatum	Pk	Z	Munro 1983d
Orthopristis chrysoptera	O	V, C, M, F	Konshina 1977
Pomadasys crocro	PB	C, F	FAO 1978
Inermiidae			
Inermia vittata	Pk	Co, Cc, A, Oi	Randall 1967
Sparidae			
Archosargus rhomboidalis	H	Vp, Va, Cc, Mg	Randall 1967
Calamus bajonado	B	Cc, A, M, Oi	Randall 1967
Calamus calamus	B	Cc, A, M, Oi	Randall 1967
Calamus penna	B	Ce, M	Randall 1967
Calamus pennatula	B	Cs, Cc, A, M, Oi	Randall 1967
Calamus proridens	B	C	Druzhinin 1976
Diplodus argenteus caudimacula	H	Va, Mg, Cc	Randall 1967
Lagodon rhomboides	O	C, M, F, V	FAO 1978
Sciaenidae			
Bairdiella batabana	B	Cs, Ca, Ci	Druzhinin 1974
Bairdiella ronchus	B	F, Cs, Cc, A, Oi, Fl	García and Nieto 1978
Bairdiella sanctaeluciae	B	Cs	FAO 1978
Equetus acuminatus	B	Cs, Ce, F, FL, Oi	Randall 1967
Equetus lanceolatus	B	Cs, Ce, A	Randall 1967
Equetus punctatus	B	Cc, Cs, A, M, Oi	Randall 1967
Larimus breviceps	B	M, C, F	Druzhinin 1974
Micropogonias furnieri	B	C, A, D	García 1979
Micropogonias undulatus	B	C, A, D	Naranjo 1956
Odontoscion dentex	B	Cs, F, Cc, Fl, Oi	Randall 1967
Umbrina coroides	B	C	Druzhinin 1974
Mullidae			
Mulloidichthys martinicus	B	A, M, Cc, Cs, Oi	Randall 1967

Continued on next page

Appendix 5.1. continued

Pseudupeneus maculatus	B	Cc, Cs, A, M, F	Randall 1967
Upeneus parvus	B	Cs, F	Cervigón 1994
Pempheridae			
Pempheris schomburgkii	Pk	A, Cc, Cs, Ce, Ca	Randall 1967
Kyphosidae			
Kyphosus incisor	H	Va	Randall 1967
Kyphosus sectatrix	H	Va, Ve	Randall 1967
Ephippidae			
Chaetodipterus faber	B	A, Oi, Va	Randall 1967
Chaetodontidae			
Chaetodon aculeatus	B	A, Ee, Cs, Co, Cot	Randall 1967
Chaetodon capistratus	B	Zo, A, G, Tu	Randall 1967; Lasker 1985
Chaetodon ocellatus	B	A, Ee, Ca, C, Va	Munro 1983b
Chaetodon sedentarius	B	A, Cs, Ca, H	Randall 1967
Chaetodon striatus	B	A, An, C, Oi	Randall 1967
Pomacanthidae			
Centropyge argi	H	Va	Randall 1967
Holacanthus ciliaris	B	S, Va, Tu, H	Randall 1967
Holacanthus tricolor	B	S, Zo, Va	Randall 1967
Pomacanthus arcuatus	B	S, Tu, Va, Zo, Oi	Randall 1967
Pomacanthus paru	B	S, Va, Zo, Tu, G	Randall 1967
Pomacentridae			
Abudefduf saxatilis	B	An, Co, Va, Tu, F	Randall 1967
Abudefduf taurus	H	Va, An, Cc, Mg	Randall 1967
Chromis cyanea	Pk	Co, Tu, Ce, Hp	Randall 1967
Chromis multilineata	Pk	Co, Tu, Ce, Oi	Randall 1967
Microspathodon chrysurus	H	Va, C, S	Randall 1967
Stegastes adustus	O	Va, A, Co, F, Mg, An	Randall 1967
Stegastes leucostictus	O	Va, A, F, Tu, Cc, Oi	Randall 1967
Stegastes planifrons	O	Va, An, C, A, E, Oi	Randall 1967
Stegastes variabilis	O	Va, A, Ci, S, H, Oi	Randall 1967
Cirrithidae			
Amblycirrhitus pinos	Pk	Co, Cs, Cc, Ci	Randall 1967
Labridae			
Bodianus pulchellus	B	C	FAO 1978
Bodianus rufus	B	Cc, Eo, Ee, Mg, Oi	Randall 1967
Clepticus parrae	Pk	Co, Cs, Cc, Oi	Randall 1967
Halichoeres bivittatus	B	Cc, Ed, A, Mg, Oi	Randall 1967
Halichoeres garnoti	B	Cc, Eo, Mg, F, Oi	Randall 1967
Halichoeres maculipinna	B	A, Co, Cs, Mg, Oi	Randall 1967
Halichoeres poeyi	B	Cc, Mg, Eo, F, Oi	Randall 1967
Halichoeres radiatus	B	Mf, Mg, Ee, Cc, Oi	Randall 1967
Lachnolaimus maximus	B	Mf, Mg, Cc, Eo	Randall 1967; Claro et al. 1989
Thalassoma bisfasciatum	B	Co, Cc, Mg, F, Oi	Randall 1967
Xyrichtys novacula	B	Mg, Mp, A, Cs,Oi	Randall 1967
Xyrichtys splendens	Pk	Co, Ca, Cs, Fe, Oi	Randall 1967
Scaridae			
Cryptotomus roseus	H	V	FAO 1978
Nicholsina usta usta	H	V	FAO 1978
Scarus coelestinus	H	Va, Vp, Oi	Randall 1967
Scarus coeruleus	H	V	FAO 1978
Scarus guacamaia	H	Va, Ve, S	Randall 1967; Dunlap and Pawlik 1998
Scarus iserti	H	Va, S	Randall 1967; Dunlap and Pawlik 1998
Scarus taeniopterus	H	Va, Ve, S	Randall 1967; Dunlap and Pawlik 1998
Scarus vetula	H	Va, Ve	Randall 1967
Sparisoma atomarium	H	V	Randall 1967
Sparisoma aurofrenatum	H	Va, Ve, S	Randall 1967; Dunlap and Pawlik 1998
Sparisoma chrysopterum	H	Va, Ve, S	Randall 1967; Dunlap and Pawlik 1998

Continued on next page

Appendix 5.1. continued

Sparisoma radians	H	Vp, Va	Randall 1967; Targett et al. 1986
Sparisoma rubripinne	H	Va, Ve	Randall 1967
Sparisoma viride	H	Va, Vp	Randall 1967
Mugilidae			
Mugil curema	D	D	González-Sansón and Alvarez-Lajonchere 1978
Mugil incilis	D	D	González-Sansón and Alvarez-Lajonchere 1978
Mugil liza	D	D	González-Sansón and Alvarez-Lajonchere 1978
Mugil longicauda	D	D	González-Sansón and Alvarez-Lajonchere 1978
Mugil trichodon	D	D	González-Sansón and Alvarez-Lajonchere 1978
Polynemidae			
Polydactylus virginicus	PB	Mc, F	Cervigón 1994
Opistognathidae			
Opistognathus aurifrons	Pk	Cc, Cs, Fe, A	Randall 1967
Opistognathus macrognathus	Pk	F, Cs	Randall 1967
Opistognathus maxillosus	Pl	Cs, Ci, F, A	Randall 1967
Uranoscopidae			
Kathetostoma cubana	O	F, Cs, Va	Cervigón 1994
Labrisomidae			
Labrisomus guppyi	B	Cc, Mg, Ci	Randall 1967
Labrisomus nuchipinnis	B	Cc, Cj, Eo, Ee, F	Randall 1967
Blenniidae			
Entomacrodus nigricans	H	Va, A	Randall 1967
Ophioblennius atlanticus	H	Va, Fe	Randall 1967
Parablennius marmoreus	H	Va, Eo, A	Randall 1967
Scartella cristata	H	Va	Randall 1967
Gobiidae			
Coryphopterus glaucofraenum		O, Va, Eo, Mp, Co	Randall 1967
Gobiosoma multifasciatum	B	Ci	Randall 1967
Acanthuridae			
Acanthurus bahianus	H	Va, Vp	Randall 1967
Acanthurus chirurgus	H	Va, Vp	Randall 1967
Acanthurus coeruleus	H	Va, Vp	Randall 1967
Sphyraenidae			
Sphyraena barracuda	P	F, Mc	Randall 1967; Valdés-Muñoz 1980
Sphyraena guachancho	PB	F, Mc	Valdés-Muñoz and Silva 1977
Sphyraena picudilla	P	F, Mc	Randall 1967
Gempylidae			
Gempylus serpens	PB	F, C, Mc	FAO 1978
Lepidocybium flavobrunneum	PB	C, Mc	FAO 1978
Ruvettus pretiosus	PB	F, C, Mc	FAO 1978
Trichiuridae			
Trichiurus lepturus	P	F, Cs	Ros and Pérez 1978
Scombridae			
Acanthocybium solandri	P	F, Mc	Naranjo 1956; Manooch and Hogarth 1983
Auxis rochei	PB	F, Cj, Ce, Mc	FAO 1978
Auxis thazard	P	F	Cervigón 1994
Euthynnus alletteratus	PB	F, Mc, A	Randall 1967
Katsuwonus pelamis	P	F, M, A	Guevara 1984b
Sarda sarda	P	F	FAO 1978
Scomber japonicus	P	F, I	FAO 1978
Scomberomorus cavalla	P	F, Mc	Randall 1967
Scomberomorus maculatus	P	F, Cs	Naranjo 1956
Scomberomorus regalis	P	F, Mc	Randall 1967
Thunnus alalunga	P	F, Mc, C	FAO 1978
Thunnus albacares	P	F, C, Mc	FAO 1978
Thunnus atlanticus	P	F, C, Mc	Guevara 1984a
Thunnus obesus	P	F	Borodulina 1974
Thunnus thynnus	PB	F, C, Mc	FAO 1978

Continued on next page

Appendix 5.1. continued

Xiphiidae			
Xiphias gladius	P	F, Mc	Naranjo 1956; Guitart Manday 1964
Istiophoridae			
Istiophorus platypterus	PB	F, Mc, C	FAO 1978
Makaira nigricans	PB	F, Mc, C	FAO 1978
Tetrapturus albidus	PB	F, Mc, C	FAO 1978
Tetrapturus pfluegeri	PB	F, Mc, C	FAO 1978
Ariommatidae			
Ariomma regulus	Pk	Z	FAO 1978
Nomeidae			
Nomeus gronovii	Pk	Z	FAO 1978
Psenes cyanophrys	Pk	Z	FAO 1978
Stromateidae			
Peprilus paru	B	F, C, A, Oi	FAO 1978
Peprilus triacanthus	B	F, C, A, Oi	FAO 1978
Scorpaenidae			
Pontinus castor	B	—	FAO 1978
Scorpaena brasiliensis	B	Cs, C, Ce, F	Randall 1967
Scorpaena grandicornis	B	Cs, C, F	Randall 1967
Scorpaena inermis	B	Cs, Ce, Cc, F	Randall 1967
Scorpaena plumieri	B	F, Cc, Cs, Mc	Randall 1967
Dactylopteridae			
Dactylopterus volitans	B	M, F, Cs, Cc	Randall 1967
Bothidae			
Bothus lunatus	P	F, Cc, Mc	Randall 1967
Bothus ocellatus	PB	F, Cc, Cs, Ca, Ce	Randall 1967
Paralichthyidae			
Syacium papillosum	PB	F, Cc, Ca, Cs	Cervigón 1994
Cynoglossidae			
Symphurus diomedeanus	B	Cc, A	FAO 1978
Symphurus plagiusa	B	Cc, A, Co, Ca	FAO 1978
Balistidae			
Balistes capriscus	B		FAO 1978
Balistes vetula	B	Ee, Cc, A, Mp, Eh, Oi, E, Va	Randall 1967; Turingan et al. 1995
Canthidermis maculata	B	Ee, Te, Z, Va	Randall 1967
Canthidermis sufflamen	B	Cl, Cc, Zo, Mg	Sierra et al. 1994
Melichthys niger	H	Va, Cc, F, Tu, Oi	Randall 1967; Turingan et al. 1995
Xanthichthys ringens	O	Z, V, An, S	Turingan et al. 1995
Monacanthidae			
Aluterus schoepfi	H	Vp, Va, Ce, Mg	Randall 1967
Aluterus scriptus	O	H, Va, G, Vp, Zo, Oi	Randall 1967
Cantherines macrocerus	B	S, H, G, Va, E, An	Randall 1967; Turingan et al. 1995
Cantherines pullus	O	Va, S, Tu, Vp, H, Oi	Randall 1967; FAO 1978
Monacanthus ciliatus	O	Va, Vp, Co, Cot, Z, A	Randall 1967
Monacanthus setifer	O	V, I	FAO 1978
Monacanthus tuckeri	Pk	Co, Ci, Cc, Ct	Randall 1967
Ostraciidae			
Acanthostracion polygonius	B	Tu, Al, S, Cs, Mg	Randall 1967
Acanthostracion quadricornis	B	S, Tu, Zo, Cc, V, G	Randall 1967
Lactophrys bicaudalis	B	Cc, Mp, A, Ee, V, Tu	Randall 1967
Lactophrys trigonus	B	Cc, Mp, A, Ee, V, Tu	Randall 1967
Lactophrys triqueter	B	A, Sp, Cc, Cs, Tu, S	Randall 1967
Tetraodontidae			
Canthigaster rostrata	O	Vp, S, Ce, M, A, E	Randall 1967
Sphoeroides nephelus	B		FAO 1978
Sphoeroides spengleri	B	Cc, M, A, E, V, Oi	Randall 1967
Sphoeroides testudineus	B		FAO 1978

Continued on next page

Appendix 5.1. continued

Diodontidae

Chilomycterus antennatus	B	Mg, Cc, Ci, Cs	Randall 1967
Diodon holacanthus	B	Mg, Mp, Ee, Cc	Randall 1967
Diodon hystrix	B	Ee, Mg, Cc, Mp	Randall 1967

[a] Feeding habits: P: piscivorous; PI: piscivorous and invertebrate predators; B: benthophagous; Pk: Planktivorous; O: omnivorous; H: herbivorous.

[b] Main food components: A: annelids; Al: alcionarians; An: anthozoans; C: crustaceans; Ca: amphipods; Cc: crabs; Ce: stomatopods; Ci: isopods; Cj: swimming crabs; Cl: crustacean larvae; Co: copepods; Cot: other crustaceans; Cr: macrurids; Cs: shrimps; Ct: tanaidaceans; D: detritus; E: echinoids; Ee: sea urchins; Eh: holothurians; Eo: ophiuroids; F: fishes; Fe: fish eggs; Fl: fish larvae; G: gorgonians; H: hydrozoans; I: invertebrates; Oi: other invertebrates; i: insects; M: mollusks; Mc: cephalopods; Mg: gastropods; Mp: pelecipods; S: sponges; Sp: sipunculids; T: turtles; Te: pteropods; Tu: tunicates; V: vegetation; Va: algae; Vp: phanerogams; Z: zooplankton; Zo: zooantharians.

Appendix 5.2. Dietary patterns of marine fishes studied in Cuba.

Family / Species	Method[b]	N[c]	Region	F	Cc	Cs	Cot	Mo	A	E	Z	V	O	D	Source
Elopidae															
Elops saurus	(V)	—	SE*	41.5	29.2	26.8	—	—	—	—	—	—	4.8	—	González-Sansón 1979
Megalops atlanticus	(V)	—	SE*	24.4	47.5	25.6	—	—	—	—	—	—	2.4	—	González-Sansón 1979
Clupeidae															
Harengula clupeola	(M)	—	NW	4.0	—	—	—	5.0	55.0	—	30.0	5.0	1.0	—	García 1976
Harengula clupeola	(W)	202	NW	—	—	—	—	—	1.3	—	98.7	—	—	—	Sierra and Díaz-Zaballa 1984
Harengula humeralis	(M)	—	NW	8.0	—	—	—	1.0	43.0	—	47.0	1.0	—	—	García 1976
Harengula humeralis	(W)	208	NW	0.3	—	—	—	—	78.4	—	21.2	—	—	—	Sierra and Díaz-Zaballa 1984
Harengula humeralis	(W)	1014	SW	4.0	—	16.1	—	—	17.1	—	53.9	0.5	8.0	—	Sierra 1987
Jenkinsia lamprotaenia	(W)	666	SW	4.0	—	—	—	—	0.7	—	98.7	0.6	—	—	Sierra et al. 1994
Engraulidae															
Anchovia clupeoides	(C)	24	SE*	—	—	—	100	—	—	—	—	41.3	6.9	62.0	González-Sansón 1979
Anchoa hepsetus	(W)	15	SW	—	—	—	—	—	—	—	100	—	—	—	Sierra et al. 1994
Hemiramphidae															
Hemiramphus brasiliensis	(W)	15	SW	—	—	—	—	—	—	—	—	100	—	—	Sierra et al. 1994
Belonidae															
Strongylura notata	(W)	11	SW	88.9	—	11.1	—	—	—	—	—	—	—	—	Sierra et al. 1994
Strongylura timucu	(V)	—	SE*	71.9	—	25.6	—	—	—	—	—	—	2.4	—	Sierra et al. 1994
Tylosurus crocodilus crocodilus	(W)	11	SW	96.6	3.4	—	—	—	—	—	—	—	—	—	Sierra et al. 1994
Tylosurus acus acus	(W)	10	SW	100	—	—	—	—	—	—	—	—	—	—	Sierra et al. 1994
Centropomidae															
Centropomus ensiferus	(V)	—	SE*	37.8	2.4	54.8	—	—	—	—	—	—	4.8	—	González-Sansón 1979
Centropomus undecimalis	(V)	—	SE*	90.2	2.4	7.3	—	—	—	—	—	—	—	—	González-Sansón 1979
Centropomus undecimalis	(V)	23	SE*	79.0	—	—	27.0	—	—	—	—	—	—	—	Alvarez-Lajonchere et al. 1982
Serranidae															
Epinephelus itajara	(W)	3	SW	25.6	8.4	—	71.5	—	—	—	—	—	—	—	Sierra et al. 1994
Epinephelus striatus	(W)	83	SW	82.0	—	—	11.7	6.3	—	—	—	—	—	—	Claro et al. 1990c
Epinephelus striatus	(W)	38	NE	87.3	—	—	9.1	3.6	—	—	—	—	—	—	Claro et al. 1990c
Epinephelus striatus	(W)	29	NW	42.9	—	—	18.9	38.2	—	—	—	—	—	—	Claro et al. 1990c
Epinephelus striatus	(O)	231	NW	53.0	22.5	—	3.5	21.0	—	—	—	—	—	—	Silva Lee 1974b
Mycteroperca bonaci	(W)	22	NW	81.3	—	—	—	—	—	—	—	—	18.7	—	Valdés-Muñoz 1980
Mycteroperca bonaci	(W)	19	SW	100	—	—	—	—	—	—	—	—	—	—	Sierra et al. 1994
Mycteroperca tigris	(W)	14	NW	100	—	—	—	—	—	—	—	—	—	—	Valdés-Muñoz 1980
Mycteroperca tigris	(W)	39	SW	100	—	—	—	—	—	—	—	—	—	—	García-Arteaga et al. 1999

Prey Components[a]

Continued on next page

Continued on next page

Appendix 5.2. continued

Taxon		n											Source
Mycteroperca venenosa	(W)	11	NW	85.1	—	—	—	14.4	—	—	0.5	—	Valdés-Muñoz 1980
Mycteroperca venenosa	(W)	13	SW	100	—	—	—	—	—	—	—	—	Sierra et al. 1994
Rachycentridae													
Rachycentron canadum	(W)	6	SW	100	—	—	—	—	—	—	—	—	Sierra et al. 1994
Carangidae													
Caranx bartholomaei	(W)	619	SW	97.1	—	—	—	1.6	0.9	—	0.4	—	Sierra et al. 1986
Caranx hippos	(V)	—	SE*	37.8	2.4	54.8	—	—	—	—	4.8	—	González-Sansón 1979
Caranx crysos	(W)	13	SW	100	—	—	—	—	—	—	—	—	Sierra et al. 1994
Caranx latus	(W)	4	SW	100	—	—	—	—	—	—	—	—	Sierra et al. 1994
Caranx ruber	(W)	1164	SW	81.7	0.1	4.9	2.3	—	9.5	—	0.9	—	Sierra and Popova 1982
Caranx ruber	(W)	369	NW	94.4	—	—	—	—	—	—	4.1	—	Sierra et al. 1994
Selar crumenophthalmus	(W)	47	SW	1.9	—	97.3	—	—	—	—	0.1	—	Sierra et al. 1994
Seriola dumerili	(W)	21	NW	99.5	—	—	—	—	—	—	0.5	—	Valdés-Muñoz 1980
Lutjanidae													
Lutjanus analis	(W)	884	NW	31.0	47.2	4.0	1.9	—	—	—	—	—	Claro 1981c
Lutjanus analis	(W)	452	SW	41.3	41.6	8.4	2.8	6.9	—	—	0.7	—	Claro 1981c
Lutjanus apodus	(V)	—	SE*	54.9	6.1	15.8	—	—	—	—	23.2	—	González-Sansón 1979
Lutjanus apodus	(W)	11	SW	88.7	6.1	—	5.2	—	—	—	—	—	Sierra et al. 1994
Lutjanus cyanopterus	(W)	10	SW	71.8	5.5	22.7	—	—	—	—	—	—	Sierra et al. 1994
Lutjanus griseus	(V)	116	SW**	81.8	5.8	1.6	7.7	0.1	—	—	2.6	—	Valdés-Muñoz and Silva Lee 1977
Lutjanus griseus	(W)	374	SW	74.1	16.5	6.6	0.6	0.1	—	—	0.1	—	Claro 1983c
Lutjanus griseus	(V)	—	SE*	25.0	46.5	22.0	—	—	—	—	6.5	—	González-Sansón 1979
Lutjanus jocu	(V)	—	SE*	9.7	64.6	13.4	—	—	—	—	12.2	—	González-Sansón 1979
Lutjanus jocu	(W)	79	SW	83.6	2.6	3.5	17.6	—	—	—	—	—	Claro et al. 1999
Lutjanus jocu	(W)	13	NE	92.1	—	1.0	3.0	3.9	—	—	—	—	Claro et al. 1999
Lutjanus jocu	(W)	13	NW	83.0	1.3	—	15.7	—	—	—	—	—	Claro et al. 1999
Lutjanus synagris	(V)	32	SW**	80.0	4.6	2.4	1.0	9.8	—	0.5	1.4	—	Valdés Muñoz and Silva Lee 1977
Lutjanus synagris	(W)	905	SW	45.8	28.4	2.8	11.9	—	2.1	1.4	2.4	—	Claro 1981b
Lutjanus synagris	(C)	207	—	31.9	—	26.6	1.4	—	12.0	2.4	26.0	—	Rodríguez Pino 1962
Ocyurus chrysurus	(W)	224	SW	49.4	13.4	1.7	11.0	7.6	5.4	11.4	—	—	Claro 1983b
Ocyurus chrysurus	(W)	534	NW	57.5	12.2	4.7	3.9	17.6	0.4	0.6	5.8	—	Claro 1983b
Ocyurus chrysurus	(V)	250	—	84.6	—	—	1.2	—	—	—	—	—	Piedra 1965
Mullidae													
Pseudupeneus maculatus	(W)	16	SW	8.3	15.3	26.1	32.5	14.4	3.1	—	0.3	—	Sierra et al. 1994

Appendix 5.2. continued

														Reference
Mulloidichthys martinicus	(W)	4	SW	—	—	3.5	35.8	—	—	14.7	—	—	—	Sierra et al. 1994
Gerreidae														
Diapterus rhombeus	(O)	44	SE*	—	—	—	100	2.0	6.1	—	61.1	77.5	49.0	González-Sansón 1979
Eucinostomus gula	(O)	28	SE*	—	—	—	100	—	19.1	—	9.5	—	38.0	González-Sansón 1979
Eugerres brasilianus	(O)	525	SE*	—	—	—	100	—	10.2	—	77.7	20.9	79.6	González-Sansón and Rodríguez-Viñas 1983
Gerres cinereus	(O)	116	SE*	—	—	—	87.2	24.8	—	29.8	58.4	5.4	—	González-Sansón and Rodríguez-Viñas 1983
Haemulidae														
Haemulon album	(O)	1351	SW	7.2	17.2	—	23.2	33.6	12.1	16.5	18.5	16.3	—	Sierra 1983
Haemulon parra	(V)	4	SW**	—	—	—	15.7	5.2	57.8	5.2	—	16.1	—	Valdés-Muñoz and Silva Lee 1977
Haemulon plumieri	(V)	126	SW**	—	—	5.6	4.4	1.1	16.8	1.4	—	67.4	—	Valdés-Muñoz and Silva Lee 1977
Haemulon plumieri	(V)	25	SW	—	14.0	1.6	36.6	0.5	4.1	0.5	—	42.5	—	Valdés-Muñoz and Silva Lee 1977
Haemulon sciurus	(V)	26	SW	—	13.4	2.0	39.7	0.7	1.6	0.3	—	41.8	—	Valdés-Muñoz and Silva Lee 1977
Haemulon sciurus	(V)	85	SW**	—	0.2	1.0	4.3	1.5	16.6	0.5	1.0	77.8	—	Valdés-Muñoz and Silva Lee 1977
Sparidae														
Archosargus rhomboidalis	(O)	15	SE*	—	—	—	100	—	13.3	—	86.6	86.6	13.3	González-Sansón 1979
Sciaenidae														
Bairdiella ronchus	(O)	447	S	17.6	6.6	2.2	34.4	5.2	4.5	0.2	—	2.9	53.5	García and Nieto 1978
Bairdiella ronchus	(O)	69	NW	2.9	—	—	47.8	5.8	4.3	—	—	—	37.6	García and Nieto 1978
Micropogonias furnieri	(W)	486	S	3.5	1.3	2.2	20.1	4.0	12.0	—	—	1.8	56.4	García 1979
Micropogonias furnieri	(W)	77	NW	9.1	31.7	2.6	19.5	32.5	2.6	—	—	10.4	22.0	García 1979
Micropogonias furnieri	(V)	—	SE*	6.1	—	36.5	—	—	—	—	—	25.6	—	González-Sansón 1979
Labridae														
Lachnolaimus maximus	(O)	—	SW	—	10.4	—	17.9	85.1	—	5.2	6.6	3.7	—	Claro et al. 1989
Lachnolaimus maximus	(W)	12	NE	63.5	63.5	—	—	50.6	—	—	—	—	—	Sierra et al. 1994
Halichoeres bivittatus	(W)	41	SW	6.2	6.2	—	0.7	65.6	0.7	—	13.8	13.1	—	Sierra et al. 1994
Sphyraenidae														
Sphyraena barracuda	(W)	37	NW	99.7	—	—	0.1	0.1	—	—	0.1	0.1	—	Valdés-Muñoz 1980
Sphyraena barracuda	(W)	—	SE*	100	—	—	—	—	—	—	—	—	—	González-Sansón 1979
Sphyraena barracuda	(W)	23	SW	100	—	—	—	—	—	—	—	—	—	Sierra et al. 1994
Sphyraena guachancho	(V)	25	SW**	30.3	—	—	0.4	55.0	—	—	—	14.2	—	Valdés-Muñoz and Silva Lee 1977
Trichiuridae														
Trichiurus lepturus	(O)	326	S	90.5	—	9.5	—	—	—	—	—	—	—	Ros and Pérez 1978

Continued on next page

Appendix 5.2. continued

	Method	N	Region	F	Cc	Cs	Cot	Mo	A	E	Z	V	O	D	Reference
Scombridae															
Katsuwonus pelamis	(V)	137	NE	84.1	—	—	—	—	—	—	8.9	7.0	—	—	Carles 1971
Katsuwonus pelamis	(V)	126	SW	81.6	—	—	—	—	—	—	1.6	16.3	—	—	Guevara 1984b
Scomberomorus regalis	(W)	5	SW	100	—	—	—	—	—	—	—	—	—	—	Sierra et al. 1994
Thunnus atlanticus	(V)	86	SE	60.3	—	—	—	—	—	—	17.0	22.4	—	—	Suárez-Caabro and Duarte-Bello 1961
Thunnus atlanticus	(V)	146	NE	49.3	—	—	—	—	—	—	17.1	33.6	—	—	Carles 1971
Thunnus atlanticus	(V)	154	SW	830	—	—	—	—	—	—	4.8	121	—	—	Guevara 1984a
Xiphiidae															
Xiphias gladius	(O)	—	NW	53.4	13.9	—	—	—	—	—	2.3	16.2	—	—	Guitart Manday 1964
Balistidae															
Canthidermis sufflamen	(W)	5	NE	—	22.3	44.3	—	—	—	—	24.5	8.8	—	—	Sierra et al. 1994

[a] Prey components: F: fishes; Cc: crabs; Cs: shrimps; Cot: other crabs; Mo: mollusks; A: annelids; E: echinoids; Z: zooplankton; V: vegetation; O: other organisms; D: detritus.

[b] Method: Data is provided according to measurement units used in the references: (O) % occurrence, (V) % volume, (W) % weight, (M) % organic matter.

[c] Number of fishes with gut contents.

* In coastal lagoons; ** On artificial reefs.

6

Growth Patterns of Fishes of the Cuban Shelf

RODOLFO CLARO AND JUAN P. GARCÍA-ARTEAGA

6.1 Introduction

Growth parameters provide key measures of the relationships between organisms and their environment. They reflect food ingestion, assimilation, and transformation within the organism and are strongly associated with both biotic and abiotic environmental factors. In fishes, natural and human-induced changes in growth rates can alter mortality rates, the timing of sexual maturation, fecundity, longevity, and the response of populations to fishing pressures. Daily and seasonal fluctuations of environmental factors can provoke changes in fish growth rates that are registered in bones, scales, and otoliths.

6.2 Growth Mark Formation

In temperate fishes, reduced winter temperatures, food supply, and hours of daylight contribute to declines in growth rate and the formation of rings or marks in some skeletal structures. In tropical waters, seasonal climatic fluctuations are less notable and many fish spawn intermittently. Nevertheless, many investigators have recorded periodic marks in bones, otoliths, and scales that reflect the dynamics of growth (Fowler 1995). Alternating hyaline and opaque marks have been observed that delineate periods of slow and fast growth. Such marks can occur annually or subannually and have been used to estimate growth parameters for many decades.

On a finer scale, daily rings have also been recorded on otoliths (Panella 1971, 1974), particularly in early life stages. Campana and Neilson (1982, 1985) provided evidence that daily marks can result from an endogenous circadian rhythm, although photoperiod, temperature fluctuation, and feeding activity might also affect deposition. Reading of daily otolith rings brought major methodological changes to the study of fish growth. The correct interpretation of daily ring

formation in otoliths is useful not only for age and growth studies, but also for investigating recruitment, early mortality, back-calculation of spawning, and other factors (Brothers et al. 1976, 1983; Struhsaker and Uchiyama 1976; Barkman 1978; Taubert and Coble 1978; Brothers 1979, 1981, 1983; Schmidt and Fabrizio 1980; Steffensen 1980; Townsend 1980; Wilson and Larkin 1980; Brothers and McFarland 1981; Miranda 1981; Campana and Neilson 1982, 1985; Geffen 1982; Campana 1983, 1984; Brothers and Thresher 1985; Keener et al. 1988; Victor 1991; Williams et al. 1994; Fowler 1995; Sponaugle and Cowen 1997; Wilson and McCormick 1997).

6.2.1 Annual Mark Formation

Most of the available data on age and growth of adult stages of Caribbean fishes are based on the study of presumably annual marks. Research in Cuba on adult life stages has emphasized the use of rings deposited on bones (particularly the urohyal), scales, and otoliths. In the bones or scales of most of the fishes aged by Cuban researchers, two annual hyaline marks have been observed, whereas otoliths typically showed a single, clear annual ring.

In the lane snapper (*Lutjanus synagris;* Olaechea and Quintana 1970), mutton snapper (*L. analis;* Pozo 1979), gray snapper (*L. griseus;* Claro 1983e), bar jack (*Caranx ruber;* García-Arteaga and Reshetnikov 1992), and hogfish (*Lachnolaimus maximus;* Claro et al. 1989) in which urohyal bones were examined, two annual marks were recorded, including an early-winter ring (Fig. 6.1) coinciding with the most abrupt water temperature change of the year (November–December; see Section 1.5). A winter ring was also observed in the redear sardine (*Harengula humeralis*) in the scale ring patterns (García-Arteaga 1993), and in mullets (white, *Mugil curema,* and fantail, *M. trichodon*) in fin spines (Alvarez-Lajonchere 1981a). This winter mark was not visible

149

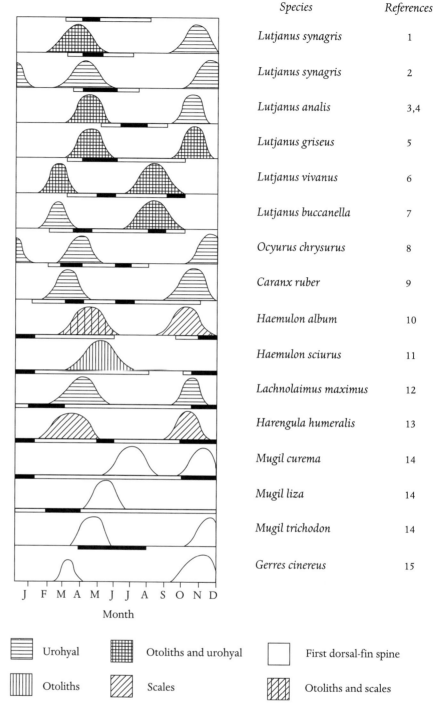

Species	References
Lutjanus synagris	1
Lutjanus synagris	2
Lutjanus analis	3,4
Lutjanus griseus	5
Lutjanus vivanus	6
Lutjanus buccanella	7
Ocyurus chrysurus	8
Caranx ruber	9
Haemulon album	10
Haemulon sciurus	11
Lachnolaimus maximus	12
Harengula humeralis	13
Mugil curema	14
Mugil liza	14
Mugil trichodon	14
Gerres cinereus	15

Fig. 6.1. Periods of mark formation on skeletal structures of fishes studied in Cuba. The horizontal bar above each graph indicates the range of the reproductive period and the dark bar is the reproductive peak. References: 1: Claro and Reshetnikov 1981; 2: Olaechea and Quintana 1970; 3: Claro 1981c; 4: Pozo 1979; 5: Claro 1983c; 6: Pozo and Espinosa 1982; 7: Espinosa and Pozo 1982; 8: Claro 1983e; 9: García-Arteaga and Reshetnikov 1992; 10: García-Arteaga 1983; 11: García-Arteaga 1992; 12: Claro et al. 1989; 13: García-Arteaga 1993; 14: Alvarez-Lajonchere 1981c; 15: Báez and Alvarez-Lajonchere 1983.

Urohyal   Otoliths and urohyal   First dorsal-fin spine

Otoliths   Scales   Otoliths and scales

in the liza (*Mugil liza*); no summer growth was detected, so there was no growth to generate a winter ring. This mark might be considered annual because it occurs regularly.

The winter mark is usually formed in the urohyal by two or more narrow rings closely apposed. Rings might be caused by the abrupt temperature drop that accompanies the passage of cold fronts during the winter (see Fig. 1.5 for an example of the intensity of temperature declines). In mutton snapper, feeding intensity and growth rate decline sharply in

early winter and then increase (Claro and Colás 1987). It has also been suggested that the winter marks in the scales of the sparid *Pagellus erythrinus* in coastal Spain are caused by decreasing temperatures in November–December rather than the lowest temperatures of winter (January–February), and that as the fishes become acclimated to the lower temperatures they resume growing (Larrañeta 1967).

The winter ring is usually visible in the urohyal and in scales, but not in the otoliths of grunts (*Haemulon* spp.;

García-Arteaga 1983, 1992a, 1992b) and shows little contrast in otoliths of the mutton (Claro 1981c), lane (Claro and Reshetnikov 1981) and yellowtail (*Ocyurus chrysurus;* Claro 1983e) snappers. This ring was not considered to be a growth mark in such species. However, this ring is apparent in gray (Claro 1983b), silk (*Lutjanus vivanus;* Pozo and Espinosa 1982; Pozo et al. 1983), and blackfin (*Lutjanus buccanella;* Espinosa and Pozo 1982; Espinosa et al. 1984) snappers, which have thinner transparent otoliths.

A second annual ring often marks the winter-summer transition period (Fig. 6.1). This ring has been detected in the urohyal and otoliths of snappers, principally during April–May, which is the reproductive peak for lane, mutton, and yellowtail snappers. However, because portions of the population reproduce after the period of ring formation, the ring might not be related to this process; for example, gray snapper spawn mainly during July–August.

This spring mark was also observed in the urohyal of bar jack (García-Arteaga and Reshetnikov 1992) and hogfish (Claro et al. 1989); in the otoliths of the margate (*Haemulon album;* García-Arteaga 1983), white grunt (*H. plumieri;* García-Arteaga 1992a), bluestriped grunt (*H. sciurus;* García-Arteaga 1992b), tiger grouper (*Mycteroperca tigris;* García-Arteaga et al. 1999), Nassau grouper (*Epinephelus striatus;* Claro et al. 1990c), and dog snapper (*Lutjanus jocu;* Claro et al. 1999); in the scales of redear sardine (García-Arteaga 1993); in the fin spines of mullets (liza, white, and fantail; Alvarez-Lajonchere 1981a); and in the otoliths and urohyals of silk and blackfin snapper (Espinosa and Pozo 1982; Pozo and Espinosa 1982; Pozo et al. 1983; Espinosa et al. 1984).

In some of the species mentioned above, the period of ring formation coincides with the reproductive season, but the reproductive season is usually extended. Consequently, the ring has been used as a reference point, or an annual ring, for determining age and growth. For most fishes, however, this ring is associated with an increase in water temperature and photoperiod (usually during April–May in Cuba), which in turn leads to increased feeding intensity (Section 5.4.2) and thus growth. In species that reproduce during the winter, the winter ring should be considered an annual ring. Ring formation in fish skeletal structures has been said to require a seasonal temperature amplitude of more than 5°C (Menon 1950). The monthly average amplitude range is 5–6°C in Cuba, and 3–4°C in Jamaica.

Environmental seasonality does not lead to the same ring pattern for all fishes (Fig. 6.1); for example, the ring formation period can be longer than the period of environmental change. Environmental changes can vary interannually over time and in intensity, as can the geographic location of the individual fish. The intensity of metabolic processes of fishes can change with age, and that will also affect the influence of environmental variation. For example, the long reproductive periods in tropical fishes can result in size differences in individuals of the same generation, which will affect their responses to the

environment. Other factors, such as the different criteria and methods used for examining skeletal structures, as well as sample sizes and sampling frequency, can help explain some of the lack of correlation noted above. Many projects may not pay sufficient attention to determining what causes ring formation, but aim only at determining age and growth rate through the annual ring pattern. Monthly sampling, in which sex and size ranges are well represented, provides more reliable data.

We found that in some species, such as margate and lane, mutton, and yellowtail snappers, some skeletal structures are more likely to register growth changes than others. This effect can be attributed to a distinct pattern of calcium accretion in different structures (bones, otoliths, scales). The urohyal, a flat bone, grows mainly in surface area, whereas the otolith grows in volume. A growth decline at the onset of winter might not provoke the same decrease in calcium accumulation as at the beginning of spring. In spring, the environmental changes coincide with an abrupt drop in feeding intensity (Section 5.4) and the start of gonadal maturation (which diverts a large part of the anabolic processes and mineral salts in the organism).

Maturation of the ovaries leads to reductions in scale calcium in tilapia (*Tilapia esculenta*) because calcium is used in gamete development (Garrod and Newell 1958). In these circumstances, the winter mark, which is visible in the urohyal and potentially in the otoliths, is masked in the otoliths by calcium deposition. This effect might not occur with the spawning mark because the calcium deficit would be clearly depicted in this structure.

The hyaline layer in the otoliths of the vermilion snapper (*Rhomboplites aurorubens*) starts to appear 4 to 6 months (November) before the scale ring (March–April) (Grimes 1978). The ring-formation period for the Brazilian mojarra (*Eugerres brasilianus*) is also different in scales and anal-fin spines (Báez et al. 1982a); the annual variation of the marginal increment width suggests that the winter scale ring does not occur in all individuals. The asynchrony of ring formation on different structures is probably related to the divergence in calcium metabolism in individuals of different sizes, particularly between adults and juveniles.

The use of data from fishes of different ages might mask the actual time of ring formation: one set of data shows that rings in red grouper (*Epinephelus morio*) are formed March–May in fishes aged 1 to 4 years, and May–July in fishes aged 5 to 10 years (Moe 1969). Clearly, selecting the appropriate structure for determining the age of a fish is critical.

We noted some overall patterns in lutjanid otoliths and urohyal bones, although each species has its own character-istics. Otoliths are generally large, easy to remove, and easy to preserve and interpret. Growth rate changes are clearly marked in the first years of life, with little interference from false rings. In some snapper species, only one hyaline ring in the otolith is formed per year and coincides, at least partially, with the spawning period; this fact facilitates data interpretation. In other species (gray, silk, and blackfin snappers) two annual rings are developed (Fig. 6.1), and less asynchrony with the reproduction period is seen. In older individuals, otoliths are whiter, with less contrast between the hyaline and opaque zones, and the rings become increasingly narrow. However, these older individuals are not common because of intense fishing pressure.

Marks on the urohyal are similar in all of the snapper species we studied. Two or more narrow annual hyaline rings were observed. False rings typically occur on this bone, which might be related to the metabolic sensitivity of growth of the urohyal bone. In older individuals, unlike on otoliths, on the urohyal the last rings are clearly marked but earlier rings are not. This is probably caused by a notable thickening of the anterior part of the urohyal as the fish grows, which covers the marks formed earlier.

Snapper scales are often difficult to read. In lane, mutton, and yellowtail snappers, growth marks on scales have little contrast and sometimes are not clearly sequential. The scales of gray snapper have more contrast and have been used by several authors for growth determination (Croker 1962; Starck 1970; Guerra and Bashirullah 1975).

Otoliths were used in growth studies of the margate, white, and bluestriped grunts (García-Arteaga 1983, 1992a, 1992b); the ring formations are similar in size, shape, and pattern to those of lane, mutton, and yellowtail snappers. The urohyal bone is strongly calcified, and opaque and hyaline zones are poorly distinct; therefore the urohyal was not used for any of these species. However, rings are clearly identifiable in grunt scales.

Both scales and otoliths of jacks (Carangidae) are tiny and difficult to manipulate, but the urohyal clearly depicts changes in growth rate. In bar jacks, two rings are formed each year, and scales seem to be more convenient for age studies (García-Arteaga and Reshetnikov 1992). In mullets (Mugilidae), spines have also been used for study (Alvarez-Lajonchere 1981a). Growth rings have been clearly seen in the scales and second anal spine of the Brazilian mojarra, particularly in the latter, although in older individuals the spine posed some difficulties for ring counting (Báez et al. 1983). A similar pattern in both structures has been observed in yellowfin mojarra (*Gerres cinereus*; Báez and Alvarez-Lajonchere 1983).

In groupers (Serranidae), otoliths are the primary skeletal structure for growth studies. They were used in studies of the red grouper in Florida (Moe 1969; Stiles and Burton 1994); graysby (*Cephalopholis cruentata*) in Curaçao (Nagelkerken 1979); yellowfin grouper (*Mycteroperca venenosa*; Thomson and Munro 1983); gag (*Mycteroperca microlepis*; McErlean 1963; Manoch and Haimovici 1978) and *Diplectrum formosum* (Bortone 1971) in Florida; warsaw and black groupers (*Epinephelus nigritus* and *Mycteroperca bonaci*; Manooch and Mason 1987) in southeastern United States; and Nassau grouper (Claro et al. 1990c), and tiger grouper (García-Arteaga et al. 1999) in Cuba. In many of these

studies, it was necessary to clear the otoliths to heighten the contrast between opaque and hyaline zones.

6.2.2   Juvenile Marks

The correct definition of the first annual ring is critical in determining age by hard structures. In many species, one or more rings are formed during the juvenile stage, before the individual attains the first year of life. These rings, which are associated with habitat or feeding shifts, can be mistaken for the first annual ring. The description of the juvenile rings is important for fish growth studies, because an error in defining the number of juvenile rings can lead to serious errors in age estimation.

Two juvenile rings were typically observed in the hard structures of individuals in their first year of life (0+ individuals) in the following species: lane, mutton, gray, dog, and yellowtail snappers, bluestriped and white grunts, margate, and bar jack. The first ring seems to be related to shifts in habitat and changes from planktivory to invertebrate feeding. Two juvenile rings have also been recorded in the bluestriped grunt in Jamaica (Billings and Munro 1974). These rings have been related to a change from diurnal planktivory to nocturnal foraging on invertebrates (Cummings et al. 1966).

The second juvenile mark observed in most of the studied species also coincides with diet changes. In snappers and bar jack, for example, the second ring occurs when the individuals start feeding on larger and more mobile prey—small fishes and crustaceans (mainly swimming crabs). Ring formation in the juvenile stage may or may not coincide with the period of winter ring formation. Regardless, a third juvenile ring can occur, as was found in many bar jack juveniles (García-Arteaga and Reshetnikov 1992). Such juvenile rings are a generalized feature in fish, both in tropical and in temperate waters. Up to seven juvenile rings were recorded in the otoliths of graysby in Curaçao (Nagelkerken 1979).

6.3   Seasonal Variations in Growth Rate

Fluctuations in the width of the space between the outermost increment and the outer margin of the skeletal structure used for aging can indicate seasonal variations in growth intensity. Fig. 6.2 summarizes such patterns for populations of various species from Cuba and neighboring areas. For some species, the curves are based on our own data. In other cases, because different methods were used, we interpreted the authors' data with a y-axis standardized for the maximum and minimum values.

Snappers show a notable, but brief, drop in growth rate during the period of transition from winter to summer (March–April). During April–May, the marginal zone increased again, and reached its maximum during September–October. This is the period of maximum feeding

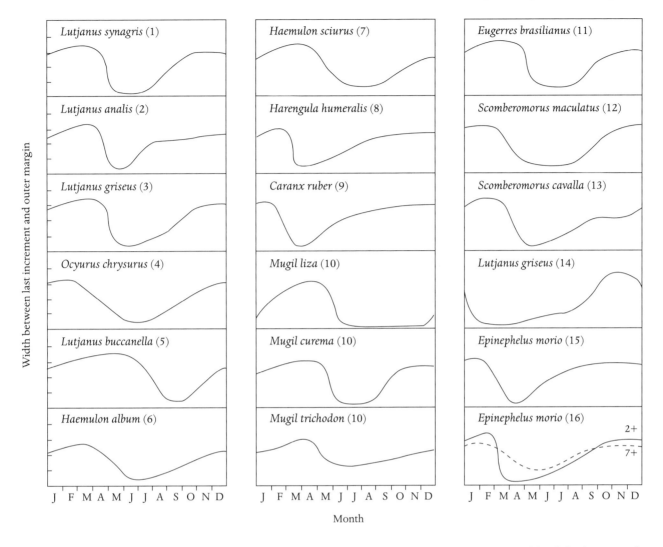

Fig. 6.2. Seasonal variation in the width of the spacing between the last increment and the outer margin of the skeletal structure for various species. This width varies during the year and is proportional to growth rate. The curves are based on an incremental scale standardized for all species (all maximum and minimum widths are at the same level), after interpreting data from other authors. 2+: juveniles at age of 2 years; 7+: adult fishes at ages up to 7 years. References: 1: Claro and Reshetnikov 1981; 2: Claro 1981c; 3: Claro 1983c; 4: Claro 1983e; 5: Espinosa and Pozo 1982; 6: García-Arteaga 1983; 7: García-Arteaga 1992; 8: García-Arteaga 1993; 9: García-Arteaga and Reshetnikov 1992; 10: Alvarez-Lajonchere 1981c; 11: Báez et al. 1983; 12: Powell 1975; 13: Beaumarriage 1973; 14: Starck 1970; 15: González 1983; 16: Moe 1969.

activity (see Section 5.4), which, in addition to increased water temperatures, might contribute to protein formation in the organism, particularly leading to an increase in muscular mass and fish length, to the detriment of fat storage (see Section 7.4). However, a temporal dip is depicted in the hotter months of the year (July–August), probably caused by a large increase in energy expenditures.

Under experimental conditions, maximum growth of subadult mutton snappers was found during May–June (Claro and Colás 1987). In adult fishes, however, the highest growth rate occurs in late summer, once the postspawning recuperation period is over (Claro 1981c). A similar pattern occurs in the marginal zone in the otoliths of lane, gray, and yellowtail

snappers (Claro and Reshetnikov 1981; Claro 1983b, 1983e). Gray snappers of Florida, like those in the Golfo de Batabanó, Cuba, grow faster during August–September (Starck 1970). Nevertheless, growth rate during the rest of the year was slower in Florida, probably because of the lower water temperatures prevalent in this more northern area.

In deep-water snappers (e.g., silk and blackfin), the greatest growth rates occur during October–December (Espinosa and Pozo 1982; Pozo and Espinosa 1982; Pozo et al. 1983; Espinosa et al. 1984), precisely when deep slope waters reach their minimum temperature in Cuba (Fig. 1.7 in Claro and Reshetnikov 1994). Evidently, the optimal growth temperature for these species is lower than for shallow-water snappers. Temperature might also be the factor that contributes to faster growth in winter (December–March) for the liza, whereas growth is almost negligible in the summer (Alvarez-Lajonchere 1981a).

Growth rates in two haemulids (margate and bluestriped grunt) reach their maximum from September to November (the period in which reproductive activity is less intense). The redear sardine grows almost uniformly most of the year; however, the growth rate declines in winter at the beginning of the reproductive period (Fig. 6.2). The bar jack grows intensely in early summer; however, juveniles were most abundant in these samples (García-Arteaga and Reshetnikov 1992).

6.4   Relationships among Different Size Measures

Taken together with the annual periodicity of ring formation, the relationship between the structure used for ring reading and fish size is another test of back-calculations of size at different ages. Claro and García-Arteaga (1994a) summarized these correlations for the marine fishes of Cuba. They also compiled correlations among total (TL), fork (FL), and standard length (SL) measures of size.

Length to weight relationships have been estimated for a considerable number of tropical marine fishes (see summaries by Bohnsack and Harper 1988; Claro and García-Arteaga 1994a; and García-Arteaga et al. 1997). Appendix 6.1 summarizes data collected for different species in Cuba. Some of these data were calculated from small samples and should be used with caution; fish weight by size might also change seasonally.

Length to weight relationships might also vary geographically within a single species. For example, mutton snapper in Cuba up to about 55 cm FL weighed more (at the same length) in the north-central region than in the northwest region (Appendix 6.1). This divergence was most notable in juveniles (< 40 cm FL). However, mutton snapper larger than about 55 cm in the north-central region weighed less than fishes of comparable length in the two western regions. Mutton snappers in Cuba exhibited greater weight at the same size than in Florida (Mason and Manooch 1986), but less than in Colombia (Echardt and Meinel 1977).

6.5   Age and Growth Rate

6.5.1   Methods Used to Evaluate Growth Rate

Fish growth is usually estimated by the length and weight of the individual and its corresponding age. Unlike growth in other vertebrates, fish growth continues throughout the life span. Growth rate at any age can be calculated by the slope of the length–time equation. The weight growth curve, although reaching an asymptotic value, exhibits a progressive increment during the early life stages up to an inflection point, after which weight growth gradually slows. A similar

pattern was observed in length–age during the first months of the fish's life. This distinct growth pattern is often not considered when overall fish growth is examined.

As already discussed, growth rings in skeletal structures of subtropical fishes are used by many researchers, but formation patterns can be difficult to define in fish populations of the lower latitudes (Munro 1983a). In such cases, modal progressions of body length are used by many authors (e.g., Thompson and Munro 1974, 1978, 1983; Munro 1976; Pauly 1980; Pauly and David 1981; Gayanilo et al. 1988). We back-calculated length with regression equations that relate the size of skeletal features (otoliths, urohyal bones, scales) to length, or with empirical equations when appropriate. To represent growth rate, we used the von Bertalanffy (1938) equation

$$L_t = L_\infty(1 - e^{-K(t-t_0)})$$

where L_t is the length at age t, L_∞ is the asymptotic length when $t = \infty$, K is the growth constant, and t_0 is the theoretical age at length 0. The L_∞ and K parameters were initially obtained by plotting a Walford (1946) line:

$$L_{t+1} = L_\infty(1 - k) + kL_t$$

where L_t is fish length at age t, and k is the Walford line slope. The slope (k) is equal to e^{-K}; so for the first estimation, K = lnk. The preliminary L_∞ values were obtained by plotting $L_{t+1} = y/(1 - k)$, where y is the intercept.

To check the accuracy of the L_∞ and K estimates, we used the method proposed by Everhart et al. (1975) and further developed by Ricker (1975). This method consists of plotting ln $(L_\infty - l_t)$ against t. By using test values with a 10-mm interval for L_∞ and taking the obtained K value, a more accurate value of L_∞, which corresponds with the highest correlation value, can be obtained. The slope of that line is the k value. Using both parameters and the y-intercept value, we calculated t_0 with the equation

$$t_0 = (y - \ln L_\infty) / K$$

The asymptotic weight (W_∞) was estimated using L_∞ and the corresponding length–weight equation for the species (Appendix 6.1).

6.5.2   Growth Parameters of Species of the Wider Caribbean

We compiled growth data for fishes of the wider Caribbean (Appendix 6.2). To compare growth parameters estimated with different methods, we followed Pauly (1980) who demonstrated that plotting the logarithm of the growth coefficient K against the logarithm of the asymptotic weight (W_∞) provides a linear regression with a slope close to a theoretical value of 0.67. Thus, log K = $\phi - 2/3$ logW_∞. Because the slope is constant, the differences in growth rate are reflected in the value of ϕ, a growth performance index.

Because data on asymptotic weight were not always available, we transformed the equations as follows to allow for the use of L_∞: $\phi' = 2 \log L_\infty + \log K$ (Pauly and Munro 1984; Pauly and Binolhan 1996).

We used this formula to estimate ϕ' from original data in the literature (see Appendix 6.2). In Cuban coastal fishes we found a logarithmic correlation between ϕ' and L_∞ (Fig. 6.3). Similar ϕ' coefficients might be characteristic of families or groups of species with similar ecological patterns (Munro 1983c). Thus, values of ϕ' can be used for comparing growth of different populations of the same species and for testing the validity of the growth estimates. We therefore calculated this and other growth parameters for species studied by us and others. When these parameters were not available, we estimated them from available data (Appendix 6.2).

Appeldoorn (1992) showed that relationships based on ϕ' are best used for intraspecific comparisons. For comparisons between species, he proposed the relationships

$$\Phi = \log k + 0.25 \log W_\infty \text{ and } \Phi' = \log k + 0.75 \log L_\infty.$$

Appeldoorn concluded that "use of ϕ with interspecific data leads to wrong conclusions; it also increases the apparent range of variation around mean values for higher taxa. Variation in Φ is greatly conserved when comparing growth parameters from within a single study, i.e., where similar methodology has been applied to different species

from the same area. Parameter estimation or verification is substantially enhanced if both ϕ and Φ are available."

Growth data are available for most commercial species of the following families: Lutjanidae, Serranidae, Haemulidae, Carangidae, Mugilidae, Gerreidae, Scombridae, Clupeidae, and Sparidae. Information on most small, noncommercial reef fishes is lacking. For these species, data obtained through tagging methods for many species in the Virgin Islands were useful (Randall 1962, 1963a). However, tagging procedures might alter growth rates.

Snappers are the most studied fishes in Cuba. The growth of the lane snapper, the most abundant coastal snapper in Cuba, has been examined by several researchers, particularly in the Golfo de Batabanó (Rodríguez Pino 1962; Olaechea and Quintana 1970; Claro and Reshetnikov 1981; Rubio et al. 1985), as well as in other areas of the tropical northwestern Atlantic (Alegría and Ferreira de Menezes 1970; Manooch and Mason 1984; Manickchand-Dass 1987; Torres and Chávez 1987). With the exception of Olaechea and Quintana (1970), who used the urohyal bone, and Torres and Chávez (1987), who used size composition, all of these studies used otoliths.

Although Olaechea and Quintana (1970) used different skeletal structures, their data for lane snapper are similar to ours (Table 6.1; Fig. 6.4). However, they differ from those provided by Rodríguez Pino (1962), whose calculated lengths for individuals 1 to 3 years of age are notably higher. This difference has been attributed to the threefold increase of commercial catches in the Golfo de Batabanó during the 1960s and the subsequent decrease in fish size (Claro 1981a). High fishery effort can reduce predation pressure on large amounts of prey, which can increase growth rates of smaller fishes. Increased numbers of individuals can overburden food supplies, which can ultimately lead to a decline in growth rates within populations (Nikolsky 1974a). This might be a cause of the decrease in growth rates of lane snapper estimated by Claro (1981a), and might result in an earlier attainment of maturity than in 1961 (Rodríguez Pino 1962). Comparing Manooch and Mason's (1984) data for southern Florida with our data for Cuba shows a great similarity in back-calculated length for the age 1-to-4 year classes. (Table 6.1). However, the Florida fish displayed slower growth and a more prolonged life span at older stages. In the Golfo de Batabanó, the oldest lane snapper sampled was 6 years old. In Florida, individuals up to 10 years are common (maximum length, 39.4 cm). This divergence might also be due to more intensive fishing pressure on the lane snapper in Cuba.

Lane snapper growth estimates in Brazil (Alegría and Ferreira de Menezes 1970), as well as in Yucatán, México (Torres and Chávez 1987), were higher than in the Golfo de Batabanó. Growth estimates are even higher in the northwest region of the Golfo de Paria (Venezuela) and the northern coast of Trinidad (Manickchand-Dass 1987),

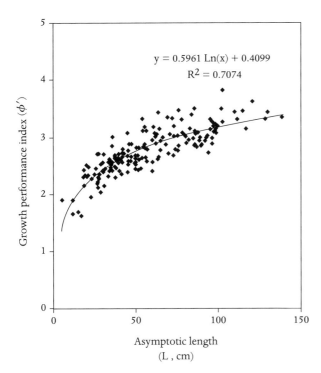

$$y = 0.5961 \, Ln(x) + 0.4099$$
$$R^2 = 0.7074$$

Fig. 6.3. Correlation between asymptotic fish length (cm) and ϕ', a growth performance index, for selected marine fishes occurring in Cuba.

Table 6.1.   Length and growth parameter estimates for the lane snapper, *Lutjanus synagris,* by different authors

	Source and Locality								
	1 Cuba	2 SW Cuba	3 SW Cuba	4 SW Cuba	5 S Florida	6 Trinidad M	 F	7 Brazil	8 Mexico
Back-calculated length (cm FL) for age (year)									
0_1	—	—	5.6	—	—	—	—	—	—
0_2	—	—	9.5	—	—	—	—	—	—
1	16.9	11.9	12.5	—	12.6	22.7	21.7	11.2	—
2	20.5	17.6	17.1	—	18.2	29.9	29.9	18.5	20.5
3	23.7	21.9	22.3	—	21.6	41.1	37.3	26.6	25.0
4	26.6	25.1	26.3	—	24.2	48.5	43.1	28.9	28.5
5	27.2	27.6	29.0	—	26.4	—	—	32.5	31.2
6	30.2	29.4	32.0	—	28.7	—	—	35.4	—
7	—	—	—	—	31.3	—	—	—	—
8	—	—	—	—	34.0	—	—	—	—
9	—	—	—	—	38.0	—	—	—	—
10	—	—	—	—	39.4	—	—	—	—
Growth parameters									
L_∞ (cm)	37.9*	35.1*	47.0*	40.1	46.2	70.5	60.3	42.0*	41.0
W_∞ (g)	823*	780*	1,758*	1,184	2,300	5,715	3,352	1,290*	1,066
K (yr)	0.20*	0.28*	0.175*	0.16	0.134	0.22	0.20	0.302*	0.25
t_0 (yr)	−1.85*	−0.52*	−0.878*	−1.83	−1.49	−0.55	−0.68	−0.026*	−1.82
ϕ'	2.46*	2.54*	2.57*	2.41*	2.51*	3.04*	2.86*	2.72*	2.61*

Sources: 1: Rodríguez Pino 1962; 2: Olaechea and Quintana 1970; 3: Claro and Reshetnikov 1981; 4: Rubio et al. 1985; 5: Manooch and Mason 1984; 6: Manickchand-Dass 1987; 7: Alegría and Ferreira de Menezes 1970; 8: Torres and Chávez 1987.

M: male; F: female.

* Values calculated in the present study using data from the original source.

where males grow faster and attain larger sizes than females. In Cuba, however, no significant differences were found in the growth rates between sexes, although females grow larger and live longer than males (Claro and Reshetnikov 1981). The ϕ' values calculated by various authors ranged from 2.41 to 3.04. Despite the wide divergence, a relationship between ϕ' and L_∞ and W_∞ was observed (Table 6.1).

Mutton snapper growth has been studied in three regions of the Cuban shelf: the southwest (Golfo de Batabanó), the northwest (Claro 1981c), and the north-central (Pozo 1979), as well as in eastern Florida (Mason and Manooch 1986). For back-calculating the length for the age of mutton snapper in the southwest and northwest Cuba, we examined the linear correlations between fish length and the otolith radius for each region separately (Claro 1981c). No significant differences between sexes were found. The growth parameters of mutton snapper in the southwest and northwest regions were L_∞ = 82 cm FL and L_∞ = 87 cm, respectively. However, the largest individuals sampled from commercial catches in the southwest and northwest regions were 72 cm FL and in the north-central region were 79 cm (Pozo 1979).

Although mutton snappers grow during the first year of life at rates similar to other snapper species, from the second year on, their growth increases substantially. In 5- to 7-year-old individuals, weight growth can reach more than 800 g per year. Length and weight growth in the northwest seems to be lower than in the southwest during the three first-year classes, but faster after the third year of life (Table 6.2; Fig. 6.4). A similar pattern was observed in the length–weight relationship for both sexes in adult individuals: they weighed more in the southwest than in the northwest at the same length. These differences might be related to the relative extent of habitat available for juveniles and for adults. The preferred habitat for juveniles (seagrass beds) is more extensive in the southwest, so this southwest population may be larger than in the northwest. However, the area of adult habitat (shelf-border reefs and patch reefs) is similar in both regions. Therefore, the amount of food available per adult individual might be higher in the northwest than in the southwest.

Mutton snapper lengths by age in north-central Cuba (Pozo 1979) were more in the first two years of life and less from the fourth year on. Pozo (1979) found three rings in the urohyal bone in the young-of-the-year, but another study considered only two of those to be 0+ rings (Claro 1981c).

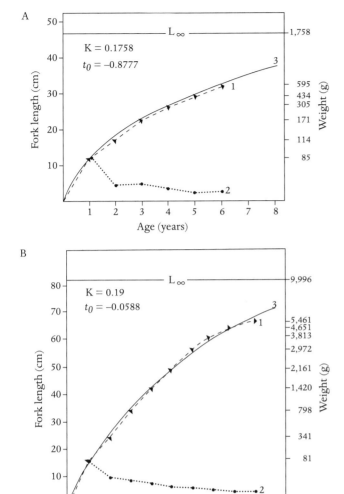

Fig. 6.4. Back-calculated values for (1) fish length, (2) mean annual growth, and (3) theoretical growth curves for (A) lane snapper (*Lutjanus synagris*) and (B) mutton snapper (*L. analis*) in southwestern Cuba.

These differences can be attributed to a difference in the definition of the first annual ring (Claro reported that the first ring was visible only in the youngest individuals). Despite the divergence between the data, adult fishes from the north-central region clearly grow more slowly than in the southwest and northwest regions. Back-calculated and theoretical values for this species in Florida (Mason and Manooch 1986) are similar to those collected by us in Cuba (Table 6.2), although slightly higher in the second and third year of life. This difference is reversed for individuals four years or older. At Isla Cubagua, Venezuela, mutton snapper growth rate was higher than that found in Florida and Cuba (Palazón and González 1986). The values of the ϕ' coefficient that we calculated from the data of different researchers are

similar, except for the coefficient from the data of Pozo (1979), which is about two standard deviations lower than for the other regions.

Gray snapper growth in the Caribbean and Florida has been studied by several authors (Croker 1962; Starck 1970; Guerra and Bashirullah 1975; Báez et al. 1980a; Claro 1983b; Manooch and Matheson 1983; Johnson et al. 1994; Burton 2000). Some differences in findings might be attributable to the different methods used. Fig. 6.5A summarizes back-calculated lengths for gray snappers from the Golfo de Batabanó, Cuba, and Table 6.3 summarizes lengths from Florida, other regions of Cuba, and Venezuela. Data from Croker (1962), Starck (1970), and Manooch and Matheson (1983) show lower annual growth during the first and second years of life in contrast to most studies from Cuba. Their back-calculated lengths coincide with the ones considered by us to be 0+ rings. If the first ring recorded by Croker and Manooch and Matheson, and the two observed by Starck are not considered to be annual rings, different growth estimates result (Table 6.3). The degree to which the subannual growth rings identified in Cuban studies apply to studies from other regions needs to be further evaluated.

The largest gray snappers recorded in five studies in Florida vary with latitude. Croker (1962) and Starck (1970), working in south Florida, did not find individuals larger than 46 cm FL. Based on extensive sampling, Manooch and Matheson (1983), Johnson et al. (1994), and Burton (2000) concluded that gray snapper attain a larger size in north Florida than in any other studied area. This difference could be due to less fishery pressure in north Florida (Burton 2000). They also eliminated the possibility that gray snapper samples in their studies were misidentified as the similar, but larger, cubera snapper (*Lutjanus cyanopterus*) (Burton 2000).

Guerra and Bashirullah (1975) and Hurtado and Bashirullah (1975) estimated high growth rates in Venezuelan gray snappers. This may be due to better feeding conditions in that region. Invertebrates were the dominant food for gray snapper in Florida (Croker 1962; Starck 1970; see Table 6.4), whereas in the Golfo de Batabanó they constituted only 26% of the food weight, and in Venezuela 9% of the volume. The proportion of fish in the diet might increase from north to south, as do species growth rates. Despite this, Croker (1962) estimated a body size at age 6+ in Florida that is similar to that reported for Cuba. Until they attained the fourth year of life, fishes appeared to grow faster in Cuba. In addition, individuals more than 4 years old were under-represented in Croker's samples. The growth rate in coastal lagoon populations might also be lower than in open shelf waters (Báez et al. 1980a). Crustaceans were consumed by gray snappers in both types of ecosystems, but they are dominant in the dietary composition of the coastal lagoon population, whereas fishes were a more important food in shelf areas.

Gray snapper growth rate during the first year of life was higher than for lane, mutton, and yellowtail snappers in

Table 6.2. Length and growth parameter estimates for the mutton snapper, *Lutjanus analis*, by different authors

	Source and Locality				
	1 E Florida	2 Venezuela	3 North-Central Cuba	4 SW Cuba	5 NW Cuba
Back-calculated length (cm FL) for age (year)					
0_1	—	—	—	6.3	3.5
0_2	—	—	—	13.0	11.4
1	14.2	21.9	20.1	15.7	15.0
2	26.8	32.8	25.0	24.7	22.4
3	35.0	42.6	34.5	34.1	31.3
4	40.2	51.0	39.5	41.9	42.6
5	44.7	57.6	44.4	49.4	50.3
6	49.0	63.3	47.2	56.0	56.2
7	52.8	67.8	49.4	60.9	62.6
8	56.2	71.8	52.5	64.4	67.3
9	59.9	—	57.7	66.4	—
10	62.7	—	—	—	—
14	70.9	—	—	—	—
Growth parameters					
L_∞ (cm)	78.6	93.7	80.7	82.0*	87.0*
W_∞ (g)	8,676	15,489	8,034*	9,996*	11,794*
K (yr)	0.1534	0.17	0.116	0.19*	0.1894*
t_0 (yr)	−0.5788	−0.62	−1.427	−0.0588*	−0.3658*
ϕ'	3.05*	3.25*	2.89*	3.07*	3.14*

Sources: 1: Mason and Manooch 1986; 2: Palazón and González 1986; 3: Pozo 1979; 4: Claro 1981c; 5: Claro 1981c.

* Values calculated in the present study using data from the original source.

the same areas. This might be related to the possibility that gray snappers eat more fishes at early stages of development (Section 5.3.3). As yellowtail snappers shift from planktivory to a more piscivorous pattern, the growth rate becomes similar to that of gray snappers.

Growth of yellowtail snappers was studied in three regions of the Cuban shelf: the southwest (Claro 1983e), northwest (Piedra Castañeda 1965; Claro 1983e), and southeast (Carrillo de Albornoz and Ramiro 1988), as well as south Florida (Johnson 1983; García 2000) and the Virgin Islands (Manooch and Drennon 1987). No differences were observed between growth of fishes in the northwest and southwest regions of Cuba. However, weight increase appeared to be higher in the southwest region for annual classes (Claro 1983e). Samples from the northwest coast were obtained from traps, whereas in other regions, samples were obtained primarily from other gear (e.g., hand lines) that often netted larger individuals. Thus, growth estimates were skewed toward smaller asymptotic lengths in northwest samples.

Data obtained by Piedra Castañeda (1965) on the age of 89 yellowtail snappers are notably different from ours, probably because she considered all vertebral rings to be annual marks, whereas we observed two rings per year in the urohyal bone and scales that might also occur in the vertebrae. Piedra Castañeda (1965) also did not consider the

first ring in the vertebrae to be a 0+ juvenile ring. Accounting for these differences, the second, fourth, sixth, and eighth rings detected by Piedra Castañeda (the values in parentheses in Table 6.5) coincide with those we observed as first, second, third, and fourth rings by Claro (1983e) for the same region. The value of the growth coefficient, ϕ', under this assumption is similar in fishes of both studies. According to the data provided by Johnson (1983) and Manooch and Drennon (1987), growth rates for yellowtail snappers in Florida and the Virgin Islands are lower than in Cuba, although the former populations were estimated to have greater longevity. Length at first year, however, is quite similar to that observed by Claro (1983e) for age 0+ fishes.

Dog snappers, which attain large sizes, appeared to have a higher growth rate during the first year than do mutton, lane, and yellowtail snappers. This could be attributed to its piscivorous feeding habits. As juveniles, dog snappers inhabit mangrove areas where competitors for food are scarce. However, as they migrate out to reef areas where large species are abundant, dog snappers might encounter more competition for food, and the growth rate could diminish (Fig. 6.6A). No growth differences were observed between dog snapper populations of the north-central and southwest regions of Cuba, although they do reach larger sizes in the north-central region. Unlike the other snappers studied, male dog snappers

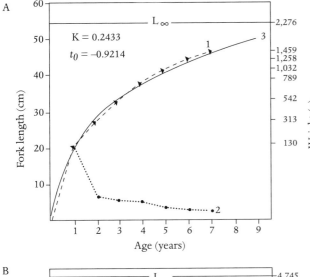

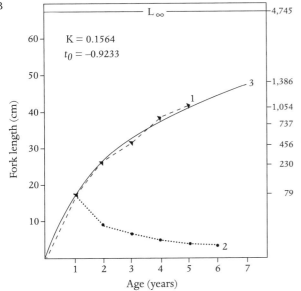

Fig. 6.5. Back-calculated values for (1) fish length, (2) mean annual growth, and (3) theoretical growth curves for (A) gray snapper (*Lutjanus griseus*) and (B) yellowtail snapper (*Ocyurus chrysurus*) in southwestern Cuba.

grow faster than females, particularly after the fifth year, and we found that they were more abundant in the 10+-year and 65-cm FL class, mainly in the north-central region (Claro et al. 1999).

Growth of silk and blackfin snappers of the southeast and north-central regions (judged by the urohyal bone and otoliths) was similar for both species (Espinosa and Pozo 1982; Pozo and Espinosa 1982; Pozo et al. 1983; Espinosa et al. 1984). The silk snapper reached a larger size and had a faster growth rate than the blackfin snapper. The K values obtained by Thompson and Munro (1983) for a small sample of these two species may be excessively high; the ϕ' values estimated from these data are not common among snappers. In South Carolina, vermillion snappers attain only 8.8 cm

during the first year (Grimes 1978), which is less than the growth displayed by other snappers during this period. However, the growth of the next year's classes is comparable to that of other species of the family. Possibly, as in other snappers, the first ring reported in vermilion snappers is formed before attaining the first year of age.

Data on Cuban snappers show that the K parameter can range from 0.09 to 0.34 (0.157 ± 0.076 SD). The ϕ' value ranged from 2.41 to 3.14, with a marked relationship between ϕ' and W_∞. When data from other areas of the wider Caribbean are included (also based on otoliths, scales, or bones), the ranges widen: 0.09 to 0.70 (0.247 ± 0.155 SD) for K, and 2.41 to 3.81 (2.89 ± 0.28 SD) for ϕ'. Comparison of these data with those summarized by Munro (1983b) for tropical and subtropical waters of the Indo-Pacific show that K parameters can be higher (0.11–0.38; 0.29 ± 0.08) than in Caribbean species.

A relationship between growth and relative species abundance was observed for grunts. The margate is the largest grunt, analogous to mutton snapper among coastal snappers (García-Arteaga 1983; Fig. 6.6B). However, this is the least numerous commercial grunt (Section 8.6). The K parameters for margates obtained by García-Arteaga (1983) in Cuba are similar to those estimated by Billings and Munro (1974) in Jamaica (using the modal progression method of age determination), although the L_∞ and ϕ' estimates were higher in the Cuban studies (Table 6.6). These differences could be due to the lack of small (< 20 cm FL) and large (> 42 cm FL) individuals in the samples from Jamaica.

The data collected on the bluestriped and white grunts (Fig. 6.7) indicate similar growth rates in southwest Cuba (García-Arteaga 1992a, 1992b). Values for these species obtained in the north-central region for white grunt are quite similar to those obtained for the southwest region (Ramos and Pozo 1984). Both of these species in Cuban waters showed K and back-calculated length values similar to those estimated for species in Jamaican waters (Billings and Munro 1974; see Table 6.6). The samples collected in southwest Cuba lacked individuals larger than 27 cm for white grunt, and 28 cm for bluestriped grunt. The sizes calculated by Billings and Munro from scale readings on bluestriped grunt are much lower than those estimated in Jamaica by the same authors using the modal shift method.

Latitudinal variations in growth are now well documented in the white grunt. Capote (1971) and Manooch (1976) found that white grunt from the Banco de Campeche and Florida attained larger size at age than in Cuba. Potts and Manooch (2001) also demonstrated that in the Carolinas, individuals attain larger size at age than in Florida, particularly the older individuals (Table 6.6).

According to unpublished data obtained by García-Arteaga, the growth rate and longevity of the tomtate (*Haemulon aurolineatum*) in the Golfo de Batabanó is lower than in North and South Carolina (Manooch and Barans 1982) and the Banco de Campeche (Sokolova 1965; Olaechea

Table 6.3.   Length and growth parameter estimates for the gray snapper, *Lutjanus griseus,* by different authors[a]

	Source and Locality										
	1		2		3	4		5	6	7	
	S. Florida		Florida Keys		Florida	Florida		SW	SE	Venezuela	
						NE	SE	Cuba	Cuba	M	F
Back-calculated length (cm)[b] for age (year)											
0_1	—	—	—	(7.8)	—	—	—	6.0	7.1	—	—
0_2	—	(8.1)	—	(14.1)	—	—	—	14.5	—	—	—
1	8.1	(18.0)	7.8	(19.6)	9.2	—	—	20.4	14.2	—	—
2	18.0	(24.1)	14.1	(25.0)	18.9	—	—	27.1	22.0	30.9	31.1
3	24.1	(29.5)	19.6	(28.8)	26.2	—	—	32.8	28.3	40.2	36.8
4	29.5	(35.2)	25.0	(32.7)	31.7	—	—	41.5	33.8	42.3	42.1
5	35.2	(43.1)	28.8	(37.0)	36.0	—	—	44.0	38.0	44.2	50.1
6	43.1	(45.6)	32.7	(42.4)	40.0	—	—	46.0	40.7	49.3	53.4
7	45.6	—	37.0	(46.4)	43.9	—	—	—	42.7	—	—
8	—	—	42.4	—	47.4	—	—	—	—	—	—
9	—	—	46.4	—	50.5	—	—	—	—	—	—
10	—	—	—	—	72.6	—	—	—	—	—	—
Growth parameters											
L_∞ (cm)	78.8*	74.0*	120*	85.0*	82.8	71.6	62.5	54.8	51.3	—	—
W_∞ (g)	6,836*	5,694*	23,248*	8,523*	9,743	—	—	2,424	1,893	—	—
K (yr)	0.16*	0.142*	0.05*	0.087*	0.101	0.17	0.13	0.243	0.24	—	—
t_0 (yr)	−0.17*	−0.793*	−0.29*	−1.8*	−0.315	−0.001	−1.33	−0.921	−0.616	—	—
ϕ'	3.00*	2.89*	2.86*	2.80*	2.90*	—	—	2.84*	2.80*	—	—

Sources: 1: Croker 1962; 2: Starck 1970; 3: Manooch and Matheson 1983; 4: Burton 2000; 5: Claro 1983b; 6: Báez et al. 1980; 7: Guerra and Bashirrullah 1975. M: male; F: female.
[a]Values in parentheses are our interpretations of the source data (see discussion of subannual rings in text).
[b]All data are in cm FL, except Burton 2000 in cm TL from NE and SE Florida.
* Values calculated in the present study using data from the original source.

and Sauskan 1974; Olaechea et al. 1975). There is evidence that French grunts in Cuba may be smaller than in other areas (Appendix 6.6 in Claro and García-Arteaga 1994a). This might be related to their high abundance in Cuban shallow-water areas (Chapter 2).

Growth of tropical jack species (Carangidae) has been little studied, possibly because it is difficult to interpret the marks in skeletal structures. Only the bar jack has been studied in the Caribbean. Growth parameters estimated through analysis of modal shifts yielded values of L_∞ = 52.0 cm FL; K = 0.24; ϕ' = 1.66 (Thompson and Munro 1974). Well-defined growth rings (two per year) were found in the urohyal bone of this species in the Golfo de Batabanó (García-Arteaga and Reshetnikov 1992). The growth

Table 6.4.   Relationships between diet composition and growth of the gray snapper, *Lutjanus griseus,* in different latitudes

	Food Items		Growth at Year 6		
Region	Invertebrates	Fishes	Length FL (cm)	Weight (g)	Source
Everglades, Florida (O)	80	34	43.1	1,203*	Croker 1962
Florida Keys (V)	67.9	31.7	39.6	942*	Starck 1970
Golfo de Batabanó, Cuba (W)	25.9	74.1	44.0	1,270	Claro 1983c
Cubagua I., Venezuela (V)	9.3	90.7	53.4* F	2,317 F	Guerra and
			49.3* M	1,930 M	Bashirrullah 1975

O: by percentage of occurrence; V: by percentage of volume; W: by percentage of weight; F: female; M: male.
* Values calculated in the present study using data from the original source.

Table 6.5.   Length and growth parameter estimates for the yellowtail snapper, *Ocyurus chrysurus,* by different authors

	Source and Locality						
1 S. Florida	2 Virgin I.	3[a] NW Cuba		4 NW Cuba	5 SW Cuba	6 SE Cuba	7 SE US
Back-calculated length (cm FL) at age (year)							
0_1	—	—	—	—	7.1	9.4	—	—
0_2	—	—	—	(12.4)	11.0	13.1	—	—
1	13.6	11.7	12.4	(17.7)	16.2	17.5	17.4	17.7
2	22.7	18.4	17.7	(25.3)	24.8	25.3	23.6	24.0
3	27.7	23.1	21.8	(31.6)	31.6	31.4	28.7	27.0
4	31.5	26.3	25.3	(36.0)	35.8	35.8	32.5	29.4
5	34.2	28.7	28.6	—	—	41.4	36.1	31.8
6	37.1	30.8	31.6	—	—	—	39.4	33.7
7	38.9	32.4	35.3	—	—	—	—	35.9
8	41.6	34.0	36.0	—	—	—	—	38.8
9	42.2	35.5	—	—	—	—	—	41.1
10	43.2	37.5	—	—	—	—	—	42.9
11	44.5	39.4	—	—	—	—	—	47.8
12	45.8	41.3	—	—	—	—	—	—
13	41.8	42.9	—	—	—	—	—	—
14	42.9	44.2	—	—	—	—	—	—
Growth parameters							
L_∞ (cm)	45.1*	50.3	51.6*	(50.0*)	47.3	68.0	69.6	41.0
W_∞ (g)	1,627*	1,687	1,885*	(1,726*)	1,497	4,145	4,768	—
K (yr)	0.279*	0.139	0.26*	(0.280*)	0.341	0.156	0.103	0.210
t_0 (yr)	−0.355*	−0.955	−0.61*	(−0.547*)	−0.236	−0.923	−1.79	−2.37
ϕ'	2.76*	2.55*	2.53*	(2.84*)	2.87*	2.87*	2.70*	2.55

Sources: 1: Johnson 1983; 2: Manooch and Drennon 1987; 3: Piedra Castañeda 1965; 4: Claro 1983e; 5: Claro 1983e; 6: Carrillo de Albornoz and Ramiro 1988; 7: García 2000.

[a] Values in parentheses are our interpretations of the source data (see discussion of subannual rings in text).

* Values calculated in the present study using data from the original source.

coefficient (K) was lower than that estimated by Thompson and Munro (1974) for Jamaica. (Fig. 6.8A).

Growth data on groupers (Serranidae) from different parts of the wider Caribbean are summarized in Appendix 6.2. The species of this family have a wide range of maximum lengths, ranging from 6 to 10 cm in species of *Serranus* to more than 200 cm in the jewfish and warsaw grouper (*Epinephelus itajara* and *E. nigritus*). Otoliths and the urohyal bone have been used in growth studies. For example, red grouper in Florida (Moe 1969) and the Banco de Campeche (Melo 1976; González 1983; Valdés Alonso and Fuentes-Castellanos 1987); graysby in Curaçao (Nagelkerken 1979); red hind (*Epinephelus guttatus*) in Puerto Rico and St. Thomas (Sadovy et al. 1994); Nassau grouper in northeast and southwest regions of Cuba (Claro et al. 1990c); gag in Florida (McErlean 1963; Manooch and Haimovicii 1978); tiger grouper in southwest Cuba (García-Arteaga et al. 1999); black grouper in the Virgin Islands and Puerto Rico (Manooch and Mason 1987); and sand perch (*Diplectrum formosum*) in Florida (Bortone 1971).

The red grouper showed a higher growth rate in Florida than in the eastern part of the Banco de Campeche

(Moe 1969; Melo 1976; González 1983). However, other authors have reported faster growth during the first two years in the Banco de Campeche, and then a similar rate for the next year classes (Valdés-Alonso and Fuentes-Castellanos 1987). The notable differences in the values obtained for the first year of life could be the result of different criteria for selecting the first annual ring. The data estimated by Nagelkerken (1979), based on length distribution using the ELEFAN program (Pauly and David 1981), show a lower growth rate for graysby than for other Serranidae. Data on the sand perch also show a low growth rate, particularly in the first year of life (Bortone 1971). The first annual ring was possibly defined as one formed before the completion of the first year; if this ring is not considered, the growth rate would be more comparable to other tropical fish species.

The K value for Nassau grouper ranged from 0.05 to 0.13 (mean, 0.09) based on $L_\infty = 90$ cm TL (Thompson and Munro 1978; using data from Randall 1962, 1963a on the recapture of 45 of 124 tagged individuals). With such data, a ϕ' value of 3.9 would be obtained, which is about four standard deviations (0.36) higher than the mean value (1.66)

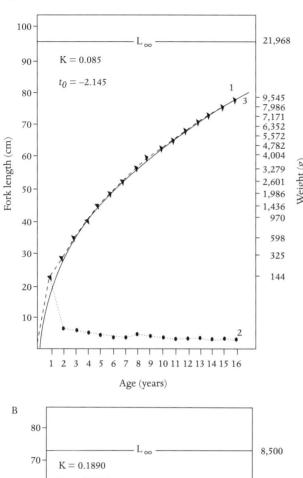

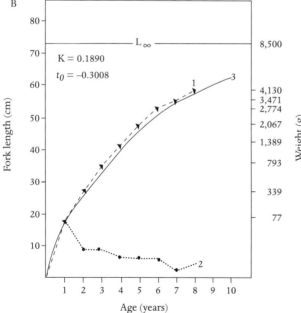

Fig. 6.6. Back-calculated values for fish length (1), mean annual growth (2), and theoretical growth curves (3) for (A) dog snapper males (*Lutjanus jocu*) and (B) margate (*Haemulon album*) in southwestern Cuba.

reported by Munro (1983c) for the Serranidae. The growth parameters calculated for this species in the Virgin Islands by Olsen and LaPlace (1979) might be more appropriate.

Growth parameters of Nassau grouper in the north-central and southwest regions of Cuba were estimated by Claro et al. (1990c) (Fig. 6.8B). The calculated lengths for age and growth rate were higher in the north-central region. Nevertheless, the asymptotic length obtained for the southwest region was higher and the individuals were older. During the first years of life, particularly during the first, the growth rate was found to be high in both regions (Appendix 6.4 in Claro and García-Arteaga 1994a). The tiger grouper (Fig. 6.9A) exhibited a slightly faster growth than the Nassau grouper in the southwest region of Cuba (García-Arteaga et al. 1999). The K and ϕ' parameters are similar to those of other serranids.

Thompson and Munro (1978) did not find any definite pattern of ring formation in red hind and coney (*Cephalopholis fulva*) by calculating growth parameters using the modal progression method. Munro (1983a) reexamined these data using the ELEFAN program (Pauly and David 1981) and recalculated the theoretical growth parameters of both species (Appendix 6.2).

Among the Labridae represented in the Cuban ichthyofauna, only the hogfish has been studied for age and growth. This is the largest species of the family and the only one with commercial significance (Claro et al. 1989). Length growth (Fig. 6.9B) is comparable to that for mutton and silk snappers and some groupers, such as the red, Nassau, and yellowfin groupers. Unlike other large predators that are mainly piscivorous, the hogfish consumes mostly mollusks and, to a lesser extent, crabs (Section 5.3.1). The relative abundance of mollusks on the Cuban shelf (Section 1.5.3) might insure that the hogfish has ample, secure food resources in this region. This might explain why all surveyed hogfishes had food in their stomachs and also high growth rates.

Growth studies of the gerreids, the yellowfin and Brazilian mojarras (*Gerres cinereus* and *Eugerres brasilianus*), in the coastal lagoon system of the Tunas de Zaza, Cuba (Báez et al. 1982a; 1983), and the striped mojarra (*Eugerres plumieri*) in the Colombian Caribbean (Rubio 1975) and Venezuela (Angell 1976) found that females have a higher growth rate than males, at least starting from the second year of life. However, the growth rate in Venezuela was lower than in Cuba, and even lower than in Colombia (Appendix 6.2).

Studies of mullet growth in the coastal lagoon system of the Tunas de Zaza (Alvarez-Lajonchere 1976, 1981a) found that liza and hospe mullet (*Mugil hospes*) females also grow faster than males. The growth rate reported for white mullet in the Tunas de Zaza is somewhat higher than that recorded in Virginia (Richards and Castagna 1976). Bustamante and Enomoto (1981) found a first-year growth rate of 90–110 g (18–20 cm FL) in pond-raised white mullet. In the liza, first-year growth in ponds was 200–270 g (25–30 cm, approximately), whereas fantail grew only 13 cm (30 g). These

Table 6.6.   Length and growth parameter estimates for three grunt species (Haemulidae) of various regions by different authors

	Haemulon album		Haemulon plumieri						Haemulon sciurus	
Source	1	2 SW	3 N & S	3 SE	4 N & S	5 Campeche	6	7 North-Central	8	9
Locality	Jamaica (FL)	Cuba (FL)	Carolina (TL)	Florida (TL)	Carolina (TL)	Bank (TL)	SW Cuba (FL)	Cuba (FL)	Jamaica (FL)	SW Cuba (FL)
Back-calculated length (cm) at age (year)										
0_1	—	6	—	—	—	—	3.7	—	5.3	5.5
0_2	—	12.1	—	—	—	—	—	—	—	—
1	20.3	18.1	20.6	20.8	9.7	—	10.2	9.2	10.5	10.2
2	25.9	30.7	25.6	23.9	18.5	—	14.9	14.9	16.8	14.9
3	33	35.3	29.4	26.0	24.4	23.6	18.6	18.7	—	18.5
4	—	41.4	32.0	27.2	31.4	26	21.8	21.1	—	21.6
5	—	48.1	34.4	28.2	39.1	28	—	22.7	—	—
6	—	51.9	36.1	29.0	36.7	30.7	—	23.7	—	—
7	—	54.2	37.8	29.3	41.4	—	—	—	—	—
8	—	58.3	39.1	29.7	43.9	—	—	—	—	—
9	—	—	40.4	29.3	46.5	—	—	—	—	—
10	—	—	41.6	30.5	49.3	—	—	—	—	—
Growth parameters										
L_∞ (cm)	65	73	59.1	32.7	64	—	36	25.6	40	34
W_∞ (g)	—	8,500	—	—	4,334	—	954	284	1,114	746
K (yr)	0.2	0.19	0.08	0.31	0.11	—	0.2	0.43	0.26	0.22
t_0 (yr)	—	-0.301	-4.21	-4.21	-1.007	—	-0.682	-0.027	—	-0.642
ϕ'	2.93*	3.00*	2.45*	2.52*	2.65*	—	2.41*	2.45*	2.62*	2.40*

Sources: 1: Billings and Munro 1974; 2: García-Arteaga 1983; 3: Potts and Manooch 2001; 4: Manooch 1976; 5: Capote 1971; 6: García-Arteaga 1992a; 7: Ramos and Pozo 1984; 8: Billings and Munro 1974; 9: García-Arteaga 1992b.

FL: fork length; TL: total length.

* Values calculated in the present study using data from the original source.

growth rates are similar to those reported by Alvarez-Lajonchere (1976, 1981a) in natural populations in the coastal lagoons of the Tunas de Zaza.

Herring and sardines (Clupeidae) are small-sized and short-lived species. According to the data obtained for scaled sardine (*Harengula jaguana*) in Florida and Brazil (Martínez and Houde 1975; Hubold and Mazzetti 1982) and for the redear sardine in Cuba (García-Arteaga 1993), their growth is similar in all the studied regions (Appendix 6.2). Both species attain more than 50% of their maximum length during the first year of life, although their growth in weight during this period is low.

We calculated the growth parameters of dwarf sardines (*Jenkinsia lamprotaenia*) in the southwest region by analyzing length distribution with the ELEFAN program. We found a K value of 3.34 and an asymptotic length of 49.8 mm FL. Both parameters are characteristic of short-lived species (not more than 10–12 months) and rapid growth. Growth is fast during the first two months and exceeds 50% of maximum length. Preliminary estimates of daily growth, using the otolith microstructures, showed results similar to those obtained using the length–frequency analysis. Among the Engraulidae, data are available only for the Atlantic anchoveta (*Cetengraulis*

edentulus) in Venezuelan waters (Simpson and Griffiths 1973). We found differences between the growth rates of males and females.

Growth studies of the king and Spanish mackerel (*Scomberomorus cavalla* and *S. maculatus*) in Florida (Klima 1959; Beaumarriage 1973; Powell 1975; Johnson et al. 1983), Brazil (Nomura 1967; Nomura and Rodrigues 1967; Carneiro Ximenes et al. 1978), Mexico (Mendoza 1968; Medina-Quej and Dominguez-Viveros 1997), and Trinidad (Sturm 1978) show notable differences (Appendix 6.2) that are difficult to interpret. Nevertheless, most studies suggest that females grow faster than males.

6.5.3   General Patterns of Growth

Data available for most tropical fishes show a high growth rate during the first year of life. During this period, an individual can reach 35–50% of its maximum length. This pattern allows fishes to avoid the high predation pressure characteristic of these regions for other taxa, and it contributes to their adaptation from early stages to a diet based on organisms with high caloric value (fish, crustaceans, mollusks, etc.). Nevertheless, this pattern is not applicable to

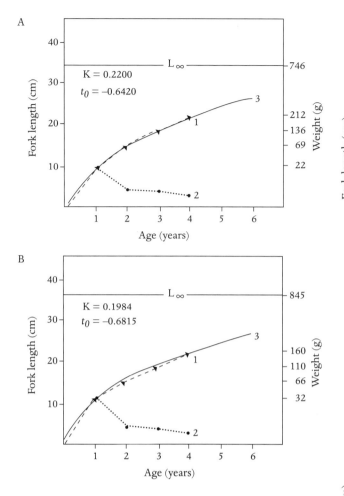

Fig. 6.7. Back-calculated values for (1) fish length, (2) mean annual growth, and (3) theoretical growth curves for (A) bluestriped grunt (*Haemulon sciurus*) and (B) white grunt (*Haemulon plumieri*) in southwestern Cuba.

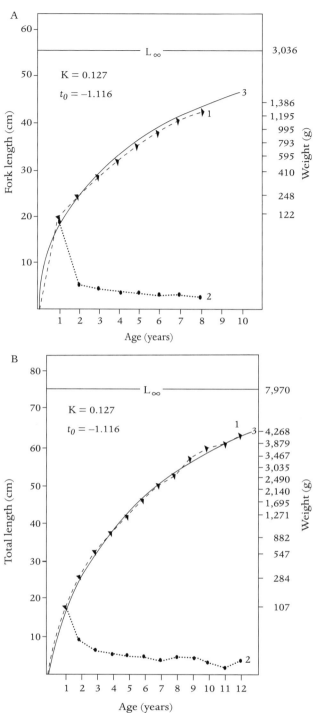

Fig. 6.8. Back-calculated values for (1) fish length, (2) mean annual growth, and (3) theoretical growth curves for (A) bar jack (*Caranx ruber*) and (B) Nassau grouper (*Epinephelus striatus*) in southwestern Cuba.

territorial reef species, whose defense mechanisms against predation are often based on the use of shelters, a strategy that limits their ability to exploit distant food sources. The body shape of nonterritorial fishes is usually fusiform and slender, more adapted to rapid swimming. Species with territorial habits more commonly are compressed and deep-bodied, a shape less correlated with rapid swimming.

Our data suggest relationships between species growth rates and maximum size, and, possibly, the comparative abundance of populations. Slow growth and small size might allow the formation of large populations, whereas high growth rates and large sizes might be associated with lower abundance (Nikolsky 1974a). In both cases, food supply is assumed to be the primary limiting factor. On the other hand, other studies have found that growth rates (K) tend to increase with stress, such as an increase in water temperature that causes an increase in oxygen consumption (Pauly 1980). Coastal zones are much more affected by environmental

changes, so higher fish growth rates and lower maximum sizes could be expected there.

These principles are clearly manifested among the snappers (Fig. 6.10) and grunts of Cuba. Among the four most abundant species, the lane snapper attains the smallest

A

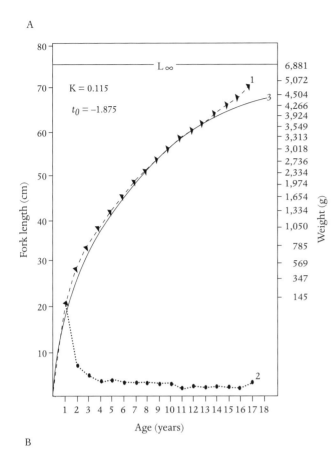

B

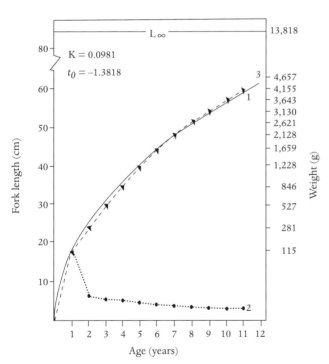

Fig. 6.9. Back-calculated values for (1) fish length, (2) mean annual growth, and (3) theoretical growth curves for (A) tiger grouper (*Mycteroperca tigris*) and (B) hogfish (*Lachnolaimus maximus*) in southwestern Cuba.

size (Fig. 6.4A), which contributes to the existence of a large population. However, small size is also associated with high mortality from predation that should be compensated for by a higher reproductive capacity. The silk snapper, on the other hand, has a high growth rate and attains a large size; its populations are much less numerous and inhabit mainly the shelf drop-off or nearby areas. Gray and yellowtail snappers occupy an intermediate position, both in growth rate and comparative abundance.

In Jamaica, unlike in Cuba and Florida where ring formations were clear, defining a pattern of ring formation in many species is difficult or impossible (Munro 1983a). Considering the difference between these regions in water temperature fluctuations, this finding suggests that the amplitude of water temperature fluctuation is an important factor in the formation of growth rings.

6.6 Methodological Considerations in Growth Investigations

The data available for growth rates of fishes of the wider Caribbean require validation of the methods used to obtain them. The most frequent problems include lack of detailed analysis of validation of annual rings; little attention to the environmental conditions that determine ring formation; difficulty in defining the correct annual ring when several rings are formed per year; and the uncertainty created by the interpretation of subannual rings as annual rings.

The variety of criteria used by different researchers could lead investigators to different conclusions. Such methodological gaps make it difficult to interpret the information published on tropical fish growth. Some researchers prefer not to assign ages to growth marks and refer instead to the "rings" recorded. This approach is still difficult because the main issue—the growth rate in a certain time period and its variation—remains unresolved. Based on our experience, we have some suggestions that could help to improve the definition of growth parameters.

1. Collect young individuals, from the smallest possible size, and characterize the biotic and abiotic factors that determine mark formation in skeletal structures. This would allow for better definition of the first annual ring. In some species, the first ring might be verified as annual by counting all preceding daily rings.
2. Use monthly sampling, if necessary, to establish the number of rings and the moment of their formation during the annual cycle, as well as their relationship with environmental changes. Annual classes should therefore be examined separately, or at least young fish should be examined separately from adults.
3. Use at least two different skeletal structures and compare the results. Other methods (tagging, modal progression, FISAT) can contribute to enhanced analyses.

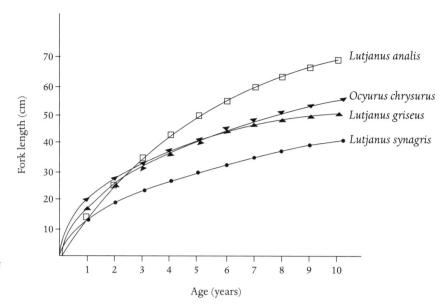

Fig. 6.10. Theoretical growth curves for mutton (*Lutjanus analis*), yellowtail (*Ocyurus chrysurus*), gray (*L. griseus*), and lane (*L. synagris*) snappers in southwestern Cuba.

6.7   Summary

Despite relatively small variations in environmental conditions on the Cuban shelf, physiological rhythms at several temporal scales are reflected in fish skeletal structures. The formation of periodic growth marks in otoliths, scales, and the urohyal bone appears to be related to two primary environmental changes: (1) decreasing temperatures at the beginning of winter, which leads to ring formation in bones and scales in most species (but not in otoliths of all species), and (2) rising temperatures and a lengthening photoperiod during summer, which increases feeding intensity and growth, as well as energy expenditure for reproductive metabolism in adults. The latter environmental change is accompanied by the mobilization of calcium from bones and scales to the gonads. This process can be reflected in the composition of the skeletal structures, especially otoliths. The combination of these factors influences the formation of one or two rings every year, depending on the species ecology, and year class, as well as on the skeletal structures affected.

When comparing growth rate estimates in several populations of the same species, differences have been observed that are apparently related to the methodological criteria for selecting the annual marks. Care is needed in defining the moment of mark formation and its relationship to environmental and fish physiological changes. This step is critical for growth studies. Knowledge of seasonal variations in the width of the marginal increment from at least two different structures is preferred, as well as establishing the percentage of individuals with a newly

formed ring in separate annual size classes, in both adults and juveniles. Rings formed before attaining the first year of life (0+) can be considered as annual marks. Evidence from various Cuban studies suggests that two or more subannual rings of this type can be formed. There is also evidence for the formation in some structures of two marks per year in subsequent years.

The information available allowed us to compare growth rates of several species in various regions of the Caribbean and Gulf of Mexico. The parameter ϕ' was useful for validating growth parameter estimates (Pauly 1980). Some species in Cuba seemed to have slower growth than in other regions (e.g., Jamaica, southeastern Florida, Banco de Campeche, and northern South America). This trend is not absolute, and other species in Cuba have somewhat higher growth rates than in Florida and northern Gulf of Mexico. These trends cannot be easily generalized because growth rates are associated with geographically variable environmental conditions that might not influence all species in the same way. Interannual variation of species growth rates was also found and might be due to fisheries exploitation.

Cuban coastal fishes are characterized by a short life span and relatively high growth rate. In the first year of life, 35–50% of maximum size can be reached. This might be an adaptive mechanism to avoid high predation pressures on younger fishes. Growth typically increases during early summer, although peak growth periods shift to later in the year in adults of some species, generally after the spawning season. Seasonality of growth is strongly related to feeding conditions and reproduction. After sexual maturity, growth increases are mainly in weight.

Appendix 6.1.   Length (L) parameters that can be used to estimate weight (W) from growth studies of marine fishes of Cuba.

Family Species	*n*	Length	$W = aL^b$ a	b	Units W/L	Locality	Source
Ginglymostomatidae							
Ginglymostoma cirratum	123	55–157 SL	2.55×10^{-5}	2.86	kg/cm	SE Cuba	Espinosa, in press a
Alopiidae							
Alopias superciliosus	15	90–215 SL	1.83×10^{-4}	3.45	kg/cm	NW Cuba	Guitart Manday 1975
Alopias superciliosus	13	135–210 SL	3.51×10^{-2}	2.44	kg/cm	NW Cuba	Quevedo and Aguilar 1984
Lamnidae							
Isurus oxyrinchus	23	156–255 FL	0.12×10^{-5}	3.46	kg/cm	NW Cuba	Guitart Manday 1975
Isurus oxyrinchus	14	150–210 FL	0.50×10^{-3}	2.32	kg/cm	NW Cuba	Quevedo and Aguilar 1984
Isurus oxyrinchus	72	60–200 SL	1.26×10^{-4}	2.58	kg/cm	NW Cuba	Espinosa, in press a
Carcharhinidae							
Carcharhinus falciformis	44	61–240 SL	0.88×10^{-5}	3.09	kg/cm	NW Cuba	Guitart Manday 1975
Carcharhinus falciformis	54	60–250 FL	0.19×10^{-4}	2.93	kg/cm	NW Cuba	Quevedo and Aguilar 1984
Carcharhinus falciformis	225	60–230 SL	1.01×10^{-5}	3.06	kg/cm	NW Cuba	Espinosa, in press a
Carcharhinus leucas	32	179–300 TL	1.75×10^{-5}	2.84	kg/cm	Cuba	Espinosa, in press a
Carcharhinus limbatus	100	122–230 TL	6.14×10^{-6}	3.01	kg/cm	Cuba	Espinosa, in press a
Carcharhinus longimanus	61	87–190 SL	0.72×10^{-4}	2.68	kg/cm	NW Cuba	Guitart Manday 1975
Carcharhinus longimanus	32	80–200 FL	0.17×10^{-4}	2.98	kg/cm	NW Cuba	Quevedo and Aguilar 1984
Carcharhinus longimanus	86	60–200 SL	4.01×10^{-5}	2.89	kg/cm	NW Cuba	Espinosa, in press a
Carcharhinus signatus	39	113–188 SL	0.30×10^{-6}	3.74	kg/cm	NW Cuba	Guitart Manday 1975
Carcharhinus signatus	13	164–239 SL	0.25×10^{-5}	3.33	kg/cm	NW Cuba	Quevedo and Aguilar 1984
Rhizoprionodon porosus	49	51–105 TL	1.14×10^{-4}	2.29	kg/cm	Cuba	Espinosa, in press a
Triakidae							
Mustelus canis insularis	55	24.5–123 TL	2.1×10^{-3}	3.12	g/cm	NE Cuba	This chapter
Negaprion brevirostris	35	49–210 SL	0.53×10^{-5}	3.16	kg/cm	SE Cuba	Espinosa, in press a
Sphyrnidae							
Sphyrna mokarran	25	187–325 TL	1.91×10^{-6}	3.16	kg/cm	Cuba	Espinosa, in press a
Squalidae							
Squalus cubensis	13	57.5–102 TL	2.2009	1.60	g/cm	NE Cuba	This chapter
Dasyatidae							
Dasyatis americana	164	30–134 DD	7.39×10^{-5}	2.81	kg/cm	SE Cuba	Espinosa, in press b
Myliobatidae							
Aetobatus narinari	14	97–159 DD	1.4392	2.09	g/cm	NE Cuba	This chapter
Elopidae							
Elops saurus	776	12–40 FL	5.60×10^{-3}	3.1	g/cm	SE Cuba	Carles 1967
Megalops atlanticus	7	110–134 FL	2.25×10^{-2}	2.36	g/cm	NE Cuba	This chapter
Albulidae							
Albula vulpes	171	24–64 FL	2.79×10^{-5}	2.89	g/cm	Cuba	This chapter
Muraenidae							
Gymnothorax funebris	16	62–155 TL	0.3×10^{-3}	3.38	g/cm	NE Cuba	This chapter
Clupeidae							
Harengula clupeola	100	3–14 FL	1.45×10^{-2}	3.05	g/cm	SW Cuba	Claro and García-Arteaga 1994a
Harengula clupeola	36	8–14 FL	2.33×10^{-2}	2.84	g/cm	NW Cuba	Claro and García-Arteaga 1994a
Harengula humeralis	100	8–16 FL	1.35×10^{-2}	3.1	g/cm	NW Cuba	García-Arteaga 1993
Harengula humeralis	100	4–17 FL	1.07×10^{-2}	3.19	g/cm	SW Cuba	García-Arteaga 1993
Harengula humeralis	23	64–90 FL	6.65×10^{-4}	2.67	g/mm	Cuba	R. González, pers. comm.
Harengula jaguana M	973	4–11 FL	1.34×10^{-2}	3.15	g/cm	NW Cuba	Suarez-Caabro et al. 1961
Harengula jaguana F	1060	4–12 FL	1.06×10^{-2}	3.25	g/cm	NW Cuba	Suarez-Caabro et al. 1961
Jenkinsia lamprotaenia	1881	1–5 FL	0.93×10^{-6}	3.62	g/mm	SW Cuba	Bustamante et al. 1992
Opisthonema oglinum	8	10.2–16 FL	1.05×10^{-2}	3.17	g/mm	NE Cuba	This chapter
Opisthonema oglinum	892	8–20 FL	1.86×10^{-2}	2.92	g/cm	SE Cuba	Valdés and Sotolongo 1983
Opisthonema oglinum	18	117–150 FL	1.74×10^{-5}	2.98	g/mm	Cuba	R. González, pers. comm.
Sardinella aurita	565	11–23 TL	3.98×10^{-6}	3.16	g/mm	Venezuela	González 1985b

Continued on next page

Appendix 6.1. continued

Engraulidae

Anchoa parva	22	5–8.4 FL	2.8×10^{-3}	3.59	g/cm	NE Cuba	This chapter
Anchoa parva	32	79–104 FL	1.45×10^{-6}	3.41	g/mm	Cuba	R. González, pers. comm.
Cetengraulis edentulus	38	126–155 FL	7.24×10^{-6}	3.15	g/mm	Cuba	R. González, pers. comm.

Batrachoididae

Opsanus beta	13	4.1–11 FL	1.35×10^{-2}	2.98	g/cm	NE Cuba	This chapter
Opsanus beta	140	2–13 FL	5.76×10^{-5}	2.64	g/mm	SW Cuba	Claro and García-Arteaga 1994a
Opsanus phobetron	19	3.1–10.6 FL	1.35×10^{-2}	3.06	g/cm	NE Cuba	This chapter
Opsanus phobetron	7	5–12 TL	2.27×10^{-5}	2.67	g/mm	SW Cuba	Claro and García-Arteaga 1994a

Ophidiidae

Lepophidium brevibarbe	39	12–24 FL	4.80×10^{-4}	2.115	g/cm	SE.Cuba	This chapter
Ogilbia cayorum	13	3.2–4.8 FL	3.8×10^{-3}	3.37	g/cm	NE Cuba	This chapter

Hemiramphidae

Hemiramphus brasiliensis	50	24–38 FL	3.2×10^{-3}	3.10	g/cm	NE Cuba	This chapter
Hemiramphus brasiliensis	27	26.5–32.3 FL	4.9×10^{-3}	2.96	g/cm	SW Cuba	This chapter
Hemiramphus brasiliensis	90	19–32 FL	1.22×10^{-3}	3.37	g/cm	SW.Cuba	This chapter

Belonidae

Strongylura notata	125	10.5–64.5 FL	1.2×10^{-3}	3.06	g/cm	SW Cuba	This chapter
Strongylura notata	34	5–28.4 FL	1.5×10^{-2}	2.16	g/cm	NE Cuba	This chapter
Strongylura timucu	4	9.4–35.8	3.0×10^{-4}	3.40	g/cm	NE Cuba	This chapter
Tylosurus crocodilus crocodilus	113	25.1–64.3 FL	5.0×10^{-4}	3.36	g/cm	SW Cuba	This chapter
Tylosurus crocodilus crocodilus	92	17–47.5 FL	6.0×10^{-4}	3.29	g/cm	NE Cuba	This chapter

Holocentridae

Holocentrus adscensionis	37	16.5–32 FL	4.52×10^{-2}	2.70		NE Cuba	This chapter
Myripristis jacobus	6	11.2–14	9.26×10^{-2}	2.53	g/cm	NE Cuba	This chapter

Syngnathidae

Hippocampus erectus	34	2.4–10.8 TL	1.23×10^{-2}	2.63	g/cm	NE Cuba	This chapter
Hippocampus punctulatus	9	6.6–11.6 TL	3.3×10^{-3}	3.22	g/cm	NE Cuba	This chapter
Syngnathus pelagicus	18	6.6–20 FL	1.0×10^{-4}	3.61	g/cm	SW-NE Cuba	Claro and García-Arteaga 1994a

Centropomidae

Centropomus undecimalis	27	31–82 FL	2.6×10^{-3}	3.32	g/cm	NE Cuba	This chapter
Centropomus undecimalis	24	34–88 TL	3.8×10^{-3}	3.29	g/cm	NE Cuba	This chapter
Centropomus undecimalis	715	18–94 FL	1.74×10^{-2}	2.86	g/cm	SE.Cuba	Alvarez-Lajonchere et al. 1982

Serranidae

Cephalopholis fulva	9	17.1–26.4 FL	3.5×10^{-3}	3.49	g/cm	NE Cuba	This chapter
Epinephelus guttatus	15	28–40 FL	1.03×10^{-2}	3.12	g/cm	NE Cuba	This chapter
Epinephelus striatus	75	15–63 TL	1.98×10^{-2}	2.98	g/cm	NE Cuba	Claro et al. 1990c
Epinephelus striatus	270	15–71 TL	5.2×10^{-3}	3.30	g/cm	SW Cuba	Claro et al. 1990c
Hypoplectrus puella	15	4–8 TL	9.0×10^{-2}	3.04	g/cm	SW Cuba	Claro and García-Arteaga 1994a
Mycteroperca bonaci	66	45–98 SL	1.73×10^{-5}	3.11	g/mm	NW Cuba	E. Valdés, pers. comm.
Mycteroperca bonaci	100	25–110 TL	8.22×10^{-3}	3.14	g/cm	SW Cuba	Claro and García-Arteaga 1994a
Mycteroperca bonaci	30	29–132 TL	1.13×10^{-2}	3.05	g/cm	NE Cuba	This chapter
Mycteroperca tigris	63	29–62 TL	1.48×10^{-5}	3.11	g/mm	NW Cuba	E. Valdés, pers. comm.
Mycteroperca tigris	145	29–74 TL	9.40×10^{-3}	3.12	g/cm	SW Cuba	Claro and García-Arteaga 1994a
Mycteroperca venenosa	8	29–57 TL	6.2×10^{-3}	3.25	g/cm	NE Cuba	This chapter
Mycteroperca venenosa	36	31–69 TL	2.82×10^{-5}	2.98	g/mm	SW Cuba	E. Valdés, pers. comm.
Mycteroperca venenosa	54	25–92 TL	1.32×10^{-2}	3.04	g/cm	SW Cuba	Claro and García-Arteaga 1994a

Apogonidae

Astrapogon stellatus	5	4.8–6 TL	3.58×10^{-2}	2.67	g/cm	NE Cuba	This chapter

Malacanthidae

Malacanthus plumieri	44	27.2–42 FL	7.1×10^{-3}	2.98	g/cm	NE Cuba	This chapter

Carangidae

Caranx bartholomaei	450	9–78 FL	3.40×10^{-2}	2.84	g/cm	SW Cuba	Sierra et al. 1986
Caranx crysos	16	17.5–31 FL	1.65×10^{-2}	3.04	g/cm	SW Cuba	This chapter
Caranx hippos	53	17.9–53 TL	5.82×10^{-2}	2.57	g/cm	NE Cuba	This chapter
Caranx latus	39	7.3–69 FL	3.83×10^{-2}	2.82	g/cm	NE Cuba	This chapter
Caranx latus	28	5–51 FL	2.10×10^{-2}	2.97	g/cm	SW Cuba	Claro and García-Arteaga 1994a

Continued on next page

Appendix 6.1. continued

Caranx ruber	17	4.5–61 FL	1.8×10^{-2}	2.99	g/cm	NE Cuba	This chapter
Caranx ruber	1862	5–50 FL	1.80×10^{-2}	2.99	g/cm	SW Cuba	García-Arteaga and Reshetnikov 1992
Chloroscombrus chrysurus	21	2.5–27.4 FL	4.3×10^{-3}	3.34	g/cm	NE Cuba	This chapter
Chloroscombrus chrysurus	80	103–203 FL	3.64×10^{-4}	2.74	g/mm	Cuba	R. González, pers. comm.
Oligoplites saurus	71	109–240 FL	4.7×10^{-3}	1.88	g/mm	NE Cuba	This chapter
Selar crumenophthalmus	10	4.9–7 FL	2.38×10^{-2}	2.48	g/cm	NE Cuba	This chapter
Selar crumenophthalmus	135	5–23 FL	7.40×10^{-3}	3.29	g/cm	SW Cuba	Claro and García-Arteaga 1994a
Selene vomer	130	9.5–31.7 FL	0.1723	2.40	g/cm	NE Cuba	This chapter
Seriola dumerili	5	62.5–108 FL	3.8×10^{-2}	2.77	g/cm	NE Cuba	This chapter
Trachinotus goodei	48	9.6–24.6 FL	0.4744	2.09	g/cm	NE Cuba	This chapter
Selene setapinnis	21	72–118 FL	1.84×10^{-4}	2.59	g/mm	NW Cuba	E. Valdés, pers. comm.
Coryphaenidae							
Coryphaena hippurus	56	50–120 FL	3.21×10^{-5}	2.67	g/cm	NW Cuba	Quevedo and Aquilar 1984
Lutjanidae							
Lutjanus analis	1154	18–68 FL	1.37×10^{-2}	3.06	g/cm	NW Cuba	Claro 1981b
Lutjanus analis	974	18–72 FL	1.52×10^{-2}	3.04	g/cm	SW Cuba	Claro 1981b
Lutjanus analis	559	—	3.54×10^{-2}	2.77	g/cm	NE Cuba	Ramos 1988
Lutjanus analis	1609	12–74 FL	1.20×10^{-2}	3.1	g/cm	SW Cuba	Ramos 1988
Lutjanus analis F	1051	23–73 FL	9.20×10^{-2}	2.59	g/cm	NE Cuba	Pozo 1979
Lutjanus analis M	1010	23–73 FL	1.14×10^{-1}	2.53	g/cm	NE Cuba	Pozo 1979
Lutjanus apodus	15	3.4–22.1 FL	2.69×10^{-2}	2.83	g/cm	NE Cuba	This chapter
Lutjanus apodus	66	4–45 FL	2.16×10^{-2}	2.92	g/cm	SW Cuba	Claro and García-Arteaga 1994a
Lutjanus buccanella	—	19–74 FL	1.75×10^{-5}	3.02	g/mm	SE Cuba	Espinosa and Pozo 1982
Lutjanus cyanopterus	28	25–99 FL	0.98×10^{-2}	3.12	g/cm	NE Cuba	This chapter
Lutjanus cyanopterus	107	33–109 FL	9.26×10^{-1}	2.88	g/cm	SW Cuba	Claro and García-Arteaga 1994a
Lutjanus griseus	25	11–40 FL	1.66×10^{-2}	2.96	g/cm	NE Cuba	This chapter
Lutjanus griseus	956	6–52 FL	2.07×10^{-2}	2.91	g/cm	SW Cuba	Claro 1983b
Lutjanus griseus	1499	15–56 FL	1.82×10^{-2}	2.94	g/cm	SW Cuba	M. Sosa, pers. comm.
Lutjanus griseus F	769	15–50 FL	1.03×10^{-4}	2.68	g/cm	SE.Cuba	Báez and Alvarez-Lajonchere 1980b
Lutjanus griseus M	519	15–43 FL	1.70×10^{-5}	3.00	g/cm	SE.Cuba	Báez and Alvarez-Lajonchere 1980b
Lutjanus jocu	11	14.5–54 FL	2.63×10^{-2}	2.82	g/cm	NE Cuba	This chapter
Lutjanus jocu	117	30–77 FL	0.85×10^{-2}	3.2	g/cm	NW/SW Cuba	Claro et al. 1999d
Lutjanus synagris	4443	9–41 FL	4.96×10^{-5}	2.8	g/mm	Cuba	Rodríguez Pino 1962
Lutjanus synagris	684	—	1.82×10^{-2}	2.9	g/cm	NE Cuba	Ramos 1988
Lutjanus synagris	1708	12–43 FL	5.17×10^{-2}	2.64	g/cm	SW Cuba	M. Sosa, pers. comm.
Lutjanus synagris	3284	6–36 FL	1.86×10^{-2}	2.97	g/cm	SW Cuba	Claro and García-Arteaga 1994a
Lutjanus synagris F	1248	13–36 FL	2.40×10^{-2}	2.89	g/cm	SW Cuba	Claro and García-Arteaga 1994a
Lutjanus synagris M	977	13–30 FL	2.17×10^{-2}	2.93	g/cm	SW Cuba	Claro and García-Arteaga 1994a
Lutjanus vivanus	—	19–56 FL	1.66×10^{-5}	3.03	g/cm	SE.Cuba	Pozo and Espinosa 1982
Ocyurus chrysurus	42	4.5–37 FL	1.15×10^{-2}	3.14	g/cm	NE Cuba	This chapter
Ocyurus chrysurus	5828	10–52 FL	7.32×10^{-2}	2.74	g/cm	NW Cuba	Piedra 1965
Ocyurus chrysurus	2393	—	0.28×10^{-1}	2.83	g/cm	SE Cuba	Carrillo de Albornoz and Ramiro 1988
Ocyurus chrysurus	564	—	2.00×10^{-2}	2.9	g/cm	NE Cuba	Ramos 1988
Ocyurus chrysurus	2524	12–54 FL	8.53×10^{-2}	2.47	g/cm	SW Cuba	M. Sosa, pers. comm.
Ocyurus chrysurus All	384	15–40 FL	3.14×10^{-2}	2.79	g/cm	NW Cuba	Claro 1983c
Ocyurus chrysurus All	378	6–43 FL	1.89×10^{-2}	2.94	g/cm	SW Cuba	Claro 1983c
Ocyurus chrysurus F	125	17–40 FL	3.80×10^{-2}	2.8	g/cm	NW Cuba	Claro 1983c
Ocyurus chrysurus F	136	16–38 FL	2.44×10^{-2}	2.86	g/cm	SW Cuba	Claro 1983c
Ocyurus chrysurus M	208	15–34 FL	2.98×10^{-2}	2.73	g/cm	NW Cuba	Claro 1983c
Ocyurus chrysurus M	100	18–43 FL	3.22×10^{-2}	2.78	g/cm	SW Cuba	Claro 1983c
Gerreidae							
Diapterus rhombeus	14	8.2–16.5 FL	6.37×10^{-2}	2.57	g/cm	NE Cuba	This chapter
Eucinostomus havana	40	83–117 FL	6.14×10^{-3}	2.77	g/mm	Cuba	R. González, pers. comm.
Eucinostomus havana	250	2–12 FL	1.23×10^{-2}	3.23	g/cm	SW Cuba	This chapter
Eucinostomus jonesi	33	85–117 FL	9.23×10^{-5}	2.65	g/mm	Cuba	R. González, pers. comm.
Eugerres brasilianus	450	1–50 FL	1.79×10^{-2}	2.91	g/cm	NE Cuba	García-Arteaga et al. 1997

Continued on next page

Appendix 6.1. continued

Eugerres brasilianus	550	13–30 FL	5.29×10^{-2}	2.82	g/cm	SE Cuba	Báez et al 1982a
Eugerres plumieri	99	13–28.5 FL	1.35×10^{-2}	3.23	g/cm	NE Cuba	This chapter
Gerres cinereus	106	13–43 FL	1.47×10^{-5}	3.12	g/cm	Cuba	García-Arteaga et al. 1997
Gerres cinereus	83	5.4–27 TL	2.07×10^{-2}	3.10	g/cm	NE Cuba	This chapter
Gerres cinereus	770	—	1.30×10^{-4}	2.69	g/mm	SE Cuba	Báez and Alvarez-Lajonchere 1983
Haemulidae							
Anisotremus virginicus	6	18–25 FL	2.5×10^{-2}	3.78	g/cm	NE Cuba	This chapter
Haemulon aurolineatum	1827	7–61 FL	1.40×10^{-2}	3.09	g/cm	SW Cuba	García-Arteaga 1983
Haemulon bonariense	300	7–15 FL	1.10×10^{-2}	3.2	g/cm	SW Cuba	This chapter
Haemulon flavolineatum	14	2.8–7.8 FL	3.41×10^{-2}	2.59	g/cm	SW Cuba	This chapter
Haemulon flavolineatum	235	3–12.4 FL	1.36×10^{-2}	3.14	g/cm	NE Cuba	This chapter
Haemulon melanurum	23	16.3–26 FL	3.12×10^{-2}	2.85	g/cm	NE Cuba	This chapter
Haemulon parra	150	13–30 FL	2.8×10^{-2}	2.8	g/cm	SW Cuba	Claro and García-Arteaga 1994a
Haemulon parra	20	5.5–25 FL	4.36×10^{-2}	2.51	g/cm	NE Cuba	This chapter
Haemulon plumieri	200	6.2–29.5 FL	2.24×10^{-2}	2.88	g/cm	NE Cuba	This chapter
Haemulon plumieri	472	—	9.81×10^{-3}	3.17	g/cm	NE Cuba	Ramos 1988
Haemulon plumieri	670	7–25 FL	1.60×10^{-2}	3.05	g/cm	SW Cuba	This chapter
Haemulon sciurus	60	2.8–19.9 FL	1.86×10^{-2}	3.00	g/cm	NE Cuba	This chapter
Haemulon sciurus	222	—	9.98×10^{-2}	2.38	g/cm	NE Cuba	Ramos 1988
Haemulon sciurus	857	8–27 FL	2.00×10^{-2}	3.01	g/cm	SW Cuba	This chapter
Orthopristis chrysoptera	6	2.6–8.5 FL	3.32×10^{-2}	2.67	g/cm	NE Cuba	This chapter
Pomadasys crocro	17	3.8–18 FL	1.84×10^{-2}	2.93	g/cm	NE Cuba	This chapter
Sparidae							
Archosargus rhomboidalis	213	5–26 FL	1.56×10^{-2}	3.19	g/cm	NE Cuba	This chapter
Calamus bajonado	14	13.8–29.5 FL	4.76×10^{-2}	2.68	g/cm	NE Cuba	This chapter
Calamus penna	—	—	1.25×10^{-2}	3.18	g/cm	NE Cuba	Ramos 1988
Calamus pennatula	41	5–28.1 FL	2×10^{-2}	3.06	g/cm	NE Cuba	This chapter
Calamus pennatula	650	7–36 FL	1.78×10^{-2}	3.11	g/cm	SW Cuba	Claro and García-Arteaga 1994a
Sciaenidae							
Bairdiella batabana	26	10–22 TL	9.0×10^{-4}	3.06	g/cm	NE Cuba	This chapter
Equetus acuminatus	23	2.4–10.3 FL	1.53×10^{-2}	3.01	g/cm	NE Cuba	This chapter
Micropogonias furnieri	180	18–47 FL	1.47×10^{-2}	2.93	g/cm	Cuba	García-Arteaga et al. 1997
Mullidae							
Mulloidichthys martinicus	64	9.5–27 FL	2.8×10^{-3}	3.65	g/cm	NE Cuba	This chapter
Pseudupeneus maculatus	11	5.5–10 FL	7.6×10^{-3}	3.30	g/cm	NE Cuba	This chapter
Pseudupeneus maculatus	118	15–23 FL	2.14×10^{-2}	2.96	g/cm	SW Cuba	Claro and García-Arteaga 1994a
Pempheridae							
Pempheris schomburgkii	82	9–13 FL	4.39×10^{-2}	2.62	g/cm	NW Cuba	This chapter
Ephippidae							
Chaetodipterus faber	109	14–38 FL	1.61×10^{-2}	2.51	g/cm	NE Cuba	This chapter
Chaetodipterus faber	27	11–34 FL	4.07×10^{-1}	2.25	g/cm	SW Cuba	Claro and García-Arteaga 1994a
Chaetodontidae							
Chaetodon capistratus	21	4–10 FL	2.58×10^{-2}	2.98	g/cm	NE Cuba	This chapter
Chaetodon capistratus	26	4–12 FL	4.70×10^{-2}	2.86	g/cm	SW Cuba	Claro and García-Arteaga 1994a
Pomacanthidae							
Pomacanthus arcuatus	9	26.9–41 FL	2.65×10^{-2}	3.04	g/cm	NE Cuba	This chapter
Pomacentridae							
Abudefduf saxatilis	18	3–13 FL	1.70×10^{-1}	3.12	g/cm	SW Cuba	Claro and García-Arteaga 1994a
Stegastes leucostictus	6	4.2–7.4 FL	4.38×10^{-2}	2.77	g/cm	NE Cuba	This chapter
Labridae							
Halichoeres bivittatus	17	4.5–17.7 FL	1.29×10^{-2}	2.93	g/cm	NE Cuba	This chapter
Halichoeres bivittatus	50	3–11 FL	9.43×10^{-3}	3.15	g/cm	SW Cuba	Claro and García-Arteaga 1994a
Halichoeres maculipinna	8	8.5–11.1 FL	4.3×10^{-2}	3.43	g/cm	SW Cuba	This chapter
Lachnolaimus maximus	20	3.6–43 FL	1.52×10^{-2}	3.11	g/cm	NE Cuba	This chapter
Lachnolaimus maximus	196	12–62 FL	2.37×10^{-2}	2.95	g/cm	SW Cuba	Claro et al 1989
Scaridae							
Scarus coelestinus	41	39–69 FL	2.01×10^{-2}	3.02	g/cm	SW Cuba	Claro and García-Arteaga 1994a

Continued on next page

Appendix 6.1. continued

Scarus guacamaia	29	39–85 FL	3.52×10^{-2}	2.88	g/cm	SW Cuba	Claro and García-Arteaga 1994a
Scarus guacamaia		—	1.22×10^{-2}	3.07	g/cm	NE Cuba	This chapter
Scarus iserti	102	3.1–22.7 FL	1.1×10^{-2}	3.20	g/cm	NE Cuba	This chapter
Scarus iserti	99	4–11 FL	2.08×10^{-5}	2.92	g/mm	SW Cuba	Claro and García-Arteaga 1994a
Sparisoma chrysopterum	80	4.5–20.7 FL	1.84×10^{-2}	2.94	g/cm	NE Cuba	This chapter
Sparisoma chrysopterum	275	7–31 FL	1.35×10^{-2}	3.1	g/cm	SW Cuba	Claro and García-Arteaga 1994a
Sparisoma radians	100	5.5–16.1 TL	2.85×10^{-2}	2.78	g/cm	NE Cuba	This chapter
Mugilidae							
Mugil curema M	—	—	3.73×10^{-2}	2.69	g/cm	T.de Zaza	Alvarez-Lajonchere 1976
Mugil curema F	—	—	2.93×10^{-2}	2.77	g/cm	T. de Zaza	Alvarez-Lajonchere 1976
Mugil curema All	—	—	3.3×10^{-2}	2.74	g/cm	T. de Zaza	Alvarez-Lajonchere 1976
Mugil curema	87	20–35 FL	2.3×10^{-5}	2.93	g/cm	NE Cuba	This chapter
Mugil hospes	946	13–18 FL	9.66×10^{-2}	2.25	g/cm	SE Cuba	Alvarez-Lajonchere 1980b
Mugil liza	871	16–65 FL	1.30×10^{-2}	2.98	g/cm	SE Cuba	Alvarez-Lajonchere 1980b
Mugil trichodon	1347	12–28 FL	2.08×10^{-2}	2.85	g/cm	SE Cuba	Alvarez-Lajonchere 1980b
Sphyraenidae							
Sphyraena barracuda	68	7.5–114 FL	5.6×10^{-3}	3.01	g/cm	NE Cuba	This chapter
Sphyraena barracuda	78	12–92 FL	4.4×10^{-3}	3.08	g/cm	SW Cuba	This chapter
Sphyraena guachancho	42	23.5–46 FL	0.6×10^{-2}	2.98	g/cm	NE Cuba	This chapter
Labrisomidae							
Malacoctenus macropus	23	3–5 TL	3.41×10^{-5}	2.72	g/mm	SW Cuba	Claro and García-Arteaga 1994a
Paraclinus fasciatus	56	2–4 TL	2.09×10^{-5}	2.84	g/mm	SW Cuba	Claro and García-Arteaga 1994a
Gobiidae							
Bathygobius soporator	7	2.6–9.2TL	4.3×10^{-2}	3.52	g/cm	SW Cuba	This chapter
Coryphopterus glaucofraenum	46	2–6 TL	3.45×10^{-5}	2.68	g/mm	SW Cuba	Claro and García-Arteaga 1994a
Lophogobius cyprinoides	22	2.8–7.8 FL	5.7×10^{-3}	3.44	g/cm	NE Cuba	This chapter
Acanthuridae							
Acanthurus bahianus	11	6.7–19 FL	2.78×10^{-2}	2.94	g/cm	NE Cuba	This chapter
Acanthurus chirurgus	17	4.4–24.2 FL	1.76×10^{-2}	3.13	g/cm	NE Cuba	This chapter
Acanthurus coeruleus	6	4.3–19 FL	1.63×10^{-2}	3.22	g/cm	NE Cuba	This chapter
Trichiuridae							
Trichiurus lepturus	393	28–105 TL	0.78	3.48	g/mm	S Cuba	Ros and Pérez 1978
Scombridae							
Katsuwonus pelamis	367	30–57 FL	1.12×10^{-2}	3.15	g/cm	Cuba	Suarez-Caabro and Duarte-Bello 1961
Katsuwonus pelamis	1612	42–60 FL	5.29×10^{-6}	3.22	g/mm	W Cuba	Carles 1971
Katsuwonus pelamis M	—	—	5.72×10^{-3}	3.34	g/cm	SW Cuba	García-Coll 1984
Katsuwonus pelamis F	—	—	4.80×10^{-3}	3.39	g/cm	SW Cuba	García-Coll 1984
Scomberomorus cavalla	311	52–97 FL	1.57×10^{-5}	2.87	g/mm	Cuba	León and Guardiola 1984
Scomberomorus regalis	262	40–66 FL	2.02×10^{-5}	2.80	g/mm	Cuba	León and Guardiola 1984
Thunnus atlanticus	13	39.5–70.4 FL	4.9×10^{-3}	3.22	g/cm	NE Cuba	This chapter
Thunnus atlanticus	875	28–60 FL	1.38×10^{-2}	3.1	g/cm	Cuba	Suarez-Caabro and Duarte-Bello 1961
Thunnus atlanticus	1760	30–56 FL	2.62×10^{-5}	2.96	g/mm	W Cuba	Carles 1971
Thunnus atlanticus M	—	—	1.31×10^{-2}	3.12	g/cm	SW Cuba	García-Coll 1984
Thunnus atlanticus F	—	—	1.54×10^{-2}	3.08	g/cm	SW Cuba	García-Coll 1984
Xiphiidae							
Xiphias gladius	242	84–254 TL	4.90×10^{-7}	3.64	kg/cm	NW Cuba	Guitart Manday 1964
Xiphias gladius	252	80–249 FL	2.30×10^{-6}	3.33	kg/cm	NW Cuba	Espinosa et al. 1988
Istiophoridae							
Makaira nigricans	73	92–220 FL	2.57×10^{-5}	2.76	kg/cm	NW Cuba	Espinosa et al. 1988
Triglidae							
Prionotus punctatus	14	3.8–14.5 FL	1.23×10^{-2}	2.85	g/cm	NE Cuba	This chapter
Prionotus punctatus	94	—	6.6×10^{-6}	3.14	g/mm	SE Cuba	This chapter
Achiridae							
Achirus lineatus	13	3.5–9 FL	1.56×10^{-2}	2.99	g/cm	NE Cuba	This chapter
Balistidae							
Balistes vetula	47	18.4–38 FL	9.89×10^{-2}	2.61	g/cm	NE Cuba	This chapter
Canthidermis sufflamen	13	34.5–53.2 FL	2.63×10^{-2}	2.04	g/cm	SW Cuba	This chapter

Continued on next page

Appendix 6.1. continued

Monacanthidae						
Aluterus schoepfi	7	13.4–30.3 FL	2.5×10^{-3}	3.30	g/cm	NE Cuba	This chapter
Monacanthus ciliatus	88	2.5–11.1 FL	2.4×10^{-2}	2.84	g/cm	NW Cuba	This chapter
Monacanthus ciliatus	163	2–7 TL	2.56×10^{-2}	2.7	g/cm	SW Cuba	Claro and García-Arteaga 1994a
Monacanthus setifer	38	3.5–14.2 FL	2.07	2.93	g/cm	NE Cuba	This chapter
Ostraciidae						
Acanthostracion quadricornis	10	9–24.5 TL	5.3×10^{-2}	2.66	g/cm	SE Cuba	This chapter
Acanthostracion quadricornis	9	9.4–24 TL	2.02×10^{-2}	2.95	g/cm	NE Cuba	This chapter
Acanthostracion quadricornis	181	5–32 TL	1.53×10^{-1}	2.25	g/cm	SW Cuba	Claro and García-Arteaga 1994a
Lactophrys trigonus	72	4.5–54.5 TL	6.53×10^{-2}	2.66	g/cm	NE Cuba	This chapter
Lactophrys trigonus	27	7–47 TL	3.75×10^{-1}	2.1	g/cm	SW Cuba	Claro and García-Arteaga 1994a
Tetraodontidae						
Canthigaster rostrata	4	5–7.5 TL	4.9×10^{-2}	2.60	g/cm	NE Cuba	This chapter
Sphoeroides spengleri	26	3.6–21 TL	0.0248	2.90	g/cm	NE Cuba	This chapter
Sphoeroides spengleri	104	2–16 TL	4.20×10^{-2}	2.61	g/cm	SW Cuba	Claro and García-Arteaga 1994a
Sphoeroides testudineus	26	3.2–21.6 TL	3.15×10^{-2}	2.83	g/cm	NE Cuba	This chapter
Diodontidae						
Chilomycterus schoepfii	30	1.8–18 TL	8.63×10^{-2}	2.69	g/cm	NE Cuba	This chapter
Diodon holacanthus	6	—	0.2799	2.28	g/cm	NE Cuba	This chapter
Diodon hystrix	25	24.5–65 TL	1.4×10^{-2}	3.14	g/cm	NE Cuba	This chapter

SL: standard length; FL: fork length; TL: total length; DD: disk diameter; M: male; F: female.

Appendix 6.2. Growth parameters of some fishes that occur on the Cuban shelf.

Family Species	G	L_∞	W_∞	K	t_0	Φ'^a	Locality	Source
Alopiidae								
Alopias vulpinus (TL)	V	650.9	—	0.10	−2.36	4.63	California	Cailliet et al. 1983
Lamnidae								
Isurus oxyrinchus M	V	321.0	369,108	0.07	−3.75	3.86	California	Cailliet et al. 1983
Isurus oxyrinchus F	V	345.0	314,345	0.20	−1.0	4.38	E US	Pratt and Casey 1983
Isurus oxyrinchus	V	302.0	446,337	0.26	−1.0	4.37	E US	Pratt and Casey 1983
Carcharhinidae								
Carcharhinus brevipinna	V	214.0	62,573	0.21	−1.94	3.98	Gulf of Mexico	Branstetter 1987
Carcharhinus limbatus	V	176.0	39,434	0.27	−1.20	3.92	Gulf of Mexico	Branstetter 1987
Carcharhinus obscurus* M	V	169.9	—	0.11	−2.65	3.50	SE US	Schwartz 1983
Carcharhinus obscurus* F	V	194.3	—	0.10	−2.33	3.58	SE US	Schwartz 1983
Carcharhinus plumbeus M	V	257.0	—	0.05	−4.5	3.52	E US	Casey et al. 1983
Carcharhinus plumbeus F	V	299.0	—	0.04	−4.9	3.55	E US	Casey et al. 1983
Prionace glauca	LF	394.0	262,743	0.13	−0.8	4.30	E US	Aasen 1966
Prionace glauca	V	423.0	332,320	0.11	−1.03	4.29	E US	Stevens 1975
Prionace glauca (TL)	V	265.5	—	0.22	−0.3	4.19	California	Cailliet et al. 1983
Rhizoprionodon terraenovae M	Sp	104.7	5,168	0.19	−1.57	3.32	Yucatán	Alvarez 1989
Rhizoprionodon terraenovae F	Sp	119.3	5,462	0.15	−1.85	3.33	Yucatán	Alvarez 1989
Sphyrnidae								
Sphyrna lewini* M	V	184.5	—	0.11	−1.68	3.57	SE US	Schwartz 1983
Sphyrna lewini* F	V	230.9	—	0.09	−1.23	3.68	SE US	Schwartz 1983
Sphyrna tiburo	LF	119.0	—	0.17	−1.9	3.38	Yucatán	Alvarez 1989
Squalidae								
Squalus acanthias F	Sp	125.0	—	0.05	−4.88	2.89	W US	Ketchen 1975
Squalus acanthias M	Sp	99.8	—	0.07	−4.7	2.84	W US	Ketchen 1975
Elopidae								
Elops saurus*	Sc	49.0	1,007	0.22	−0.54	2.72	SE Cuba	Carles 1967
Megalops atlanticus M	O	206.0	—	0.08	0.17	3.53	Brazil	Ferreira de Menezes and Pinto 1966
Megalops atlanticus F	O	263.0	—	0.06	0.17	3.62	Brazil	Ferreira de Menezes and Pinto 1966
Megalops atlanticus F	O	181.8	64,621*	0.10	−1.410	3.51*	S Florida	Crabtree et al. 1995
Megalops atlanticus M	O	156.7	41,481*	0.12	−1.575	3.47*	S Florida	Crabtree et al. 1995
Albulidae								
Albula vulpes*	Sc	66.3	5,230	0.22	−0.93	2.99	Florida	Bruger 1974
Clupeidae								
Harengula humeralis	Sc	18.5	198	0.41	−0.79	2.15	SW Cuba	García-Arteaga 1993
Harengula jaguana*	LF	18.2	136	0.6	0.06	2.30	Brazil	Hubold and Mazzetty 1982
Harengula jaguana	LF	21.0	87	0.48	−0.64	2.33	Yucatán	Arreguín-Sánchez et al. 1989
Jenkinsia lamprotaenia	LF	4.9	1.3	3.34	−0.07	1.90	SW Cuba	Bustamante et al. 1992
Opisthonema oglinum	LF	21.5	73	0.65	−0.55	2.48	Yucatán	Arreguín-Sánchez et al. 1989
Sardinella aurita	O	27.4	201	0.25	−1.66	2.27	Venezuela	González 1985b
Engraulidae								
Cetengraulis edentulus* M	LF	14.9	34	0.22	−4.7	1.69	Venezuela	Simpson and Griffiths 1973
Cetengraulis edentulus* F	LF	16.9	51	0.15	−4.7	1.63	Venezuela	Simpson and Griffiths 1973
Centropomidae								
Centropomus undecimalis* M	O	76.3	4,213	0.49	0.73	3.46	Florida	Volpe 1959
Centropomus undecimalis* F	O	85.7	5,373	0.42	0.47	3.49	Florida	Volpe 1959
Holocentridae								
Holocentrus adscensionis	T	26.1	—	0.23	—	2.19	Jamaica	Munro 1999
Holocentrus rufus	T	18.8	—	0.48	—	2.23	Jamaica	Munro 1999
Serranidae								
Cephalopholis cruentata (TL)	O	41.5	1,174	0.13	−0.94	2.35	Curaçao	Nagelkerken 1979
Cephalopholis cruentata (TL)	LF	34.0	690	0.34	—	2.59	Jamaica	Munro 1983a

Continued on next page

Appendix 6.2. continued

Cephalopholis cruentata (TL)	O	45.1	1,683	0.12	−1.24	2.39	SE US	Potts and Manooch 1999
Cephalopholis fulva (TL)	LF	35.5	669	0.56	—	2.85	Jamaica	Munro 1983a
Cephalopholis fulva (TL)	O	37.2	934	0.32	0.20	2.65	SE US	Potts and Manooch 1999
Cephalopholis fulva (TL)	T	31.0	378	0.14	—	2.14	US Virgin I.	Munro 1983a
*Diplectrum formosum**	O	27.1	211	0.27	0.18	2.30	Florida	Bortone 1971
Epinephelus itajara (TL)	—	200.6	161,880*	0.126	−0.49	3.71	Gulf of Mexico	Bullock et al. 1992
Epinephelus adscensionis (TL)	T	49.9	1,900	0.11	—	2.44	US Virgin I.	Randall 1962 in Munro and Williams 1985
Epinephelus adscensionis (TL)	O	49.9	2,479	0.17	−2.49	2.62	SE US	Potts and Manooch 1995
Epinephelus guttatus (TL)	O	60.1	3,093*	0.07	−4.69	2.40	St. Thomas	Sadovy et al. 1992
Epinephelus guttatus (TL)	O	51.5	2,040*	0.10	−0.44	2.42	Puerto Rico	Sadovy et al. 1992
Epinephelus guttatus (TL)	LF	54.5	2,453	0.22	−2.944	2.82	Jamaica	Munro 1983a
Epinephelus guttatus (TL)	—	50.7	—	0.18	−0.44	2.67	Bermuda	Burnett-Herkes 1975 in Sadovy et al. 1992
Epinephelus guttatus (TL)	T	56.8	2,470	0.12	—	2.58	US Virgin I.	Randall 1962 in Munro and Williams 1985
Epinephelus guttatus (TL)	O	47.1	1,748	0.2	−2.39	2.65	SE US	Potts and Manooch 1995
Epinephelus morio (SL)	O	67.2	13,370	0.18	−0.45	3.05	Florida	Moe 1969
Epinephelus morio (TL)	O	92.8	13,869	0.11	0.09	2.98	Banco de Campeche	Melo 1976
Epinephelus morio (TL)	U	86.0	17,056	0.10	−1.5	2.87	Banco de Campeche	González 1983
Epinephelus morio (TL)	U	96.3	14,489	0.10	−1.9	2.97	Banco de Campeche	Valdés-Alonso and Fuentes-Castellanos 1987
Epinephelus morio (TL)	U	89.1	9,353	0.12	—	2.98	Banco de Campeche	Rodriguez 1986
Epinephelus morio (TL)	O	80.2	6,796	0.16	—	3.01	Banco de Campeche	Doi et al. 1981
Epinephelus morio (TL)	U	86.0	8,400	0.10	—	2.87	Banco de Campeche	Valdés-Muñoz and Padron 1980
Epinephelus morio (TL)	U	87.5	8,853	0.10	—	—	Banco de Campeche	González et al. 1974
Epinephelus morio (TL)	—	93.6	10,862	0.12	—	3.02	Banco de Campeche	Guzman 1986
Epinephelus morio (TL)	O	82.1	7,296	0.13	—	2.94	Banco de Campeche	Salazar 1988
Epinephelus morio (TL)	O	92.8	10,582	0.11	—	2.98	Banco de Campeche	Muhlia 1976
Epinephelus morio (TL)	LF	98.5	12,681	0.17	—	3.22	Banco de Campeche	Arreguín-Sánchez 1987
Epinephelus morio (TL)	LF	92.0	10,308	0.10	—	2.93	Banco de Campeche	Arreguín-Sánchez 1992
Epinephelus morio (TL)	O	93.8	—	0.15	−0.099	3.12	SE US	Stiles and Burton 1994
Epinephelus nigritus (TL)	O	239.4	245,682	0.05	−3.61	3.46	SE US	Manooch and Mason 1987
Epinephelus niveatus (TL)	O	125.5	—	0.07	−1.92	3.04	N Carolina	Matheson and Huntsman 1984
Epinephelus niveatus (TL)	O	111.7	—	0.12	−1.409	3.17	N and S Carolina	Wyanski et al. 2000
Epinephelus niveatus (TL)	O	132	—	0.08	−1.01	3.14	Florida Keys	Moore and Labinsky 1984
Epinephelus striatus (TL)	LF	90	12,900	0.09	—	2.86	US Virgin I.	Randall 1962
Epinephelus striatus (TL)	T	110.0	22,500	0.22	—	3.43	US Virgin I.	Pauly 1978
Epinephelus striatus (SL)	LF	97.4	16,839	0.18	0.49	3.23	US Virgin I.	Olsen and LaPlace 1979
Epinephelus striatus (TL)	O	94	16,553	0.06	−3.27	2.72	SW Cuba	Claro et al. 1990c
Epinephelus striatus (TL)	O	76	7,970	0.12	−1.11	2.84	NE Cuba	Claro et al. 1990c
Mycteroperca bonaci (TL)	O	135.2	37,890	0.11	−0.93	3.30	SE US	Manooch and Mason 1987
Mycteroperca bonaci (TL)	O	130.6	—	0.17	0.77	3.46	Florida	Crabtree and Bullock 1998
*Mycteroperca microlepis**	O	98.7	17,729	0.14	−0.55	3.13	Florida	McErlean 1963
Mycteroperca microlepis (TL)	O	129.0	38,809	0.12	−1.13	3.30	SW Cuba	Manooch and Haimovici 1978
Mycteroperca microlepis (TL)	LF	130.0	40,226	0.16	—	3.42	Gulf of Mexico	Saloman and Fable 1981
Mycteroperca microlepis (TL)	O	109.2	—	0.19	−1.33	3.35	SE US	Harris and Collins 2000
Mycteroperca tigris (TL)	O	74.0	6,881	0.11	−1.875	2.78	SW Cuba	García-Arteaga et al. 1999
Mycteroperca venenosa (TL)	O	86.0	12,000	0.17	—	3.10	Jamaica	Munro 1983a
Mycteroperca venenosa (TL)	T	89.5	11,300	0.09	—	2.84	US Virgin I.	Munro and Williams 1985
Paranthias furcifer	O	37.2	—	0.22	−0.25	2.48	Gulf of Mexico	Nelson et al. 1985
Paranthias furcifer	LF	31.4	263	0.28	—	2.44	Puerto Rico	Posada and Appeldoorn 1996
Rachycentridae								
*Rachycentron canadum** M	Sc	121.0	21,504	0.28	−0.06	3.61	E US	Richards 1967
*Rachycentron canadum** F	Sc	164.0	55,086	0.22	−0.08	3.77	E US	Richards 1967

Continued on next page

Appendix 6.2. continued

Carangidae

Caranx ruber	LF	52.0	—	0.24	—	2.81	Jamaica	Thompson and Munro 1974
Caranx ruber	U	56.0	3,036	0.14	−1.73	2.64	SW Cuba	García-Arteaga and Reshetnikov 1992
Seriola dumerili	O	151.4	37,163	0.11	−1.17	5.40	SE US	Manooch and Potts 1997a
Seriola dumerili	O	110.9	19,078	0.22	−0.72	3.43	Gulf of Mexico	Manooch and Potts 1997b

Coryphaenidae

Coryphaena hippurus	Sc	162.5	64,500	0.73	−0.14	4.29	Florida	Beardsley 1967

Lutjanidae

Apsilus dentatus F LT	Sc	61.6	2,617	0.30	—	3.06	Jamaica	Thompson and Munro 1983
Apsilus dentatus M LT	Sc	63.8	2,927	0.65	—	3.42	Jamaica	Thompson and Munro 1983
Etelis oculatus LT	LF	102.0	23,300	0.29	—	3.48	Saint Lucia	Murray and Moore 1992
Etelis oculatus LT	LF	103.0	23,900	0.61	—	3.81	Saint Lucia	Murray et al. 1992
Lutjanus analis	U	80.7	8,034	0.12	−1.42	2.89	NE Cuba	Pozo 1979
Lutjanus analis	O	88.0	12,563	0.15	−0.35	3.07	SW Cuba	Claro 1981c
Lutjanus analis	O	117.0	30,087	0.10	−0.43	3.14	NW Cuba	Claro 1981c
Lutjanus analis (TL)	O	86.2	8,676	0.15	−0.57	3.05	Florida	Mason and Manooch 1986
Lutjanus analis (TL)	U	102.8	15,487	0.17	−0.62	3.25	N Venezuela	Palazón and González 1986
Lutjanus apodus	LF	57.0	3,800	0.18	—	2.77	Jamaica	Munro 1983a
Lutjanus apodus	T	57.0		0.21	—	2.83	Jamaica	Munro 1999
Lutjanus bucannella	LF	46.0	1,890	0.35	—	2.87	Jamaica	Munro 1983a
Lutjanus buccanella	LF	54.0	3,200	0.70	—	3.31	Jamaica	Munro 1983a
Lutjanus buccanella	U	60.2	4,339	0.10	−3.16	2.56	NW Cuba	Espinosa et al. 1984
Lutjanus buccanella	U	63.5	5,098	0.10	−2.05	2.61	SE Cuba	Espinosa and Pozo 1982
*Lutjanus campechanus**	O	64.4	4,228	0.33	−0.29	3.14	Florida	Futch and Bruger 1976
*Lutjanus campechanus**	—	117.0	22,572	0.18	−0.21	3.39	Gulf of Mexico	Wakeman et al. 1979
*Lutjanus campechanus**	O	97.0	15,485	0.16	−0.1	3.18	Florida	Nelson and Manooch 1982
Lutjanus campechanus (TL)	LF	95.4	15,730	0.22	—	3.30	US	Munro 1983b
Lutjanus campechanus (TL)	LF	100.0	12,700	0.15	—	3.18	Banco de Campeche	Leonce-Valencia and Monroy 1993
Lutjanus campechanus (TL)	—	98.7	12,200	0.12	—	3.07	Banco de Campeche	Rodríguez Castro 1992
Lutjanus campechanus (TL)	—	97.4	11,700	0.13	—	3.09	Banco de Campeche	González 1988
Lutjanus campechanus (TL)	O	95.5	12,198	0.15	0.18	3.12	SE US	Manooch and Potts 1997c
Lutjanus griseus	O	54.8	2,424	0.23	−1.06	2.84	SW Cuba	Claro 1983b
*Lutjanus griseus**	Sc	120.0	23,248	0.05	−0.29	2.86	Florida	Starck 1970
*Lutjanus griseus**	O	51.3	1,893	0.24	−0.62	2.80	SE Cuba	Báez et al. 1980
Lutjanus griseus (TL)	O	89.0	9,182	0.10	−0.31	2.90	E Florida	Manooch and Matheson 1983
Lutjanus griseus (TL)*	Sc	78.8	6,836	0.16	0.17	3.00	Florida	Croker 1962
Lutjanus griseus (TL)	O	792.0	6,325	0.078	−3.897	3.69	SE US	Johnson et al. 1995
Lutjanus griseus (TL)	T	46.4	—	0.61	—	3.12	Jamaica	Munro 1999
Lutjanus griseus (TL)	O	62.5	—	0.13	−1.33	2.70	SE Florida	Burton 2000
Lutjanus griseus (TL)	O	71.6	—	0.17	−0.001	2.94	NE Florida	Burton 2000
Lutjanus jocu F	O	85.4	13,000	0.10	−2	2.86	SW and NE Cuba	Claro et al 1999
Lutjanus jocu M	O	96.4	21,938	0.08	−2.14	2.87	SW and NE Cuba	Claro et al. 1999
Lutjanus purpureus	O	85.0	8,960	0.13	−0.86	2.97	Trinidad	Manickchand and Phillip 1996
Lutjanus synagris	—	42.0	1,111	0.30	−0.03	2.72	Brazil	Alegría and Ferreira de Menezes 1970
Lutjanus synagris	O	47.0	—	0.17	−0.88	2.57	SW Cuba	Claro and Reshetnikov 1981
Lutjanus synagris		30.7	444	0.53	−0.95	2.70	Mexico	Ayala 1984
Lutjanus synagris	O	40.1	1,184	0.16	−1.83	2.41	SW Cuba	Rubio et al. 1985
Lutjanus synagris	O	70.5	5,715	0.22	−0.55	3.04	Trinidad	Manickchand-Dass 1987
Lutjanus synagris	LF	41.0	1,066	0.24	−1.84	2.61	Mexico	Torres and Chávez 1987
Lutjanus synagris	O	454.0	3,175	0.126	−4.255	2.46	N Gulf of Mexico	Johnson et al. 1994
Lutjanus synagris	LF	42.8	1,195	0.28	−0.07	2.71	Yucatán	Mexicano and Arreguín-Sánchez 1989a
Lutjanus synagris	LF	45.0	1,209	0.23	—	2.67	Puerto Rico	Acosta and Appeldoorn 1992
Lutjanus synagris	LF	35.2	—	0.26	—	2.51	Campeche Sound	Rivera-Arriaga et al.1996

Continued on next page

Appendix 6.2. continued

Species							Location	Reference
Lutjanus synagris (TL)	O	50.1	2,300	0.13	−1.49	2.51	Florida	Manooch and Mason 1984
Lutjanus synagris★	O	37.9	823	0.20	−1.85	2.46	Cuba	Rodríguez Pino 1962
Lutjanus synagris★	O	35.1	780	0.28	−0.52	2.54	SW Cuba	Oleachea and Quintana 1970
Lutjanus synagris	T	39.0	—	0.37	—	2.75	Jamaica	Munro 1999
Lutjanus vivanus	U	75.7	8,735	0.10	−2.08	2.76	SE Cuba	Pozo and Espinosa 1982
Lutjanus vivanus	U	72.9	7,837	0.09	−2.64	2.68	NE Cuba	Pozo et al. 1983
Ocyurus chrysurus	V	49.2	6,836	0.14	−1.19	2.53	NW Cuba	Piedra 1965
Ocyurus chrysurus	Sc	53.0	23,248	0.10	−2.86	2.45	Yucatán	Cantarell 1982
Ocyurus chrysurus	LF	60.0	3,600	0.25	—	2.95	Jamaica	Munro 1983a
Ocyurus chrysurus	O	47.3	1,497	0.33	−0.27	2.87	NW Cuba	Claro 1983e
Ocyurus chrysurus	O	68.1	4,745	0.16	−0.85	2.87	SW Cuba	Claro 1983e
Ocyurus chrysurus	O	69.9	4,768	0.1	−1.79	2.70	SE Cuba	Carrillo and Ramiro 1988
Ocyurus chrysurus	O	50.2	1,687	0.14	−0.95	2.55	Virgin I.	Manooch and Drennon 1987
Ocyurus chrysurus	Sc	45.3	1,715	0.16	−1.89	2.52	Yucatán	Mexicano and Arreguín-Sánchez 1989b
Ocyurus chrysurus★	O	45.1	1,627	0.28	−0.36	2.76	Florida	Johnson 1983
Ocyurus chrysurus	O	41.0	—	0.21	−2.37	2.55	E Florida	García 2000
Rhomboplites aurorubens	O	62.7	2,966	0.20	0.13	2.89	N Carolina	Grimes 1978
Rhomboplites aurorubens	O	56.2	2,183	0.20	−0.11	2.80	SE US	Zhao et al. 1997
Rhomboplites aurorubens	O	65.0	3,398	0.14	−0.23	2.77	SE US	Potts et al. 1998
Rhomboplites aurorubens (TL)	O	29.8	—	0.25	−3.9	2.35	E Gulf of Mexico	Hood and Johnson 1999
Gerreidae								
Eugerres brazilianus★ F	Sp	30.8	831	0.29	−1.79	2.44	SE Cuba	Báez et al. 1983
Eugerres brazilianus★ M	Sp	27.9	629	0.28	−1.92	2.34	SE Cuba	Báez et al. 1983
Eugerres plumieri★ F	Sc	39.7	1,669	0.32	−0.27	2.70	Colombia	Rubio 1975
Eugerres plumieri★ M	Sc	33.9	1063	0.32	−0.56	2.57	Colombia	Rubio 1975
Eugerres plumieri★ F (SL)	Sc	28.8	667	—	—	—	Venezuela	Angell 1976
Eugerres plumieri★ M (SL)	Sc	22.9	343	0.17	−4.79	1.95	Venezuela	Angell 1976
Gerres cinereus	T	24.7	—	0.33	—	2.30	Jamaica	Munro 1999
Gerres cinereus★	Sp	28.0	493	0.65	0.00	2.71	SE Cuba	Báez and Alvarez-Lajonchere 1983
Haemulidae								
Haemulon album	LF	65.0	5,300	0.20	—	2.93	Jamaica	Billings and Munro 1974
Haemulon album	O	73.0	8,500	0.19	0.30	3.01	SW Cuba	García-Arteaga 1983
Haemulon album	—	62.3	5,563	0.196	—	2.88	Cuba	Baisre and Páez 1981
Haemulon aurolineatum★	LF	23.0	188	0.35	—	2.27	—	Hartruijker 1982
Haemulon aurolineatum★	LF	37.6	1,030	0.138	—	2.29	Banco de Campeche	Baisre and Páez 1981
Haemulon aurolineatum★	LF	27.1	424	0.184	—	2.13	Banco de Campeche	Olaechea et al. 1975
Haemulon aurolineatum★	Sc	26.7	284	0.21	−0.94	2.18	Banco de Campeche	Sokolova 1965
Haemulon aurolineatum★	Sc	27.1	164	0.18	−1.75	2.12	E Banco de Campeche	Olaechea et al. 1975
Haemulon aurolineatum (TL)	Sc	31.0	430	0.22	−1.28	2.21	N and S Carolina	Manooch and Barans 1982
Haemulon aurolineatum	Sc	20.0	—	0.506	−1.71	2.31	Brazil	Ximenes-Carvalho and Fonteles-Filho 1996
Haemulon bonariense	—	45.7	126	0.183	—	2.58	Banco de Campeche	Baisre and Páez 1981
Haemulon flavolineatum M	LF	25.0	374	0.35	—	2.34	Jamaica	Hartsuiyker 1982
Haemulon flavolineatum F	LF	26.9	476	0.179	—	2.11	Jamaica	Hartsuiyker 1982
Haemulon flavolineatum	LF	35.0	993	0.24	—	2.47	Puerto Rico	Dennis 1988
Haemulon melanurum	Sc	35.0	767	0.32	−0.1	2.59	N Gulf of Mexico	Nelson et al. 1985
Haemulon parra	O	38.8	1,093	0.24	−0.27	2.56	SW Cuba	This chapter
Haemulon plumieri	LF	42.0	1,360	0.35	—	2.79	Jamaica	Billings and Munro 1974
Haemulon plumieri	LF	39.8	1,159	0.275	—	2.64	Jamaica	Munro 1983a
Haemulon plumieri	LF	45.8	1,752	0.157	—	2.52	Jamaica	Hartsuiyker 1982
Haemulon plumieri	LF	48.0	1,206	0.328	—	2.88	Banco de Campeche	Capote 1971
Haemulon plumieri	LF	47.3	—	0.35	—	2.89	Banco de Campeche	Dominguez-Viveros and Avila-Martinez 1996
Haemulon plumieri	LF	41.6	—	0.21	—	2.56	Banco de Campeche	Dominguez-Viveros and Avila-Martinez 1996

Continued on next page

Appendix 6.2. continued

Haemulon plumieri	LF	37.6	998	0.34	—	2.68	Puerto Rico	Dennis 1988
Haemulon plumieri (TL)	Sc	64.0	4,334	0.11	−1.01	2.65	Carolinas	Manooch 1976
*Haemulon plumieri**	LF	25.6	284	0.43	−0.02	2.45	NE Cuba	Ramos and Pozo 1984
Haemulon plumieri	O	36.0	954	0.2	−0.69	2.41	SW Cuba	García-Arteaga 1992a
*Haemulon plumieri**	T	59.1	3,400	0.08	−4.21	2.44*	SW Florida	Potts and Manooch 2001
*Haemulon plumieri**	T	32.7	464	0.31	−4.21	2.52	SE Florida	Potts and Manooch 2001
Haemulon sciurus	LF	40.0	1,114	0.24	—	2.58	Jamaica	Billings and Munro 1974
Haemulon sciurus	O	34.0	746	0.22	−0.64	2.41	SW Cuba	García-Arteaga 1992b
Haemulon sciurus	—	49.7	1,377	0.189	—	2.67	Banco de Campeche	Baisre and Páez 1981
Haemulon sciurus	LF	37.1	948	0.3	—	2.62	Puerto Rico	Dennis 1988
Sparidae								
Archosargus rhomboidalis	O	20.0	320	0.81	0.07	2.51	Mexico	Chavence et al. 1986
Archosargus rhomboidalis	LF	30.4	418	0.27	−0.08	2.40	Yucatán	Arreguín-Sánchez et al. 1989
Calamus bajonado	Sc	75.6	9,044	0.18	−0.12	3.01	E Banco de Campeche	Olaechea et al. 1975
Calamus proridens	Sc	35.3	802	0.3	0.16	2.57	E Banco de Campeche	Olaechea et al. 1975
Diplodus holbrooki (TL)	O	33.2	579	0.19	−1.50	2.32	N Carolina	Manooch and Potts 1996
Lagodon rhomboides	LF	19.3	104	0.59	−0.27	2.34	Yucatán	Arreguín-Sánchez et al. 1989
Sciaenidae								
Micropogonias furnieri	Sc	50.7	3,305	0.25	−0.14	2.81	S Brazil	Vazzoler 1962
*Micropogonias furnieri**	O	85.0	8,283	0.13	−0.86	2.97	Trinidad	Manickchand-Heileman and Phillip 1996
Mullidae								
Mulloidichthys martinicus M	LF	29.0	360	0.4	—	2.53	Jamaica	Munro 1976
Mulloidichthys martinicus F	LF	30.0	520	0.4	—	2.56	Jamaica	Munro 1976
Mulloidichthys martinicus	T	25.0	—	1.07	—	—	Jamaica	Munro 1999
Pseudupeneus maculatus M	LF	27.0	360	0.7	—	2.71	Jamaica	Munro 1976
Pseudupeneus maculatus F	LF	25.0	260	0.35	—	2.34	Jamaica	Munro 1976
Pseudupeneus maculatus	T	24.4	—	1.14	—	—	Jamaica	Munro 1999
Pomacentridae								
Stegastes planifrons F	LF	11.6	36	0.58	—	1.89	Jamaica	Pauly and Ingles 1981
Stegastes planifrons M	LF	11.6	35	0.33	—	1.65	Jamaica	Pauly and Ingles 1981
Labridae								
Lachnolaimus maximus	U	85.0	13,818	0.098	−1.38	2.83	SW Cuba	Claro et al. 1989
Mugilidae								
Mugil curema	Sp	36.0	—	0.78	0.06	3.00	Virginia USA	Richards and Castagna 1976
Mugil curema	Sp	53.2	1,519	0.10	−5.9	2.45	NW Cuba	Alvarez-Lajonchere 1976
*Mugil hospes** M	Sp	30.3	275	0.28	−0.37	2.41	SE Cuba	Alvarez-Lajonchere 1981a
*Mugil hospes** F	Sp	51.8	1,483	0.15	−0.57	2.60	SE Cuba	Alvarez-Lajonchere 1981a
*Mugil liza** M	Sp	55.4	2,080	0.24	−1.41	2.87	SE Cuba	Alvarez-Lajonchere 1981a
*Mugil liza** F	Sp	88.8	8,361	0.11	−3.23	2.94	SE Cuba	Alvarez-Lajonchere 1981a
*Mugil trichodon**	Sp	28.6	300	0.13	−3.4	2.03	SE Cuba	Alvarez-Lajonchere 1981a
Scaridae								
Sparisoma aurofrenatum	T	22.3	—	0.48	—	2.38	Jamaica	Munro 1999
Sparisoma chrysopterum	T	40.0	—	0.79	—	3.99	Jamaica	Munro 1999
Sparisoma viride	T	54.9	—	0.27	—	3.75	Jamaica	Munro 1999
Sphyraenidae								
Sphyraena barracuda M	Sc	128.8	11,326	0.15	−0.68	3.40	Florida	DeSylva 1963
Sphyraena barracuda F	Sc	126	18,973	0.17	−0.38	3.43	Florida	DeSylva 1963
Scombridae								
*Katsuwonus pelamis** F	Sp	78.4	10,675	0.19	−4.54	3.07	N Carolina	Batts 1972
*Katsuwonus pelamis** M	Sp	87.2	15,289	0.15	−4.71	3.06	N Carolina	Batts 1972
*Katsuwonus pelamis**	V	139	68,817	0.11	−1.97	3.33	W Cuba	Carles 1974
*Katsuwonus pelamis** F	Sp	62.5	5,697	0.29	−2.5	3.05	SW Cuba	García-Coll 1984
*Katsuwonus pelamis** M	Sp	62.1	5,752	0.26	−2.6	3.00	SW Cuba	García-Coll 1984
*Scomberomorus cavalla** M	—	116.0	11,828	0.18	−0.22	3.38	Brazil	Nomura and Rodrígues 1967
*Scomberomorus cavalla** F	—	137.0	18,973	0.15	−0.13	3.45	Brazil	Nomura and Rodrígues 1967
*Scomberomorus cavalla** M	O	84.0	6,520	0.35	−2.5	3.39	Florida	Beaumariage 1973

Continued on next page

Appendix 6.2. continued

*Scomberomorus cavalla** F	O	115.0	18,959	0.21	−2.4	3.44	Florida	Beaumarriage 1973
*Scomberomorus cavalla**	O	124.9	14,573	0.18	−1.8	3.45	Brazil	Carneiro et al. 1978
Scomberomorus cavalla M	—	96.5	6,903	0.20	−1.17	3.27	SE US	Johnson et al. 1983
Scomberomorus cavalla F	—	152.9	27,769	0.14	−2.08	3.51	SE US	Johnson et al. 1983
Scomberomorus cavalla	LF	114.6	10,629	0.16	−0.25	3.32	Yucatán	Cabrera-Vázquez and Arreguín-Sánchez 1989
Scomberomorus maculatus M	O	60.8	2,966	0.40	0.14	3.17	Florida	Klima 1959
Scomberomorus maculatus F	O	72.0	4,406	0.40	0.27	3.32	Florida	Klima 1959
Scomberomorus maculatus	O	70.5	4,249*	0.36	0.36	3.25	SE Mexico	Medina-Quej and Dominguez-Viveros 1997
Scomberomorus maculatus M	O	83.0	7,498	0.20	−0.4	3.14	Brazil	Nomura 1967
Scomberomorus maculatus F	O	111.7	17,497	0.16	−0.12	3.30	Brazil	Nomura 1967
Scomberomorus maculatus	O	133.3	14,351	0.07	−3.26	3.09	Mexico	Mendoza 1968
Scomberomorus maculatus M	O	55.5	2,260	0.48	−1.12	3.17	Florida	Powell 1975
Scomberomorus maculatus F	O	69.4	3,926	0.45	−0.78	3.34	Florida	Powell 1975
*Scomberomorus maculatus**	O	73.0	3,043	0.29	−0.55	3.19	Trinidad	Sturm 1978
*Thunnus atlanticus**	V	78.0	9,225	0.33	−1.57	3.30	W Cuba	Carles 1974
*Thunnus atlanticus** F	Sp	54.3	3,387	0.43	−1.7	3.10	SW Cuba	García-Coll 1984
*Thunnus atlanticus** M	Sp	56.3	3,794	0.41	−1.7	3.11	SW Cuba	García-Coll 1984
Xiphiidae								
Xiphias gladius F	O	277.2	312,023	0.07	−3.94	3.73	E US	Radtke and Hurley 1983
Xiphias gladius M	O	266.7	274,691	0.12	−1.68	3.93	E US	Radtke and Hurley 1983
Istiophoridae								
Istiophorus platypterus F	Sp	183.0	28,800	0.16	3.12	3.73	Florida	Hedgepeth and Jolley 1983
Istiophorus platypterus M	Sp	147.0	54,100	0.3	1.96	3.81	Florida	Hedgepeth and Jolley 1983
Stromateidae								
*Peprilus triacanthus**	O	18.3	133	0.8	0.12	2.43	Virginia	Du Paul and McEachran 1973
Balistidae								
Balistes vetula	LF	45.0	1,950	0.57	—	3.06	Virgin I.	Randall 1962
Balistes vetula (TL)	LF	53.8	3,286	0.15	0.0	2.64	Brazil	Ferreira 1985
Balistes vetula	Sp	41.5	1,561	0.30	−0.6	2.71	Virgin I.	Manooch and Drennon 1987
Acanthuridae								
Acanthurus bahianus	T	28.0	—	0.36	—	2.45	Jamaica	Munro 1999
Acanthurus chirurgus	T	31.3	—	0.56	—	3.55	Jamaica	Munro 1999
Acanthurus coeruleus	T	30.0	—	0.20	—	3.15	Jamaica	Munro 1999

[a] All Φ′ values were calculated from data in original references.

Lengths are expressed as fork length unless noted otherwise (TL: total length; SL: standard length).

* Values calculated from data in original reference.

G: method used to estimate growth; V: vertebrae; Sc: scales; O: otoliths; LF: length frequency; Sp: spines; U: urohyal; T: tagging/recaptures; M: male; F: female.

7

Ecophysiology of Cuban Fishes

GEORGINA BUSTAMANTE, RODOLFO CLARO, AND MIJAIL I. SHATUNOVSKY

7.1 Introduction

Several physiological and biochemical indicators have been used successfully to describe the physiological condition of fishes. Research on the ecophysiological characteristics of marine and freshwater fishes increased during the 1970s and 1980s, particularly regarding physiological cycles and their relationship to environmental factors and life history processes. However, most of these investigations were conducted on European freshwater and marine fishes (Everson 1970; Shulman 1972, 1985; Kozlov 1978; Shatunovsky 1980; Lapina et al. 1984; Gershanovich et al. 1987). Herring (*Clupea harengus harengus*), cod (*Gadus morhua*), and halibut (*Platichthys flesus*) are the most well-known. The European sprat (*Sprattus sprattus*), European pilchard (*Sardina pilchardus*), European anchovy (*Engraulis encrasicholus*), and Mediterranean horse mackerel (*Trachurus mediterraneus*) are also relatively well-studied species. On the other hand, biochemical and physiological studies on fishes of the tropical western Atlantic are scarce and have a limited ecological perspective (Tornes et al. 1971; Nunes et al. 1976; Gutiérrez 1976). We have interpreted the potential ecological implications of the patterns found in Cuban fishes based largely on the temperate literature because there have been few studies on these parameters in tropical reef fishes.

Various indicators, such as hepatosomatic (HSI), gonadosomatic (GSI), and mesenteric fat (MFI) indices, as well as condition factor (K), have been used to describe physiological conditions of fishes from natural populations. These indices reflect the weight of liver and gonads, and fat in the body cavity relative to the gutted fish weight, as well as the robustness of the individual fish (the gutted fish weight divided by length cubed). In addition, lipid, protein, and water in muscle, liver, and gonads can be estimated; these are termed *biochemical indicators* in this chapter. These indices provide information about the intensity of fish metabolism through-

out different periods of the life cycle (e.g., Shulman 1972, 1985; Love 1980; Shatunovsky 1980; Lapina and Lapin 1982; Lapina et al. 1984; Gershanovich et al. 1987).

Of these indicators, fat content in tissues has typically been used in ecophysiological studies of fishes because lipid content in fishes varies markedly with metabolic changes. The intense motor activity of swimming in fishes demands the expenditure of large amounts of energy, and fat is an economic source of energy because of its high caloric content and the ease with which it is deposited in and mobilized from reserves (Shulman 1972).

This chapter summarizes research performed in Cuba on the seasonal and ontogenetic patterns of various physiological and biochemical indicators for four snapper species (Lutjanidae), and for redear sardine (*Harengula humeralis*), bar jack (*Caranx ruber*), margate (*Haemulon album*), and dwarf herring (*Jenkinsia lamprotaenia*). These studies allowed us to determine the relationship between changes in fish physiological conditions and variation in temperature, as well as seasonal patterns of reproduction, feeding, and growth. More detailed information can be found in Claro and Lapin (1971, 1973), Claro (1976, 1983a, 1983c), Bustamante (1983, 1987a, 1987b, 1988, 1989), and Bustamante et al. (1981, 1988, 1994).

7.2 Fish Fat Content

Data from several studies (Tornes et al. 1971; Nunes et al. 1976; Table 7.1 in Bustamante et al. 1994) show that most fish species of the wider Caribbean are "lean," according to the classification of Stansby (1963), because most have less than 5% body fat. This classification was based on data from fishes inhabiting polar and temperate latitudes that typically have higher fat content than species in the tropics and subtropics. However, body fat of tropical fish species, typically lower in absolute values, has been recorded with ranges as high as

Table 7.1.   Physiological and biochemical indicators in female bar jack, *Caranx ruber*, at two stages of sexual development (from Bustamante 1989)

Indicator[a]	Resting or Developing (Stage II-III) [b]	Early Maturation or Resting (Stage II) [b]	t-value	p
n	8	10		
Fish total weight (g)	495 ± 48	314 ± 33	3.1	0.01
Gonadosomatic index	0.40 ± 0.05	0.16 ± 0.02	4.5	0.001
Mesenteric fat index	0.74 ± 0.11	0.28 ± 0.04	3.8	0.01
Hepatosomatic index	1.46 ± 0.24	0.82 ± 0.06	2.6	0.05
Condition factor	1.60 ± 0.06	1.55 ± 0.03	0.75	0.25
% of lipids in muscles	3.78 ± 1.18	1.13 ± 0.60	2.0	0.05
% of lipids in liver	5.71 ± 0.40	3.55 ± 0.35	4.1	0.001

[a] Indicators were calculated using gutted fish weight.

[b] Mean ± standard error.

those of cooler area species. For example, muscular lipid content in redear sardine can increase 14 to 16 times over the amount present during seasons of minimum lipid content (Bustamante 1987b). The same seems to occur in the round sardinella (*Sardinella anchovia*) in Venezuelan waters (Tornes et al. 1971).

The lower proportion of body fat in tropical species is associated with various factors. One of the most important factors is that fishes in the tropics feed throughout the year, whereas fishes in temperate and polar regions have a relatively short feeding season (3–4 months for Arctic fishes and 6–8 months for temperate fishes; see Lapina et al. 1984). The shorter feeding period means storage of a larger amount before the fasting period. Another important factor is the depression of lipid synthesis that seems to occur when temperatures reach 30–31°C (see Section 7.3).

Data from various studies suggest that differences in fat content of the same species from populations on the continental shelves of Venezuela and Brazil compared with those in Cuban waters correlate with different levels of biological productivity (Tornes et al. 1971; Gutiérrez 1976; Nunes et al. 1976; and Bustamante et al. 1994). Both Venezuela and Brazil have major sources of coastal productivity (e.g., the drainage of large rivers and the occurrence of upwellings). Such features are limited in Cuba. Similarly, differences in the growth rates of lane and gray snappers (*Lutjanus synagris* and *L. griseus*) in Trinidad and Venezuela (Manickchand-Dass 1987 and Guerra and Bashirullah 1975, respectively) and Cuban populations can be attributed to differential biological productivity (see Chapter 6).

Based on available evidence, pelagic species have the highest fat content (Bustamante et al. 1994). Some active swimmers, such as jacks (Carangidae) and herrings (Clupeidae), necessarily have higher fat content than demersal species with more limited movement (e.g., lutjanids and haemulids). This difference is associated with two major functions of body fat storage: hydrostatic and energetic

(Shulman 1972). Hydrostatic storage refers to the contribution of lipids to buoyancy; energetic storage is associated with the biochemical capacity of lipids as an energy source for fish motor activity.

Among our study species, the liver functioned as an organ for storing fat only in grunts (margate, *Haemulon album;* bluestriped grunt, *H. sciurus;* white grunt, *H. plumieri;* and tomtate, *H. aurolineatum*). In some margates, lipids reached 40–50% of the organ weight. This high fat content gave the liver a white color. In other fish families, musculature and the body cavity are the main sites of fat storage. Biochemical studies of different types of tissues in redear sardine of Cuban waters (Wittenberger et al. 1969) showed that the fat content of the red lateral muscle is six times higher than that of the liver. These studies indicate that the red muscle of the redear sardine has a higher percentage of lipid, glycogen, and sarcoplasmatic proteins than the white muscles. Wittenberger et al. (1969) concluded that red muscle is the main storage area for energy reserves and that this tissue plays a major metabolic role. Because reproduction influences metabolic processes, the seasonal patterns of physiological condition of juvenile and adult portions of populations are presented in the next section.

7.3   Seasonal Patterns of Physiological and Biochemical Indicators

7.3.1   Juvenile Stages

Bar jack and margate are the only species whose physiological cycles have been investigated separately for juveniles and adults in Cuba (Bustamante et al. 1981, 1988, 1994; Bustamante 1983, 1987a, 1988). Juvenile bar jacks showed a marked seasonal physiological pattern that showed some similarities to the pattern of immature margate (Bustamante 1983; Bustamante et al. 1994). Two peaks in fat content (MFI) were evident: May–June and December–January (Fig. 7.1). Toward the end of the dry season (March–April), part of the

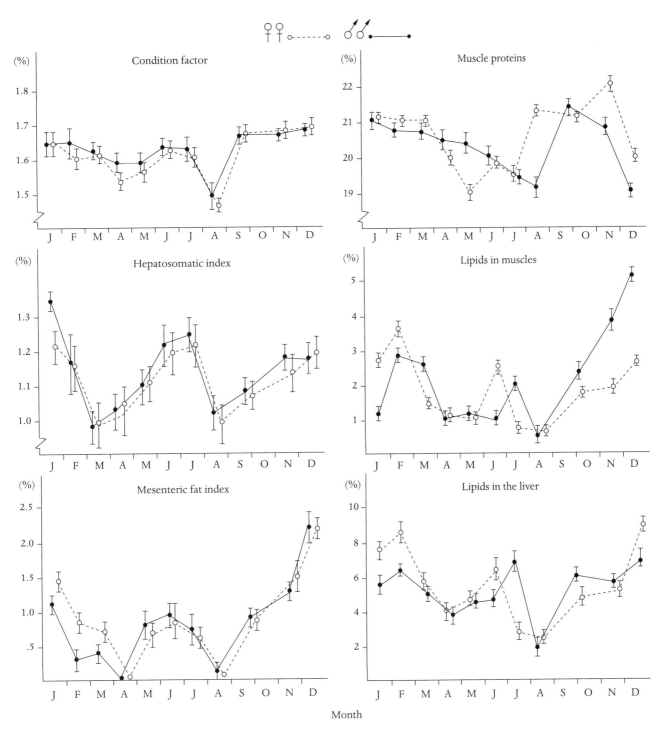

Fig. 7.1. Seasonal variations of physiological and biochemical parameters in juvenile bar jack (*Caranx ruber*) of southwestern Cuba. Mean ± standard error is shown (from Bustamante et al. 1981).

fat stored in organs and tissues was expended, which was reflected in all the indicators studied.

In Cuban shelf waters, when temperatures reach the highest values (30–31°C in July–August; see Lluis-Riera 1983c), body fat in bar jack and margate seems to be depleted (Bustamante 1983; Bustamante et al. 1994).

However, unlike what is seen in temperate and polar fishes, in Cuban fishes, the proportion of proteins in the musculature also declines during the summer (Fig. 7.1). This suggests that high temperatures provoke a decrease in anabolic processes and probably a reduction in growth rate (both in length and weight as reflected in the decline of the condition factor), followed by a recovery as water temperatures decline during the autumn. These data coincide with growth studies conducted on these and other fish species in Cuba. Growth intensity of fishes appeared to decrease during July–August and to increase during September–October (see Section 6.3).

The probable increase of respiratory rate with the elevation of temperature is another factor reducing anabolic processes during late summer. Information is lacking concerning respiratory metabolism for bar jack and margate, but experimental data collected by Claro and Colás (1987) on mutton snapper (*Lutjanus analis*) showed that from February to August respiratory metabolism increased from 18.8 to 28.5 cal/kg/day. In contrast, the indicators suggested an intensification of lipid synthesis and accumulation of reserves during the fall and winter.

7.3.2   Adult Stages

In adults, reproduction is one of the most critical biological processes affecting physiological condition. Fluctuations of physiological and biochemical indicators can reflect profound changes that occur in fish metabolism with successive phases of the reproductive cycle. Variations related to body fat are the most important because fat plays a major metabolic role as a source of energy and materials for gamete formation and spawning.

The level of fat reserves determines the beginning of sexual maturation. Table 7.1 provides comparative data for two groups of 3-year-old female bar jacks (captured in the months before the beginning of the reproductive season) that are in slightly different stages of gonad maturation. The first group (Stage II-III), with a slightly more advanced stage of sexual maturation, were larger and had higher levels of fat reserves in organs and tissues than the group at Stage II (before vitellogenesis, see Chapter 4). For each individual, the metabolic processes of sexual maturation and spawning require use of previously accumulated fat reserves. Table 7.2 shows a reduction of the condition factor (K) and marked reductions of muscular and mesenteric fat (MFI) in adult bar jack. These trends coincide with gonad maturation and spawning (data were collected during the second annual peak of the reproductive cycle).

In lane snappers, we recorded the expenditure of mesenteric fat in females during days of peak spawning (Fig. 7.2). During the 7- to 10-day period in which females released various portions of ripe eggs, the amount of fat stored in the body cavity declined by approximately two-thirds. During the trophoplasmic growth of oocytes (see Chapter 4 for terms related to gonad development), a rapid accumulation of nutritive substances (lipids and proteins) occurs. These substances later form the embryo and the prelarval nutritional reserve. Testis development, on the

Table 7.2.   Gonadosomatic index (GSI), condition factor (K), fat content in the body cavity or mesenteric fat index (MFI), and the percentage of lipids in muscles of adult bar jack, *Caranx ruber*, (32–37 cm FL) during the reproductive cycle (from Bustamante 1989)

| | | Month and Stage of Maturation Mean (Sample Size) (Range) | | |
| | | June | July | August |
Indicator	Sex	Stage II [a]	Stage IV [b]	Stage VI-II [c]
GSI	Female	0.40 (6) (0.21–0.61)	1.71 (7) (1.01–3.40)	0.47 (5) (0.32–0.72)
	Male	0.47 (4) (0.17–0.76)	1.64 (6) (0.86–2.41)	0.10 (1)
K	Female	1.72 (6) (1.61–1.78)	1.51 (7) (1.34–1.63)	1.49 (5) (1.41–1.57)
	Male	1.73 (4) (1.66–1.75)	1.47 (6) (1.39–1.55)	1.49 (1)
MFI	Female	2.80 (6) (1.31–6.50)	0.15 (7) (0–0.95)	0 (5)
	Male	2.73 (4) (0.10–4.90)	0 (6)	0 (1)
Lipids in muscles	Female	7.90 (3) (4.20–9.10)	0.79 (6) (0.21–1.27)	0.94 (4) (0.50–1.11)
	Male	9.39 (1)	0.88 (6) (0.39–1.33)	—

[a] Immature; [b] ripe; [c] spent and immature, see Table 4.2.

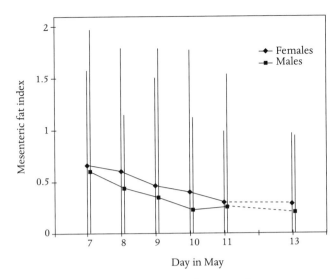

Fig. 7.2. Mesenteric fat index (MFI) of spawning lane snapper (*Lutjanus synagris*) during the spawning peak of May 1972 in southwestern Cuba (Claro 1983a). Ranges around mean are depicted for females (left bars) and males (right bars).

contrary, does not require the accumulation of such a large amount of material because maturation occurs mostly with a proliferation of cells (spermatogonia) and the redistribution of genetic material.

Table 7.3 shows the composition of the gonads of bar jack and margate at different stages of maturation. During vitellogenesis (up to Stage IV), the weight of the ovaries increased an average of sevenfold in bar jack as a result of a ninefold increase in protein content, and gonad weight

increased 15-fold in margate as a result of a 25-fold increase in protein content. Oocyte development occurred differently in these species. In the margate, maturation of oocytes was more asynchronous than in the bar jack (García-Cagide 1985, 1986a). Thus, in margate, individuals with partially spent ovaries (Stage VI-IV) still had vitellogenic oocytes in Phase E (García-Cagide 1986a; see also Chapter 4 for stages and terms). These types of ovaries had an intermediate weight (Stage VI-IV individuals, with mean protein weight of 4.27 g; see Table 7.3) but a relatively high proportion of protein (20.6%). On the other hand, lipid content and percentage of protein increased as lipids accumulated in the oocytes during development. In both species, the spent ovaries (Stage VI) had an extremely high water content (80–82%).

In males, gonad biomass grows (on average) 23-fold in bar jack and twofold in margate. The limited increase of testes weight in the latter might be associated with a much longer extension of the spawning period. The production of mature spermatozoa in the male margate seems to occur continuously, so that ripe testes do not exhibit large weights. In the bar jack, although ripe testes have a mean weight similar to that of the ovaries, the protein and lipid content (15.4 and 2.1%, respectively) are much lower in the testes than in the ovaries (25.3 and 8.1%, respectively). In both bar jack and margate, protein percentage does not change with maturation (15.1–15.4%, and 11.9–11.5%, respectively), despite increases in gonad weight.

In all species studied, marked seasonal variations in indicator values were associated with annual fluctuations of water temperature and breeding intensities within populations (Claro 1976, 1983a, 1983f, 1985; Bustamante et

Table 7.3. Protein, lipid, and water content in ovaries and testes of adult bar jack, *Caranx ruber* (30–44 cm FL), and margate, *Haemulon album* (45–62 cm FL), at different gonadal stages

Species Sex, Gonad Stage	*n*	Protein g	Protein %	Lipids g	Lipids %	Water g	Water %	Total g
Caranx ruber								
Female, II [a]	28	0.44	18.2	0.08	3.3	1.89	78.4	2.41
Female, III [b]	5	1.36	20.0	0.39	5.8	5.05	74.3	6.80
Female, IV [c]	14	4.15	25.3	1.33	8.1	10.91	66.6	16.39
Female, VI [d]	5	0.66	18.9	0.03	0.9	2.55	80.2	3.18
Male, II and VI	8	0.11	15.1	0.05	6.8	0.57	78.1	0.73
Male, IV	12	2.54	15.4	0.35	2.1	13.85	82.4	16.49
Haemulon album								
Female, II	8	1.27	15.1	0.18	2.1	6.96	82.9	8.40
Female, IV	16	32.42	26.1	6.88	5.5	84.90	68.4	124.2
Female, VI-IV [e]	8	4.27	20.6	1.15	5.6	15.31	74.0	20.63
Female, VI	16	1.87	15.5	0.20	1.7	9.95	82.2	12.02
Male, II and VI	14	0.69	11.9	0.17	2.9	4.87	83.9	5.73
Male, IV	5	1.53	11.5	0.18	1.3	11.71	87.4	13.42

[a] Immature; [b] mature (early development); [c] ripe; [d] spent; [e] spent, but with mature oocytes, see Table 4.2.

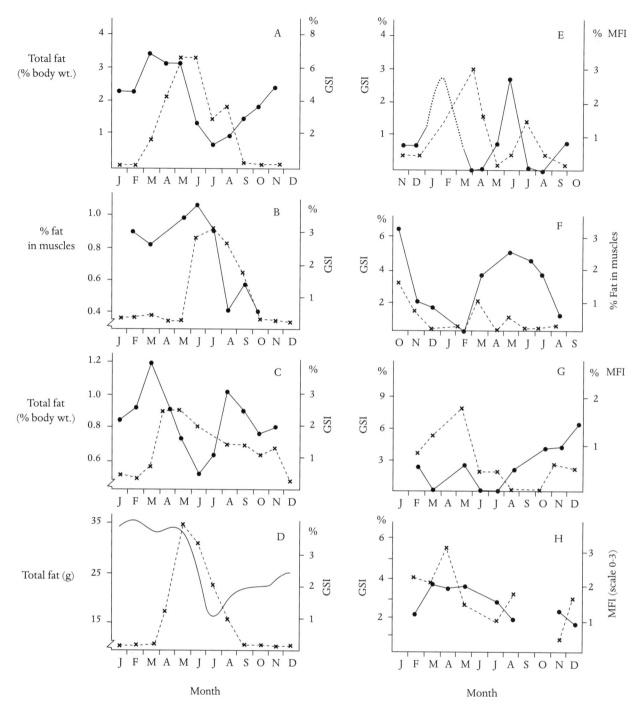

Fig. 7.3. Seasonal variations of different indicators of fat content, and the gonadosomatic index (GSI) of adult females of various species of southwestern Cuba. (A) lane snapper, *Lutjanus synagris* (Claro 1983a); (B) gray snapper, *L. griseus* (Claro 1985); (C) yellowtail snapper, *Ocyurus chrysurus* (Claro 1976), (D) mutton snapper, *L. analis* (Claro 1983e); (E) bar jack, *Caranx ruber* (Bustamante et al. 1981; and Bustamante 1989); (F) margate, *Haemulon album* (Bustamante 1983, 1986); (G) redear sardine, *Harengula humeralis* (Bustamante 1987b); (H) dwarf herring, *Jenkinsia lamprotaenia* (Bustamante et al. 1992). In H, the right-hand scale is the proportion of external stomach lining covered by fat: 1: <25%; 2: 25 to 50%; and 3: >50%. Broken line: mean GSI; solid line: different indicators of fat content; MFI: mesenteric fat index.

al. 1981; Bustamante 1987a, 1987b, 1989). Fig. 7.3 shows seasonal variations in body fat content measured through various indicators, as well as the gonadosomatic index (GSI),

for adults of several coastal fish species of Cuba. Typically, an increase of body fat was noted from winter until March–May, when the reproductive season began or peaked in some species, such as mutton, lane, gray, and yellowtail snapper (*Ocyurus chrysurus*), redear sardine, and dwarf herring. In bar jacks (Fig. 7.3E), as in snappers, body fat content increased from August on, at the end of the reproductive season. In March–April, when breeding started, ripe fishes had the lowest level of body fat content; about 0.7% of lipid content was in the muscles and nonmesenteric fat (Bustamante 1989).

The absence of adult bar jacks in January and February trawl surveys during three consecutive years suggests that adults move to other areas of the shelf during this period. This, together with data from Sierra and Popova (1982), who found a marked decline of prey consumption by bar jack larger than 30 cm fork length (FL), suggests a winter migration to areas where their main prey (parrotfishes and small wrasses) are most abundant. Such feeding migrations might lead to nutritional and physiological conditions necessary for the building of fat reserves for gonadal maturation. A new period of fat accumulation occurred during May–June, before the second annual spawning peak (June–July) of the species in the study area (Fig. 7.3E).

In the adults of margate, whose breeding season extends throughout the year, muscular lipid content declined markedly, paralleling the decrease in GSI (Fig. 7.3F). In the redear sardine, the highest levels of fat content were found during November–December, at the beginning of the spawning season, and before peak spawning during February–May (Fig. 7.3G). In the other small clupeid, the dwarf herring, the most mesenteric fat was noted during March–May, the primary period of egg production.

In adults, as in juvenile fishes, the body fat content of all species studied declined markedly during July–August. On average, adult bar jack inhabiting the eastern Golfo de Batabanó expended almost all mesenteric fat and 60–90% of the lipids stored in the muscular tissue from March–April to July–August. Fishes spawning late in the season showed lower body fat than those breeding in early spring. This pattern is probably associated with the influence of high summer temperatures on fish metabolism–higher respiratory rates and lower lipid production.

Lipids in ripe ovaries of the studied species also changed seasonally (Table 7.4). In the lane snapper, this body fat content dropped by approximately one-half from April to August, whereas in the bar jack the reduction was about one-third. Except for the mean value seen in the ripe ovaries of margate in December (because they were in Stage VI-IV, partially spawned), the proportion of protein seemed to change only slightly.

Data suggest that seasonal changes in the size of oil droplets in eggs occur with the progress of the reproductive season. Specifically, droplet size decreases during July–August when temperature is at a maximum. However, this pattern might not directly affect larval survivorship. High temperatures could also shorten the period during which larvae feed on internal reserves, counterbalancing the reduction of the nutritional reserve in late summer. This physiological aspect of larval survivorship requires further study because of its importance to the seasonality of recruitment in tropical fishes that display long breeding periods. This is also important for the artificial production of larvae in fish culture.

As in fish species of European seas (Shatunovsky and Krivobok 1976; Maslennikova 1978; Shatunovsky 1980), liver weight and its composition could serve as good indicators of the physiological condition of fishes of the Caribbean. In species with relatively little movement, the liver serves as a reserve for lipids and glycogen (Shatunovsky 1980).

Table 7.4. Composition of ripe ovaries of lane snapper, *Lutjanus synagris* (Claro 1983a), bar jack, *Caranx ruber* (Bustamante 1989), and margate, *Haemulon album* (Bustamante 1986), during the reproductive season

Species	Month	Protein (%)	Lipids (%)	Water (%)
Lutjanus synagris	April	22.0	6.8	71.2
	May	28.0	4.6	67.4
	June	25.5	4.7	69.8
	July	27.8	4.2	70.0
	August	24.2	3.6	72.7
Caranx ruber	April	23.0	7.5	69.5
	June	22.7	6.2	71.1
	July	29.0	4.9	66.1
Haemulon album	March	26.7	6.1	67.2
	May–June	26.2	7.4	66.4
	July	28.2	4.0	67.8
	September	26.2	5.1	68.7
	December	21.3	4.5	74.2

Storozhuk (1975) found that in the cod (*Pollachius virens*), lipids can make up 90% of the liver weight of some individuals. Among the species studied in Cuban waters, only grunts (*Haemulon* spp.) exhibited such a pattern, with some individuals having up to 50% fat in their liver. All margate with fatty livers were adults, larger than 40 cm FL, and with a high body fat content. This finding suggests that in this species, the liver functions for lipid storage. It can be considered a fat depot, although the amount of lipids contained in the liver never exceeded the amount contained in the muscles or the body cavity.

In all species studied, the most substantial change in the HSI occurred during gonad maturation. A notable increase in liver weight during vitellogenesis was observed in lane snapper (Claro 1983a), gray snapper (Claro 1985), mutton snapper (Claro 1983f), yellowtail snapper (Claro 1976), margate (Bustamante 1983), bar jack (Bustamante et al. 1981; Bustamante 1989), redear sardine (Bustamante 1987b), and dwarf herring (Bustamante et al. 1992), regardless of the species' reproductive season. The HSI of ripe female lane snappers was considerably higher than in immature and spent individuals (Fig. 7.4). The production of nutritive substances for gamete maturation requires more functional work in the female liver than in males. In all species, increases in liver activity generated a notable increase in weight, mostly in females, during the production of ovovitelline and lipids for oocyte maturation.

An increase in liver weight during vitellogenesis has also been recorded in many other temperate and polar fish species (Lapin 1976; Reshetnikov 1976; Shatunovsky 1976, 1980; Maslennikova 1978; Lapina et al. 1984), and the increase is associated with intensified synthesis and transformation of organic compounds in the liver during gamete development. In species of polar and temperate waters, with the completion of vitellogenesis, a reduction of liver weight occurs as production of lipids and ovovitelline for the formation of the oil droplets and yolk in the oocytes is completed. The diminished liver weight is caused by the lessened metabolic work of this organ. Rainbow trout (*Onchorhynchus mykiss*) display such a pattern; histological analyses indicate that during yolk formation, liver weight increases 100% because of increased cell and nucleus size. When yolk formation is complete, the liver returns to its previous weight (Zahnd and Clavert 1960).

No reduction of liver weight was detected in Cuban fish species before ovulation. This might be associated with the fact that in tropical fishes the period between the end of vitellogenesis and the release of eggs is relatively short (a few hours). In fishes of higher latitudes females can stay in an advanced stage of maturation for months before spawning. In the snappers, bar jack, margate, redear sardine, and dwarf herring, the liver weight of female spawners diminished only during spawning days or immediately thereafter. For example, in a group of bar jack spawners surveyed during the first spawning peak of the year (April), the mean HSI was 1.9 in females in Stage IV (GSI = 3.1), but in partially spent females with oocytes in early stages of development (Stage VI-III, see García-Cagide 1985), HSI was notably lower, approximately 1.0 (Bustamante 1989).

In June, female lane snappers in Stages VI (spent) and VI-II (resting) had a notably lower HSI than ripe females (Fig. 7.4). Except for late-spawners (those spawning in July–August), percentages of proteins in the livers of ripe females were considerably higher than in immature adults. This difference highlights the importance of the liver in ovovitelline production during yolk formation. Consequently, the proportion of water in the liver of just-spent females (in June–August) is high.

In addition to the lane snapper, all adults of the studied species (mutton, gray, and yellowtail snappers, bar jack, margate, redear sardine, and dwarf herring) showed marked seasonal changes in HSI and liver composition. This process is illustrated for bar jack in Fig. 7.5. The liver weight of mature bar jacks drops in July–August because of a decline in generative metabolism as water temperature rises. This pattern was also shown by decreased absolute contents and percentages of proteins and lipids. Lipid content in the livers of adult individuals in June (before the second spawning peak of the year in July) were high (Fig. 7.5). After July, HSI, liver protein, and lipid content increased gradually as a result of increased metabolism. This was a period of active growth and accumulation of fat for the next reproductive season.

Although the species investigated in Cuba have an extended reproductive season, a marked decline in the intensity of gamete production and lipid content in ripe oocytes toward late summer was evident. This is likely related to increased respiratory metabolism and the decline of lipid synthesis as temperatures rise in July and August. Subsequently, in September and October, intensified feeding activity (Chapter 5) and dropping temperatures, with a likely decline in respiratory metabolism, might contribute to increased lipid synthesis in preparation for the next reproductive cycle.

These physiological rhythms are different in different species according to the season in which reproduction takes place, and according to feeding habits and other factors related to the ecological requirements of fish populations within different habitats. More detailed investigations on the physiological aspects of Caribbean fish populations will lead to identifying the factors that control the dynamics of fish biological productivity in the tropics.

7.4 Ontogenetic Dynamics of Physiological and Biochemical Indicators

The relationship between fishes and their environment changes with age. Several authors have described ontogenetic variations in fish physiology from polar and temperate areas (Hatanaka et al. 1956a, 1956b; Lasker 1970; Shulman 1972; Daan 1975; Shatunovsky 1976, 1978, 1980; Maslennikova

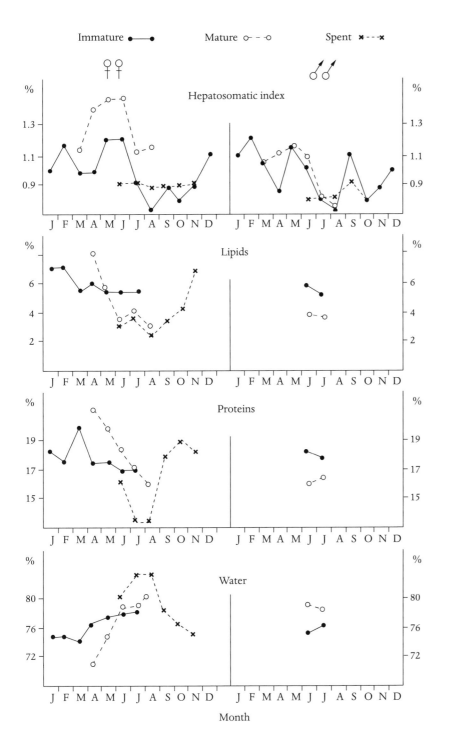

Fig. 7.4. Seasonal variations of the hepatosomatic index, and lipid, protein, and water content in liver tissues of male and female lane snapper (*Lutjanus synagris*) in southwestern Cuba (from Claro 1983a).

1978). Our investigations on Cuban fishes established that as juveniles, the lane, gray, yellowtail, and mutton snappers, as well as margate, bar jack, and redear sardine have no fat in the body cavity during the first months of development. In this period, anabolic metabolism is directed toward rapid growth. This physiological process might have adaptive significance: as fishes reach a larger size, they are more capable of avoiding the strong predation pressures existing in reef–seagrass habitats.

We investigated the physiological status during the first year of life in the bar jack using monthly data on size distributions, sex ratio, ring formation in the urohyal bone, feeding, and physiological indicators (Bustamante et al. 1988). Fig. 7.6 summarizes several biological and environmental factors influencing the life history of the young bar jack (more specific data can be found in Bustamante et al. 1988). The growth and fat content of juveniles spawned during both peaks of the breeding season in the

Golfo de Batabanó (March–April and June–July) showed a strong relationship between physiological condition, growth, feeding habits, and variations of habitat and water temperature. Beginning in May, the juveniles (3–10 cm FL) start to switch habitat from the drifting *Sargassum* algae to shallow water areas. This change in habitat leads to the formation of one mark in the urohyal bone at about 10 cm FL (García-Arteaga and Reshetnikov 1992). Active feeding and intense somatic metabolism contribute to the rapid growth of juveniles during spring. In June, fishes that were spawned at the spring spawning peak (Group 1 of Fig. 7.6) recruited to the trawl fisheries in the eastern part of

the Golfo de Batabanó at a mean size of 15 cm FL (34% of the maximum length of this species).

The relatively low daily ration of bar jack during July–August (approximately 1.8% of body weight) and the likely elevation of respiratory metabolism in the summer might cause reduced growth and the formation of a new mark on the urohyal bone in young recruits in July. An acceleration of growth followed by an accumulation of body fat was recorded beginning in August. Fish weight and condition factor, as well as liver weight and lipid content, increased substantially in the fall (Bustamante et al. 1988). Small fishes first appeared in the stomachs of juveniles of

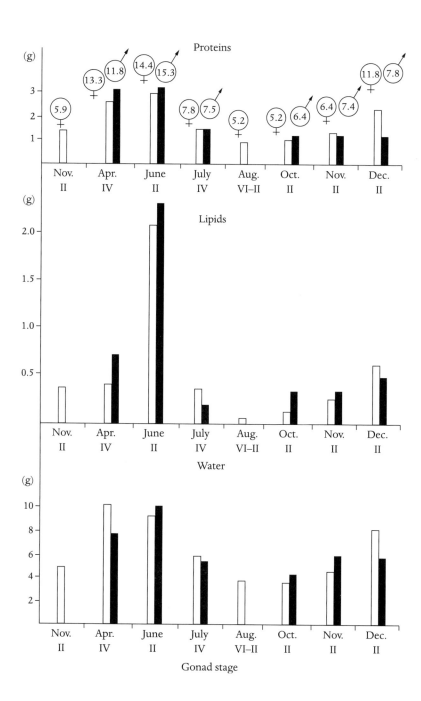

Fig. 7.5. Seasonal variations of liver weight (circled), and its protein, lipid, and water content in adult male (solid bars) and female (open bars) bar jack (*Caranx ruber*) in southwestern Cuba (from Bustamante 1989).

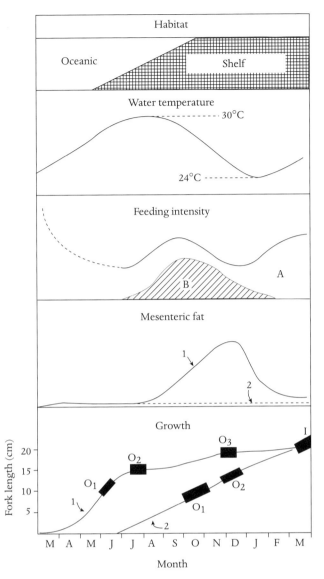

Fig. 7.6. Seasonal relationships among habitats, coastal water temperatures, feeding intensity, and food composition of 0+ bar jack (*Caranx ruber*), and mesenteric fat content and growth for cohorts born in the first (1) and second (2) spawning peaks. 0_1, 0_2, and 0_3 are marks on the urohyal bone formed before the first (I) annual ring. (A) planktonic crustacean and small shrimps; (B) juvenile fishes (from Bustamante et al. 1988).

Group 1 during the summer; zooplankton had previously been the primary food item. (The capture of large prey is more bioenergetically efficient for fishes than is picking small planktonic organisms; see Weatherley and Gill 1987.) This change in feeding habits might contribute to the development of gonads, because in October it was possible to determine the sex of 90% of the fish surveyed; the proportion was lower in the August sample.

A second group of recruits appeared in the November samples (taken from trawl fisheries). These recruits were probably produced during the second annual spawning peak in June–July. The two groups had different physiological features. Body fat content was consistently higher in Group 1. From December to March, both groups fed exclusively on zooplankton. For older individuals (4 to 5 cm larger than fishes from Group 2), the increase in energetic costs of capturing smaller prey (zooplankton) might cause a loss of body fat content and slower growth (see Fig. 7.6). For smaller fishes (Group 2), ingesting smaller prey could cost less bioenergetically (the ratio of predator size to prey size is lower). Thus, Group 2 fishes grew faster from December to March than Group 1 fishes. The two juvenile length groups in trawl fisheries gradually merge by March (Fig. 7.6). The difference in length increments of different groups within one generation is termed compensatory growth by Mina (1967). In the case of bar jack in the Golfo de Batabanó, such a process could explain why young-of-the-year generated from the two spawning peaks attained similar sizes by March.

Fig. 7.7 shows ontogenetic variations of mesenteric fat in different species. In the redear sardine (Fig. 7.7G) the completion of fat deposit occurred by the end of year one. A similar pattern was observed in the lane and yellowtail snappers (Fig. 7.7, C and E). In species such as the mutton snapper and margate, both of which attain large sizes, the ontogenetic pattern was different (Fig. 7.7, B and F). In these species, mesenteric fat was limited, particularly in the first years of life. Substantial deposits were present only in adults.

A marked decline in mesenteric fat content occurred during the second and third years for most species (Fig. 7.7). In the bar jack, this process is associated with the temporary lowering of daily food rations during the second year of life (Chapter 5) and the attainment of sexual maturity of part of the population during the second and third years (Fig. 7.8; see Bustamante 1988 and Bustamante et al. 1994 for more details). The decrease in feeding activity is probably associated with the occurrence of dietary shifts.

Gradual ontogenetic changes in bar jack feeding habits occurred with growth. As individuals attained 30 cm FL and three years of age, they switched from a combined diet of zooplankton, shrimp, and fishes to a diet based exclusively on fish. However, morphological (gape size) and behavioral (predation ability) changes might occur slowly. As a result, the food ration declined as did the condition factor, HSI, and body fat content (Bustamante 1988; Bustamante et al. 1994). This depletion process was also reflected in an abrupt slowing of growth, from 20 cm in the first year of life to 4 cm in the second. This nutritional shift was completed as bar jack became totally piscivorous; the food ration increased, as well as the body fat content, which in turn contributed to the attainment of maturity of most of the year-3 generation. In lane snapper, declines of mesenteric fat content in the second and third years of life also paralleled a shift to piscivory.

Data for the redear sardine shows that expenditure of fat reserves during the reproductive season is higher in larger individuals (Bustamante 1987b). At the beginning of the

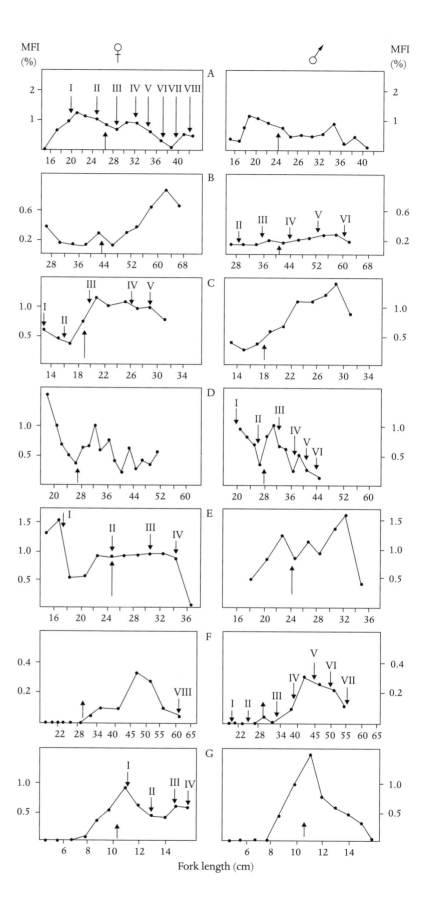

Fig. 7.7. Ontogenetic changes of the mesenteric fat index (MFI) with size in male and female fishes of the following species: (A) bar jack, *Caranx ruber;* (B) mutton snapper, *Lutjanus analis;* (C) lane snapper, *L. synagris;* (D) gray snapper, *L. griseus;* (E) yellowtail snapper, *Ocyurus chrysurus;* (F) margate, *Haemulon album;* (G) redear sardine, *Harengula humeralis.* Arrows pointing upward indicate the approximate size at first reproduction; roman numerals indicate the estimated age at that size for fishes of both sexes.

breeding season (November–December), the proportion of lipids in muscles increases with size (Fig. 7.9). At the end of the breeding season (June–July), the larger the adults (>10 cm), the lower the fat content. The proportion of protein in muscles shows a similar pattern. These results suggest that energy (fat) expenditures for reproduction are higher in larger individuals (Fig. 7.9). Ontogenetic variation in mesenteric fat content in all studied species (Fig. 7.7 through 7.9) reflected a reduction of synthetic activity in the largest individuals. This pattern has also been observed in fishes of polar and temperate waters and was associated with the decreased efficiency of anabolic processes in older individuals (Shatunovsky 1980).

7.5   Summary

Marine fishes of tropical areas typically have lower fat content (< 5% of body weight is lipids) than polar and

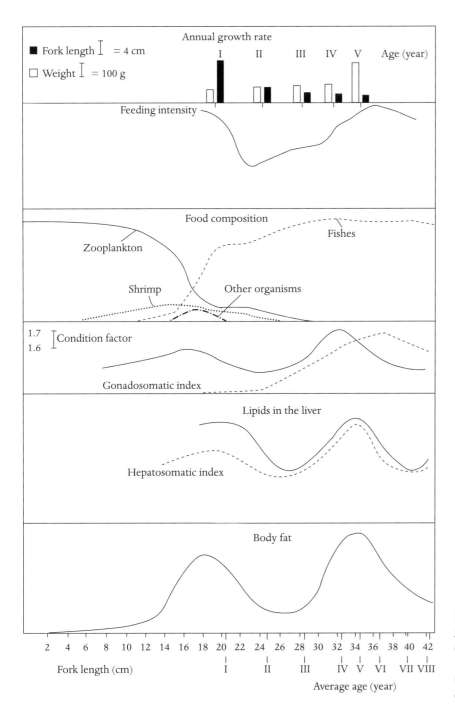

Fig. 7.8. Ontogenetic variation in annual growth increments, feeding intensity, food composition (from Sierra and Popova 1982, 1988), and physiological and biochemical indicators in the bar jack (*Caranx ruber*) (from Bustamante 1988).

temperate species. Only some pelagic families such as clupeids, engraulids, and scombrids exhibited high levels of body fat. Lower reserves are an adaptation to less severe environmental conditions than in higher latitudes. Tropical conditions allow fishes to feed all year, whereas polar and temperate fishes have a limited vegetative season during which they eat intensively and accumulate a large amount of fat, which is then expended during reproduction and winter fasting. Metabolic depression of lipid synthesis during the warmest period of the year (July–August in Cuban waters) is

another important factor influencing the low lipid content of tropical fishes. The high fat content of pelagic fishes in comparison with demersal species is attributable to the double function (hydrostatic and energetic) of fat reserves: they save metabolic energy by aiding buoyancy as well as providing an energy source.

Changes in the proportions of proteins and lipids in organs and tissues, and fluctuations of physiological indicators (HSI, GSI, mesenteric fat content, condition factor) allowed us to examine seasonal variations of physiology in several species

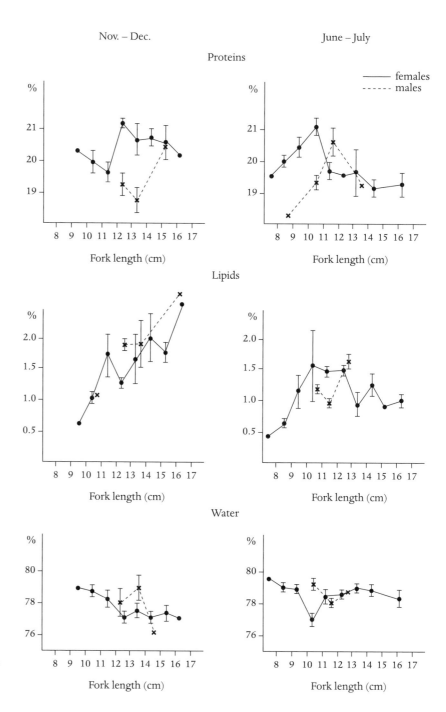

Fig. 7.9. Seasonal and size-specific variations in the relative percentage of lipids, proteins, and water content in the muscle of redear sardine (*Harengula humeralis*) at the beginning (November–December) and end (June–July) of the reproductive season (Bustamante 1987b). Bars represent standard error around the mean.

from the Cuban shelf. These patterns are related to primary biological processes (feeding, growth, reproduction) and annual variations in temperature. Marked ontogenetic variations of fat content in all species were also recorded. In juvenile stages, fat levels are low because anabolism is mostly directed to produce proteins that assure fast growth during the first months of life. This fast growth helps fishes avoid predation, which is intense in reef areas. Size-specific variations of these indicators show that near the completion of the first year of life, fat reserves reach a maximum in all the studied species except margate and mutton snapper, in which reserves are a low percentage of body weight.

The trend toward declining body fat during the second and third years of life seems to be associated with a nutritional shift that leads to reduced food ingestion. This transition appears to result from slow modifications of morphology and behavior that favor the capture of larger and more mobile prey. The bioenergetic cost of predation gradually goes down as the individual becomes more skillful at catching prey and the gape increases. When bar jacks reach 30 cm FL, food rations increase and become exclusively composed of fishes; body fat content also rises and fishes attain sexual maturity. A similar process probably takes place in the lane snapper, but that process has not been well documented. In the redear sardine, a continuous spawner, an increase in fat expenditure for reproduction was recorded.

In bar jack and margate juveniles, seasonal variations in temperature, feeding intensity, and diet composition contribute to two seasonal peaks in body fat: May–June and November–December. During July–August, high water temperatures (30–31°C) depress protein and lipid production and increase metabolism, which leads to slower growth in length and weight. In the fall, synthetic metabolic processes are activated, provoking accelerated growth and accumulation of fat reserves. During March–April, declines in feeding activities and probably increases in energetic costs for food capture contribute to a new period of depleted reserves.

In adult fishes, an additional process, reproduction, greatly influence physiological dynamics. As lipids are expended during both gonad maturation and spawning, the building of fat reserves typically peaks during the winter. Liver weight increases as a result of this metabolic activity. Although water temperatures fluctuate little throughout the year, marked variations were observed in the physiological condition of spawners. The lowest lipid content in ripe ovaries occurs during the highest temperatures of the year. This pattern is probably associated with reduced capacity for lipid synthesis combined with elevated respiration. It is likely that the reduction in reproductive activity of Cuban fishes in August is a result of the decline of anabolic processes in adult fishes during this warmest period.

8

Cuban Fisheries: Historical Trends and Current Status

RODOLFO CLARO, JULIO A. BAISRE,
KENYON C. LINDEMAN, AND JUAN P. GARCÍA-ARTEAGA

8.1 Introduction

Until the 1960s, most Cuban fisheries were focused on the insular shelf using artisanal fishing gear and small boats (3–11 m long), most of which were without engines. Larger vessels (20–25 m) were restricted to the tuna fishery and a few shrimp trawlers. Some other vessels fished high-priced species near the continental shelves of the Bahamas, Florida, and the Banco de Campeche. Total catches were less than 30,000 metric tons (t) per year. By 1977, Cuban fishery catches had increased to more than 200,000 t per year worldwide. This growth was partly due to the expanded activities of Cuban fleets in international fishing areas of the Atlantic and Pacific—the Gulf of Mexico Fleet, the Cuban High-Seas Fishing Fleet, and the Tuna Fleet. Since the 1980s, fisheries activities outside of the Cuban Exclusive Economic Zone (EEZ) have been seriously curtailed by the new Law of the Sea provisions and by the high cost of fuel and machinery.

Within the EEZ, fisheries entered a growth phase in 1955 after a 20-year period during which landings were less than 10,000 t per year (Baisre 1985a, 1985b, 1993). From about 1962 to 1965, motorboats replaced sailboats, fishing cooperatives were formed, better supplies became available, prices for catches rose, and more efficient fishing gear was developed. Landings in the EEZ expanded considerably, from about 22,500 t in 1959 to 79,000 t in the mid-1980s (Fig. 8.1). However, steadily increasing fishing effort resulted in the overexploitation of some stocks and reduced total landings in the late 1970s. In 1981, regulatory efforts were increased to protect overfished species and production efforts focused on underused species, such as stingrays, swimming crabs, and clams. These changes allowed for increased total catches in subsequent years, which

reached a record 78,989 t in 1985. Several analyses have concluded that many existing fishery resources were near maximum sustainable yield throughout the 1980s (Baisre 1981, 1985a, 1985b, 1993; Baisre and Páez 1981).

Baisre (2000a) estimated fishery trends from a 1978–1995 time series of landings for 21 species groups. By 1995, about 40% of these species groups were in the senescent phase of fishery development, about 50% were in the mature phase with a high exploitation level, and only 10% were in a developing phase with some potential for growth (terms based on Caddy 1984 and analyses of Grainger and García 1996). The trend of total landings was negative but not significant for most species groups. Baisre considered significant negative correlations for Nassau grouper, mullet, gray snapper, land crab, and stone crab landings particularly noteworthy. Less important, but also significant were the negative trends in the landings of sharks, grunts, and mangrove oysters (Baisre 2000a). Currently, the Cuban shelf has an exploitation level that can be increased only by changing fishery development strategies based on careful analyses of all factors influencing fishery production.

In this chapter we summarize the primary characteristics of Cuban fishery resources and expand on earlier analyses. We also summarize information on historical catch trends within the four fishery statistical zones of the Cuban shelf: Cabo San Antonio to Punta Hicacos (the northwest zone), Punta Hicacos to Punta de Maisí (the northeast zone, including the Archipiélago Sabana-Camagüey of north-central Cuba), Punta de Maisí to Bahía Cienfuegos (the southeast), and Bahía Cienfuegos to Cabo San Antonio (the southwest) (Figs. 1.1 and 1.3). All statistical data from 1959 to 1976 were obtained from *Serie Cronológica de la Actividad Extractiva* (Junta Central de Planificación 1975–1977). The

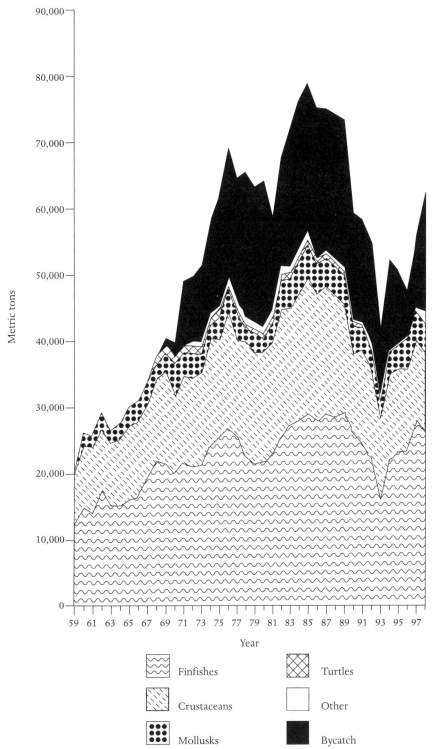

Fig. 8.1. Summary of fisheries landing data from the Cuban Exclusive Economic Zone, 1959–1998.

Ministry of Fishing Industries provided data totals for Cuba for 1977 through 1998. Data from associations that fish in more than one statistical zone were obtained directly from those associations. The Gulf Fleet catches in the EEZ were incorporated into their respective zones.

8.2   Fishery Resources of the Cuban Exclusive Economic Zone

The four statistical zones constitute relatively independent fishing areas that may represent somewhat discrete popula-

Table 8.1.   Mean annual catches (in metric tons) and proportion of total catch for major groups of marine organisms in the Cuban Exclusive Economic Zone during three five-year periods

Group or Species	1981–1985 t	%	1986–1990 t	%	1991–1995 t	%
Finfishes	25,600.5	37.2	28,557.8	38.8	22,063.8	42.5
Crustaceans (total)	17,647.6	25.7	16,241.0	22.6	12,956.2	25.0
Lobsters	11,897.5	17.3	10,690.0	14.9	9,364.9	18.0
Shrimps	4,666.1	6.8	3,841.1	5.3	2,436.8	4.7
Other Crustacea	1,084.0	1.6	1,710.0	2.4	1,154.5	2.2
Mollusks (total)	4,221.1	6.1	4,089.0	5.7	3,254.7	6.3
Oysters	2,521.0	3.7	2,353.2	3.2	1,656.0	3.2
Queen conch	552.9	0.8	189.0	0.3	56.3	0.1
Clams	1,347.2	2.0	1,547.0	2.2	1,533.2	3.0
Sponges	49.0	0.1	49.1	<0.1	49.4	0.1
Turtles	968.5	1.4	789.0	1.1	298.0	0.6
Bycatch	19,166.5	27.9	21,735.4	31.3	13,114.1	25.3
Other	1,090.5	1.6	284.8	0.4	126.2	0.2
Total catch	68,743.8		71,746.1		51,862.4	

tion units for some harvested species. Among the most notable fishery characteristics within the EEZ are the wide diversity of exploited species and the wide variety of artisanal fishing gear. The most economically valuable group of organisms are the crustaceans, which represented 23–26% of the landings from 1981 through 1995 (Table 8.1). Within this group, the spiny lobster (*Panulirus argus*) fishery represents approximately 15% of the total catch (Fig. 8.2) and is the most valuable fishery species in Cuba. Lobster landings increased from 6,500 t in 1959 to 13,578 t in 1985. Peak landings were typically followed by declines. From 1986 to 1990, lobster landings remained stable at more than 10,000 t per year, but catches since 1990 have decreased, fluctuating from 9,000 to 9,900 t per year (Cruz Izquierdo 1999). Evidently, expanding the fishing areas by installing artificial habitats contributed to the increases in the 1980s. Regulations on the lobster fishery were also enacted in 1977, including closures of 45 to 90 days during the reproductive season and control over the landing of individuals under 69 mm carapace length (Baisre 1985b, 1993, 2000b). It has been calculated that changing the size limit to 77 mm would increase the catch from 890 to 1,200 t per year (Cruz Izquierdo et al. 1991).

In general, more than 60% of the national lobster catch is from the southwest shelf and approximately 20–25% from the southeast shelf. Lobsters are less abundant in the northeast (15%) and northwest (2–3%) regions. Approximately 10 fishing associations, 240 vessels (27% of the boats working the shelf), and 1,300 fishermen work in the lobster fishery. In the 1980s the number of participating vessels and fishermen was higher, but high catch levels are now maintained by increased efficiency.

The shrimp fishery (*Penaeus notialis* and *P. schmitti*) is the second most valuable in Cuban waters. It is mainly concentrated on the southeast shelf, where nearshore estuarine conditions in large lagoons favor shrimp production. Shrimp landings increased at a steady rate from 1953 (1,900 t) to 1977 (5,864 t). However, landings have declined since 1977, signaling overexploitation of the fishery (Table 8.1). With implementation of new regulations in 1983 and other measures, effort was sharply reduced in the southeast (Baisre and Zamora 1983). These changes have contributed to a partial recovery of shrimp fishery resources. Shrimp landings leveled off at 4,000–4,500 t per year, then dropped abruptly in the late 1980s because key juvenile nursery areas were degraded by factors that included reduced river outflow after dam construction, a prolonged drought from 1983 to 1988, and the extremely hot winter of 1986. These factors appear to have brought about extremely low recruitment (Baisre 1993). Catches fell even lower in subsequent years (to about 2,000 t per year from 1995 to 1998).

Mollusks constituted 6% of the fishery landings from 1980 through 1997. Within this group, the predominant species (about 50%) was the mangrove oyster (*Crassostrea rhizophorae*). Despite large fluctuations in landings in some regions because of overexploitation, landings averaged near 2,500 t annually from 1981 to 1990 and have represented the same relative percentage of the national catch into the 1990s (Table 8.1). A significant part of this production has been accomplished through oyster farming.

Harvesting of the turkey wing clam (*Arca zebra*) began in 1975 in separate natural shoals between the ports of Guayabal and Santa Cruz del Sur in southeast Cuba. Between 1986 and 1990, harvest levels reached 1,800 to 1,900 t. Other

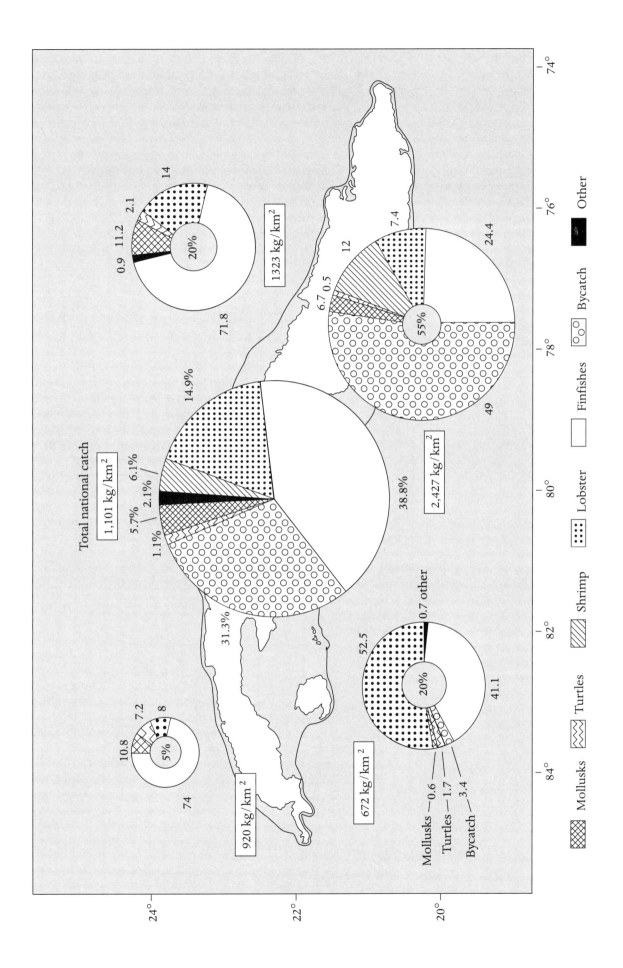

Fig. 8.2. Composition of average catch (%) from 1986–1990: country total and four statistical reporting zones. Production for each zone was calculated on the basis of catches and total area per zone (kg/km²).

Mollusks   Turtles   Shrimp   Finfishes   Bycatch   Other

Total national catch

1,101 kg/km²

5.7% 2.1% 6.1%
1.1%
14.9%
31.3%
38.8%

2,427 kg/km²

1323 kg/km²

0.9 11.2 2.1 14
71.8
20%

55%
12 6.7 0.5 7.4
24.4
49

920 kg/km²

10.8 7.2 8
74
5%

672 kg/km²

0.7 other
52.5
41.1
20%
Mollusks — 0.6
Turtles — 1.7
Bycatch — 3.4

clams, such as *Pecten laurentis* and *Laevicardium laevicardium,* are collected as bycatch from the shrimp fishery. In the 1984 catch, clams constituted approximately 5% of the biomass obtained in shrimp trawls (Fernández 1984).

The queen conch (*Strombus gigas*) has traditionally been used for consumption, but it has also been used for bait, and the shells are collected for handicrafts. The intense harvesting of this mollusk in some regions led to overexploitation, for which regulatory measures have been adopted. Since 1992, conch landings have been largely prohibited; the exceptions are small-quota catches permitted under special authorization. Nonetheless, queen conch is still used as bait. Sponges are collected in the southwest and northeast regions, but with variable fishing effort because of changing demand. Sponge landings have averaged 50 t per year (Ministerio de la Industria Pesquera, unpublished data).

Sea turtle landings increased after an expansion of fishing effort from 1968 to 1975. Landings ranged from 800 to 1,000 t per year until 1992, when the fishery was largely closed. Currently, only two small groups of fishermen, at Isla de la Juventud and Nuevitas, carry licenses to take turtles.

Finfish landings account for the largest fishery biomass within the EEZ. Although these fisheries take more than 140 fish species, only about 40 species contribute to significant landings. Of these, a small group with high economic value constitute more than 50% of the annual finfish landings. A detailed breakdown of the main fisheries by region is provided in Section 8.6.

8.3   Ecological Subsystems

Baisre (1985a, 1985b, 1993) has identified three ecological subsystems from a fisheries perspective in Cuban waters: the littoral estuarine, seagrass beds and reefs, and the oceanic water complex. The littoral estuarine subsystem covers approximately 8,500 km² and, during the 1980s, had a fisheries yield of approximately 1.4 t/km². Environmental conditions in this subsystem have deteriorated as a result of pollution, sedimentation, the irregular influx of freshwater, and nutrient loading. The largest estuarine systems are concentrated in the southeast region and, to a lesser extent, in north-central Cuba, but estuarine conditions are also found in Ensenada de la Broa and in most coastal bays throughout Cuba. Shrimp represent approximately 50% of the catch in this subsystem.

Baisre's (1985a) second ecological subsystem includes seagrass beds and coral reefs. This complex subsystem covers approximately 45,000 km², has a fisheries yield of 0.4 t/km², and is characterized by greater environmental stability. We suggest that the ichthyofauna inhabiting the mangroves of the keys should be included in this category. Many mangrove species are generally concentrated in areas close to coral reefs (Section 2.3.2) and make diurnal migrations to feed on seagrass beds and adjacent sand plains (Chapter 5). Collectively, this subsystem contributes more than 60% of the national catch and has a higher economic value than

estuarine areas because it includes lobster and some of the more valuable demersal fishes. The fishery yield per area is higher than indicated by Baisre (1985a) because the actual fishing grounds do not include all of the area encompassed by these habitats. For instance, the fishing regions for finfishes in the Golfo de Batabanó cover less than 50% of its area. Large areas lack sufficient shelter (rocks, corals, etc.) to provide adequate habitat for many coastal fishes.

Baisre's third ecological subsystem is the oceanic water complex. This subsystem covers approximately 13,800 km² with a yield of 0.24 t/km², if only the fringe areas adjacent to the shelf (where most fisheries are conducted) are considered. This habitat is characterized by relatively high environmental stability because it covers a large area and the influence of shelf waters is typically low. Skipjack (*Katsuwonus pelamis*) and blackfin (*Thunnus atlanticus*) tuna account for more than 75% of the catch from oceanic waters.

8.4   Fishing Gear

Until the late 1950s, fisheries efforts were focused on large, valuable fish species in the deeper waters of the shelf using selective gear, such as hook and line. By 1988, the shelf fleet consisted of approximately 1,850 boats focused on short-distance fishing operations in shallow and oceanic waters. Approximately 840 of these boats were dedicated to catching finfishes; by 1998 that number had dropped to 400 boats. Most of these 15- to 20-m long cement, wood, or steel motorboats were equipped with only one type of gear for shallow-water fisheries. Currently, the fishing fleets are adding fiberglass-hulled boats equipped with two or more types of gear. The most common gear used for shelf finfishes are traps, seines, and set nets with a wide variety of shapes and uses depending on shelf physiography, local tradition, and the species targeted. In oceanic waters (primarily for sharks), the most common gear are pelagic longlines or bottom longlines. For small tunas (skipjack and blackfin tuna) rod fishing with lures or live bait (mostly dwarf herring) has been the primary method for the last 60 years. The great changes occurring in the last 35 years in types, features, and use of fishing gear, as well as in boats, make it difficult to assemble standardized estimates of fishing effort and historic variations.

Most fishing gear is built in an artisanal manner. Some gear, such as set nets, are highly productive because of the expertise of the fishermen, the relatively high fishing power, and favorable markets. Some gear is harmful to habitats because no regulations control their use. To increase the production of some fishery resources, more advanced technologies will be required, particularly for deep-water species and small tuna fisheries.

8.4.1   Nets
Nets are used mainly in shallow-water shelf areas and coastal lagoons. The most common nets are seines or trawls, set

nets, and gill nets. The common trawl is usually 800 to 1,000 m long, and 8 to 10 m high. These trawls are used in broad flat areas (seagrass meadows and sandy areas) where they are pulled by two 15- to 20-m boats for 2 to 3 hours. Recently, some fishermen have increased the length of the seine to 2,000 m. The net can be retrieved from the main boat after a netted "wall" is placed between the two seine wings to prevent fishes from escaping. This gear is used mainly in the Sabana-Camagüey area.

In the south, particularly in the Golfo de Batabanó, a *chinchorro de boliche* trawl is more common (Fig. 8.3). It is smaller (200–250 m long) and can be used to fish small coral outcrops and, especially, artificial habitats often made of tires or mangrove branches tied together. Two small boats (7–12 m)

tow the *chinchorro* over the area while divers keep the net from becoming entangled. The two wings are pulled together through two long poles tied with a rope at the bottom and operated by a fisherman standing in a small skiff. The wings close until the cod end reaches the poles. Once the cod end is closed, the net is easily lifted. When the *chinchorro* is used to fish on artificial habitats, the tires or other artificial structures beneath the net are removed, and the mangrove branch piles can be lifted to the surface so that the wings can be joined (Claro and García-Arteaga 1999). This technique is usually used for snappers, jacks, and grunts, although other species can also be caught. A similar gear, the *chinchorro bolapié*, is used primarily for clupeids, but it is deployed differently (as a beach seine, *playero*).

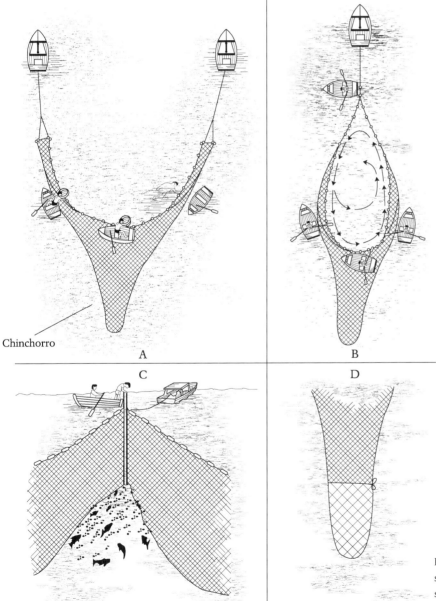

Chinchorro

A   B

C   D

Fig. 8.3. Four steps in the deployment of a small trawl (*chinchorro de boliche*). The net surrounds the habitat, which may or may not be removed after initial net closure. See text for details.

The rapid introduction of large trawler fisheries in the 1960s, along with new boat types, is a primary cause of the large increase in fisheries production in Cuban waters. However, heavy use of this gear has also contributed to the decline of Cuban fishery resources. Trawl catches are increased with a small mesh size (20–30 mm in the wings and 10–14 mm in the cod end; Ramos and Obregón 1983) that results in high bycatch levels. This problem was partially solved with the use of a device called the selective crown (Obregón and Pedroso 1984), a netted attachment added to the end of the cod end that can be opened after the fishing operation is completed to allow smaller fishes to escape.

Set nets or pens (*tranques* or *corrales*) are usually deployed perpendicular to the coast to intercept migrating spawning aggregations. On the shelf, fishermen place these nets in channels between the keys that are the usual routes used by fish to reach their spawning areas near the shelf edge (Fig. 8.4). In coastal lagoons, set nets are deployed in channels, entirely spanning the passage between inshore and offshore waters. Most of the spawners are therefore caught before they reproduce, and those that escape or do not enter the traps often undergo oocyte resorption and fail to spawn

(Section 4.3.3). Set nets are used mainly for snappers, particularly lane snapper during spawning runs. Set nets are also used for mullet, mojarras, and drums in shallow-water areas (1–3 m). These nets are made with about 20-mm mesh hexagonal wire or synthetic net. In some regions, a large trap (*jaulón*) is added to the set net system (Fig. 8.4). The indiscriminate use of set nets during spawning migrations has resulted in severe overfishing of lane snapper in the Golfo de Batabanó (Section 8.6.2) and mullets in the coastal lagoons of the south-central coast.

Gill nets have a more limited use than seines and set nets. They are used mainly for catching rays, sharks, jacks, mullets, bonefish, and mackerels. Nevertheless, the misuse of bottom or submarine gill nets is a serious concern because they remain negligently submerged longer than necessary, or are abandoned, leading to the useless death of many fishes and sometimes turtles and dolphins.

8.4.2   Fish Traps
Fish traps play a large role in reef fisheries of the entire Caribbean because most other fishing gear is hard to operate

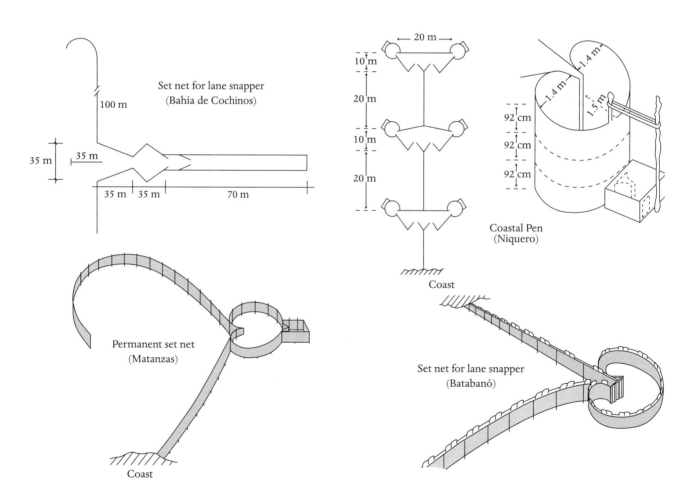

Fig. 8.4. Important set net types used in four different areas of the Cuban shelf (area of use in parentheses).

and inefficient on reefs. In waters 4- to 10-m deep, shallow-water traps (*nasas de placer*) are primarily used. They are often deployed on seagrass meadows next to patch and reef crests. Although not baited, these traps are effective in catching fishes of many sizes, including many species that are not directly targeted. Fishermen shade traps by covering them with branches of *yanilla* (*Suriana maritima;* a plant abundant in the keys). The traps are often deployed in strings, 50–60 units tied together with a separation of 30–40 m between each trap. They are typically lifted every 24–48 hours or more. Habitat damage can occur when traps are used in such arrays. Similar traps, without shade but with bait, are used on the slope reefs (15–30 m deep).

Trap sizes and shapes differ according to local tradition, but the most common traps are rectangular or cylindrical (Fig. 8.5). Traps are usually built with a frame of steel bars covered with wire mesh. These traps have been used traditionally for fishing small shallow-water species, mainly lane and gray snappers and grunts. Deep-water traps (*nasa del alto*), larger than those for shallow-water areas, are used for fishing deep-water snappers and groupers on 120- to 300-m deep rocky bottoms (Fig. 8.5). They are usually built with double strings of cane and are baited with lobster heads, queen conch, and noncommercial species (moray eel, sardines, and smaller grunt species).

In some instances, fish traps may be less damaging than seines and trawls. Fishermen can throw back live undersized individuals. In addition, Cuban fisheries regulations have established a minimum mesh size (40 mm) for trap walls or entrances, which, according to some information, allows the escape of 90% of snappers and grunts smaller than 18 cm fork length (FL) (Artamendi 1970). However, in recent years, the appropriate size of wire mesh is not always available and smaller mesh is frequently used. The extremely high catch

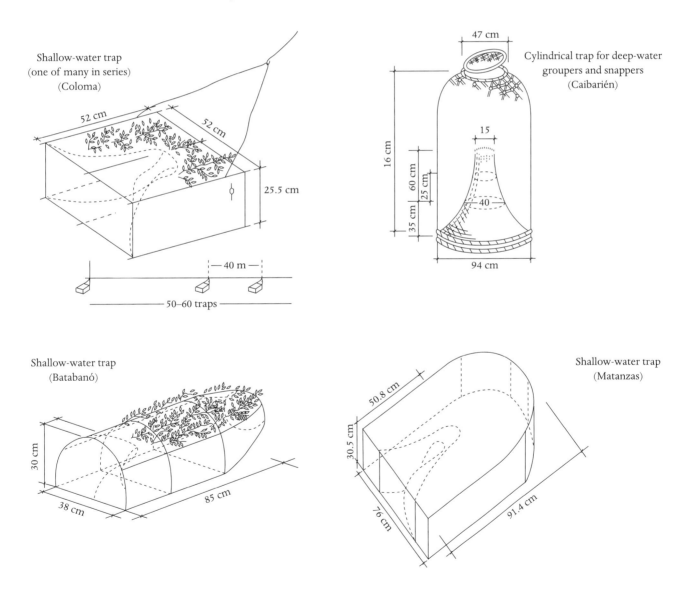

Fig. 8.5. Important trap types used in four different areas of the Cuban shelf (area of use in parentheses).

effectiveness of traps is problematic in terms of sustainability and many fisheries throughout the Caribbean show growth overfishing, in part due to heavy use of traps. Ghost fishing, when traps are lost but continue to catch fish over long periods of time, can also damage fish stocks. This problem can be avoided by making biodegradable escape doors, but these are not currently used in Cuba.

8.4.3 Longlines

Longlines are typically used in shelf slope or oceanic fisheries. Two main types are distinguished: the bottom longline for demersal fishes and sharks and the drifting longline for pelagic species such as billfish, swordfish, tuna, and sharks (Fig. 8.6). Bottom longlines can be used in shallow water for fishing coastal sharks; deep-water longlines are used for bathyal sharks, groupers, and others. The two types of longlines differ primarily in the length of the vertical lines and the distance between the hooks. Drifting longlines are generally deployed across the current axis for billfishes, swordfish, sailfish, and oceanic sharks. This fishing gear has a low productivity.

8.5 Fisheries Infrastructure

Commercial fisheries are conducted from 43 main fishing ports that belong to 15 fishing associations (Fig. 8.7) within Cuba's 14 provinces and the Isla de la Juventud, a special municipality. These associations are state-owned and under the general control of the Cuban Ministry of Fishing Industries. Most associations have their own processing plants. The associations in Pinar del Río, Caibarién, Batabanó, and Santa Cruz del Sur are the most important and collectively account for about 50% of the total catch.

Except for the lobster and shrimp fisheries, the associations do not have demarcated operating areas. In some areas, more than four units fish simultaneously, which makes it difficult to assess fishery stocks, as well as to establish and enforce regulations for stock protection. The advantages of regionalized fishery activities and limited entry have been demonstrated in the lobster fishery (Baisre 2000b) where fishing areas are demarcated for each association. This fosters spatially discrete catch statistics and better compliance with regulations, and thus highly productive and relatively stable lobster stocks.

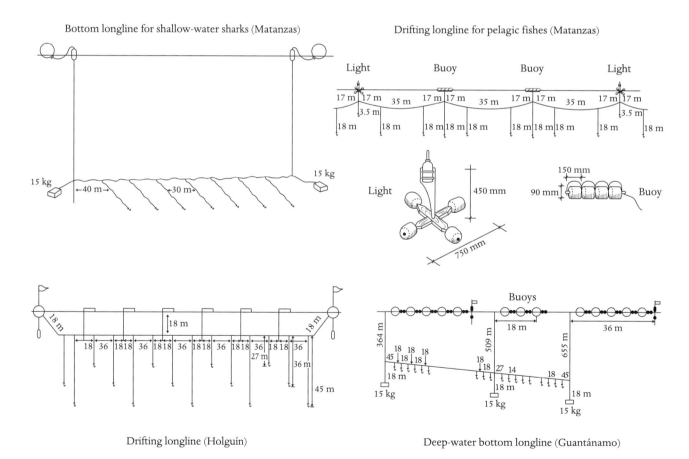

Fig. 8.6. Types of longlines used in billfish, marlin, swordfish, and shark fisheries (area of use in parentheses; numbers without units are in meters).

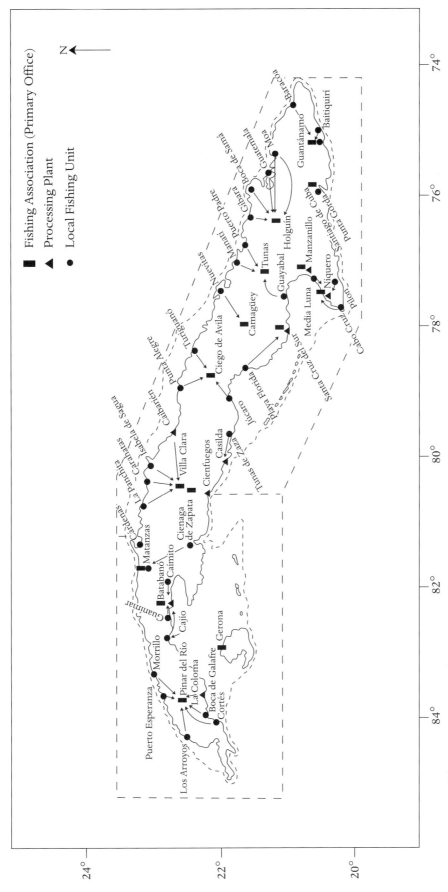

Fig. 8.7. Location of fishing associations and their components. Arrows show administrative connections between associations and local fishing units.

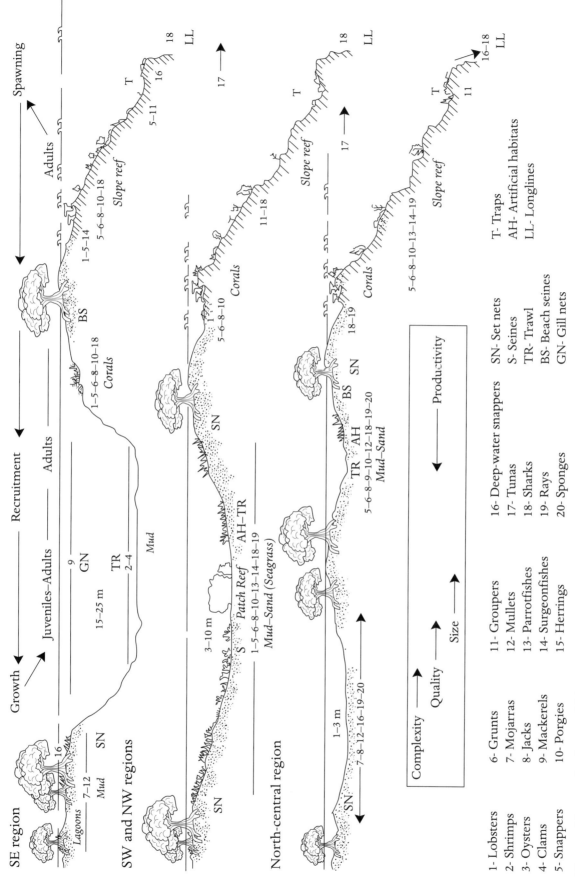

Fig. 8.8. Schematic distribution of fishing activities, habitats, and life stages of fishes in several regions of the Cuban shelf.

Without fisheries regionalization, many boats operate in areas far from their ports, which is economically inefficient. This issue has been addressed in a recent proposal for new fishing regions, and a complex nationwide fisheries regionalization program is under development. Enhanced fishery regionalization could also aid implementation of a coordinated, large-scale artificial reef program (Claro and García-Arteaga 1999).

Fig. 8.8 summarizes the zonation of fishing operations, primary fishing methods, and the dynamics of the main life history stages of principal commercial fishes. A marked gradient of complexity and fish size runs from shallow-water to deep slope areas. However, in the last 25 years, fishing has been focused on the shallower areas with a massive artisanal fishing effort.

8.6 Catches by Species and Region

From 1986 to 1990, the total catches of seafood products in the Cuban EEZ averaged 71,746 t per year; 28% were invertebrates and 39% were finfishes (Table 8.1). Since 1990, however, Cuban fishery activities have been seriously

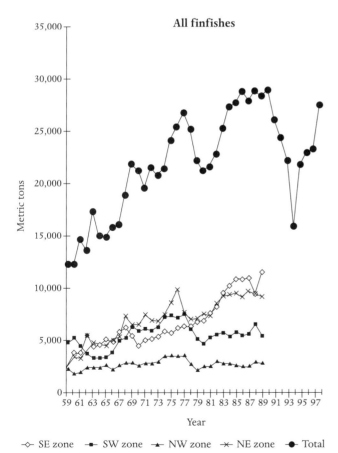

Fig. 8.9. Regional and total catches for all finfishes in Cuba, 1959–1998.

disrupted (except for lobsters) by the economic crisis and continued difficulties in obtaining parts and supplies, with many negative consequences, including a lack of fuel and deterioration of the fleet. Catch levels dropped to 41,343 t in 1993, followed by a period of partial recovery two years later, reaching 62,114 t in 1998 (Fig. 8.1). Catch statistics for the 1990s reflect neither the full potential of the fishing effort nor the available biological resources. Consequently, we have primarily used pre-1990 information in our analysis.

Cuban fisheries statistics include all edible fishes under "finfish." Seven quality and price categories include about 90 species; a few species in each category have commercial significance. Fig. 8.9 shows the history of total finfish landings and the contribution of each fishing zone. Finfish catches increased from 1959 to 1976, after which they decreased because of overfishing of the main finfish resources. However, finfish production increased again to 29,720 t (38% of total catch) in 1987 after adoption of new management measures. Overall fishing effort and catch has declined since 1990 as a result of the economic crisis.

The largest finfish catches have often been in the northeast zone (30–35% of the nation's total), although in the southeast zone, catches increased to 35% of the total by the mid-1980s. If we include the fishes caught as bycatch in shrimp fisheries (not shown in Fig. 8.9), the southeast zone contributes more than 50% of the overall finfish catch (Fig. 8.2). Yields per unit area within the two northern zones are almost twice those of the southern zones, although a higher biomass of crustaceans (mainly lobster and shrimps) is obtained on the south coast. Table 8.2 shows mean catches per species or group of species and the approximate proportion that each contributes to total marine finfish production.

8.6.1 Snappers

Snappers (Lutjanidae) are the principal finfish fishery group in Cuba. This group is composed mainly of five shallow-water species: lane (*Lutjanus synagris*), mutton (*L. analis*), gray (*L. griseus*), cubera (*L. cyanopterus*), and yellowtail (*Ocyurus chrysurus*) snappers, as well as several deep-water species, among which the silk snapper (*L. vivanus*) is the most important. Because of their high commercial value, snappers are recorded separately in fisheries statistics. Snappers constituted about 30% of the total finfish catch until 1975, when this percentage began to decrease. From 1976 through 1980, the mean annual catch was 5,196 t (22% of the total finfish); from 1986 to 1990, it reached about 5,371 t (18.8%).

Lane snapper is one of the most numerous commercial fishes of the Cuban shelf. Until 1960, lane snapper catches fluctuated from 515 to 1,030 t per year. At that time, fishing was mainly done with hook and line or trap, with some seine boats in the southwest zone. A subsequent increase in fishing effort, however, saw shallow-water traps, set nets, and seines gradually replace the hook and line. For many years after

Table 8.2.   Mean annual catches (in metric tons) and proportion of total catch for major finfish groups during three five-year periods

Group or Species	1981–1985		1986–1990		1991–1995	
	t	%	t	%	t	%
Lane snapper	1,793.5	7.0	1,589.0	5.6	1,508.8	6.8
Mutton snapper	928.3	3.6	1,145.5	4.0	661.3	3.0
Cubera and gray snapper	1,077.3	4.2	843.6	3.0	451.5	2.0
Yellowtail snapper	864.2	3.4	848.5	3.0	594.3	2.7
Grunts	1,849.5	7.2	1,988.0	7.0	1,731.4	7.8
Nassau grouper	362.8	1.4	256.2	0.9	95.5	0.4
Billfishes, swordfish	310.3	1.2	208.3	0.7	106.0	0.5
Skipjack, blackfin tuna	1,885.1	7.4	1,890.0	6.6	1,366.8	6.2
Mackerels	411.2	1.6	614.0	2.2	433.9	2.0
Jacks	389.7	1.5	494.6	1.7	358.7	1.6
Sharks	2,767.8	10.8	2187.0	7.7	1,247.2	5.7
Rays	1,801.9	7.0	2,942.1	10.3	1,819.1	8.2
Sparids	383.1	1.5	447.4	1.6	342.7	1.6
Mullets	765.3	3.0	778.9	2.7	246.7	1.1
Atlantic thread herring	2,071.1	8.1	2,230.6	7.8	1,710.8	7.8
Other herrings	436.6	1.7	509.3	1.8	834.1	3.8
Mojarras	1,166.5	4.6	1,248.5	4.4	1,957.4	8.9
Other fishes	6,336.4	24.8	8,336.3	29.2	6,597.8	29.9
Total finfishes	25,600.5	100	28,557.8	100	22,063.8	100

that, lane snapper ranked first in finfish catches; catches reached 4,366 t in 1969 and remained above 3,090 t until 1977 (Fig. 8.10). Before 1977, lane snapper catches in the Golfo de Batabanó accounted for more than 50–60% of the nation's total catch of this species.

From 1973 to 1977, the indiscriminate use of set nets across migratory spawning routes was, in our opinion, the main cause of the collapse of the lane snapper fishery in the Golfo de Batabanó. Elimination of the minimum-size regulation also contributed to the collapse. In the southeast zone, the total catch from 1976 to 1980 was approximately 50% of the previous five-year period. A decline also occurred in the other fishing zones.

Starting in 1978, fisheries regulations were established for lane snapper in the Golfo de Batabanó, including the prohibition of fishing during the reproductive season. From 1978 to 1990, a low catch quota was also established in the southwest. Recovery signs have been evident only since 1995. Fisheries scientists expected a faster recovery because of the lane snapper's relatively high reproductive rate and short life span. Claro (1991) has suggested that the reduction in the number of lane snapper from overfishing may have allowed grunts such as the tomtate (*Haemulon aurolineatum*), as well as the gray snapper, to replace the ecological position of the lane snapper.

Before the capture of spawning migrations was prohibited, lane snapper catch data showed a marked seasonality because the species was fished primarily during the spawning season. In the eastern part of the Golfo de

Batabanó, close to the Golfo de Cazones, 60 to 75% of the total yearly catch of the species was obtained during the breeding season (May–August), most of it during the 10–12 days of peak spawning. During these days, fishermen would catch 40–68% of their annual total. Such seasonal patterns were strengthened from 1970 to 1975 when set nets were used intensively during the spawning migrations. Maximum catches in the Golfo de Batabanó and the northwest zone were obtained during spawning peaks in the fifth lunar cycle, typically occurring in May. In the two eastern zones, however, peak spawning and catches appear to occur one month later (Rodriguez Pino 1962; Claro 1982).

Obregón et al. (1990) reported annual figures for maximum sustainable yield of lane snapper in the northeast zone as 220 t using the production models of Fox (1975), and as 250 t using the production models of Csirke and Caddy (1983). Using the model of Beverton and Holt (1957), Obregón et al. (1990) found a maximum yield-per-recruit (39 g) at an age of first capture of 1.84 years, approximating a fork length of 16 cm. Since 1987, without increased fishing effort, catches in this region have exceeded predictions, but they fell abruptly in 1997, possibly as result of habitat fragmentation in Bahía de Buenavista and Bahía San Juan de los Remedios, the most important areas for recruitment and fishing, respectively, in the northeast fishing zone (Claro et al. 2000).

The mutton snapper is also one of the most valuable demersal fishes of Cuba and is caught mostly by seine. However, during the spawning season, hook and line and set

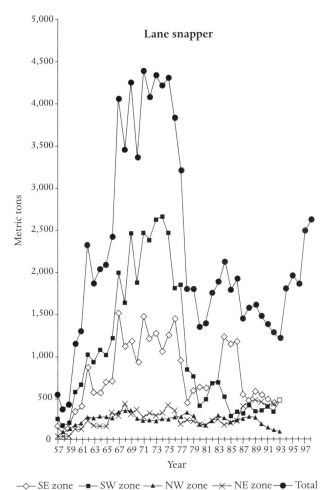

Fig. 8.10. Regional and total catches of lane snapper (*Lutjanus synagris*) in Cuba, 1957–1998.

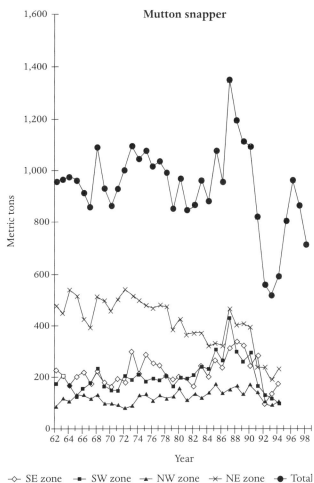

Fig. 8.11. Regional and total catches of mutton snapper (*Lutjanus analis*) in Cuba, 1962–1998.

nets are also used. Mutton snapper was one of the top fishery stocks until 1962, when its proportional catch diminished with the increase of other finfish fisheries. The mutton snapper now accounts for 4% of the nation's total finfish catches and 15% within the snapper family. Until 1986, catches remained relatively stable in the four zones (Fig. 8.11) with levels close to the estimated maximum yield, which is about 1,000 t per year (Ministerio de la Industria Pesquera 1980) for the whole country. Catches increased substantially from 1986 to 1990, particularly in the southwest zone (because of set nets used during spawning runs). During the period of economic crisis in the early 1990s, reported mutton snapper landings decreased more than most other species, partly because of increased illegal fishing.

The largest mutton snapper catches occur in the northeast zone; they typically represent more than 35% of the nation's total. Maximum catches are obtained in May–June and are focused on the spawners. In the southern zones, the maximum catch occurs primarily in May (Claro 1981c). Unfortunately, immature individuals (smaller than

40–45 cm FL) constitute more than 50% of the catch in the southwest zone, more than 80% in the northwest zone (Claro 1981c), and 90% in the northeast zone (Pozo 1979). Adults constitute the majority of catches only during the spawning season (May–June), but these individuals are often caught before they are able to spawn.

Because the size of sexual maturation for mutton snapper is approximately 50 cm FL (approximately 5 years old), establishing a minimum catch size that allows individuals to spawn at least once is difficult. In addition, fishing gear for catching mutton snapper is nonselective and is used in multispecies fisheries in which smaller fishes dominate. Mutton snapper is also one of the most prized recreational fishes. Although no statistics are available for this fishery in Cuba, the mutton snapper recreational fishery could have considerable potential, particularly along the northern coast.

Until 1981, cubera and gray snappers were grouped in fisheries statistics as "cubera." Most fishermen do not distinguish one species from the other because of similar pigmentation and shape. Gray snapper are caught mainly with traps and seines in shallow areas of the shelf (< 6 m depth). Cubera snapper are occasionally fished with seine and, during spawning runs, with set nets, traps, and hook and line, although in lower numbers than the gray snapper. Gray snapper constituted about 95% by weight and more than 98% of the individuals in this group in the southwest and northeast zones (Claro 1983b). This proportion is apparently similar in most Cuban fishing zones. However, in the Tunas de Zaza coastal lagoons of the southeast zone, 84% gray snapper and 16% cubera snapper were recorded (Báez and Alvarez-Lajonchere 1980a). In both the Golfo de Batabanó (Claro 1983b) and the northwest zone (Radakov et al. 1975), gray snapper individuals of 20–30 cm FL (73%) and 140–400 g weight (1–3 years of age) prevail in catches. The modal class ranges throughout the year from 26–30 cm FL, although during the spawning season (July–August) the dominant size increases to 38–40 cm FL (Claro 1983b). One-year-old individuals constitute about 50% of the catches; most have not attained sexual maturity at that age.

Landings of gray and cubera snappers increased somewhat from 1961 to 1970 (Fig. 8.12), then increased dramatically from 1973–1981 as a result of greater fishing effort and efficiency. From the late 1970s to the mid-1980s, the annual catch exceeded 1,000 t. After the decline of lane snapper catches in the Golfo de Batabanó, fishing effort shifted to gray snapper, which became the principal finfish resource of the area. High catches of gray snapper also decreased by the late 1980s as a result of decreased fishing effort due to poor economic conditions and fewer resources. The decline shown in Fig. 8.12 also reflects a significant change in statistical reporting; since 1982 cubera and gray snappers smaller than one pound are included in the "Other Fishes" statistical category. The largest catches of cubera and gray snapper typically occur during August in the southwest, but in the northwest, where there is a different spawning peak, they occur during July (Claro 1983b). The capture of gray snappers in spawning aggregations with set nets increased in the 1980s and early 1990s. Since 1995 a one-month closed season has been established during the

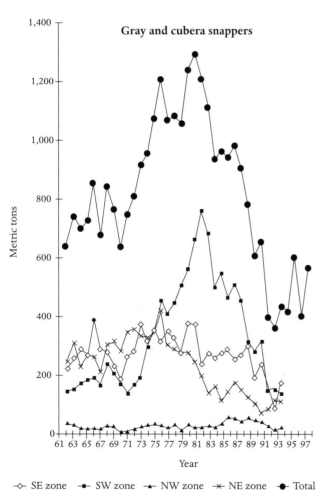

Fig. 8.12. Regional and total catches of gray (*Lutjanus griseus*) and cubera (*L. cyanopterus*) snappers in Cuba, 1962–1998.

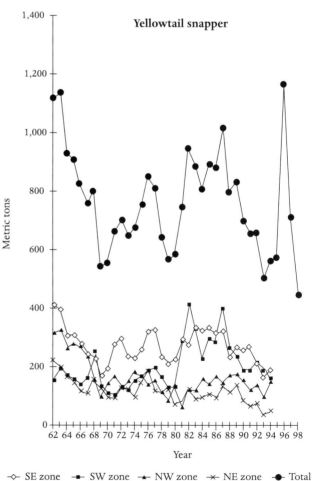

Fig. 8.13. Regional and total catches of yellowtail snapper (*Ocyurus chrysurus*) in Cuba, 1962–1998.

different spawning peaks: July in the northwest and August in the southwest.

Yellowtail snapper catches decreased gradually from 1959 to 1970 (Fig. 8.13) with a reduction in traditional nightfishing near the shelf edge. This fishery was conducted with hook and line, assisted by lights and chumming. The individuals caught with this method were larger than those taken with seine and traps in the shallow-water areas, now the most common fishing method for yellowtail snapper in Cuba. From 1971 to 1973, 75% of the yellowtail snappers caught with seine and traps were small individuals, 1 to 3 years old. Because this species first reproduces at a size larger than 25 cm FL, a large portion of the population never reproduced (Claro 1983d). Subsequent catches have fluctuated, and the notable annual fluctuations from 600 to 1,000 t correspond to changes in fishing effort. It has been suggested that catches could be maintained at 1,000 t (Ministerio de la Industria Pesquera 1980) by increasing fishing effort in deeper waters along the shelf edge.

Like the other snapper species, the yellowtail snapper is caught in larger proportions during the spawning season. Peak catches occur in the southeast zone in May–June; in the northeast zone, the catch peak occurs mostly in June. In the northwest and southwest zones, the largest catches are obtained in April. In all fishing zones, the largest catches are during warmer months and the lowest in December–February. As with lane snapper, the spawning peak in both western zones occurs one month earlier than in the eastern.

Several species of deep-water snappers inhabit the slope at depths of 70–400 m. The silk snapper is the dominant species of this group (80% of the catch) (García-Rodriguez 1978). Other species, such as blackfin snapper (*Lutjanus buccanella*), vermilion snapper (*Rhomboplites aurorubens*), and cardinal snapper (*Pristipomoides macrophthalmus*), are also fished. After exploratory investigations in the 1970s (García-Rodriguez et al. 1976; García-Rodriguez 1978; García-Rodriguez and Miranda 1979a, 1979b), proposals were made to promote deep-water reef fisheries (which had practically disappeared after 1960). In 1987, annual deep-water snapper catches reached 217 t. Since 1995, the catch has been less than 50 t yearly because of low fishing effort. Other species are caught incidentally in this type of fishery, such as the misty (*Epinephelus mystacinus*) and yellowedge (*E. flavolimbatus*) groupers. Total landings of these two species reached 49 t in 1983, but currently remain at less than 10 t per year. Misty and yellowedge groupers are fished mainly in the northeast and southwest zones. In general, deep-water snappers and groupers are underfished in Cuba.

8.6.2 Groupers

The Nassau grouper (*Epinephelus striatus*) is a highly valued species in Cuba. Catches have gradually declined from an annual mean of 1,476 t (1959–1963) to less than 100 t (1992–1998) (Fig. 8.14). The largest catches are in the southeast (more than 40% of the total) and northeast zones.

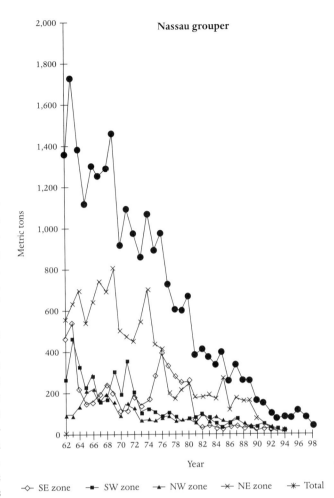

Fig. 8.14. Regional and total catches of Nassau grouper (*Epinephelus striatus*) in Cuba, 1962–1998.

Catch statistics from the northeast until 1969 included a considerable number of Nassau grouper from the Bahamas (not quantified by country). Termination of Cuban fisheries in that country in the 1970s led to a substantial decrease of Nassau grouper landings in Cuban fisheries. Landings from the Cienfuegos Association (southeast zone) showed a constant increase in fishing effort throughout the 1980s, accompanied by a gradual reduction of yields (Fernández 1984).

Compared with Nassau grouper, other grouper species are less significant in Cuba and most are grouped in the statistics as "Other Fishes." Grouper overfishing in Cuba, as well as in other regions of the Caribbean, is a dramatic example of how unsustainable fishing practices can jeopardize important economic resources (Appeldoorn and Meyers 1993; Sadovy and Eklund 1999). These species are particularly vulnerable because they often gather in discrete spawning aggregations where they can be caught with nonselective methods. They are also typically hermaphroditic with a slow growth rate. Overfishing therefore results in differential mortalities between sexes that can complicate fishery recovery.

8.6.3 Grunts

Grunts represent about 8% of the shelf finfish catches, although their commercial value is lower than for snappers and groupers. This group includes white grunt (*Haemulon plumieri*), bluestriped grunt (*H. sciurus*), margate (*H. album*), sailor's choice (*H. parra*), and other less important species. Most grunt species except the margate are relatively small. They are abundantly caught in the seagrass–reef complex and on the shelf slope. Because of their small size, many are released or escape from seines equipped with selective crowns, large-mesh traps, and pens. In the eastern part of the Golfo de Batabanó, margate constitute about 15–20% of total grunt catches from seagrass with patch reef areas. On artificial refuges in the same region, this number can reach 50%. However, on artificial habitats deployed on the inner part of the gulf, white grunts prevail in catches. The

bluestriped grunt is less abundant in catches, perhaps because it may be more territorial, which makes it less vulnerable to fishing gear. Ramos and Pozo (1984) estimated that white grunt catches of 500 t per year were possible.

From 1978 to 1980, grunts constituted only 10% of the commercial seine catches in the Golfo de Batabanó (southwest zone), but the smaller individuals constituted 73% of the noncommercial part of the seine catch (Bustamante et al. 1982). In surveys of seine catches, individuals of noncommercial size (mainly white grunt and tomtate) were twice as abundant as all individuals of commercial size and four times the total number of snappers. A similar pattern was observed in the northeast (Claro and García-Arteaga, unpublished data).

According to fisheries statistics, grunt catches increased gradually until 1976 (Fig. 8.15). After 1976, grunt fisheries declined in the northeast zone (the most important zone for this fishery, 35–45% of the total); catches in the southwest and the southeast zones remained relatively high. Catch fluctuations seem to be the result of changing fishing effort. In the Golfo de Batabanó, margate catches from 1978 to 1980 consisted principally of 18–24 cm FL juveniles (50–60% of total number). This species matures at a larger size (45–50 cm FL on average). Large individuals are caught only in areas near the shelf slope.

Surveys conducted on commercial landings showed that 75–85% of the white and bluestriped grunt catches consisted of 12- to 18-cm individuals, most of which were adults (see Chapter 4). In addition, both species show prolonged spawning of multiple batches. Grunts are caught year-round, although catches are higher from September to early summer, the main reproductive period for many of these species in Cuba.

8.6.4 Mojarras

Mojarra (Gerreidae) species in Cuba include striped, Brazilian, and yellowfin mojarras (*Eugerres plumieri, E. brasilianus,* and *Gerres cinereus*). The main mojarra fisheries are in the northeast (65–70% of the total) and southeast (30%) zones. Annual catches have increased steadily since 1976, surpassing 2,000 t from 1992 to 1995 (Fig. 8.16). Mojarra catches have increased as catches of larger species have declined (e.g., lane snapper, gray snapper, cubera snapper, Nassau grouper).

In the 1970s, at the Tunas de Zaza estuarine system (southeast zone), Brazilian mojarras represented more than 85% of the finfish catch and yellowfin mojarras only 6% (Báez and Alvarez-Lajonchere 1980b), but striped mojarras were dominant in catches in the northeast (Claro and García-Arteaga, unpublished data). Like mullets, mojarras are typical inhabitants of coastal lagoons. Yellowfin mojarras are also found in the mangrove areas of the keys and islets that fringe the Cuban shelf and in reef areas where euhaline conditions prevail.

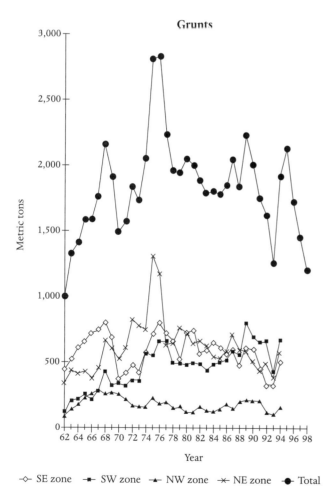

Fig. 8.15. Regional and total catches of grunts (*Haemulon* spp.) in Cuba, 1962–1998.

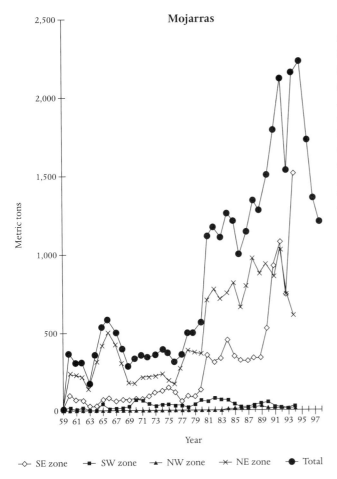

Fig. 8.16. Regional and total catches of mojarras (*Eugerres* spp., *Gerres cinereus*) in Cuba, 1959–1998.

8.6.5 Sardines and Herring

Sardines and herrings, in the family Clupeidae, are the only entirely planktivorous fish species with commercial significance in Cuba. Clupeids constitute approximately 10% of the national finfish catch, the majority being Atlantic thread herring (*Opisthonema oglinum*) (Table 8.2). This fishery also includes, the redear sardine (*Harengula humeralis*), scaled sardine (*H. jaguana*), false pilchard (*H. clupeola*), and Spanish sardine (*Sardinella aurita*). Clupeid schools often comprise two or more species. Almost all of the catch is used as bait or animal feed. Catches increased rapidly from less than 206 t in 1964 to more than 1,854 t in 1977. Clupeid catches declined abruptly after 1980 in all regions, mainly because of market changes (Baisre 1985b). The main fisheries traditionally operated in the southwest zone (more than 60% of total), followed by the northeast zone. Clupeids are fished exclusively in shallow-water areas with seines. Migrations and seasonal patterns make it difficult to locate schools during some parts of the year, but their gregarious behavior makes it easy to catch large volumes, which can compensate for time spent in locating the schools.

Catches of Atlantic thread herring have undergone market fluctuations similar to other clupeids. This species has two annual catch peaks: early winter and spring. It is used mostly as bait and is also canned for human consumption, although production of the latter has dropped since 1990. Annual catches reached more than 2,000 t after 1977 and continued to increase through the 1980s. Catches dropped in the 1990s as a result of habitat alteration and reduced fishing effort in the northeast zone. The southeast zone, where estuarine conditions prevail and the species is particularly abundant, now accounts for approximately 80% of the catch. The dwarf herring (*Jenkinsia lamprotaenia*) is of importance as bait in the skipjack and blackfin tuna fisheries. Information on the volume of these catches is not available, but García-Coll (1984) suggested that population decreases in the Golfo de Batabanó were a result of high fishing effort. Claro and García-Arteaga (1993) considered the species to be more abundant in the Archipiélago Sabana-Camagüey, where it is the preferred bait for important tuna fisheries.

8.6.6 Mullets

Various mullet species (Mugilidae) have fisheries importance in Cuba: liza (*Mugil liza*), white (*M. curema*), hospe (*M. hospes*), and fantail (*M. trichodon*) mullet. Other less significant mullets also occur in some catches: rabúa (*M. longicauda*), Parassi mullet (*M. incilis*), mountain mullet (*Agonostomus monticola*), and bobo mullet (*Joturus pichardi*) (Alvarez-Lajonchere 1978a; Ministerio de la Industria Pesquera 1980).

Fisheries statistics include all mullets in the same group with two size categories: *lisas* (large mullets) and *lisetas* (small and medium-sized mullets) (Alvarez-Lajonchere 1978a). Statistics show that until 1973 the whole mullet group accounted for 2–3% of the total finfish catch, but that has declined to about 1% since 1995. Catch surveys in Tunas de Zaza (southeast zone, one of the most important mullet fisheries grounds of Cuba) showed that large mullet represented 76%, whereas small mullet accounted for 20% of the total catch (Alvarez-Lajonchere 1978a). Of the large mullet, 86% were liza and 10% were white mullet. Of the small mullet, 55% were fantail and 39% were hospe mullet.

Mullet are fished almost exclusively in estuaries and coastal lagoons. The most important catches are obtained in the southeast and northeast zones. In the western zones, mullet capture is incidental, despite their occurrence in most estuaries and bays. Gill nets are used most commonly for mullet fishing in Cuba (Alvarez-Lajonchere 1978a), but in some locations fishermen prefer set nets, principally during spawning aggregations. The cast net has been used less frequently.

Until 1966, mullet catches in Cuba fluctuated from 500 to 900 t per year (Fig. 8.17). Increased fishing effort raised catches to 1,605 t in 1971, after which the fisheries started to decline. In 1990, the decrease intensified, and by 1995–1996 the catch was less than 100 t. Intense fishing effort, lack of

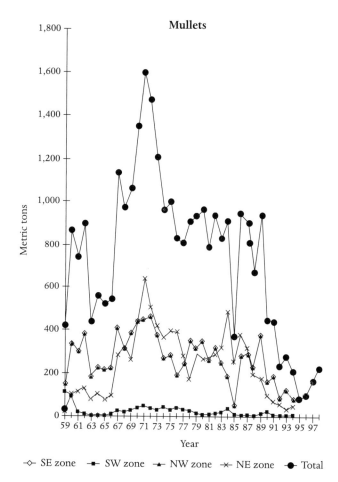

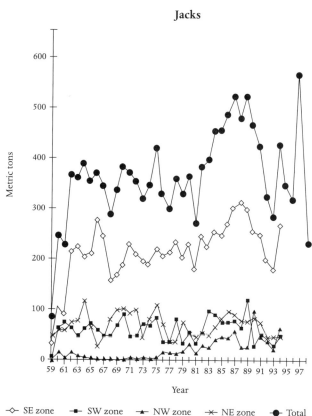

Fig. 8.17. Regional and total catches of mullets (*Mugil* spp.) in Cuba, 1959–1998.

Fig. 8.18. Regional and total catches of jacks (Carangidae) in Cuba, 1959–1998.

regulations until the 1980s, and river modifications and their downstream effects are among the factors affecting mullet abundance. The massive use of set nets during spawning runs could be the primary cause of the dramatic decline (Baisre 1981). Currently, habitat fragmentation is widespread and might represent an obstacle to stock recovery.

Mullet fisheries are markedly seasonal. The largest catches are from September through January, with a peak in November at the beginning of spawning migrations. Set nets block the passage of spawners from estuarine systems to the open sea, so mullets are caught before they reach the spawning grounds. In the coastal lagoons of the Tunas de Zaza, liza recruit to fisheries during the spring–summer period at a length of 16–18 cm (0+ age). Most individuals caught in the lagoons are not adults, so they have not been able to reproduce even once before harvest (Alvarez-Lajonchere 1978b).

8.6.7   Jacks

The jack (Carangidae) fishery includes bar (*Caranx ruber*), yellow (*C. bartholomaei*), and horse-eye (*C. latus*) jacks. Other carangid species, such as the crevalle jack (*C. hippos*), blue

runner (*C. crysos*), bigeye scad (*Selar crumenophthalmus*), and pompano (*Trachinotus* spp.) are included in the generic "Other Fishes" catch category.

Jacks are caught mainly with seine or set nets; to a lesser extent they are also taken in fish traps. Bar and yellow jacks are caught consistently with seine nets. These two species accounted for approximately 79% of the jacks caught by boats using seine nets in the Golfo de Batabanó in 1982 (Pérez-Pérez and Alvarez-Conesa 1983). The horse-eye jack is caught less often, although it forms large schools.

After a notable increase in jack catches from 1959 to 1962, catches have varied from approximately 300 to 560 t per year since (Fig. 8.18). Approximately 60% of the total is from the southeast zone. The catches in the northeast and southwest zones are similar (15–20% of the total), but vary from year to year. Catches in the northwest, despite an increase over the past decades, are low and have rarely exceeded 100 t.

The potential annual catch for this group has been estimated at 412 t (Ministerio de la Industria Pesquera 1980), but to achieve that, a better understanding of the ecology of these species is needed, particularly their migratory patterns. Installation of floating aggregation devices could make higher catches possible in certain areas. Reports of human poisoning (ciguatera) from some jack species have negative implications for the commercial value of this group.

Juveniles (16–24 cm FL) account for 65–70% of the bar jacks caught by drag-net in the Golfo de Batabanó. The adults are caught predominantly in shelf waters near the drop-off. From 1978 to 1980, the average size in the commercial catches surveyed was 21.8 cm FL. Because the average length of the bar jack at sexual maturity is 30 cm, it is evident that this fishery is based on individuals who were not yet able to reproduce.

The same fishery also catches yellow jacks primarily in the 20–36 cm size range (78%). The yellow jack takes longer than the bar jack to reach sexual maturity; the smallest sexually mature female individuals measured 32 cm FL and males measured 30 cm FL (Chapter 4). There were not enough sexually mature individuals in the surveys to establish average length at sexual maturity, but the data indicate that it might be near 40 cm FL. Munro (1983a) suggested a length of 45 cm for sexual maturity. We lack information on the sizes and lengths at sexual maturity for the other jacks in the Cuban catches, but Munro (1983a) gave values of 37 cm for horse-eye jack and 20 cm for blue runner in Jamaican waters.

8.6.8 Tunas

Skipjack and blackfin tuna are the current principal oceanic fishery resources of the Cuban archipelago. Of all fish catches, they represent the second-most important group by volume of catch. A high market value (15% of the total catch value of finfishes) also sets this group apart. Most fishing for these species traditionally occurred along a strip of oceanic water abutting the western end of the shelf, where until 1975, 95% of the catches had been obtained. Since 1976, this fishery has intensified off the Archipiélago Sabana-Camagüey, where approximately 30% of the catches have occurred in recent years. The tuna fishery primarily uses the ancient Japanese method of rod fishing, with mainly dwarf herring as live bait. The method is artisanal but quite productive; one vessel can rapidly land more than a ton of fish. Nonetheless, the benefits are somewhat limited by the time and effort spent searching for live bait and targeting schools of fishes.

Generally, both skipjack and blackfin tuna are found in the same schools. Blackfin tuna were especially prevalent in catches until the 1980s (Suárez-Caabro and Duarte-Bello 1961; Carles and Hirtenfield 1978; Valle and Carles 1983), accounting for 66% of the total catches (García-Coll 1984). Since 1983 the proportions have changed and the skipjack is now more prevalent in the catch (60–80%). Groups of mackerel (*Auxis thazard*) are also occasionally caught, primarily in the southeast. Yellowfin tuna (*Thunnus albacares*) also occur in these catches to a lesser extent. It has been suggested that some schools of skipjack and blackfin tuna remain in near-shelf waters for relatively long periods (Suárez-Caabro and Duarte-Bello 1961; Carles 1975); they are consequently sometimes referred to as "coastal residents."

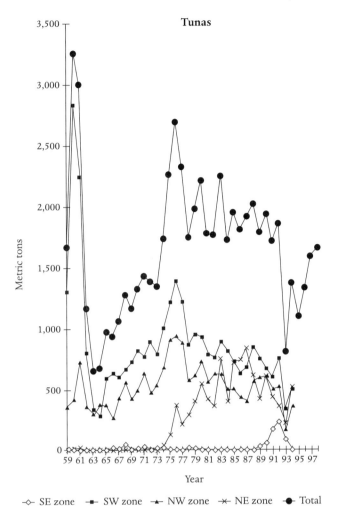

Fig. 8.19. Regional and total catches of skipjack (*Katsuwonus pelamis*) and blackfin (*Thunnus atlanticus*) tunas in Cuba, 1959–1998.

Because tuna have been among the most important fisheries since the 1950s, they hold a traditional value in Cuba. In the early 1960s, catches were higher than 3,200 t per year, but they declined in subsequent years as a result of notably decreased fishing effort (Fig. 8.19). After 1965, catch levels began to slowly rise again, reaching a level of 2,694 t in 1976, only to decline once more. García-Coll (1984) attributed this last decline to the increased effort required to find and attain live bait. From 1967 to 1977 the number of active fishing vessels doubled, which brought with it increased competition and increased search times for bait (García-Coll 1984). However, other factors might be involved (Baisre 1985b). For example, fluctuations in catch levels of skipjack tuna might be due to changes in accessibility to the schools in the areas where the fishing fleet operates and not to changes in abundance (Valle and Carles 1983). Catches of blackfin tuna, on the other hand, remained relatively stable and approximated the estimated catch potential (Valle 1983). Considering that both species are caught in the same fishery, it is probable that decreased accessibility to the blackfin tuna schools, rather than the availability of live bait, was responsible for the decrease in catch per unit effort (Carles 1972; Valle and Carles 1983).

From 1979 to 1982, the average size of skipjack tuna caught by the La Coloma fishing unit in the southwest zone ranged from 43 to 50 cm, and 50% of the individuals were between 40 and 54 cm (García-Coll 1984). In the same period, the average length of blackfin tuna was 43 to 49 cm. The median size at sexual maturity for the female and male skipjack tuna in the Atlantic Ocean has been reported to be 42 and 45 cm, respectively (Cayre and Farrugio 1986). Estimates of the fishery potential of these tuna species appear to overestimate their abundance (Baisre 1985b). Nonetheless, catch levels could be increased through improved fishing methods, expanded operating areas for the fleets, and better understanding of the environmental factors affecting these fish populations. Estimates of 1,350 t of blackfin tuna and 700 t of skipjack tuna for the western regions (Carles 1975) are comparable to estimates given by others (García-Coll 1984).

8.6.9 Mackerels

Three species of scombrids—cero mackerel (*Scomberomorus regalis*), Spanish mackerel (*S. maculatus*), and king mackerel (*S. cavalla*)—are generally mixed together in catch reporting. The main fishery is in the southeast zone (85% of the total); Spanish and king mackerel constitute 95% of this catch. These species are highly prized as food, but their yield is limited. The fishing potential has been estimated as more than 1,000 t per year (Ministerio de la Industria Pesquera 1980), which is more than double the average annual catch during most of the last 20 years (Fig. 8.20). Mackerels are usually fished from small boats by trolling with live bait, but this system is being replaced by gill nets, which are limited in regions where shrimp fishing boats work (Baisre 1981).

8.6.10 Billfishes

Four species of billfishes are commercially important in Cuba: white marlin (*Tetrapturus albidus*), blue marlin (*Makaira nigricans*), swordfish (*Xiphias gladius*), and sailfish (*Istiophorus platypterus*). Specimens of the longbill spearfish

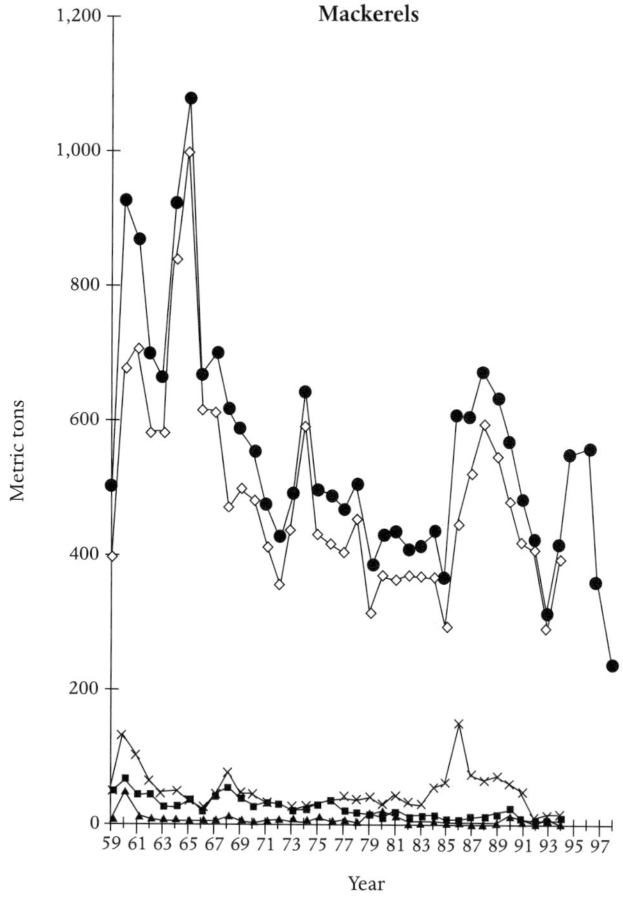

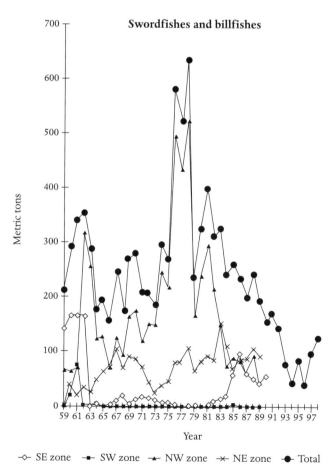

Fig. 8.20. Regional and total catches of mackerels (*Scomberomorus* spp.) in Cuba, 1959–1998.

Fig. 8.21. Regional and total catches of swordfishes and billfishes (*Xiphias gladius, Makaira nigricans, Tetrapturus albidus, T. pfluegeri,* and *Istiophorus platypterus*) in Cuba, 1959–1998.

(*Tetrapturus pfluegeri*) are occasionally caught, but their weight in catches is not significant (Guitart Manday 1975). Billfish fisheries use drifting longlines in the oceanic areas adjacent to the shelf. Catches are seasonal and related to the spawning migrations of the species. The Greater Antilles is an important reproductive area for these fishes. Since 1985, the relative catch among individual species has been highly variable, although dominated in general by sailfishes (32–76%) (unpublished data, Ministerio de la Industria Pesquera).

From 1963 to 1975, the national catch was less than 309 t; it subsequently increased to 638 t by 1978, only to decline to less than 100 t for much of the 1990s (Fig. 8.21). These fluctuations took place primarily in the northwest zone, which provides more than 70% of the national catch. From 1971 to 1973, billfishes accounted for 59% of the oceanic pelagic catches in this zone; of this, 26% were swordfish and 33% were blue marlin and sailfish (Guitart Manday 1975). Recreational fisheries also focus on these species; the potential income from international tourism could exceed the value of the species to commercial fisheries.

8.6.11 Sharks

The primary sharks in commercial catches include whitetip (*Carcharhinus longimanus*), night (*Carcharhinus signatus*), silky (*Carcharhinus falciformis*), shortfin mako (*Isurus oxyrinchus*), and dusky (*Carcharhinus obscurus*). These species represented more than 80% of the pelagic sharks caught by drifting longlines in the northwest zone (Guitart Manday 1975). The bigeye thresher shark (*Alopias superciliosus*), tiger shark (*Galeocerdo cuvier*), bignose shark (*Carcharhinus altimus*), hammerheads (*Sphyrna* spp.), and blue shark (*Prionace glauca*) are caught in smaller proportions.

Most of the catches were in areas near the border between the shelf and oceanic waters. Offshore, billfishes were the main fishery target because of their high commercial value, although, in terms of biomass, sharks constituted the major catch. Starting in the 1970s, the sixgill shark (*Hexanchus griseus*) has been caught with longlines in deep waters of the slope (160–300 m). Both pelagic and deep-bottom fisheries have nonetheless declined in recent years. Major fishing effort is also focused on sharks that inhabit coastal waters, such as the nurse shark (*Ginglymostoma cirratum*), which is frequently caught in shallow shelf-waters.

In general, shark catches increased until 1981 and have been variable since (Fig. 8.22). The yearly average was 98 t from 1959 to 1961, and 2,870 t from 1981 to 1985. Since 1985, a substantial decline in catches of some species has been observed, although other species have replaced them (Espinosa, in press a). For example, decreased catches in the northwest zone for the silky shark and the oceanic whitetip shark have been reported (Pol et al. 1990). Variations in the fishing effort and changes in the fishery make it

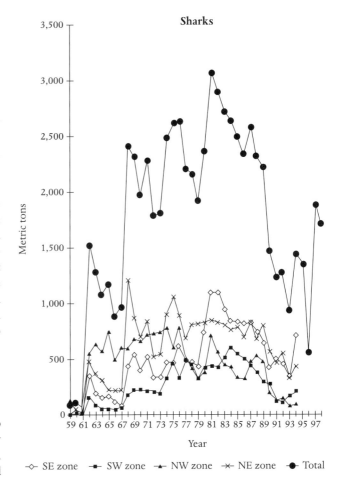

Fig. 8.22. Regional and total catches of sharks in Cuba, 1959–1998.

difficult to assess the present condition of this resource, but since 1981 there has been a tendency toward decline.

8.6.12 Other Fishes

Fishing statistics lump more than 60 species (many caught in small quantities) into an "Other Fishes" category, but since 1970, some of these species, such as stingrays (*Dasyatis* spp.) and blackedge cusk-eel (*Lepophidium brevibarbe*), have reached significant levels in the catch. The rays, mainly the southern stingray (*Dasyatis americana*) and spotted eagle ray (*Aetobatus narinari*), have become an important fishing target. In the 1990s, their yields and sizes diminished at some locations (Espinosa, in press b), which might be evidence of overfishing (also influenced by notable variations in effort at different locations).

This category also includes hogfish (*Lachnolaimus maximus*), black grouper (*Mycteroperca bonaci*), red hind (*Epinephelus guttatus*), parrotfishes (*Sparisoma chrysopterum*, *Scarus guacamaia*, and *S. coelestinus*), cobia (*Rachycentron canadum*), the little tunny (*Euthynnus alletteratus*), and some

Table 8.3.   Composition of shrimp trawl bycatch (percentage of total bycatch by weight) in the Golfo de Guacanayabo (data from the Manzanillo Fishing Association) and Golfo de Ana Maria (from Fernández 1984)

Fish Species or Invertebrate Group	Bycatch (%)	
	Golfo de Guacanayabo	Golfo de Ana María
Chondrichthyes	3.65	0.14
Osteichthyes	—	—
Clupeidae	0.04	2.32
Synodontidae	3.42	1.17
Congridae	—	2.22
Syngnathidae (*Hippocampus* spp.)	—	0.53
Centropomidae	0.45	—
Serranidae (*Diplectrum* & *Rypticus* spp.)	3.57	2.17
Carangidae	0.30	1.19
Caranx latus, Selar crumenophthalmus, Chloroscombrus chrysurus	4.74	1.80
Lutjanidae (*Lutjanus synagris*)	4.84	2.05
Gerreidae	—	—
Diapterus rhombeus	7.68	6.18
Eucinostomus spp.	10.07	5.76
Haemulidae (*Haemulon* spp.)	—	1.81
Sparidae (*Archosargus rhomboidalis, Calamus* sp.)	—	2.60
Sciaenidae (*Micropogonias furnieri*)	—	0.89
Ophidiidae (*Lepophidium brevibarbe*)	6.67	2.51
Acanthuridae (*Acanthurus* spp.)	—	1.06
Scombridae (*Scomberomorus* spp.)	0.06	—
Triglidae (*Prionotus* sp.)	6.99	1.83
Bothidae + Cynoglossidae	3.34	1.18
Balistidae	0.20	0.70
Tetraodontidae (*Sphoeroides* spp.)	0.63	—
Diodontidae	—	0.10
Ostraciidae	—	0.24
Ogcocephalidae (*Ogcocephalus* sp.)	—	1.66
Other fishes	0.88	—
Total fishes	60.45	51.07
Crustaceans	23.29	24.79
Mollusks	6.43	4.15
Sponges	3.14	7.97
Ascidiacea	—	1.73
Sea urchins	—	6.53
Coelenterata	—	3.76

jacks (*Selar crumenophthalmus* and *Trachinotus* spp.). Catches in this category increased from 4,000–5,000 t in 1962–1965 to a mean annual catch of 8,336 t from 1986–1990.

8.6.13   Bycatch
A large group of species are classified as bycatch or *morralla* and are used for producing animal food. These fishes are taken either directly by finfish fisheries that take commercially low-value species or indirectly as a bycatch of shrimp fishing, the primary source. The use of this catch for animal food started in 1969 and grew rapidly; by 1980 it surpassed 20,600 t a year. Of this total, the shrimp fisheries in the southeast accounted for 75–80%; that proportion has since

increased to almost 100%. Since 1990, these catches overall have decreased because of the diminishing shrimp fishery.

Table 8.3 summarizes the bycatch composition of shrimp trawls in the Golfo de Guacanayabo and Golfo de Ana Maria. The data is from 250 samples taken between May 1984 and May 1985 in the Golfo de Guacanayabo (from the Manzanillo Fishing Association), and from the shrimp boats of the Cienfuegos Fishing Association in the Golfo de Ana Maria and in the western part of the Golfo de Guacanayabo (Fernández 1984). Finfish represented approximately 60% and 51% of the shrimp trawl bycatch in samples from the Golfo de Guacanayabo and Golfo de Ana Maria, respectively. In general, the diversity of species caught as bycatch is high.

Gerreids, especially *Diapterus* and *Eucinostomus* species, are very abundant among the finfish bycatch. Large numbers of juvenile lane snapper and other species are also included in these catches; this could affect total recruitment to several fisheries.

8.7 Fishery Productivity

The maximum annual catch obtained on the Cuban shelf and adjacent oceanic areas was 78,989 t in 1985 (Fig. 8.1). Production on the shelf alone (an area of 53,126 km^2), excluding oceano-pelagic species, was 1,420 kg/km^2. The average production for 1986 to 1990 was 1,301 kg/km^2. Due to the economic crisis of the 1990s and limited availability of fuel and parts, fishing effort decreased substantially, making comparisons with earlier decades difficult.

From 1986 to 1990, annual catches from the southeast zone of the Cuban shelf were 2,427 kg/km^2, higher than any other region in Cuba (Fig. 8.2). Large portions of the southeast zone are estuarine and highly productive, but reef fisheries are also important. The southeast zone contributed 55% of the national catch, but almost 50% of the catch in this region was incidental fishes. Omitting bycatch, the productivity for the area was 1,274 kg/km^2, 595 kg corresponding to fishes. This figure is less than that reported for the northeast statistical zone (which includes the Archipiélago Sabana-Camagüey of north-central Cuba), where 950 kg/km^2 of fish was obtained from a total catch of 1,323 kg/km^2 (72%) over the same period. This difference is due, in great part, to differences in fishing effort: 28 boats/1,000 km^2 in the northeast zone versus 17 boats/1,000 km^2 in the southeast (data from 1981 to 1985; the number of boats has decreased, but the proportions remain similar). In the southeast zone, invertebrates are an important element of production (28%). On the north coast, fishes are the dominant group. The northeast zone is the most intensely exploited region of the Cuban shelf.

In the southwest zone, production of 672 kg/km^2 was obtained (20% of the total production from 1986 to 1990). Of that total, 52% was lobster, the most valuable fishery species in Cuba, whereas the fish production was only 276 kg/km^2, relatively low compared with other zones. Currently, fisheries in this zone are concentrated in areas that occupy less than 50% of the Golfo de Batabanó. Fish density may be low in the remaining areas because of limited habitats. The northwest zone produced 920 kg/km^2 from 1986 to 1990 (74% fish). These productivity figures are similar to those obtained for the previous five years (Claro et al. 1994).

Several studies (Wijkstrom 1974; Carpenter 1977; Munro 1977; Murdy and Ferraris 1980) project productivity of 4–6 t/km^2 per year for regions with coral reefs, taking into consideration the area, the reefs, seagrass beds, sandy bottoms, and all other habitats from the shore to the shelf edge. Studies in American Samoa (Hill 1978; Wass 1982) estimated that potential catches could be 8–19 t/km^2 in areas with active coral growth; however, data from catches in highly exploited areas are lower. Munro (1983a) estimated the fishing productivity for several regions of the Caribbean on the basis of the maximum catches in these areas. In Jamaica, catches of 1,720 kg/km^2 have been estimated. These are the largest catches reported in the Caribbean and are evidence of ecosystem overfishing (Munro 1983b, 1983d; Aiken 1993). Evidence of overfishing in the Greater Antilles also exists from Puerto Rico and Hispaniola (Appeldoorn and Lindeman 1985; Appeldoorn and Meyers 1993; Leon et al., in press). However, despite localized overfishing of some species and areas, there is no clear evidence of ecosystem-scale overfishing on the massive shelf of Cuba.

Assuming an estimated productivity of 4–6 t/km^2 (Wijkstrom 1974; Carpenter 1977; Munro 1977; Murdy and Ferraris 1980), the Cuban shelf should produce 212,000–329,600 t yearly. Even with good management of the fisheries, these values seem unrealistic. This suggests that earlier estimates of potential catches need to be reconsidered. The calculations might be based on estimates from islands where the area covered by corals is much higher than in Cuba and the wider Caribbean. Fishing productivity in the Golfo de Batabanó (southwest zone) is currently less than 0.7 t/km^2. With better fishery management, productivity could reach a level of about 1.5 t/km^2, which is less than that of Jamaica and the southeast zone of Cuba, and slightly higher than the Cuban national average in 1984 (1.3 t/km^2). On the Cuban shelf, most coral reefs are concentrated at the outer edge of the shelf, and more than 80–90% of the shelf is covered by seagrass beds, mud and sand bottoms, and estuaries (see Section 1.6.1). It is therefore unlikely that Cuban shelf reefs could reach the highest production estimates from other areas.

If we assume an average productivity value of 1.5 t/km^2 for the shelf, a logical estimate of potential annual catches would be 80,000 t, which coincides with Baisre's (1985a, 1985b) estimate. Recently, Baisre (2000a) calculated a potential maximum production of about 57,000 t (not including bycatch), which, according to the generalized fisheries model described by Caddy (1984), fits previous estimates (Baisre 1985a, 1985b; Claro et al. 1994). If we had good records from the recreational fishery, this number might be realized. Nevertheless, the present deterioration of habitats, diffuse areas of high production, and lack of regulation of fishing effort will make it difficult to reach such a level of fisheries productivity.

8.8 Use of Artificial Reefs

The use of artificial reefs has increased over the last 30 to 40 years, although such systems have been used for centuries

(Steimle and Stone 1973). In Japan and Taiwan, artificial reefs play an important role in the commercial shelf fisheries. The United States, Australia, and Mexico, among other countries, have focused artificial reef programs on recreational rather than commercial fisheries (Bohnsack and Sutherland 1985). The optimal placement of artificial reefs could increase fish catches and contribute to the rehabilitation of overexploited populations (Beets and Hixon 1994). However, there can also be ecological and geographic constraints on the actual production value of artificial reefs (Bohnsack et al. 1994; Grossman et al. 1997).

Large seagrass bed areas are a potentially important resource for maintaining fishery resources. In the extensive seagrass beds and sandy bottoms of the Golfo de Batabanó, fish catches are obtained in less than 50% of the total area, but the colonization of artificial lobster habitats placed in seagrass areas has been dramatic. Many of these large underpopulated areas could be converted to productive areas for the catch of lobsters and fishes by installing artificial habitats (Claro and García-Arteaga 1999).

A geographic framework for an artificial reef system for the Cuban shelf has been developed (Claro and García-Arteaga 1999). The authors consider strategic use of artificial reefs as a potential tool to elevate fishing production on the shelf. However, if haphazardly placed, isolated artificial structures to concentrate fishes can increase the vulnerability of stock to overfishing. The proposal for the Cuban shelf (Claro and García-Arteaga 1999) identifies appropriate regions for locating the artificial reefs, as well as the quantity and types of refuges for each region and general habitat type. More than 13,500 km^2 of shelf area have been identified as appropriate sites where at least 850 reef systems could be constructed, each reef providing up to 200 fish shelters.

8.9   Conservation and Management

Cuban fisheries are administered under a system not commonly found in other Latin American countries. Access to fishery resources is controlled in a relatively centralized system of administration, extraction, and commercialization through the Ministry of Fishing Industries. Decree Law No. 164, approved in May 1996, established several new concepts, including fishery access through fishing licenses and permitting for both commercial and recreational practices. This law also grants the Ministry of Fishing Industries the authority to set minimum size regulations and catch quotas, create closed seasons for species or regions, regulate fishing gear and methods, establish commercialization and conservation methods, and inspect food quality.

Permits provide a way to control fishing effort relative to the status of the resource. The Ministry of Fishing Industries has recently created an inspection system that enforces existing regulations. In a short period this system has yielded notable results, especially in controlling illegal fishing. However, because both conservation and production are managed by the same entity, steps should be taken to avoid an overemphasis on fishery production at the expense of sustainability.

More than 12% of Cuba's terrestrial land is classified under various protected area categories. However, because of limited resources, particularly for enforcement, there are few marine reserves. Implementing an integrated network of marine reserves is an important new initiative of the National Center for Protected Areas and the Institute of Oceanology of the Ministry of Science, Technology, and Environment. Previously, the Ministry of Fishing Industries, with other national entities, has established special regulations for the protection and limited use of approximately eight marine reserves. Most of these areas correspond to areas of high tourism development or high biodiversity. In general, fishing for everything except lobster (which is conducted under specific regulations) is largely prohibited in these reserves, although enforcement efficiency is variable. The design concepts behind the reserves often include goals other than fishery resource protection. The interplay among fisheries objectives and the different objectives of other user groups is a complicating factor in the design process for reserves in many areas (Plan Development Team 1990; Shorthouse 1990; Man et al. 1994; Roberts 1995; Pauly and Christensen 1996; Russ and Alcala 1996, 1998; Chiappone and Sullivan Sealey 2000; Walters 2000).

Until recently, fishing was the principal marine economic activity in Cuba, but fishing is currently experiencing strong competition from tourism development, which is also closely associated with ocean resources. Considering the coastal effects of terrestrial activities (pollution, deforestation, damming, and urban development), all of which constitute serious threats to fisheries resources as well as tourism, it is clear that integrated coastal zone management approaches are needed. The national administrative structure is fostering integrated management approaches where resources allow. The Global Environmental Fund (GEF)–United Nations Development Program (UNDP) Project for Integrated Management of the Sabana-Camagüey Archipelago, coordinated by the Ministry of Science, Technology, and the Environment, is one recent example (CUB/92/631 1999).

8.10   Summary

Total catches from the EEZ increased from 22,500 t in 1959 to 79,000 t in 1985. Commercial fishes and crustaceans constitute approximately 39% and 24% of the total national catch, respectively. Bycatch is high, accounting for approximately 30% of the total catch. Finfish catches increased from 12,730 to 29,720 t by 1987, but since 1990 have declined, in part because of reduced fishing effort associated with the economic crisis. Catches have improved since 1995, but remain at a lower level, perhaps as a result of habitat degradation (Claro et al. 2000).

From 1970 to 1975, increased fishing effectiveness (e.g., use of trawls, seines, and set nets) resulted in overfishing of several important species (lane snapper, mullet, Nassau grouper, stone crab, and shrimp in some regions). Overfishing also resulted from the use of unselective fishing gear, the indiscriminate use of set nets during spawning migrations, and limited enforcement of some fishery regulations (e.g., minimum legal size, closures during spawning periods). In 1981, a new fishery administration policy (first adopted in 1978) was implemented and marked the beginning of the modern Cuban fishery management period (Baisre 1993).

As a result of intensive development of the fisheries industries since the 1970s, the Cuban shelf is close to maximum exploitation. The deterioration of coastal habitats has also contributed to the reduction of some resources (shrimp, mullet). Decreased yields of some resources (lobster, sharks, batoids) and the replacement of species of high quality and size by others of lower value has occurred. In some parts of Cuba, these conditions suggest overfishing, but the situation is not fully comparable with other areas of the Caribbean, where access to fisheries can be less regulated.

Despite the high level of exploitation of the Cuban shelf, some resources are not being used to their full potential (e.g., sardines, deep-water snappers and groupers, grunts, yellowtail snapper, mackerels, and tunas). For some species, the processing industries will need to develop better technologies. For other underutilized species (e.g., certain groupers, jacks, and barracuda), efficient ways to detect *ciguatera* poisoning could improve their commercial value.

Most Cuban fisheries are multispecies fisheries; it is difficult to obtain specific catch information and to evaluate resource status and the effectiveness of regulations. This emphasizes the need for more focus on ecological interactions in fishery research and ecosystem management. Most fisheries are markedly seasonal and are often tied to spawning migrations, which constitute direct threats to fishery sustainability. Regulations that establish a minimum size for all species are not always adequate and some species are caught in large quantities before reaching the size of first spawning (e.g., mutton, lane, gray, and yellowtail snappers, margate, hogfish, mullet, Nassau grouper, and jacks). Reserve networks that consider larval connectivity and inclusive habitat protection around the massive Cuban insular shelf are logical tools for sustaining fishery production; inter-agency efforts to build a marine reserve network in Cuba are underway.

Biological productivity on the Cuban shelf appears to be lower ($1.1–1.5$ t/km^2) than estimates for some other tropical regions, and science-based changes in fishing effort and management will be required to achieve sustainability. These changes include the development of low-impact mariculture, the well-designed use of artificial reefs, enhanced regulation of fishing gear and catch parameters, regionalized fishing effort, establishment of fisheries reserves, and increased integration of decision-making protocols among agencies responsible for the management of both coastal lands and fisheries.

REFERENCES

Aasen, O. 1966. Blahaien, *Prionace glauca* (Linnaeus, 1758). Fisken Havet, 1:1–15.

Abraham, B. 1963. A study of the oogenesis and egg resorption in the mullets *Mugil cephalus* and *Mugil capito* in Israel. Proc. Fish. Counc. Medit., 7:435–453.

Abraham, M., N. Blanc, and A. Yashouv. 1966. Oogenesis in five species of mullets (Teleostei, Mugilidae) from natural and landlocked habitats. Isr. J. Zool., 15:155–172.

Academia de Ciencias de Cuba. 1965. *Area de Cuba*. Instituto de Geografia, La Habana, 181 p.

Academia de Ciencias de Cuba and Academia de Ciencias de la URSS. 1970. Atlas Nacional de Cuba. Dirección Nacional de Geodesia y Cartografia, Consejo de Ministros de la URSS, Moscú, 132 p.

Acero, A. 1980. Observaciones ecológicas de la ictiofauna de una pradera de *Thalassia* en la Bahía de Nenguange (Parque Nacional Tayrona, Colombia). Bol. Inst. Oceanogr. São Paulo, 29(3):5–8.

Acosta, A., and R. S. Appeldoorn. 1992. Estimation of growth, mortality and yield per recruit for *Lutjanus synagris* (Linnaeus) in Puerto Rico. Bull. Mar. Sci., 50(2):282–291.

Aguilar-Perera, A., and W. Aguilar-Dávila. 1996. A spawning aggregation of Nassau grouper *Epinephelus striatus* (Pisces:Serranidae) in the Mexican Caribbean. Envir. Biol. Fish., 45:351–361.

Aiken, K. 1983a. The biology, ecology and bionomics of the butterfly and angelfishes, Chaetodontidae. In: J. L. Munro, ed. Caribbean Coral Reef Fishery Resources. Intern. Center Living Aquatic Resources Management (ICLARM) Stud. Rev., 7:155–165.

Aiken, K. 1983b. The biology, ecology and bionomics of the triggerfishes, Balistidae. In: J. L. Munro, ed. Caribbean coral reef fishery resources ICLARM Stud. Rev., 7:191–215.

Aiken, K. 1993. Jamaica. In: Marine fishery resources of the Antilles. FAO Fish. Tech. Pap. 326. Rome, Italy, pp. 159–180.

Alcalá, A. 1988. Effects of marine reserves on coral fish abundance and yields of Philippine coral reefs. AMBIO 17(3):194–199.

Alcolado, P. M. 1981. Zonación de los gorgonaceos someros de Cuba y su posible uso como indicadores comparativos de la tensión hidrodinámica sobre los organismos del bentos. Informe Cient.-Tec., Inst. Oceanol., Acad. Cien. Cuba, 187:1–43.

Alcolado, P. M. 1990. Aspectos ecológicos de la macrolaguna del Golfo de Batabanó, con especial referencia al bentos. In: P. M. Alcolado, ed. El bentos de la Macrolaguna del Golfo de Batabanó. Editorial Academia, La Habana, pp. 129–157.

Alcolado, P. M., C. Jiménez, J. Espinosa, D. Ibarzábal, J. C. Martínez, R. del Valle, N. Martínez, A. Hernández, M. Abreu, L. Vega, and E. Ramírez. 1990a. Aspectos ecológicos del acuatorio del noreste de la provincia de Villa Clara. In: J. Pérez, ed. Estudios de los Grupos Insulares y Zonas Litorales del Archipiélago Cubano con Fines Turísticos Cayos Frances, Cobo, Las Brujas, Ensenachos y Santa María. Editorial Cient. Téc., La Habana, pp. 86–97.

Alcolado, P. M., C. Jiménez, J. Espinosa, D. Ibarzábal, J. C. Martínez, R. del Valle, N. Martínez, A. Hernández, M. Abreu, L. Vega, and E. Ramírez. 1990b. Ecología marina. In: L. Fernández, R. Cañizares, and H. Gómez, eds. Estudios de los Grupos Insulares y Zonas Litorales del Archipiélago Cubano con Fines Turísticos Cayos Megano Grande, Cruz, Romano y Guajaba. Editorial Cient. Téc., La Habana, pp. 99–109.

Alcolado, P. M., A. Herrera-Moreno, and N. Martínez-Estalella. 1994. Sessile communities as environmental bio-monitors in Cuban coral reefs, 1993. In: R. N. Ginsburg, ed. Proc. Colloquium and Forum on Global Aspects of Coral Reefs: Health, Hazards, and History. Univ. Miami, Miami, FL, pp. 27–33.

Alcolado, P. M., J. Espinosa, D. Ibarzabal, R. del Valle, M. Abreu, J. C. Martínez, A. Hernández, N. Martínez-Estalella, P. García, and G. Menéndez, 1996. Informe sobre el Inventario Extensivo del Bentos (Zoobentos) del Archipiélago Sabana-Camaghey. Inst. Oceanol. 44 p.

Alcolado, P. M., R. Claro, G. Menéndez, and B. Martínez-Daranas. 1997. General status of Cuban coral reefs. Proc. 8th Int. Coral Reef Symp., 1:341–344.

Alcolado, P. M., R. Claro, B. Martínez-Daranas, K. Cantelar, G. Menéndez-Macía, M. Hernández, T. García, J. Espinosa, and R. del Valle. 1999a. Evaluación Diagnóstica del Estado del Arrecife de María la Gorda (SE de la Ensenada de Corrientes, Cuba). Inst. Oceanol., 10 p.

Alcolado, P. M., R. Claro, B. Martínez-Daranas, G. Menéndez-Macía, P. García-Parrado, E. Perigó, K. Cantelar, M. E. Miravet, G. M. Lugioyo, R. del Valle, N. Melo, J. F. Montalvo, D. Enríquez, T. García-Díaz, M. Hernández-González, J. L. Hernández-López and R. Nuñes. 1999b. Evaluación Diagnóstica Preliminar de los Arrecifes Coralinos del oeste de Cayo Largo del Sur: 1998–1999. Inst. Oceanol. (informe), 37 p.

Alcolado, P. M., R. Claro, G. Menéndez, P. García-Parrado, B. Martínez-Daranas, and M. Sosa. 1999c. The Cuban coral reefs. Archivo Instituto de Oceanología, 20 pp.

Alcolado, P. M., R. Claro, B. Martínez-Daranas, G. Menéndez-Macías, P. García-Parrado, K. Cantelar-Ramos, J. Espinosa-Sáez, R. del Valle-García, J. C. Martínez-Iglesias, and T. Neff. 2000. Estado General de los Arrecifes Coralinos de Cuba y Propuestas de Manejo Ambiental. Informe Final del Proyecto. Inst. Oceanol., 66 p.

Aldenhoven, J. M. 1984. Social organization and sex change in an angelfish, *Centropyge bicolor*, on the Great Barrier Reef. Ph.D. Diss., Macquarie University, North Ryde, Australia.

Alegre, B., and J. Alheit. 1986. Un nuevo método para la determinación de la fecundidad de la sardina peruana (*Sardinops sagax*). Bol. Inst. Mar. Perd., 10(3):61–90.

Alegría, J. R., and M. Ferreira de Menezes. 1970. Edad y crecimiento del ariacó, *Lutjanus synagris* (Linnaeus), en el nordeste del Brasil. Arq. Cienc. Mar., 10(1):65–68.

Alheit, J., B. Alegre, V. H. Alarcón, and B. J. Macewicz. 1983. Batch fecundity and spawning frequency of various anchovy (Genus: *Engraulis*) populations from upwelling areas and their use for spawning biomass estimates. FAO Fish. Rep., 291(3):977–985.

Alvarez, J. H. 1989. Análisis de la pesquería de cazón en la Península de Yucatán, México. Unpublished, Centro de Investigaciones y Estudios Avanzados del Instituto Politécnico Nacional, Unidad Mérida, México, 54 p., 22 tabs., 43 figs.

Alvarez-Lajonchere, L. 1976. Contribución al ciclo de vida de *Mugil curema* Valenciennes in Cuvier et Valenciennes, 1836 (Pisces: Mugilidae). Ciencias, Ser. 8, Invest. Mar., 28:1–130.

Alvarez-Lajonchere, L. 1978a. La pesca de lisas (Pisces: Mugilidae) en Cuba. Ciencias, Ser. 8, Invest. Mar., 35:1–15.

Alvarez-Lajonchere, L. 1978b. Las pesquerías de lisas (Pisces: Mugilidae) en Tunas de Zaza, Cuba. Ciencias, Ser. 8, Invest. Mar., 36:1–86.

Alvarez-Lajonchere, L. 1979. Algunos aspectos sobre la reproducción de *Mugil liza* (Pisces: Mugilidae) en Tunas de Zaza, Cuba. Rev. Cub. Invest. Pesq., 4(2):25–61.

Alvarez-Lajonchere, L. 1980a. Algunos datos adicionales sobre la reproducción y las relaciones largo-peso, de *Mugil curema* (Pisces: Mugilidae) en Cuba. Rev. Invest. Mar., 1(1):75–91.

Alvarez-Lajonchere, L. 1980b. Determinación de la edad y el crecimiento de *Mugil hospes* y *Mugil trichodon* (Pisces: Mugilidae) en Tunas de Zaza, Cuba. Rev. Invest. Mar., 1(2–3):61–88.

Alvarez-Lajonchere, L. 1980c. Estudio sobre algunos aspectos de la reproducción de *Mugil trichodon* (Pisces: Mugilidae) en Tunas de Zaza, Cuba. Rev. Invest. Mar., 1(2–3):3–28.

Alvarez-Lajonchere, L. 1981a. Determinación de la edad y el crecimiento de *Mugil liza*, *M. curema*, *M. hospes* y *M. trichodon* (Pisces: Mugilidae) en aguas cubanas. Rev. Invest. Mar., 2(1):142–162.

Alvarez-Lajonchere, L. 1981b. Estudios de algunos aspectos sobre la reproducción de *Mugil hospes* (Pisces: Mugilidae) en Tunas de Zaza, Cuba. Rev. Invest. Mar., 2(3):101–128.

Alvarez-Lajonchere, L. 1981c. Estudio morfométrico y merístico de *Mugil trichodon* y *M. liza* (Pisces: Mugilidae) en Tunas de Zaza, Cuba. Rev. Invest. Mar., 2(3):129–173.

Alvarez-Lajonchere, L., M. Báez, and G. Gotera. 1982. Estudio de la biología pesquera del robalo de ley *Centropomus undecimalis* (Bloch) (Pisces: Centropomidae) en Tunas de Zaza, Cuba. Rev. Invest. Mar., 3(1):159–200.

Amorin Borges, G. de. 1966. Notas preliminares a biologia e pesca do xaréu préto *Caranx lugubris* Poey, no nordeste Brasileiro. Bol. Est. Pesca, 6(2):9–30.

Andreu, B. 1956. Observaciones sobre el ovario de merluza (*Merluccius merluccius* L.) y características del mecanismo de la puesta. Invest. Pesq. (Barc.), 4:49–66.

Andreu, B., and J. dos Santos Pinto. 1957. Características histológicas y biométricas del ovario de sardina (*Sardina pilchardus* Walb.) en la maduración, puesta y recuperación. Origen de los ovocitos. Invest. Pesq. (Barc.), 6:3–38.

Angell, C. 1976. Una contribución a la biología de la mojarra *Eugerres plumieri*. Mem. Soc. Cienc. Nat. La Salle, 36(105): 297–310.

Appeldoorn, R. S. 1992. Interspecific relationships between growth parameters, with application to haemulid fishes. Proc. 7th Int. Coral Reef Symp., Guam, 2:899–904.

Appeldoorn, R. S., and K. C. Lindeman. 1985. Multispecies assessments in coral reef fisheries using higher taxonomic categories as unit stocks, with an analysis of an artisanal haemulid fishery. Proc. 5th Int. Coral Reef Congr. 5:507–514.

Appeldoorn, R. A., and S. Meyers. 1993. Puerto Rico and Hispaniola. In: Marine Fishery Resources of the Antilles. FAO Fish. Tech. Pap. 326, Rome, Italy, pp. 99–158.

Appeldoorn, R. A., D. A. Hensley, D. Y. Shapiro, S. Kioroglou, and B. G. Sanderson. 1994. Egg dispersal in a Caribbean coral reef fish, *Thalassoma bifasciatum*. II. Dispersal off the reef platform. Bull. Mar. Sci., 54(1):271–280.

Arnold, L. V., and K. R. Fortunatova. 1937. Experimental study of the feeding of fishes of the Black Sea [in Russian]. Dokl. AN S.S.S.R., 15(3):505–508.

Arreguín-Sánchez, F. 1987. Estado actual de la explotación del mero (*Epinephelus morio*) del Banco de Campeche. 25 Aniv. Inst. Nacl. Pesca CRIP, Yucaltepén, INP, México.

Arreguín-Sánchez, F. 1992. An approach to the study of catchability coefficient with application to the red grouper (*Epinephelus morio*) fishery from the continental shelf of Yucatán, México. Ph.D. thesis, Cent. Invest. Estudios Avanzados del INP, México. 222 p.

Arreguín-Sánchez, F., E. Chávez, C. Martínez, J. Sánchez, M. Olvera, and P. Castañeda. 1989. Análisis integral de la pesquería de chinchorro del puerto de Celestún, Yucatán. Cent. Invest. Estudios Avanzados del Inst. Politécnico Nacl., Unidad Mérida, México, 2 p.

Arreguín-Sánchez, F., M. Contreras, V. Moreno, R. Burgos, and R. Valdés. 1996. Population dynamics and stock assessment of red grouper (*Epinephelus morio*) fishery on Campeche Bank, Mexico. In: F. Arreguín-Sánchez, J. L. Munro, M. C. Balgos, and D. Pauly, eds. Biology, Fisheries and Culture of Tropical Groupers and Snappers. ICLARM Conf. Proc. 48, p. 202–217.

Arreguín-Sánchez, F., J. L. Munro, M. C. Balgos, and D. Pauly, eds. 1996. Biology, Fisheries and Culture of Tropical Groupers and Snappers. ICLARM Conf. Proc. 48. 449 p.

Artamondi, F. 1970. Pesquerías del Golfo de Batabanó. Un estudio económico y estadístico de la zona 8. Inst. de Oceanol. Acad. Cienc. Cuba, 33 p.

Atz, J. W. 1964. Intersexuality in fishes. In: C. N. Armstrong, and A. J. Marshall, eds. Intersexuality in Vertebrates Including Man. Academic Press, pp. 145–232.

Atz, J. W. 1965. Hermaphroditic fish. Science 150:789–797.

Ault, J. S., J. A. Bohnsack, and G. Meester. 1998. A retrospective (1979–1996) multispecies assessment of coral reef fish stocks in the Florida Keys. Fish. Bull. 96(3):395–414.

Austin, H. M., and S. Austin. 1971. The feeding habits of some juvenile marine fishes from the mangroves in western Puerto Rico. Carib. J. Sci., 11(3–4):171–178.

Avello, O. 1979. Biogenic components of the Cuban shelf sediments [in Russian]. Ph.D. diss., Inst. Oceanol. Acad. Sci. USSR, Moscow, 157 p.

Ayala, D. L. 1984. Determinación de algunos parámetros poblacionales y de la biología pesquera de la biajaiba *Lutjanus synagris* (Linneo), 1758 (Pisces: Lutjanidae). Tesis de licenciatura, E.N.E.P.I., UNAM, México.

Báez, M., and L. Alvarez-Lajonchere. 1980. La pesquería de gérridos (Pisces: Gerridae) en Tunas de Zaza, Cuba. Rev. Invest. Mar., 1(2–3): 89–134.

Báez, M., and L. Alvarez-Lajonchere. 1983. Edad, crecimiento y reproducción de la mojarra *Gerres cinereus* (Walbaum 1972) en Tunas de Zaza, Cuba. Rev. Invest. Mar., 4(3):41–76.

Báez, M., L. Alvarez-Lajonchere, and B. Pedroso. 1980. Edad y crecimiento del caballerote, *Lutjanus griseus* (Linné), en Tunas de Zaza, Cuba. Rev. Invest. Mar., 1(2–3):135–159.

Báez, M., L. Alvarez-Lajonchere, and J. Gómez. 1982a. Edad y crecimiento del patao (*Eugerres brasilianus* Cuvier) (Pisces: Gerreidae) en Tunas de Zaza, Cuba. Rev. Invest. Mar., 3(3):117–152.

Báez, M., L. Alvarez-Lajonchere, and E. Ojeda Serrano. 1982b. Reproducción del caballerote, *Lutjanus griseus* (Linnaeus) en Tunas de Zaza, Cuba. Rev. Invest. Mar., 3(1):43–86.

Báez, M., L. Alvarez-Lajonchere, and J. Yonazi. 1983. Reproducción de *Eugerres brasilianus* (Pisces: Gerreidae) en Tunas de Zaza, Cuba. Rev. Invest. Mar., 4(3):21–40.

Bagnis, R., S. Chanteau, E. Chungui, J. M. Hurtel, T. Yamamoto, and A. Inoue. 1980. Origins of ciguatera fish poisoning: A new dinoflagelate, *Gambierdiscus toxicus* Adachi and Fukuyo, definitively involved as a causal agent. Toxicon, 18(2):199–208.

Baisre, J. A. 1981. Comportamiento de las Pesquerías Nacionales. Cent. Invest. Pesq., Ministerio de la Industria Pesquera, La Habana, 51 p.

Baisre, J. A. 1985a. Los complejos ecológicos de pesca: Definición e importancia en la administración de las pesquerías cubanas. FAO Fish. Report No. 327(Supl.):252–272.

Baisre, J. A. 1985b. Los recursos pesqueros marinos de Cuba: Fundamentos ecológicos y estrategias para su utilización. Tesis doctoral en Ciencias Biológicas. 189 p.

Baisre, J. A. 1993. Marine fishery resources of the Antilles: Lesser Antilles, Puerto Rico and Hispaniola, Jamaica, Cuba. FAO Fish. Tech. Pap., No. 326, Part 4, Cuba, Rome, Italy, pp. 181–235.

Baisre, J. A. 2000a. Chronicle of Cuban marine fisheries (1935–1995). Trend analysis and fisheries potential. FAO Fish. Tech. Pap. 394, 26 p.

Baisre, J. A. 2000b. The Cuban lobster fishery. In: B. F. Phillips and J. Littaka, eds. Spiny Lobsters: Fisheries and Culture. Fishing News Books, Oxford. pp. 135–152.

Baisre, J. A., and J. Páez. 1981. Los recursos pesqueros del archipiélago cubano. Estudios WECAF, 8:1–79.

Baisre, J. A., and A. Zamora. 1983. Las pesquerías de camarón: Antecedentes, situación actual y perspectivas [publ. especial]. Cent. Invest. Pesq., Ministerio de la Industria Pesquera, La Habana, 64 p.

Bakus, G. I. 1966. Energetics and feeding in shallow marine waters. Int. Rev. Gen. Exp. Zool., 4:275–369.

Bannerot, S., W. W. Fox, Jr., and J. E. Powers. 1987. Reproductive strategies and the management of snappers and groupers in the Gulf of Mexico and Caribbean. In: J. J. Polovina, and S. Ralston, eds. Tropical Snappers and Groupers. Biology and Fisheries Management. Westview Press, Boulder, CO, and London, pp. 561–603.

Bara, W. A. 1960. Histological and cytological changes in the ovaries of the mackerel, Scomber scomber L., during the annual cycle. Rev. Fac. Sci. Univ. Istanbul, 15:49–91.

Bardach, J. E. 1958. On the movements of certain Bermuda reef fishes. Ecology, 39(1):139–146.

Bardach, J. E. 1959. The summer standing crop of fish on a shallow Bermuda reef. Limnol. Oceanogr. 4:77–85.

Barkman, R. C. 1978. The use of otolith growth rings to age young Atlantic silversides, Menidia menidia. Trans. Am. Fish. Soc., 107:790–792.

Barlow, G. W. 1974. Hexagonal territories. Anim. Behav., 22:876–878.

Barlow, G. W. 1975. On the sociobiology of four Puerto Rican parrotfishes (Scaridae). Mar. Biol., 33:281–293.

Barlow, G. W. l981. Patterns of parental investment, dispersal and size among coral-reef fishes. Environ. Biol. Fish., 6(1):65–86.

Bashirullah, A. K. M. 1975. Biology of Lutjanus griseus of the Cubagua Island, Venezuela. I. Length-weight, body length–gut length relationships and condition factor. Bol. Inst. Oceanogr. Univ. Oriente Cumaná, 14(1):101–107.

Basu, A. K., E. Perigó, and G. Suárez. 1975a. Características contaminantes de algunas destilerías y cervecerías en Cuba. Resúmenes Invest. Cent. Invest. Pesq., Cuba, 2:211–214.

Basu, A. K., E. Perigó, and G. Suárez. 1975b. Observaciones sobre la índole y el aporte de cargas contaminantes por efluentes de algunos centrales en Cuba. Resúmenes Invest. Cent. Invest. Pesq., Cuba, 2:205–210.

Basu, A. K., G. Suárez, B. Vázquez, E. Perigó, and J. Gómez-Quintero. 1975c. Prospección de algunos parámetros que influyen sobre la contaminación en la zona suroccidental de la plataforma cubana. Resúmenes Invest. Cent. Invest. Pesq., Cuba, 2:238–240.

Batista, J. L. 1974. Isolíneas del escurrimiento medio anual. Voluntad Hidráulica, 32:13–15.

Batts, B. S. 1972. Age and growth of the skipjack tuna, Katsuwonus pelamis (Linnaeus) in North Carolina waters. Chesapeake Sci., 13(4):237–244.

Beardsley, G. L. 1967. Age, growth and reproduction of the dolphin, Coryphaena hippurus in the straits of Florida. Copeia, 2:441–451.

Beaumarriage, D. S. 1973. Age, growth and reproduction of king mackerel, Scomberomorus cavalla, in Florida. Fla. Mar. Res. Publ., 1:1–46.

Becebnov, L. H. 1960. Some data on the ichthyofauna of floating plants in the Pacific Ocean [in Russian]. Trudy Inst. Okeanol., 41:192–197.

Beebe, W., and J. Tee-Van. 1928. The fish of Port-au-Prince Bay, Haití with a summary of the known species of marine fish of the island of Haiti and Santo Domingo. Zoological (N.Y.), 10(1):1–279.

Beets, J. 1997. Effects of a predatory fish on the recruitment and abundance of Caribbean coral reef fishes. Mar. Ecol. Prog. Ser., 148:11–21.

Beets, J., and M. A. Hixon. 1994. Distribution, persistence, and growth of groupers (Pisces: Serranidae) on artificial and natural patch reefs in the Virgin Islands. Bull. Mar. Sci., 55(2–3):470–483.

Behety, P. A. 1975. Nuevos reportes de gorgonáceos (Coelenterata) para Cuba. Serie Oceanológica, 33:1–9.

Berry, F. H. 1959. Young jack crevalles (Caranx species) of the southeastern Atlantic coast of the United States. U.S. Fish. Wild. Serv., Fish. Bull., 59(152):417–535.

Bertalanffy, L. von. 1938. A quantitative theory of organic growth (inquiries on growth laws II). Human Biol., 10(2):181–213.

Bessonov, N. M., and O. González. 1971. On the causes of variation of the bioproductivity of the Gulf of Mexico [in Russian]. In: A. S. Bogdanov, ed. Joint Soviet-Cuban Fisheries Investigations, Pishevaia Promishlennost, Moscow, vol. 3, pp. 54–63.

Bessonov, N. M., A. A. Elizarov, and O. González. 1971. Main features of the oceanographic conditions of the Campeche Bank [in Russian]. In: A. S. Bogdanov, ed. Joint Soviet-Cuban Fisheries Investigations Pishevaia Promishlennost, Moscow, vol. 3, pp. 14–32.

Beverton, R. J., and S. J. A. Holt. 1957. Dynamics of Exploited Fish Population. Fish Invest., Ser. 2, 533 p.

Bigelow, H. B., and W. C. Shroeder. 1953. Fishes of the Gulf of Maine. U.S. Fish Wild. Serv., Fish. Bull., 74:1–577.

Billings, V. C., and J. Munro. 1974. The biology, ecology and bionomics of Caribbean reef fishes: Pomadasydae (Grunts). Part 5. Res. Rep. Zool. Dep. Univ. West Indies, 3:1–128.

Birkeland, C., and S. Neudecker. 1981. Foraging behaviour of two Caribbean chaetodontids: Chaetodon capistratus and Chaetodon aculeatus. Copeia, 1:169–178.

Blázquez-Echandi, L., and E. Romeu. 1982. Contribución al estudio de la circulación general en el Golfo de Batabanó, Zona B. Rep. Invest. Inst. Oceanol. Acad. Cienc. Cuba, 1:1–33.

Blázquez-Echandi, L., J. P. Rodríguez-Portal, I. Rosabal-Campos, and R. Calderón-Hebra. 1988. Mediciones de corriente en el Golfo de Batabanó. Rep. Invest. Inst. Oceanol. Acad. Cienc. Cuba, 14:1–36.

Boesch, D. F. 1977. Application of Numerical Classifications in Ecological Investigations of Water Pollution. Ecol. Res. Ser., EPA-GOO/3-77-033, 115 p.

Bogorov, V. G., and L. A. Zenkevich. 1947. A Primer for Hydrobiological Surveys in the Sea (Plankton and Benthos). Glavsermorputi, Moscow, 126 p.

Böhlke, J. E., and C. C. Chaplin. 1993. Fishes of Bahamas and Adjacent Tropical Waters. 2nd ed., Univ. Texas Press, Austin, TX, 771 p.

Böhlke, J. E., and J. E. Randall. 1963. The fishes of the western Atlantic serranoid genus Gramma. Proc. Acad. Nat. Sci. Phil., 115(2):33–52.

Bohnsack, J. A., and D. E. Harper. 1988. Length-Weight Relationships of Selected Marine Fishes from the United States and the Caribbean. NOAA Tech. Memo., NMFS-SEFC-215.

Bohnsack, J. A., and D. L. Sutherland. 1985. Artificial reef research: A review with recommendations for future priorities. Bull. Mar. Sci., 37:11–39.

Bohnsack, J. A., D. E. Harper, D. B. McClellan, D. L. Sutherland, and M. W. White. 1987. Resource survey of fishes within Looe Key National Marine Sanctuary. NOAA Tech. Mem. NOS MEMD 5. 108 p.

Bohnsack, J. A., D. E. Harper, D. B. McClellan, and M. Hulsbeck. 1994. Effects of reef size on colonization and assemblage structure of fishes at artificial reefs off southeastern Florida, U.S.A. Bull. Mar. Sci., 55(2&3):796–823.

Bolden, S. K. 2000. Long-distance movement of a Nassau grouper (*Epinephelus striatus*) to a spawning aggregation in the Central Bahamas. Fish. Bull., 98(3):642–645.

Borodulina, O. D. 1972. On the feeding habits of mesopelagic predators of the open ocean [in Russian]. Vopr. Ikhtiol., 12(4): 757–768.

Borodulina, O. D. 1974. Feeding habits of the big-eye tuna, *Thunnus obesus* (Lowe), in the Gulf of Guinea and its role in the pelagic trophic system [in Russian]. Vopr. Ikhtiol., 14(5):881–892.

Bortone, S. A. 1971. Studies on the biology of the sand perch, *Diplectrum formosum* (Perciformes: Serranidae). Fla. Dep. Nat. Resour. Mar. Res. Lab., Tech. Ser., 65:1–27.

Bortone, S. A. 1977. Revision of the sea basses of the genus *Diplectrum* (Pisces: Serranidae). NOAA Tech. Rep. NMFS, 404: 1–49.

Bowers, A. B., and F. G. T. Hollyday. 1961. Histological changes in the gonad associated with the reproductive cycle of the herring (*Clupea harengus* L.). Mar. Res. Indones., 5:3–16.

Branstetter, S. 1987. Age and growth validation of newborn sharks held in laboratory aquaria with comments on the life history of the Atlantic sharpnose shark *Rhizoprionodon terraenovae*. Copeia, 2:291–299.

Breder, C. M. 1948. Field Book of Marine Fishes of the Atlantic Coast from Labrador to Texas. G. P. Putnam's Sons, New York and London, 332 p.

Breder, C. M. 1959. Studies on social groupings in fishes. Bull. Am. Mus. Nat. Hist., 117(6):398–482.

Breder, C. M., and R. F. Nigrelli. 1938. The significance of differential locomotor activity as an index to the mass physiology of fishes. Zoologica (N.Y.), 23(1):1–29.

Breder, C. M., Jr., and P. M. Bird. 1975. Cave entry by schools and associated pigmentary changes of the marine clupeid *Jenkinsia*. Bull. Mar. Sci., 25(3):377–386.

Brett, J. R., and T. D. D. Groves. 1979. Physiological energetics. In: W. S. Hoar, D. J. Randall, and J. R. Brett, eds. Fish Physiology. Academic Press, London, vol. 8, pp. 290–344.

Briggs, J. C. 1995. Global Biogeography. Developments in Palaeontology and Stratigraphy No. 14. Elsevier, New York, NY, 452 p.

Brothers, E. B. 1979. Age and growth studies on tropical fishes. In: S. B. Saila and P. M. Rhoede, eds. Proc. Int. Workshop Trop. Small Scale Fish. Stock Assessment, Univ. Rhode Island, pp. 119–136.

Brothers, E. B. 1981. What can otolith microstructure tell us about daily and subdaily events in the early life history of fish? Rapp. P-V. Réun. CIESM, 178:393–394.

Brothers, E. B. 1983. Summary of round table discussion on age validation. NOAA Tech. Rep. NMFS, 8:35–44.

Brothers, E. B., and W. McFarland. 1981. Correlation between otolith microstructure, growth, and life history transitions in newly recruited french grunts (*Haemulon flavolineatum* (Desmarest) Haemulidae). Rapp. P-V. Reun. CIESM, 178:369–374.

Brothers, E. B., and R. E. Thresher. 1985. Pelagic duration, dispersal and distribution of Indo-pacific coral reef fishes. In: M. L. Reaka, ed. The Ecology of Coral Reefs Symposia Series Undersea Research. NOAA Undersea Res. Prog., 3(1):53–69.

Brothers, E. B., P. Mathews, and R. Lasker. 1976. Daily growth increments in otolith from larval and adult fishes. Fish. Bull., 74(1):1–8.

Brothers, E. B., D. M. Williams, and P. F. Sale. 1983. Length of larval life in twelve families of fishes at One Tree Lagoon, Great Barrier Reef, Australia. Mar. Biol., 76:319–324.

Brownell, W. N., and W. E. Rainey. 1971. Research and development of deepwater commercial and sport fisheries around the Virgin Islands Plateau. Virgin Islands Ecol. Res. Sta., Contr., 3:1–88.

Bruger, G. E. 1974. Age, growth, food habits, and reproduction of bonefish, *Albula vulpes*, in south Florida waters. Fla. Mar. Res. Publ., 3:1–20.

Brulé, T. T., N. J. del Maldonado, and G. Mexicano. 1991. Datos preliminares sobre la reproducción del mero, *Epinephelus morio* (Pisces:Serranidae) del Banco de Campeche. In: M. Dailey, and H. Bertoch, eds. Mem.s VIII Simp. Biol. Mar., 1990, pp. 19–30.

Brulé, T. T., C. Deniel, T. Colas-Marrufo, and M. Sanchez-Crespo. 1999. Red grouper reproduction in the southern Gulf of Mexico. Trans. Am. Fish. Soc., 128(3):385–402.

Buckman, N. S., and J. C. Ogden. 1973. Territorial behaviour of the striped parrotfish, *Scarus croicensis* (Bloch) (Scaridae). Ecology, 54(6):1377–1382.

Buesa, R. J. 1970. La biajaiba y su trama alimentaria. Mar y Pesca, 54:24–29.

Bullock, L. H., and M. D. Murphy. 1994. Aspects of the life history of the yellowmouth grouper, *Mycteroperca interstitialis* in the eastern Gulf of Mexico. Bull. Mar. Sci. 55:30–45.

Bullock, L. H., and G. N. Smith. 1991. Seabasses (Pisces: Serranidae). In: Memoirs of the Hourglass Cruises, vol VIII, part II, 243 p.

Bullock, L. H., M. D. Murphy, M. F. Godcharles and M. E. Mitchill. 1992. Age, growth and reproduction of jewfish, *Epinephelus itajara*, in the eastern Gulf of Mexico. Fish. Bull., 90(2):243–249.

Burch, R. K. 1979. The greater amberjack, *Seriola dumerili*: It's biology and fishery off southeastern Florida. M.S. thesis, Univ. Miami, Coral Gables, FL. 113 p.

Burnett-Herkes, J. 1975. Contribution to the biology of red hind, *Epinephelus guttatus*, a commercially important serranid fish from the tropical Western Atlantic. Ph.D. diss., Univ. Miami, Coral Gables, FL, 154 p.

Burton, M. L. 2000. Age, growth, mortality, reproductive seasonality and population status of gray snapper, *Lutjanus griseus*, from the east coast of Florida. M.S. thesis. North Carolina State Univ., Raleigh, NC.

Bustamante, G. 1983. Variaciones estacionales de algunos índices morfofisiológicos y bioquímicos del jallao, *Haemulon album*, en la plataforma suroccidental de Cuba. Rep. Invest. Inst. Oceanol. Acad. Cienc. Cuba, 10:1–17.

Bustamante, G. 1986. Características ecofisiológicas de algunos peces de la plataforma SW de Cuba: Cibí carbonero, *Caranx ruber* (Bloch) (Carangidae); jallao, *Haemulon album* (Cuvier) (Haemulidae); y sardina de ley, *Harengula humeralis* (Cuvier) (Clupeidae). Tesis doctoral, Inst. Oceanol. Acad. Cienc. Cuba, La Habana, 102 p.

Bustamante, G. 1987a. Aspectos ecofisiológicos de los cambios en diferentes años de la adiposidad corporal del cibí carbonero, *Caranx ruber* (Pisces: Carangidae) del SW de Cuba, en relación con el régimen hidroclimático. In: Trabajos del Primer Congreso de Ciencias del Mar, La Habana, Junio 1987.

Bustamante, G. 1987b. Características ecofisiológicas de la sardina de ley, *Harengula humeralis* (Cuvier, 1829), en aguas cubanas. Rep. Invest. Inst. Oceanol. Acad. Cienc. Cuba, 56:1–17.

Bustamante, G. 1988. Ecofisiología de la variación ontogénica del metabolismo lipídico del cibí carbonero, *Caranx ruber* (Bloch) (Pisces: Carangidae), en la plataforma SW de Cuba. Rep. Invest. Inst. Oceanol. Acad. Cienc. Cuba, 13:1–12.

Bustamante, G. 1989. Aspectos ecofisiológicos de la reproducción del cibí carbonero, *Caranx ruber* (Bloch) (Pisces: Carangidae), en la región suroeste de Cuba. Trop. Ecol., 30(2):205–217; Rep. Invest. Inst. Oceanol. Acad. Cienc. Cuba, 7:1–20.

Bustamante, G., and Y. Enomoto. 1981. Cultivo experimental de lisas (*Mugil curema, M. liza* y *M. trichodon*) en estanques. Inf. Cient.-Téc., Acad. Cienc. Cuba, 158:3–15.

Bustamante, G., and M. Schwartzberg. [unpublished]. The effect of Cyclone Frederick on the ecophysiological cycle of bar jack, *Caranx ruber* (Pisces:

Carangidae) in Cuban waters. 125th Ann. Meet. Amer. Fish. Soc., Tampa, FL.

Bustamante, G., M. I. Shatunovsky, and J. E. García-Jorge. 1981. Variación de los índices morfofisiológicos y bioquímicos del civil, *Caranx ruber,* en la plataforma suroccidental de Cuba. Cienc. Biol., 6:81–92.

Bustamante, G., J. E. García-Jorge, and J. P. García-Arteaga. 1982. La pesca con chinchorro en la región oriental del Golfo de Batabanó y algunos datos sobre las pesquerías en la plataforma cubana. Rep. Invest. Inst. Oceanol. Acad. Cienc. Cuba, 4:1–31.

Bustamante, G., L. M. Sierra, and J. P. García-Arteaga. 1988. Ecofisiología del crecimiento, la alimentación y el metabolismo lipídico de los juveniles (0+) de cibí carbonero, *Caranx ruber* (Bloch), en el Golfo de Batabanó (plataforma suroccidental de Cuba). Rep. Invest. Inst. Oceanol. Acad. Cienc. Cuba, 2:1–28.

Bustamante, G., A. García-Cagide, L. Sierra, and J. P. García-Arteaga. 1992. Características biológicas de la manjúa, *Jenkinsia lamprotaenia* (Pisces: Clupeidae) en la plataforma cubana. Unpublished, Inst. Oceanol., CITMA, Habana, Cuba. 29 p., 5 tabs., 10 figs.

Bustamante, G., R. Claro, and M. I. Shatunovsky. 1994. Ecofisiología. In: R. Claro, ed. Ecología de los Peces Marinos de Cuba. Inst. Oceanol. Cent. Invest. Quintana Roo, México, pp. 403-434.

Bustamante, G., M. Chiappone, F. X. Geraldes, E. Pugibet, E. Schmitt, R. Sluka, K. M. Sullivan-Sealey, R. E. Torres, M. Vega, Y. Rodriguez, J. Alarcon, and Y. Lichtensztajn. 1998. Reef fish assemblages and fisheries in Parque Nacional del Este, Dominican Republic. Proc. Gulf Caribb. Fish. Inst. 50:919-937.

Butler, M. J., IV, J. H. Hunt, W. F. Herrnkind, M. J. Childress, R. Bertelsen, W. Sharp, T. Matthews, J. M. Field, and H. G. Marshall. 1995. Cascading disturbances in Florida Bay, U.S.A.: Cyanobacteria blooms, sponge mortality, and implications for juvenile spiny lobsters *Panulirus argus.* Mar. Ecol. Prog. Ser. 129:119–125.

Cabrera-Vázquez, M., and F. Arreguín-Sánchez. 1989. Dinámica de la pesquería de carito (*Scomberomorus cavalla* [Cuvier, 1829]) en la costa norte de la Península de Yucatán. Unpublished, Cent. Invest. y Estudios Avanzados del Inst. Politécnico Nacl., Unidad Mérida, México, 16 p., 4 tabs., 5 figs.

Caddy, J. F. 1984. An alternative to equilibrium theory for management of fisheries. FAO Fish. Rep. No. 289, Suppl. 2. Rome, Italy, 214 p.

Cailliet, G., L. Martin, J. Harvey, D. Kusher, and B. Welden. 1983. Preliminary studies on the age and growth of blue *Prionace glauca,* common thresher *Alopias vulpinus,* and short fin mako *Isurus oxyrinchus* sharks from California waters. NOAA Tech. Rep. NMFS, 8:179–188.

Caley, M. J., M. H. Carr, M. A. Hixon, T. P. Hughes, G. P. Jones, and B. A. Menge. 1996. Recruitment dynamics and the local dynamics of open marine populations. Ann. Rev. Ecol. Syst. 27:477-500.

Campana, S. E. 1983. Feeding periodicity and the production of daily growth increments in otoliths of steelhead trout *Salmo gairdnieri* and starry flounder (*Platichthys stellatus*). Can. J. Zool., 61:1591–1597.

Campana, S. E. 1984. Microstructural growth patterns in the otoliths of larval and juvenile starry flounder, *Platichthys stellatus.* Can. J. Zool., 62:1507–1512.

Campana, S. E., and J. D. Neilson. 1982. Daily growth increments in otoliths of starry flounder (*Platichthys stellatus*) and the influence of some environmental variable in their production. Can. J. Fish. Aquat. Sci., 39:937–942.

Campana, S. E., and J. D. Neilson. 1985. Microstructure of fish otoliths. Can. J. Fish. Aquat. Sci., 42(5):1014–1032.

Campos, A. G., and A. K. M. Bashirullah. 1975. Biología del pargo *Lutjanus griseus* (Linn.) de la isla de Cubagua, Venezuela. II. Maduración sexual y fecundidad. Bol. Inst. Oceanogr. Univ. Oriente, Cumaná, 14(1):109–116.

Cantarell, E. 1982. Determinación de la edad y el ritmo de crecimiento del pargo canané (*Ocyurus chrysurus,* Bloch 1871), en el estado de Yucatán. Tesis de licenciatura E.N.E.P.I., UNAM, México.

Capote, A. M. 1971. On the age of commercially significant benthic fishes of the Campeche Bank [in Russian]. In: A. S. Bogdanov, ed. Joint Soviet-Cuban Fisheries Investigations Pishevaia Promishlennost, Moscow, vol. 3, pp. 77–81.

Carles, C. 1967. Algunos datos de la biología del banano, *Elops saurus* L. (Teleostomi: Elopidae). Cent. Invest. Pesq., Inst. Nac. Pesca Cuba, Contrib., 27:1–53.

Carles, C. 1971. Características biológico-pesqueras del bonito (*Katsuwonus pelamis*) y la albacora (*Thunnus atlanticus*) en la costa nororiental de Cuba. Cent. Invest. Pesq., Inst. Nacl. Pesca Cuba, Contrib., 32:1–51.

Carles, C. 1972. Situación actual de la pesca del bonito en Cuba. In: Tercera Reunión de balance de trabajo del Centro de Investigaciones Pesqueras, Inst. Nacl. Pesca, La Habana, vol. 3, no. 1, pp. 87–100.

Carles, C. 1974. Edad y crecimiento del bonito (*Katsuwonus pelamis*) y la albacora (*Thunnus atlanticus*) en la parte occidental de Cuba. Resúmenes Invest. Cent. Invest. Pesq., Cuba, 1(1):122–126.

Carles, C. 1975. Evaluación de la pesquería del bonito en la zona occidental de Cuba. Resúmenes Invest. Cent. Invest. Pesq., Cuba, 1(1):73–96.

Carles, C., and A. Hirtenfield. 1978. Resultados de la pesca del bonito (*Katsuwonus pelamis*) y la albacora (*Thunnus atlanticus*) en Cuba durante 1977. Rev. Invest. Cent. Invest. Pesq., Inst. Nac. Pesca Cuba, 3(2):62–78.

Carneiro Ximenes, M. O., M. Ferreira, and A. A. Fonteles-Filho. 1978. Idade e crescimento da cavala, *Scomberomorus cavalla* (Cuvier), no Estado do Ceará (Brasil). Arq. Cienc. Mar., 8(1–2):73–81.

Carneiro Ximenes, M. O., and A. A. Fonteles-Filho. 1988. Estudo da idade e crescimento do pargo, *Lutjanus purpureus* Poey (Pisces: Lutjanidae), no norte e nordeste do Brasil. Arq. Cienc. Mar, 27:69–81.

Carpenter, K. E. 1977. Philippine coral reef fisheries resources. Philipp. J. Fish., 17:95–125.

Carr, W. E., and C. Adams. 1973. Food habits of juvenile marine fishes ocupying seagrass bed in the estuarine zone near Cristal River, Florida. Trans. Am. Fish. Soc., 102:511–540.

Carrillo de Albornoz, C., and M. E. Ramiro. 1988. Estudio biológico de la rabirrubia (*Ocyurus chrysurus*) en el oeste de la plataforma sureste de Cuba. I. Edad y crecimiento. Rev. Invest. Mar., 9(1): 9–24.

Carrodeguas, C., G. Arencibia, N. Capetillo, and M. García. In press. Decoloración de corales en el Archipiélago Cubano. Rev. Invest. Pesq.

Carter, J., and D. Perrine. 1994. A spawning aggregation of dog snapper, *Lutjanus jocu* (Pisces: Lutjanidae) in Belize, Central America. Bull. Mar. Sci., 55(1):228–234.

Carter, J., G. J. Marrow, and V. Pryor. 1994. Aspects of the ecology and reproduction of Nassau grouper, *Ephinephelus striatus,* off the coast of Belize, Central America. Proc. Gulf Carib. Fish. Inst., 43:65–111.

Carvajal, R. J. 1975. Contribución al conocimiento de la biología de los robalos *Centropomus undecimalis* y *C. poeyi* en la Laguna de Términos, Campeche, México. Bol. Inst. Oceanogr. Univ. Oriente Cumaná, 14(1):51–71.

Carvalho, M. O. X., and A. A. Fonteles-Filho. 1995. Age and growth of fishes of genus *Haemulon* (Pisces: Pomadasyidae) in Ceara State, Brazil. Arquivos do Ciencia do Mar, Fortaleza, 29(1):14–19.

Caselles-Osorio, A., and P. A. Acero. 1996. Reproducción de *Anchovia clupeoides* and *Anchoa parva* (Pisces: Engraulidae) en dos ciénagas del caribe colombiano. Rev. Biol. Trop. 44(2-B):781–793.

Casey, J. G., H. L. Pratt, Jr., and C. E. Stillwell. 1983. Age and growth of the sandbar shark (*Carcharhinus plumbeus*) from the western north Atlantic. Can. J. Fish. Aquat. Sci., 42(5):963–975.

Cayre, P., and H. Farrugio. 1986. Biologie de la reproduction du listao (*Katsuwonus pelamis*) de l'Océan Atlantique. In: Actas de la Conferencia ICCAT sobre el Programa del Año Internacional del Listado, Madrid, 1983, pp. 252–273.

Centro de Investigaciones Pesqueras (CIP). 1978. Sinopsis sobre los recursos pesqueros. Segunda parte. Cent. Invest. Pesq., La Habana, Cuba. 112 p.

Cervigón, F. 1994. Los Peces Marinos de Venezuela. Estación de Investigaciones Marinas de Isla Margarita, Fundación La Salle de Ciencias Naturales, Caracas, vols. 1 and 2, 951 p.

Chavence, P., A. Yáñez-Arancibia, D. Flores-Hernández, A. L. Lara-Domínguez, and F. Amezcua. 1986. Ecology, biology and population dynamics of *Archosargus rhomboidalis* (Pisces: Sparidae) in tropical coastal lagoon system, southern Gulf of México. Annu. Inst. Cienc. Mar. Limnol. Univ. Nacl. Autón. México, 13(2):11–30.

Chiappone, M., and R. Sluka. 1996. Fishes and Fisheries. Volume 6: Site Characterization for the Florida Keys National Marine Sanctuary. Farley Court Publishers, Zenda, WI. 86 p.

Chiappone, M., and K. M. Sullivan Sealy. 2000. Marine reserve design criteria and measures of success: Lessons learned from the Exuma Cays Land and Sea Park, Bahamas. Bull. Mar. Sci., 66(3):691–706.

Chiappone, M., R. Sluka, K. M. Sullivan, E. Schmitt, G. Bustamante, J. Kelly, M. Vega, E. Pugibet, F. X. Geraldes, and R. E. Torres. 1998. Comparison of grouper assemblages in northern areas of the wider Caribbean: A preliminary assessment. Proc. Gulf and Carib. Fish. Inst. 50:427–451.

Choat, J. H., and D. R. Robertson. 1975. Protogynous hermaphroditism in fishes of the family Scaridae. In: R. Reinboth, ed. Intersexuality in the Animal Kingdom. Heidelberg: Springer-Verlag, Berlín, Germany, pp. 263–283.

Clark, E. 1959. Functional hermaphroditism and self-fertilization in serranid fish. Science, 129:215–216.

Clarke, R. D. 1977. Habitat distribution and species diversity on chaetodontid and pomacentrid fishes near Bimini, Bahamas. Mar. Biol., 40:277–289.

Clarke, T. A. 1971. Territorial boundaries, courtship and social behavior in the Garibaldi, *Hypsypops rubicunda* (Pomacentridae). Copeia, 2:205–299.

Clarke, T. A. 1978. Diel feeding patterns of 16 species of mesopelagic fishes from Hawaiian waters. Fish. Bull., 76(3):495–521.

Clarke, T. A. 1980. Diets of fourteen species of vertically migrating mesopelagic fishes in Hawaiian waters. Fish. Bull., 78:619–640.

Claro, R. 1976. Ecology and dynamics of biological indicators on lutjanid fishes in Cuba [in Russian]. Ph.D. diss., Inst. Evolutionary Morphology and Animal Ecology, Soviet Acad. Sci., Moscow, 149 p.

Claro, R. 1981a. Ecología y ciclo de vida de la biajaiba, *Lutjanus synagris* (Linnaeus), en la plataforma cubana. II. Biología pesquera. Inf. Cient.-Téc., Acad. Cienc. Cuba., 177:1–53.

Claro, R. 1981b. Ecología y ciclo de vida de la biajaiba, *Lutjanus synagris*, en la plataforma cubana. III. Nutrición. Cienc. Biol., 6:93–110.

Claro, R. 1981c. Ecología y ciclo de vida del pargo criollo, *Lutjanus analis* (Cuvier), en la plataforma cubana. Inf. Cient.-Téc., Acad. Cienc. Cuba., 186:1–83.

Claro, R. 1982. Ecología y ciclo de vida de la biajaiba, *Lutjanus synagris* (Linnaeus), en la plataforma cubana. IV. Reproducción. Rep. Invest. Inst. Oceanol. Acad. Cienc. Cuba, 5:1–37.

Claro, R. 1983a. Ecología y ciclo de vida de la biajaiba, *Lutjanus synagris*, en la plataforma cubana. V. Dinámica estacional de algunos indicadores morfofisiológicos. Rep. Invest. Inst. Oceanol. Acad. Cienc. Cuba, 16:1–24.

Claro, R. 1983b. Ecología y ciclo de vida del caballerote, *Lutjanus griseus* (Linnaeus), en la plataforma cubana. II. Edad y crecimiento, estructura de las poblaciones, pesquerías. Rep. Invest. Inst. Oceanol. Acad. Cienc. Cuba, 8:1–28.

Claro, R. 1983c. Ecología y ciclo de vida del caballerote, *Lutjanus griseus* (Linnaeus), en la plataforma cubana. I. Identidad, distribución y hábitat, nutrición y reproducción. Rep. Invest. Inst. Oceanol. Acad. Cienc. Cuba, 7:1–30.

Claro, R. 1983d. Ecología y ciclo de vida de la rabirrubia, *Ocyurus chrysurus* (Bloch), en la plataforma cubana. I. Identidad, distribución, hábitat, reproducción y alimentación. Rep. Invest. Inst. Oceanol. Acad. Cienc. Cuba, 15:1–34.

Claro, R. 1983e. Ecología y ciclo de vida de la rabirrubia, *Ocyurus chrysurus* (Bloch), en la plataforma cubana. II. Edad y crecimiento, estructura de poblaciones y pesquerías. Rep. Invest. Inst. Oceanol. Acad. Cienc. Cuba, 19:1–33.

Claro, R. 1983f. Dinámica estacional de algunos indicadores morfofisiológicos del pargo criollo, *Lutjanus analis* (Cuvier), en la plataforma cubana. Rep. Invest. Inst. Oceanol. Acad. Cienc. Cuba, 22:1–14.

Claro, R. 1985. Ecología y ciclo de vida del caballerote, *Lutjanus griseus*, en la plataforma cubana. III. Dinámica de algunos indicadores morfofisiológicos. Rep. Invest. Inst. Oceanol. Acad. Cienc. Cuba, 36:1–14.

Claro, R. 1991. Changes in fish assemblage structure by the effects of intense fisheries activity. Trop. Ecol. 32(1):36–46.

Claro, R., ed. 1994a. Ecología de los Peces Marinos de Cuba. Inst. Oceanol. Acad. Cien. Cuba y Cent. Invest. Quintana Roo, Mexico. 525 p.

Claro, R. 1994b. Características generales de la ictiofauna. In: R. Claro, ed. Ecología de los peces marinos de Cuba. Inst. Oceanol. Acad. Cien. Cuba y Cent. Invest. Quintana Roo, Mexico, pp. 55–142.

Claro, R., and G. Bustamante. 1977. Edad y crecimiento del caballerote, *Lutjanus griseus* (Linnaeus), en la plataforma suroccidental de Cuba. Inf. Cient.-Téc., Inst. Oceanol. Acad. Cienc. Cuba, 12:1–11.

Claro, R., and I. Colás. 1987. Algunos elementos del balance energético del pargo criollo, *Lutjanus analis* (Cuvier, 1828), en condiciones experimentales. Rep. Invest. Inst. Oceanol. Acad. Cienc. Cuba, 58:1–19.

Claro, R., and J. P. García-Arteaga. 1993. Estructura de las comunidades de peces asociados a los manglares del grupo insular Sabana-Camaguey, Cuba. Revista Oceanol. Ecol. Trop. Avicennia, 0:60–82.

Claro, R., and J. P. García-Arteaga. 1994a. Crecimiento. In: R. Claro, ed. Ecología de los Peces Marinos de Cuba. Inst. Oceanol. Cent. Invest. Quintana Roo, Mexico, pp. 321–402.

Claro, R., and J. P. García-Arteaga. 1994b. Estructura de las comunidades de peces en los arrecifes del grupo insular Sabana-Camaguey, Cuba. Revista Oceanol. Ecol. Trop. Avicennia, 2:83–107.

Claro, R., and J. P. García-Arteaga. 1999. Perspectives on an Artificial Habitat Program for Fishes of the Cuban Shelf. G. Bustamante, and K. C. Lindeman (translators). SGEB-49 Florida Sea Grant College Program, Univ. Florida, 38 p.

Claro, R., and E. Giménez. 1989. Desarrollo y estado actual de las pesquerías de peces en la región este del Golfo de Batabanó. Cienc. Biol., 21–22:150–160.

Claro, R., and V. I. Lapin. 1971. Algunos datos sobre la alimentación y la dinámica de las grasas en la biajaiba, *Lutjanus synagris* (Linnaeus), en el Golfo de Batabanó, plataforma sur de Cuba. Ser. Oceanol. Acad. Cienc. Cuba, 10:1–16.

Claro, R., and V. I. Lapin. 1973. Variabilidad de algunos parámetros bioquímicos en órganos y tejidos de la biajaiba (*Lutjanus synagris* [L.]), durante el proceso de maduración gonadal. Ser. Oceanol. Acad. Cienc. Cuba, 17:1–18.

Claro, R., and Y. S. Reshetnikov. 1981. Ecología y ciclo de vida de la biajaiba, *Lutjanus synagris* (Linnaeus), en la plataforma cubana. I. Formación de las marcas de crecimiento en sus estructuras. Inf. Cient.-Téc., Acad. Cienc. Cuba, 174:1–28.

Claro, R., and Y. S. Reshetnikov. 1994. Condiciones de hábitat. In: R. Claro, ed. Ecología de los peces marinos de Cuba. Inst. Oceanol. Cent. Invest. Quintana Roo, Mexico, pp. 13–54.

Claro, R., A. García-Cagide, and R. Fernández de Alaiza. 1989. Características biológicas del pez perro, *Lachnolaimus maximus* (Walbaum), en el Golfo de Batabanó, Cuba. Rev. Invest. Mar., 10(3):239–252.

Claro, R., J. P. García-Arteaga, E. Valdés-Muñoz, and L. M. Sierra. 1990a. Alteraciones de las comunidades de peces en el Golfo de Batabanó, en relación con la explotación pesquera. In: R. Claro, ed. Asociaciones de Peces en el Golfo de Batabanó. Editorial Academia, La Habana, pp. 50–66.

Claro, R., J. P. García-Arteaga, E. Valdés-Muñoz, and L. M. Sierra. 1990b. Características de las comunidades de peces en los arrecifes del Golfo de Batabanó. In: R. Claro, ed. Asociaciones de Peces en el Golfo de Batabanó. Editorial Academia, La Habana, pp. 1–49.

Claro R., A. García-Cagide, L. M. Sierra, and J. P. García-Arteaga. 1990c. Características biológico-pesqueras de la cherna criolla *Epinephelus striatus* (Bloch) (Pisces: Serranidae) en la plataforma cubana. Cienc. Biol., 23:23–42.

Claro, R., J. Baisre, and J. P. García-Arteaga. 1994. Evolución y manejo de los recursos pesqueros. In: R. Claro, ed. Ecología de los peces marinos de Cuba. Inst. Oceanol. Acad. Cien. Cuba y Cent. Invest. Quintana Roo, Mexico, pp. 435–492.

Claro, R., J. P. García-Arteaga, Y. Bouchon-Navarro, M. Louis, and C. Bouchon. 1998. La estructura de las comunidades de peces en los arrecifes de las Antillas Menores y Cuba. Revista Oceanol. Ecol. Trop. Avicennia, 8:69–86.

Claro R., L. M. Sierra, and J. P. García-Arteaga. 1999. Biología del jocu, *Lutjanus jocu* (Pisces:Lutjanidae) en las NE y SW de la plataforma Cubana. II. Alimentación, edad y crecimiento. Rev. Invest. Mar. 20(1-3):45–53.

Claro, R., K. Cantelar-Ramos, F. Pina-Amargós, and J. P. García-Arteaga. 2000a. Biodiversidad y manejo de la ictiofauna del Archipielago Sabana-Camagüey. Inst. Oceanol., 89 p.

Claro, R., R. G. Gilmore, C. R. Robins, and J. E. McCosker. 2000b. Nuevos registros para la ictiofauna marina de Cuba. Revista Oceanol. Ecol. Trop. Avicennia, 12/13:19–24.

Claro, R., J. P. García-Arteaga, and F. Pina-Amargós. In press. La ictiofauna de los fondos blandos del Archipielago Sabana-Camagüey, Cuba. Rev. Invest. Mar.

Claro, R., J. P. García-Arteaga, and F. Pina-Amargós. In press. Nuevos registros de peces marinos para Cuba. Revista Oceanol. Ecol. Trop. Avicennia.

Clifton, K. E., and L. M. Clifton. 1998. A survey of fishes from various coral reef habitats within the Cayos Cochinos Marine Reserve, Honduras. Rev. Biol. Trop. 46(4):109–124.

Coblentz, B. E. 1995. Reproductive biology of the dwarf herring (*Jenkinsia lamprotaenia*) in the Virgin Islands. Bull. Mar. Sci., 56(2):602–608.

Coleman, F. C., C. C. Koenig, A. Eklund, and C. B. Grimes. 1999. Management and conservation of temperate reef fishes in the grouper-snapper complex of the southeastern United States. Am. Fish. Soc. Symp., 23:233–242.

Colin, P. L. 1973. Comparative biology of the gobies of the genus *Gobiosoma*, subgenus *Elacatinus*, in the western North Atlantic. Ph.D. Diss., Univ. Miami. Coral Gables, FL. 247 p.

Colin, P. L. 1982. Spawning and larval development of the hogfish, *Lachnolaimus maximus* (Pisces: Labridae). Fish. Bull., 80(4):853–862.

Colin, P. L. 1992. Reproduction of the Nassau grouper, *Epinephelus striatus* (Pisces: Serranidae) and its relationship to environmental conditions. Environ. Biol. Fish., 34:357–377.

Colin, P. L., and I. E. Clavijo. 1988. Spawning activity of fishes producing pelagic eggs on a shelf edge coral reef, southwestern Puerto Rico. Bull. Mar. Sci., 43:249–279.

Colin, P. L., D. Y. Shapiro, and D. Weiler. 1987. Aspects of the reproduction of two groupers, *Epinephelus guttatus* and *E. striatus,* in the West Indies. Bull. Mar. Sci., 40(2):220–230.

Colin, P. L., W. A. Laroche, and E. B. Brothers. 1997. Ingress and settlement in the Nassau grouper, *Epinephelus striatus* (Pisces: Serranidae), with relationship to spawning occurrence. Bull. Mar. Sci., 60(4):656–667.

Collette, B. B., and F. H. Talbot. 1972. Activity patterns of coral reef fishes with emphasis on nocturnal-diurnal change over. Nat. Hist. Mus. Los Angeles County Sci. Bull., 14:98–124.

Collins, L. A., A. G. Johnson, C. C. Koenig, and M. S. Baker, Jr. 1998. Reproductive patterns, sex ratio, and fecundity in gag, *Mycteroperca microlepis* (Serranidae), a protogynous grouper from the northeastern Gulf of Mexico. Fish. Bull., 96(3):415–427.

Collins, M. R., C. W. Waltz, W. A. Rosemillat, and D. L. Stubbs. 1987. Contribution to the life history and reproductive biology of gag, *Mycteroperca microlepis* (Serranidae) in the South Atlantic Bight. Fish. Bull., 85(3):648–653.

Corral, J. I. del. 1940. El geosinclinal cubano. Editora de Libros y Folletos, La Habana, 141 p.

Corvea, A., M. Abreu, and P. M. Alcolado. 1990. Características de las comunidades de equinodermos de la macrolaguna del Golfo de Batabanó. In: P. M. Alcolado, ed. El Bentos de la Macrolaguna del Golfo de Batabanó. Editorial Academia, La Habana, pp. 90–99.

Cowey, C. B., and J. R. Sargent. 1979. Nutrition. In: W. S. Hoar, D. J. Randall and J. R. Brett, eds. Fish Physiology. 8. Bioenergetics and Growth. Academic Press, New York. pp. 1–70.

Crabtree, R. E., and L. H. Bullock. 1998. Age, growth, and reproduction of black grouper, *Mycteroperca bonaci,* in Florida waters. Fish. Bull., 96(4):735–753.

Crabtree, R. E., E. C. Cyr, and J. M. Dean. 1995. Age and growth of tarpon, *Megalops atlanticus,* from south Florida waters. Fish Bull. 93:619–628.

Crabtree, R. E., C. Stevens, D. Snodgrass, and F. J. Stengard. 1998. Feeding habitats of bonefish, *Albula vulves,* from the waters of the Florida Keys. Fish. Bull. 96(4):754–766.

Croker, R. A. 1962. Growth and food of the gray snapper, *Lutjanus griseus,* in Everglades National Park. Trans. Am. Fish. Soc., 91:379–383.

Cruz, R. 1978. Algunas consideraciones sobre las pesquerías de biajaiba (*Lutjanus synagris* Linné 1758) en la plataforma suroccidental de Cuba. Rev. Cub. Invest. Pesq., 3(3):51–82.

Cruz Izquierdo, R. 1999. Reclutamiento, procesos oceánicos y su relación con el pronóstico de la pesquería de la langosta, *Panulirus argus* (Latreille, 1804) en Cuba. Tesis doctoral, Univ. de la Habana, La Habana.

Cruz Izquierdo, R., R. Sotomayor, M. E. de León, and R. Puga. 1991. Impacto en el manejo de la pesquería de langosta (*Panulirus argus*) en el Archipiélago cubano. Rev. Invest. Mar., 12(1–3):246–253.

Csirke, J., and J. Caddy. 1983. Production modelling using mortality estimates. Can. J. Fish. Aquat. Sci., 40(1):43–51.

CUB/92/G31. 1999. Protecting biodiversity and establishing sustainable development in the Sabana-Camagüey ecosystem. P. M. Alcolado, E. E. García, and N. Espinosa, eds. Global Environmental Facility (GEF) / United Nations Development Program (UNDP) Project Sabana-Camagüey, Cuba, 145 p.

Cummings, W. C., B. D. Brany, and J. Y. Spires. 1966. Sound production, schooling and feeding habits of the margate, *Haemulon album* Cuvier, off North Bimini, Bahamas. Bull. Mar. Sci. Gulf Carib., 16(3):620–639.

Cushing, D. H. 1972. The production cycle and the numbers of marine fish. Symp. Zool. Soc. Lond., 29:213–232.

Daan, N. 1975. Consumption and production in North Sea cod, *Gadus morhua*. An assessment of the ecological status of the stock. Neth. J. Sea Res., 9(1):24–55.

Damant, G. C. C. 1921. Illumination of plankton. Nature, 108:42–43.

Damas, T., M. Borrero, N. Millares, and E. González. 1979. Fecundidad de la biajaiba, *Lutjanus synagris* (L. 1758). Rev. Cub. Invest. Pesq., 4(3):18–31.

D'Ancona, U. 1950. Détermination et différentiation du sexe chez les poissons. Arch. Anat. Microsc. Morphol. Exp., 39(3):274–292.

Danilowicz, B. S., and P. F. Sale. 1999. Relative intensity of predation on the French grunt, *Haemulon flavolineatum*, during diurnal, dusk and nocturnal periods on a coral reef. Mar. Biol., 133:337–343.

Davis, C. 1976. Biology of the hogfish in the Florida Keys. M.S. thesis, Univ. Miami, 86 p.

Davis, G. 1989. Designated harvest refugia: The next stage in marine fishery management in California. CaLCOFL Rep., 30:53–58.

Davis, W. P., and R. S. Birdsong. 1973. Coral reef fishes which forage in the water column. A review of their morphology, behavior, ecology and evolutionary implications. Helgol. Meeresunters, 24(1–14):292–306.

Dejhnik, T. V., M. Juárez, and D. Salabarría. 1976. Distribution of fish pelagic eggs and larvae in Cuban waters [in Russian]. In: V. A. Vodyanitski, ed. Central American Seas Research. Naukova Dumka, Kiev, vol. 1, pp. 131–170.

den Hartog, C. 1970. The seagrasses of the world. North Holland Publication Co., 275 p.

Dennis, G. D. 1988. Commercial catch length-frequency data as a tool for fisheries management with an application to the Puerto Rico trap fishery. Mem. Soc. Cienc. Nat. La Salle, t. 48, suppl. 3:289–310.

Dennis, G. D. 1991. The validity of length-frequency derived growth parameters from commercial catch data and their application to stock assessment of the yellowtail snapper (*Ocyurus chrysurus*). Proc. Gulf Carib. Fish. Inst., 40:126–138

Dennis, G. D. 1992a. Island mangrove habitats as spawning and nursery areas for commercially important fishes in the Caribbean. Proc. Gulf Carib. Fish. Inst., 41(B):205–225.

Dennis, G. D. 1992b. Resource utilization by members of a guild of benthic feeding coral reef fish. Ph.D. diss., Univ. Puerto Rico, Mayaguez, Puerto Rico.

De Sylva, D. P. 1963. Systematics and life history of the great barracuda *Sphyraena barracuda* (Walbaum). Stud. Trop. Oceanogr., 1:1–179.

De Sylva, D. P. 1994. Distribution and ecology of ciguatera fish poisoning in Florida, with emphasis on the Florida Keys. Bull. Mar. Sci., 54(3):944–954.

Díaz-Briquets, S., and J. Pérez-López. 2000. Conquering Nature: The Environmental Legacy of Socialism in Cuba. University of Pittsburgh Press. 328 p.

Díaz-Piferrer, L. 1972. Las algas superiores y fanerógamas marinas. In: Ecología Marina, Fundación La Salle de Ciencias Naturales, Caracas, pp. 272–307.

Díaz-Ruíz, A., Aguirre-León, C. Macuitl, and O. Pérez. 1996. Seasonal patterns of distribution and abundance of snappers in the Mexican Caribbean. In: F. Arreguín-Sánchez, J. L. Munro, M. C. Balgos, and D. Pauly, eds. Biology, Fisheries and Culture of Tropical Groupers and Snappers. ICLARM Conf. Proc. 48, pp. 43–50.

Dietz, R. S., and J. C. Holden. 1970. Reconstruction of Pangaea: Breakup and dispersion of continents, Permian to present. J. Geophys. Res., 75(26):4939–4956.

Disler, N. N. 1950. Development of the sense organs of the lateral-line system in the perch and the scorpionfish [in Russian]. Tr. Inst. Morfol. Zhibotnij Akad. Nauk SSSR, 2:85–139.

Dobzhansky, T. 1950. Evolution in the tropics. Am. Sci., 38:209–221.

Dodrill, J. 1993. Food and feeding behavior of adult snowy grouper, *Epinephelus nivealus* (Valenciennes) (Pisces: Serranidae), collected off the central North Carolina coast with ecological notes on major food groups. Brimleyana, 19:101–135.

Doherty, P. J. 1991. Spatial and temporal patterns of recruitment. In: P. F. Sale, ed. The Ecology of Fishes of Coral Reefs. Academic Press. San Diego, CA, pp. 261–293.

Doherty, P. J., and T. Fowler. 1994. An empirical test of recruitment limitation in a coral reef fish. Science 263:935–939.

Doherty, P., and J. McIlwaun. 1996. Monitoring larval fluxes through the surf zones of Australian coral reefs. Mar. Freshwater Res., 47:383–390.

Doherty, P. J., and D. McB. Williams. 1988. The replenishment of coral reef fish populations. Oceanogr. Mar. Biol. Ann. Rev., 26: 487–551.

Doi, T., D. Mendizábal, and M. Contreras. 1981. Análisis preliminar de la población de mero *Epinephelus morio* (Valenciennes) en el Banco de Campeche. Cienc. Pesq., Mexico (1):1–16.

Domeier, M. L. 1994. Speciation in the serranid fish *Hypoplectrus*. Bull. Mar. Sci. 54(1):103–141.

Domeier, M. L., and P. L. Colin. 1997. Tropical reef fish spawning aggregations: Defined and reviewed. Bull. Mar. Sci. 60(3):698–726.

Domeier, M. L., C. C. Koenig, and F. C. Coleman. 1996. Reproductive biology of the gray snapper (*Lutjanus griseus*), with notes on spawning for other western Atlantic snappers (Lutjanidae). In: F. Arreguín-Sánchez, J. Munro, M. C. Balgos, and D. Pauly, eds. Biology Fisheries and Culture of Tropical Groupers and Snappers. ICLARM Conf. Proc. 48:189–201, 449 p.

Dominguez-Viveros, M., and J. G. Avila-Martínez. 1996. Preliminary diagnostics of the white grunt (*Haemulon plumieri*, Lacepede, 1802) fishery of Campeche Bank, based on length-frequency analysis. Proc. Gulf Carib. Fish. Inst. 44:747–758.

Domm, S. B., and A. J. Domm. 1973. The sequence of appearance at dawn and disappearance at dusk of some coral reef fishes. Pac. Sci., 27(2):128–135.

Donnelly, T. W. 1988. Geologic constraints on Caribbean biogeography. In: J. K. Liebherr, ed. Zoogeography of Caribbean Insects. Comstock, Cornell Univ. Press, Ithaca, NY, pp. 15–37.

Dooley, J. K. 1972. Fishes associated with the pelagic Sargassum complex, with a discussion of the Sargassum community. Contrib. Mar. Sci., 16:1–32.

Druzhinin, A. D. 1974. Distribution, biology and fisheries of drums [in Russian]. Pishevaia Promishlennost, Moscow, 119 p.

Druzhinin, A. D. 1976. The porgy fishes of the world oceans [in Russian]. Pishevaia Promishlennost, Moscow, 195 p.

Duarte-Bello, P. P. 1959. Catálogo de peces cubanos. Lab. Biol. Mar., Univ. Sto. Tomás de Villanueva, La Habana, vol. 6, 208 p.

Duarte-Bello, P. P., and R. J. Buesa. 1973. Catálogo de peces cubanos. Primera revisión. Ciencias, Ser. 8, Invest. Mar., 3:1–255.

Dubovistky, A. A. 1973. Composition of the porgy species of the genus *Calamus* (fam. Sparidae), in the Campeche Bank [in Russian]. Tr. VNIRO, 93:230–244.

Dubovistky, A. A. 1974. Sexual ratio and population size of *Calamus bajonado* (Bloch and Schneider, 1801), in the Campeche Bank [in Russian]. In: A. S. Bogdanov, ed. Joint Soviet-Cuban Fisheries Investigations Pishevaia Promishlennost, Moscow, vol. 4, pp. 177–181.

Dubovistky, A. A. 1977. Distribution, migration and some biological features of little-head porgy (*Calamus proridens,* Jordan and Gilbert, 1884) family Sparidae, of the Gulf of Mexico. FAO Fish. Rep., 200:123–144.

Dunlap, M., and J. R. Pawlik. 1998. Spongivory by parrotfish in Florida mangrove and reef habitats. Mar. Ecol., 19(4):325–337.

Du Paul, W. D., and J. D. McEachran. 1973. Age and growth of butterfish, *Peprilus triacanthus,* in the Lower York river. Chesapeake Sci., 14(3):205–207.

Durbin, A. G. 1979. Food selection by plankton feeding fishes. In: Predator-Prey Systems in Fisheries Management. Sport Fishing Inst., Washington, DC, pp. 203–218.

Durham, J. W. 1985. Movement of the Caribbean plate and its importance for biogeography in the Caribbean. Geology, 13: 123–125.

Echardt, H., and W. Meinel. 1977. Contribution to the biology of *Lutjanus analis* (Cuvier-Valenciennes, 1828) (Lutjanidae: Perciformes: Pisces) on the Colombian Atlantic coast. Inst. Rev. Ges. Hidrobiol., 62(1):161–175.

Edwards, A. W. F. 1960. Natural selection and sex ratio. Nature, 188:960.

Edwards, R. R., J. H. Blaxter, U. H. Gapalon, C. V. Mathew, and A. M. Finlayson. 1971. Feeding, metabolism, and growth of tropical flatfish. J. Exp. Mar. Biol. Ecol., 6:279–300.

Eggleston, D. B., J. J. Grover, and R. N. Lipcius. 1998. Ontogenic diet shifts in Nassau grouper: Trophic linkages and predatory impact. Bull. Mar. Sci., 63(1):111–126.

Eklund, A. M., D. B. McClellan, and D. E. Harper. 2000. Black grouper aggregations in relation to protected areas within the Florida Keys National Marine Sanctuary. Bull. Mar. Sci., 66(3):721–728.

Elizarov, A. A., and J. A. Gómez. 1971. On the water dynamics around Cuba [in Russian]. In: A. S. Bogdanov, ed. Joint Soviet-Cuban Fisheries Research. Pishevaia Promishlennost, Moscow, vol. 3, pp. 9–13.

Emery, A. R. 1968. Preliminary observations on coral reef plankton. Limnol. Oceanogr., 13(2):293–302.

Emilson, I. 1968. Investigaciones sobre la hidrología de la Ensenada de la Broa, con vista a su posible transformación en un embalse de agua dulce. Ser. Transformación de la Naturaleza., Acad. Cienc. Cuba, 5:1–45.

Emilson, I., and J. Tápanes. 1971. Contribución a la hidrología de la plataforma sur de Cuba. Ser. Oceanol. Acad. Cienc. Cuba, 9:1–31.

Erdman, D. S. 1977. Spawning patterns of fish from the North-eastern Caribbean. FAO Inf. Pesca, 200:145–170.

Eschmeyer, W. N., ed. 1990. Catalog of the genera of recent fishes. California Acad. Sci. i-v + 1–697.

Eschmeyer, W. N., ed. 1998. Catalog of Fishes. Special Pub. No.1 of the Center for Biodiversity Research and Information, California Acad. Sci., San Francisco, CA, vols. 1–3, 2905 p.

Escorriola, J. I. 1991. Age and growth of the gray triggerfish, *Balistes capriscus,* from the southeastern United States. M.S. thesis, Univ. North Carolina, Wilmington, NC.

Espinosa, J., P. M. Alcolado, C. Jiménez, D. Ibarzábal, J. C. Martínez, R. del Valle, N. Martínez, A. Hernández, M. Abreu, L. Vega, E. Ramírez, and J. Gonzáles. 1990. Ecología marina. In: L. Fernández, E. García, M. Aguiar, G. Rodríguez, and M. Páez, eds. Estudios de los Grupos Insulares y zonas litorales del Archipiélago Cubano con fines turísticos. Cayos Guillermo, Coco y Paredón Grande. Editorial Cient.-Téc., La Habana, pp. 101–112.

Espinosa, L. 1979. Aspectos biológico-pesqueros de la cherna criolla en la plataforma cubana y regiones adyacentes. In: Segundo Foro Cient. Cent. Invest. Pesq., Resúmenes de Investigaciónes, La Habana, p. 61.

Espinosa, L. In press a. Los tiburones de Cuba, biología, pesquería, importancia comercial, manejo del recurso y estrategia de investigación. Rev. Invest. Pesq., MIP, Cuba.

Espinosa, L. In press b. Las pesquerías de batoideos en la plataforma cubana, administración, importancia de su explotación y potencialidad económica. Rev. Invest. Pesqueras, MIP, Cuba.

Espinosa, L., and E. Pozo. 1982. Edad y crecimiento del sesí (*Lutjanus buccanella* [Cuvier, 1828]) en la plataforma suroriental de Cuba. Rev. Cub. Invest. Pesq., 7(1):80–100.

Espinosa, L., E. Pozo, and R. Quevedo. 1984. Edad y crecimiento del sesí (*Lutjanus buccanella* [Cuvier, 1828]) en la plataforma nororiental de Cuba. Rev. Cub. Invest. Pesq., 9(1):9–20.

Espinosa, L., M. Sosa, S. Moreno, and R. Quevedo. 1988. Aspectos biológicos de los peces de pico en la region noroccidental de Cuba. ECCAT, Col. Doc. Cienc., 28(1):321–338.

Everhart, W. H., A. W. Eipper, and W. D. Youngs. 1975. Principles of Fishery Science. Cornell Univ. Press, Ithaca and New York, NY, 288 p.

Everson, J. 1970. The population dynamics and energy budget of *Nothothenia neglecta* Nybelin at Sighny Island, South Orkney Islands. Br. Antarct. Surv. Bull., 23:25–49.

Fabian, G., G. Molnar, and I. Torg. 1963. Comparative data and enzyme kinetic calculation in change caused by temperature in duration of gastric digestion on some predatory fishes. Acta Biol. Acad. Sci. Hungary, 14(2):123–129.

Fabré, S. 1985. Distribución cuantitativa del zooplancton en la región noroccidental de la plataforma cubana. Rep. Invest. Inst. Oceanol. Acad. Cienc. Cuba, 31:1–27.

FAO (Food and Agricultural Organization). 1978. Species identification sheets for fishery purposes, western central Atlantic, Fishing Area 31. W. Fisher, ed. FAO, Rome, Italy, vols. 1-4 (unpaged).

Feddern, A. M. 1966. Cleaning symbiosis in the marine environment. In: S. Mark Henry, ed. Symbiosis. Academic Press, New York, NY, vol. 1, pp. 327–386.

Fedoriako, B. I. 1982. The Langmuir circulation and a possible mechanism of formation of associations of fishes around floating objects [in Russian]. Okeanologia, 22(2):314–319.

Fernández, F. 1984. Estudio preliminar de la fauna acompañante del camarón en el Combinado Pesquero Industrial de Cienfuegos. Unpublished, Ministerio de la Industria Pesquera, Cienfuegos, 3 p., 1 tab.

Fernández-Vila, L., O. Leiva, and N. Romero. 1990. Características del régimen termohalino y circulación de las aguas de las bahías San Juan de los Remedios y Buenavista. Unpublished rep., GeoCuba, 25 p.

Ferreira de Menezes, M. 1985. Biología pesqueira do cangulo, *Balistes vetula* Linnaeus: Reproducao e crescimento. Arq. Cienc. Mar., 24:53–59.

Ferreira de Menezes, M., and M. dos Santos Pinto. 1966. Notes on the biology of tarpon, *Tarpon atlanticus* (Cuvier et Valenciennes) from coastal waters of Ceará, Santa Fé, Brazil. Arq. Cienc. Mar., 6(1): 83–98.

FishBase. 2000. The FishBase Project. Intern. Center Living Aquatic Resources Management (ICLARM), Makati City, Philippines. Searchable web database: *http://www.fishbase.org.*

Fishelson, L., D. Popper, and A. Avidor. 1974. Biosociology and ecology of pomacentrid fishes around the Sinai Peninsul (northern Red Sea). J. Fish. Biol., 6(2):119–133.

Fontana, A. 1969. Etude de la maturité sexuelle des sardinelles *Sardinella eba* (Val.) et *Sardinella aurita* (C. et V.) de la région de Pointe Noire. Cah. Ser. Oceanogr. ORSTOM, 7(2):101–114.

Formoso, M. 1975. Distribución cuantitativa de la macrofauna alimenticia en la región oriental del Banco de Campeche. Resúmenes Invest. Cent. Invest. Pesq. Cuba, 2:110–112.

Fortunatova, K. R. 1961. Methods for feeding studies on predator fishes of the Volga Delta. (In Russian), Nauka, Moscow, pp. 137–188.

Fortunatova, K. R., and O. A. Popova. 1973. Feeding and feeding relationship of predatory fishes of the Volga Delta. Nauka, Moscow, 296 p.

Fowler, A. J. 1995. Annulus formation in otoliths of coral reef fish: A review. In: D. Secor, J. M. Dean, and S. E. Campana, eds. Recent Developments in Fish Otolith Research. Univ. South Carolina Press, Columbia, SC, The Belle W. Baruch Library in Marine Science, vol. 19, pp. 45–63.

Fox, W. W., Jr. 1975. Fitting the generalized stock production model by least-squares and equilibrium approximation. Fish. Bul., 73(1):23–36.

Franks, J. S., and K. E. VanderKooy. 2000. Feeding habits of juvenile lane snapper *Lutjanus synagris* from Mississippi coastal waters, with comments on the diet of gray snapper, *Lutjanus griseus*. Gulf Carib. Res. 12:11–17.

Fuentes, D., R. Valdés, C. Zetina, S. Nieto, G. V. Rhos, C. Monroy, M. Contreras, and V. Moreno. 1989. Informe de investigaciones conjuntas Mexico-Cuba sobre el mero (*Epinephelus morio* [Valenciennes, 1828]) en el Banco de Campeche. CREPY, INP-CIP, Document interno. IPN, México.

Furrazola-Bermúdez, G., C. Khudoley, M. I. Mijailovskaia, Y. S. Miroliubov, I. P. Novokjatsky, A. Núñez-Jiménez, and J. B. Solsona. 1964. Geología de Cuba. Editora Nacional de Cuba, La Habana, 239 p.

Futch, C. R., and G. E. Bruger. 1976. Age, growth and reproduction of red snapper in Florida waters. Fla. Sea Grant College Progr. Rep., 17:165–184.

Gallegos A., I. Victoria, J. Zabala, M. Fernández, and I. Penié. 1998. Hidrología de los estrechos del mar Caribe noroccidental. Rev. Invest. Mar. 9(1):1–37.

García, A., and G. Bustamante. 1981. Resultados preliminares del desove inducido de lisa (*Mugil curema* Valenciennes) en Cuba. Inf. Cient.-Téc., Inst. Oceanol. Acad. Cienc. Cuba, 158:17–26.

García, E. R. 2000. Age and growth of yellowtail snapper, *Ocyurus chrysurus*, collected from the southeastern United States. M.S. thesis, East Carolina University, Greenville.

García, T. 1976. Alimentación natural de dos especies de sardinas, *Harengula humeralis* (Cuvier) y *Harengula clupeola* (Cuvier). Ciencias, Ser. 8, Invest. Mar., 24:1–16.

García, T. 1979. Study of the family Sciaenidae (Pisces) in Cuban waters [in Romanian]. Ph.D. diss., Univ. Bucharest, Romania, 164 p.

García, T., and E. Nieto. 1978. Alimentación de *Bairdiella ronchus* en dos áreas de la plataforma cubana. Ciencias, Ser. 8, Invest. Mar., 38:13–28.

Garcia-Abad, M. C., A. Yáñez-Arancibia, P. Sánchez-Gil, and M. Tapia-García. 1998. Distribución, abundancia y reproducción de *Opisthonema oglinum* (Pisces: Clupeidae) en la plataforma continental del sur del Golfo de México. Rev. Biol. Trop. 46(2):257–266.

García-Arteaga, J. P. 1983. Edad y crecimiento del jallao, *Haemulon album* en la plataforma SW de Cuba. Rep. Invest. Inst. Oceanol. Acad. Cienc. Cuba. Revista Oceanol. Ecol. Trop. 14:1–25.

García-Arteaga, J. P. 1992a. Edad y crecimiento del ronco arará, *Haemulon plumieri* (Lacépède) (Pisces: Haemulidae) en el Golfo de Batabanó, plataforma suroccidentalde Cuba. Arch. Inst. Oceanol., 14 p., 5 tab., 5 illust.

García-Arteaga, J. P. 1992b. Edad y crecimiento del ronco amarillo, *Haemulon sciurus* (Shaw) (Pisces: Haemulidae) en la plataforma suroccidental de Cuba. Cienc. Biol., 25:104–116.

García-Arteaga, J. P. 1993. Edad y crecimiento de la sardina de ley, *Harengula humeralis* (Pisces: Clupeidae) en la plataforma suroccidental de Cuba. Revista Oceanol. Ecol. Trop. Avicennia, 0:14–26.

García-Arteaga, J. P., and R. Claro. 1998. Estructura de las comunidades de peces en la pendiente arrecifal del Golfo Batabano. Arch. Inst. Oceanol., 6 p., 4 tabs, 4 illust.

García-Arteaga, J. P., and Y. S. Reshetnikov. 1992. Edad y crecimiento del cibí carbonero, *Caranx ruber* (Bloch) en las costas de Cuba. Rep. Invest. Inst. Oceanol. Acad. Cienc. Cuba, 1:1–21.

García-Arteaga, J. P., R. Claro, L. M. Sierra, and E. Valdés-Muñoz. 1990. Características del reclutamiento a la plataforma de los juveniles de peces neríticos en la región oriental del Golfo de Batabanó. In: R. Claro, ed.

Asociaciones de peces en el Golfo de Batabanó Editorial Academia, La Habana, pp. 96–122.

García-Arteaga, J. P., R. Claro, and S. Valle. 1997. Length-weight relationships of Cuban marine fishes. NAGA, ICLARM Q. 20(1):38–43.

García-Arteaga, J. P., L. M. Sierra, and R. Claro. 1999. Características biológicas del bonací gato, *Mycteroperca tigris* (Pisces: Serranidae) en la plataforma SW de Cuba. II. Alimentación, edad y crecimiento. Rev. Invest. Mar. 20(1–3):30–36.

García-Cagide, A. 1985. Características de la reproducción del civil, *Caranx ruber*, en la plataforma suroccidental de Cuba. Rep. Invest. Inst. Oceanol. Acad. Cienc. Cuba, 34:1–36.

García-Cagide, A. 1986a. Características de la reproducción del jallao, *Haemulon album*, en la plataforma suroccidental de Cuba. Rep. Invest. Inst. Oceanol. Acad. Cienc. Cuba, 47:1–33.

García-Cagide, A. 1986b. Características de la reproducción del ronco amarillo, *Haemulon sciurus*, en la región oriental del Golfo de Batabanó, Cuba. Rep. Invest. Inst. Oceanol. Acad. Cienc. Cuba, 48:1–28.

García-Cagide, A. 1987. Características de la reproducción del ronco arará, *Haemulon plumieri* (Lacépède) en la región oriental del Golfo de Batabanó, Cuba. Rev. Invest. Mar., Ser. 8, 3:39–55.

García-Cagide, A. 1988. Particularidades de la reproducción de la sardina de ley, *Harengula humeralis* (Cuvier, 1829), en la región oriental del Golfo de Batabanó, Cuba. Rep. Invest. Inst. Oceanol. Acad. Cienc. Cuba, 12:1–15.

García-Cagide, A., and R. Claro. 1983. Datos sobre la reproducción de algunos peces comerciales del Golfo de Batabanó. Rep. Invest. Inst. Oceanol. Acad. Cienc. Cuba, 12:1–20.

García-Cagide, A., and R. Claro. 1985. Characteristics of the reproduction of the neritic fishes of Cuba [in Russian]. In: B. V. Koshelev, ed. Characteristics of the reproductive cycle of fishes of different geographic areas. Nauka, Moscow, pp. 45–54.

García-Cagide, A., and L. Espinosa. 1991. Algunas características histológicas de la cherna criolla, *Epinephelus striatus* (Bloch) (Pisces: Serranidae). Cienc. Biol., 24:134–138.

García-Cagide, A., and T. García. 1996. Reproducción de *Mycteroperca bonaci* y *M. venenosa* en la plataforma cubana. Rev. Biol. Trop., 44(2):771–780.

García-Cagide, A., R. Claro, and B. V. Koshelev. 1983. Peculiaridades de los ciclos reproductivos de los peces de diferentes latitudes. Rep. Invest. Inst. Oceanol. Acad. Cienc. Cuba, 21:1–14.

García-Cagide, A., B. V. Koshelev, R. Claro, and Y. S. Reshetnikov. 1985. Reproduction, spawning time and periodicity. In: V. E. Sokolov and G. García, eds. Ecology of fishes of the Cuban shelf [In Russian], Nauka, Moscow, pp. 174–205.

García-Cagide, A., R. Claro, and B. V. Koshelev. 1994. Reproduccion. In: R. Claro, ed. Ecología de los peces marinos de Cuba. Inst. Oceanol. Acad. Crinc. Cuba y Cent. Invest. Quintana Roo, Mexico, pp. 187–262.

García-Cagide, A., R. Claro, and J. P. García-Arteaga.1999a. Caracterí sticas biológicas del bonací gato, *Mycteroperca tigris* (Pisces: Serranidae) en la plataforma SW de Cuba. I. Características generales, reproducción. Rev. Invest. Mar., 20(1–3):8–14.

García-Cagide, A., R. Claro, and J. P. García-Arteaga. 1999b. Biología del jocú, *Lutjanus jocu* (Pisces: Lutjanidae) en las zonas NE y SW de la plataforma cubana I. Distribución, hábitat, reproducción y dinámica de los indicadores morfofisiológicos. Rev. Invest. Mar., 20(1–3):22–29.

García-Coll, I. 1984. Determinación de la edad y el crecimiento del bonito, *Katsuwonus pelamis* y la albacora *Thunnus atlanticus* en las regiones suroccidental y nororiental de Cuba y evaluación de las pesquerías de estas especies en La Coloma. Tesis doctoral, Univ. la Habana, 104 p.

García Díaz, C. 1981. Temperatura de las aguas oceánicas de Cuba. II. Aguas sub-superficiales. Rev. Cub. Invest. Pesq., 6(2):16–35.

García Díaz, C., A. L. Chirino, and J. P. Rodríguez. 1991. Corrientes geostróficasa en la ZEE al sur de Cuba. Rev. Invest. Mar., 12(1–3): 29–38.

García-Parrado, P., and P. M. Alcolado-Menendez. 1998. Nuevos registros de octocorales para Cuba. Rev. Oceanol. Ecol. Trop. Avicennia, 8–9:105–108.

García Rodríguez, J. 1978. Resultados de la pesca exploratoria del alto en la región central de la costa N de la plataforma cubana. In: Primer Foro Cient. Cent. Invest. Pesq., La Habana, 8 p., 4 tabs., 14 figs.

García Rodríguez, J., and L. Miranda. 1979a. Resultados de la pesca exploratoria del alto de la plataforma suroriental de la Isla de Cuba. Unpublished, Cent. Invest. Pesq., La Habana, 10 p.

García Rodríguez, J., and L. Miranda. 1979b. La pesca del alto en Cuba. Unpublished, Cent. Invest. Pesq., La Habana, 11 p., 8 figs.

García Rodríguez, J., L. Miranda, and R. de la Rúa. 1976. Resultados preliminares de la pesca exploratoria del alto en la costa norte de la provincia de Matanzas. Unpublished, Cent. Invest. Pesq., La Habana, 2 p., 1 tab., 2 figs.

García-Rubies, A., and M. Zabala. 1990. Effects of total fishing prohibition on the rocky fish assemblages of Medes Islands marine reserve (NW Mediterranean). Fish structure in marine reserves. Sci. Mar., 54(4):317–328.

Garrick, J. A. 1967. Revision of sharks of the genus Isurus with description of a new species (Galeoidea, Lamnidae). Proc. U.S. Natl. Mus., 118(3537):663–690.

Garrido, O. H., and E. Valdés-Muñoz. 1984. Nuevos registros de peces para Cuba, y comentarios sobre su comportamiento, distribución y coloración en vida. Rep. Invest. Inst. Oceanol., Acad. Cienc. Cuba, 28:1–11.

Garrod, D. J., and B. S. Newell. 1958. Ring formation in the scales of Tilapia sculenta. Nature, 181:1411–1412.

Gayanilo, F. C. Jr., M. Soriano, and D. Pauly. 1988. A draft guide to the complete ELEFAN. ICLARM Software 2, 65 p.

Geffen, A. J. 1982. Otolith ring deposition in relation to growth rate in herring (Clupea harengus) and turbot (Scophthalmus maximus) larvae. Mar. Biol., 71(3):317–326.

Gershanovich, A. D., V. A. Pegasov, and M. I. Shatunovsky. 1987. Ecology and physiology of juvenile sturgeons [in Russian]. Agropromizdat, Moscow, 214 p.

Gibbs, R. H., Jr., and B. B. Collette. 1959. On the identification, distribution and biology on the dolphins, Coryphaena hippurus and C. equisetis. Bull. Mar. Sci. Gulf Carib., 9(2):117–152.

Gilmore, R. G., Jr. 1995. Environmental and biogeographic factors influencing ichthyofaunal diversity: Indian River Lagoon. Bull. Mar. Sci., 57(1):153–170.

Gilmore, R. G. Jr., and R. J. Jones. 1992. Color variation and associated behavior in the epinepheline groupers Mycteroperca microlepis and M. phenax. Bull. Mar. Sci. 51:83–103.

Giménez, E., R. Tizol, and M. L. Hernández. 1986. Uso y distribución de los arrecifes artificiales en Cuba. Arch. Cent. Invest. Pesq., MIP, Cuba, 12 p.

Ginsburg, J. 1952. Fishes of the family Carangidae of the northern Gulf of Mexico and three related species. Publ. Inst. Mar. Sci. Univ. Texas, 2:43–117.

Gladfelter, W. B., and E. H. Gladfelter. 1978. Fish community structure as a function of habitat structure on West Indian patch reefs. Rev. Biol. Trop. Univ. Costa Rica, 26(1):65–84.

Gobert, B. 1990. Production relative des pecheries cotières en Martinique. Centre ORSTOM de Fort-de-France Cedex (France), 3:181–191.

Goldberg, S. R., and D. W. K. Au. 1983. The spawning schedule of skipjack tuna from southeastern Brazil as determined from histological examination of ovaries, with notes on spawning in the Caribbean. In: Actas de la Conferencia del ICCAT sobre el Año Internacional del Listado, Madrid, 18 p.

Goldman, B., and F. H. Talbot. 1976. Aspects of the ecology of coral reef fishes. In: O. A. Jones, and R. Endean, eds. Biology and Geology of Coral Reefs. Academic Press, New York, NY, vol. 3, pp. 125–154.

Gómez, F. E., and A. K. M. Bashirullah. 1984. Relación longitud-peso y hábitos alimenticios de Rhizoprionodon porosus Poly, 1861 (Familia Carcharhinidae) en el oriente de Venezuela. Bol. Inst. Oceanogr. Univ. Oriente Cumaná, 23(1–2):49–54.

Gómez, G. J., R. A. Guzmán, and L. A. Marcano. 1996. Biological aspects of the yellow eye snapper (Lutjanus vivanus) (Pisces: Lutjanidae) from the Los Hermans Islands, Eastern Venezuela. In: F. Arreguín-Sánchez, J. L. Munro, M. C. Balgos, and D. Pauly, eds. Biology, Fisheries and Culture of Tropical Groupers and Snappers. ICLARM Conf. Proc. 48, pp. 51–58.

Gómez, J. A. 1979. Corrientes geostróficas alrededor de Cuba. Rev. Cub. Invest. Pesq., 4(3):89–102.

Gómez, O., D. Ibarzábal, and A. Silva. 1980. Evaluación cuantitativa del bentos en la región suroccidental de Cuba. Inf. Cient.-Téc., Inst. Oceanol. Acad. Cienc. Cuba, 149:1–15.

González, E., T. Damas, N. Millares, and M. Borrero. 1979. Desove inducido en el caballerote (Lutjanus griseus Linné, 1758), en condiciones de laboratorio. Rev. Cub. Invest. Pesq., 4(1):43–64.

González, M. 1995. Principales fuentes contaminantes de Cuba. Inst. Unpublished, Geografia Tropical, Cuba.

González, M. E. 1988. Estudio de la pesqueria del huachinango (Lutjanus campechanus Poey) del Banco de Campeche. Tesis maestría, CINVESTAV Unidad Merida.

González, P. 1983. Age determination and back calculation of growth of the red grouper, Epinephelus morio, of the Campeche Bank, Gulf of Mexico [in Russian]. Sistematika y Ekologia, 5:57–66.

González, P. D., S. Zupanovic, and H. E. Ramis. 1974. Biología pesquera de la cherna americana en el Banco de Campeche. INP/CIP Cuba Res. Invest. 1:107–111.

González, W. L. 1985a. Relación longitud-peso y factor de condición de la caitipa Diapterus rhombeus (Cuvier, 1829) (Pisces: Gerreidae) del sur-oeste de la isla de Margarita Venezuela. Bol. Inst. Oceanogr. Univ. Oriente Cumaná, 24(1–2):15–22.

González, W. L. 1985b. Determinación de edad y crecimiento de la sardina Sardinella aurita Valenciennes, 1847 (Pisces: Clupeidae) de la región nororiental de Venezuela. Bol. Inst. Oceanogr. Univ. Oriente Cumaná, 24(1–2):111–128.

González-Sansón, G. 1979. Fish feeding of the coastal lagoons in a mangrove area of Cuba with special emphasis on the ecosystem food web [in German]. Ph.D. diss., Univ. Wilhelm-Pieck, Rostock, Germany, 155 p.

González-Sansón, G. 1984. Ecología de las lagunas costeras de la región suroriental de Cuba. Rev. Invest. Mar., 5(1):127–171.

González-Sansón, G., and C. Aguilar Betancourt. 1983. Estudio comparativo de la estructura de las comunidades de peces de las lagunas costeras de la región suroriental de Cuba. Rev. Invest. Mar., 4(2):91–124.

González-Sansón, G., and L. S. Alvarez-Lajonchere. 1978. Alimentación natural de Mugil liza, M. curema, M. trichodon y M. hospes (Pisces: Mugilidae) en las lagunas coste ras de Tunas de Zaza, Cuba. Ciencias, Ser. 8, Invest. Mar., 41:1–40.

González-Sansón, G., and R. Lalana-Rueda. 1982. Aporte de materia orgánica del manglar al ecosistema acuático de lagunas costeras en Cuba. Rev. Invest. Mar., 3(1):3–32.

González-Sansón, G., and L. Rodríguez-Viñas. 1983. Alimentación natural de Eugerres brasilianus (Cuvier) y Gerres cinereus (Walbaum) (Pisces: Gerreidae) en las lagunas costeras de Tunas de Zaza, Cuba. Rev. Invest. Mar., Ser. 8, 4(1):91–134.

González-Sansón, G., L. S. Alvarez-Lajonchere, and M. Báez. 1978. Lista preliminar de peces presentes en las lagunas costeras de Tunas de Zaza, Cuba. Ciencias, Ser. 8, Invest. Mar., 32:1–15.

Gooding, R. M., and J. J. Magnuson. 1967. Ecological significance of a drifting object to pelagic fishes. Pac. Sci., 21(4):486–497.

Goodwin, J. M., IV, and J. H. Finucane. 1985. Reproductive biology of blue runner (*Caranx crysos*) from the eastern Gulf of Mexico. Northeast Gulf Sci., 7(2):139–146.

Gorbunova, N. M. 1954. Reproduction and growth of mintai, *Thelagra chalcogramma* (R.) [in Russian]. Tr. IOAN, 11:132–195.

Gorbunova, N. M. 1965. Periods and conditions of reproduction of scombroid fishes (Pisces: Scombroidei) [in Russian]. Tr. IOAN, 80:36–61.

Gorelova, T. A. 1974. Experimental data on the digestion rate in fingerlings of the snapper, *Lutjanus peru*. J. Ichthyol., 19(6):161–164.

Graber, R. C. 1974. Food intake patterns in captive juvenile lemon sharks, *Negaprion brevirostris*. Copeia, 2:554–556.

Grainger, R. J. R., and S. M. García. 1996. Chronicles of fishery landings (1950–1994). Trend analysis and fisheries potential. FAO Fish. Tech. Pap. 359, 51 p.

Grimes, C. B. 1978. Age, growth and length-weight relationship of vermillion snapper, *Rhomboplites aurorubens*, from North Carolina and South Carolina waters. Trans. Am. Fish. Soc., 107(3):454–456.

Grimes, C. B. 1979. Diet and feeding ecology of the vermilion snapper, *Rhomboplites aurorubens*(Cuvier) from North Carolina and South Carolina waters. Bull. Mar. Sci. 29(1):53–61.

Grimes, C. B. 1987. Reproductive biology of the Lutjanidae: A review. In: J. J. Polovina, and S. Ralston, eds. Tropical Snappers and Groupers: Biology and Fisheries Management. Westview Press, Boulder, CO, pp. 239–294.

Grossman, G. D., G. P. Jones, and W. Seaman, Jr. 1997. Do artificial reefs increase regional fish production? A review of existing data. Fisheries, 22(4):17–23.

Grover, J. J. 1993. Trophic ecology of pelagic early-juvenile Nassau grouper, *Epinephelus striatus*, during and early phase of recruitment into demersal habits. Bull. Mar. Sci., 53:1117–1125.

Grover, J. J., B. L. Olla, and R. I. Wicklund. 1992. Food habit of nassau grouper (*Epinephelus striatus*) juveniles in three habitats in the Bahamas. Proc. Gulf Carib. Fish. Inst., 42:247.

Grover, J. J., D. B. Eggleston, and J. M. Shenker. 1998. Transition from pelagic to demersal phase in early-juvenile Nassau-grouper, *Epinephelus striatus*: Pigmentation, squamation, and ontogeny of diet. Bull. Mar. Sci., 62(1):97–113.

Gruber, S. H. 1982. Role of the lemon shark, *Negaprion brevirostris* (Poey), as a predator in the tropical marine environment: A multidisciplinary study. Fla. Sci., 45(1):46–75.

Guardiola, M., and M. E. León. 1988. Selectividad en redes de enmalle de la sierra (*Scomberomorus cavalla*) en la zona suroriental de Cuba. ICCAT vol. Sci. Pap., 38:303–308.

Gudger, E. W. 1929. On the morphology, coloration, and behavior of seventy teleostean fishes of Tortugas. Fla. Pap. Tortugas Lab., 26:149–204.

Guerra, A., and A. Bashirullah. 1975. Biología del pargo *Lutjanus griseus* (Linn.) de la isla de Cubagua, Venezuela. II. Maduración sexual y fecundidad. Bol. Inst. Oceanogr. Univ. Oriente Cumaná, 14(1):109–116.

Guevara, E. C. 1984a. Alimentación de la albacora *Thunnus atlanticus* en la región suroccidental de Cuba. Rev. Invest. Mar., 5(3):37–45.

Guevara, E. C. 1984b. Alimentación del bonito, *Katsuwonus pelamis* en la región suroccidental de Cuba. Rev. Invest. Mar., 5(2):9–22.

Guitart Manday, D. 1959. Gorgonias del litorial de la costa norte de Cuba. Publ. Acuario Nacional, Serie Técnica 1:1–24.

Guitart Manday, D. 1964. Biología pesquera del emperador o pez de espada, *Xiphias gladius* Linnaeus (Teleostomi: Xiphiidae) en las aguas de Cuba. Poeyana Ser. B, 1:1–37.

Guitart Manday, D. 1974–1978. Sinopsis de los peces marinos de Cuba. Inst. Oceanol., Acad. Cienc. Cuba, 4 t., 881 p.

Guitart Manday, D. 1975. Las pesquerías pelágico-oceánicas de corto radio de acción en la región noroccidental de Cuba. Ser. Oceanol. Acad. Cienc. Cuba, 31:1–26.

Guitart Manday, D. 1981. Análisis de las pesquerías deportivas de agujas (Pisces: géneros Istiophorus, Tetrapturus, Makaira) en la región noroccidental de Cuba. Cienc. Biol., 6:125–143.

Guitart Manday, D. 1983. Los Tiburones. Editorial Científico-Técnica., La Habana, 57 p.

Gutiérrez, P. B. 1976. Composición aproximada de algunos peces de Venezuela. Soc. Cienc. Nat. La Salle, 36(104):1-6.

Guzmán, E. 1986. Contribución al conocimiento de la pesquería del mero (*Epinephelus morio*, Valenciennes) de las costas de Yucatán. Tesis profesional, ENEP-Iztacala, UNAM, México.

Halstead, B. W. 1967. Poisonous and venemous marine animals of the world. U.S. Gov. Printing Office, Washington, DC, vol. 2, 1070 p.

Harmelin-Vivien, M. L. 1981. Trophic relationships of reef fishes in Tulear (Madagascar). Oceanol. Acta, 4(3):365–374.

Harris, P. J., and M. R. Collins. 2000. Age, growth and age at maturity of gag, *Mycteroperca microlepis*, from the southeastern United States during 1994–1995. Bull. Mar. Sci., 66(1):105–117.

Hartley, P. H. T. 1948. Food and feeding relationships in a community of freshwater fishes. J. Anim. Ecol., 17:1–14.

Hartline, A. C., P. H. Hartline, A. M. Szmant, and A. O. Flechsig. l972. Escape response in a pomacentrid reef fish, *Chromis cyaneus*. Nat. Hist. Mus. Los Angelas County, Sci. Bull., 14:93–97.

Hartruijker, L. 1982. Trap fishing survey of Pedro Bank (Jamaica), 2nd phase: A re-assessment of the stocks of reef fish on Pedro Bank, Jamaica. Fish Div., Tech. Rep. 4, 43 p.

Hatanaka, M., M. Kosaka, and G. Sato. 1956a. Growth and food consumption in plaice. I. *Kareius bicoloratus* (Basilievsky). Tohoku J. Agric. Res., 7(2):163–174.

Hatanaka, M., M. Kosaka, and G. Sato. 1956b. Growth and food consumption in plaice. I. *Limanda yokohamae* (Gunther). Tohoku J. Agric. Res., 7(2):151–162.

Head, W. D., W. O. Watanabe, S. C. Ellis, and E. P. Ellis. 1996. Hormone-induced multiple spawning of captive Nassau grouper broodstock. Prog. Fish. Cult. 58(1):65–69.

Heald, E. J., and W. E. Odum. 1970. The contribution of mangrove swamps to Florida fisheries. Proc. Gulf Carib. Fish. Inst., 22:130–135.

Hedgepeth, H., and J. Jolley. 1983. Age and growth of sailfish, *Istiophorus platypterus*, using cross sections from the fourth dorsal fin spine. NOAA Tech. Rep. NMFS 8:131–135.

Helfman, G. S. 1978. Patterns of community structure in fishes: Summary and overview. Environ. Biol. Fish., 3(1):129–148.

Helfman, G. S., J. L. Meyer, and W. N. McFarland. 1982. The ontogeny of twilight migration patterns in grunts (Pisces: Haemulidae). Anim. Behav., 30:317–326.

Hernández-Corujo, C. 1975. Algunos aspectos preliminares del desove de tres especies del género *Calamus* (Familia Sparidae) en el Banco de Campeche. Unpublished rep., Res. Invest. Cent. Invest. Pesq., MIP, La Habana, Cuba.

Herrera, A., and N. Martinez. 1987. Efectos de la contaminación sobre las comunidades de corales escleractineas al oeste de la Bahia de La Habana. Rep. Invest. Inst. Oceanol. Acad. Cien.Cuba, 62:1–29.

Hettler, W. F., Jr. 1989. Food habits of juveniles of spotted sea trout and gray snapper in western Florida Bay. Bull. Mar. Sci., 44(1):155–162.

Hiatt, K. O. 1979. Feeding strategy. In: W. S. Hoar, D. J. Randall, and J. R. Brett, eds. Fish Physiology. 8. Bioenergetics and Growth Academic Press, New York, San Francisco and London, pp. 71–120.

Hiatt, R. W., and D. W. Strasburg. 1960. Ecological relationships of the fish fauna on coral reefs of the Marshall Islands. Ecol. Monogr., 30(1):65–127.

Hill, R. B. 1978. The use of nearshore marine life as food resources by American Samoans. Pac. Is. Stu. Progr. Univ. Hawaii, 170 p.

Hixon, M. A. 1991. Predation as a process structuring coral reef fish communities In: P. F. Sale, ed. The Ecology of Fishes on Coral Reefs. Academic Press, San Diego, CA, pp. 475–508.

Hixon, M. A., and M. H. Carr. 1997. Synergistic predation, density dependence, and population regulation in marine fish. Science, 277:946–949.

Hoar, W. S. 1957. The gonads and reproduction. In: M. L. Brown, ed. Fish Physiology. Vol. 4, Metabolism. Academic Press, New York. pp. 287–322.

Hobson, E. S. 1974. Feeding relationships of teleostean fishes on coral reef in Kora, Hawaii. Fish. Bull., 72(4):915–1030.

Hobson, E. S. 1991. Trophic relationships of fishes specialized to feed on zooplankters above coral reefs. In: P. Sale, ed. The Ecology of Fishes on Coral Reefs. Academic Press Inc., San Diego, CA, pp. 69–95.

Hood, P. B., and A. K. Johnson. 1999. Age, growth, mortality, and reproduction of vermilion snapper, *Rhomboplites aurorubens,* from the eastern Gulf of Mexico. Fish. Bull., 97(4):828–841.

Hood, P. B., and R. A. Schlieder. 1992. Age, growth and reproduction of gag, *Mycteroperca microlepis* (Pisces: Serranidae), in the eastern gulf of Mexico. Bull. Mar. Sci., 51(3):337–352.

Hourigan, T. F., and C. D. Kelly. 1985. Histology of the gonads and observations on the social behavior of the Caribbean angelfish *Holacanthus tricolor.* Mar. Biol., 88:311–322.

Hubold, G., and M. V. Mazzetti. 1982. Growth, morphometry and aspects of the life history of the scaled sardine *Harengula jaguana* Poey, 1865 in the Guanabara Bay (Brazil). Meeresforschung, 2:80–88.

Hughes, T. P., and J. H. Connell. 1999. Multiple stressors on coral reefs: A long-term perspective. Limnol. Oceanogr., 44(3, part 2):932–940.

Humphries, C. J., and L. R. Parenti. 1999. Cladistic biogeography: Interpreting patterns of plant and animal distributions, 2nd ed. Oxford Univ. Press, New York, NY. 187 p.

Hunter, J. R. 1968. Fishes beneath flotsam. Sea Front., 14(5):280–288.

Hunter, J. R., and S. R. Goldberg. 1980. Spawning incidence and batch fecundity in northern anchovy, *Engraulis mordax.* Fish. Bull., 77:641–652.

Hunter, J. R., and B. J. Macewicz. 1980. Sexual maturity, batch fecundity, spawning frequency, and temporal pattern of spawning for the northern anchovy, *Engraulis mordax* during the 1979 spawning season. California Coop. Oceanic Fish. Invest. Rep., 21:139–149.

Hunter, J. R., and C. T. Mitchell. 1967. Association of fishes with flotsam in the offshore waters of Central America. U.S. Fish Wildl. Serv., Fish. Bull., 66(1)13–29.

Hunter, J. R., N. H. Lo, and R. Leong. 1984. Batch fecundity in multiple spawning fishes. An egg production method for estimating spawning biomass of pelagic fish. In: R. Lasker, ed. Application to the northern anchovy, *Engraulis morda.* NOAA Tech. Rep. NMFS, 366:67–77.

Huntsman, G. R., and W. E. Schaaf. 1994. Simulation of the impact of fishing on reproduction of a protogynous grouper, the graysby. N. Am. J. Fish. Manage. 14:41–52.

Hurtado, G. P., and A. K. M. Bashirullah. 1975. Biología del pargo, *Lutjanus griseus* de la Isla de Cubagua, Venezuela. III. Análisis del contenido estomacal. Bol. Inst. Oceanogr. Univ. Oriente Cumaná, 15:191–205.

Ibarzábal-Bombalier, D. 1982. Evaluación cuantitativa del bentos en la región noroccidental de la plataforma de Cuba. Cienc. Biol., 8:57–80.

Ibarzábal-Bombalier, D. 1990. Características de la macroinfauna de la macrolaguna del Golfo de Batabanó. In: P. M. Alcolado, ed. El Bentos de la Macrolaguna del Golfo de Batabanó. Editorial Academia, La Habana, pp. 113–128.

ICGC (Instituto Cubano de Geodesia y Cartografia). 1978. Atlas de Cuba, 20 años del Triunfo de la Revolución Cubana, La Habana, 143 p.

Ida, H., Y. Hiyama, and T. Kusaka. 1967. Study of fishes gathering around floating seaweed. I. Abundance and species composition. Bull. Jap. Soc. Sci. Fish., 33(10):923–929.

Ionin, A. S., Y. A. Pavlidis, and O. Avello-Suárez. 1977. Geology of the Cuban Platform [in Russian]. Nauka, Moscow, 215 p.

Iturralde-Vinent, M. A. 1972. Principal characteristics of Oligocene and Lower Miocene stratigraphy of Cuba. Am. Assoc. Petrol. Geol. Bull., 56(12):2369–2379.

Iturralde-Vinent, M. A. 1975. Problemas en la aplicación de dos hipótesis tectónicas modernas a Cuba y la región del Caribe. Rev. Tec., 13(1):46–63.

Iturralde-Vinent, M. A. 1977. Los movimientos tectónicos de la etapa de desarrollo platafórmico en Cuba. Inf. Cient.-Tec., Inst. Geol. Paleontol. Acad. Cienc. Cuba, 20:1–24.

Iturralde-Vinent, M. A. 1981. Nuevo modelo interpretativo de la evolución geológica de Cuba. Cienc. Tierra Espacio, 3:51–89.

Iturralde-Vinent, M. A. 1982. Aspectos geológicos de la biogeografía de Cuba. Cienc. Tierra Espacio, 5:1–16.

Ivlev, V. E. 1955. Experimental feeding ecology of fishes [in Russian]. Pishevaia Promishlennost, Moscow, 255 p.

Jackson, J. B. C. 1997. Reefs since Columbus. Coral Reefs, 16:S23–S32.

Jennings, S., and N. V. C. Polunin. 1997. Impacts of predator depletion by fishing on the biomass and diversity of non-target reef fish communities. Coral Reefs, 16:71–82.

Jiménez, C., and D. Ibarzábal. 1982. Evaluación cuantitativa del mesobentos en la plataforma nororiental de Cuba. Cienc. Biol., 7:53–69.

Jiménez Domínguez, C., and P. M. Alcolado. 1990. Características del macrofitobentos del Golfo de Batabanó. In: P. M. Alcolado, ed. El benthos de la macrolaguna del Golfo de Batabanó, Editorial Academia, La Habana, pp. 4–7.

Johannes, R. E. 1978. Reproductive strategies of coastal marine fishes in the tropics. Environ. Biol. Fish., 3(1):54–84.

Johnson, A. 1983. Age and growth of yellowtail snapper from South Florida. Trans. Am. Fish. Soc., 112:173–177.

Johnson, A., W. Fable, M. Williams, and L. Barger. 1983. Age, growth and mortality of king mackerel *Scomberomorus cavalla,* from the southeastern United States. Fish. Bull., 81(1):97–106.

Johnson, A. J., L. A. Collins, and C. P. Keim. 1994. Age-size structure of gray snapper from the southeastern United States: A comparison of two methods of back-calculating size at age from otolich data. Proc. Annu. Conf. SE Assoc. Fish and Wild. Agencies, 48:592–600.

Johnson, A. J., L. A. Collins, J. Dahl, and M. S. Baker, Jr. 1995. Age, growth, and mortality of lane snapper from the northern Gulf of Mexico. Proc. Annu. Conf. SE Assoc. Fish and Wild. Agencies, 49: 178–186.

Johnson, H. H. 1932. The problem of self-fertilization in teratologically hermaphroditic fishes. Copeia, 1:21–26.

Jolley, J. W., Jr. 1977. The biology and fishery of Atlantic sailfish *Istiophorus platypterus,* from southeast Florida. Fla. Mar. Res. Publ., 28:1–31.

Jones, G. P. 1991. Postrecruitment processes in the ecology of coral reef fish populations: A multifactorial perspective. In: P. F. Sale, ed. The Ecology of Fishes on Coral Reefs. Academic Press, San Diego, CA, pp. 294–330.

Jones, G. P., D. J. Ferrell, and P. F. Sale. 1991. Fish predation and its impact on the invertebrates of coral reefs and associated sediments. In: P. F. Sale, ed. The Ecology of Fishes on Coral Reefs. Academic Press, San Diego, CA, pp. 156–182.

Juanes, F. 1994. What determines prey size selectivity in piscivorous fishes? In: D. Stouder, K. Fresh and R. Feller, eds. Theory and Application in Fish

Feeding Ecology. Columbia, South Carolina, University of South Carolina Press, pp. 45–59.

Junta Central de Planificación (Juceplan); Cuba. 1975–1977. Serie cronológica de la actividad extractiva, 1959–1973. Dirección Central de Estadísticas, Junta Central de Planificación, La Habana, 7 t.

Kabanova, Y. G., and L. López-Baluja. 1973. Producción primaria en la región meridional del Golfo de México y cerca de la costa noroccidental de Cuba. Ser. Oceanol. Acad. Cienc. Cuba, 16:1–34.

Kabanova, Y. G., V. V. Siniukov, L. López-Baluja, N. Borrero, and M. Lluis-Riera. 1968. Resultados obtenidos en el estudio de la producción primaria y su relación con la concentración de elementos biogénicos en las aguas cubanas. Unpublished rep., Inst. Oceanol., Acad. Cienc. Cuba, 208 p.

Karpevich, A. F., and E. N. Bokova. 1937. Rate of digestion of marine fishes [in Russian]. Zool. Zh., 16(1):28–44.

Keast, A., and D. Webb. 1966. Mouth and body form relative to feeding ecology in the fish fauna of a small lake Opinion, Ontario. J. Fish. Res. Board Can., 23(12):1845–1874.

Keener, P., G. D. Johnson, B. W. Stender, E. B. Brothers and H. R. Beatty. 1988. Ingress of postlarval gag, *Mycteroperca microlepis* (Pisces: Serranidae), through a South Carolina barrier island inlet. Bull. Mar. Sci. 42(3):376–396.

Kelly, C. D., A. Moriwake, G. Miyamoto, V. Nicol, and W. Watanabe. 1994. The use of LHRH-a for induced spawning of five different species of marine teleost fishes. Abstract in World Aquaculture 94. World Aquaculture Soc., New Orleans, LA, 14–18 Jan., 1994, 147 p.

Ketchen, K. S. 1975. Age and growth of dogfish, *Squalus acanthias*, in British Columbia waters. J. Fish. Res. Board Can., 32:43–59.

Klima, E. F. 1959. Aspects of the biology and the fishery for Spanish mackerel, *Scomberomorus maculatus* (Mitchill), of southern Florida. Fla. Board Conserv. Tech. Ser., 27:1–39.

Koenig, C. C., F. C. Coleman, L. A. Collins, Y. Sadovy, and P. L. Colin. 1996. Reproduction in gag (*Mycteroperca microlepis*) (Pisces: Serranidae) in the eastern Gulf of Mexico and the consequences of fishing spawning aggregations. In: F. Arreguin-Sanchez, J. L. Munro, M. C. Balgo, and D. Pauly, eds. Biology, Fisheries and Culture of Tropical Groupers and Snappers. ICLARM Conf. Proc. 48:307–323, 449 p.

Koenig, C. C., F. C. Coleman, C. B. Grimes, G. R. Fitzhugh, K. M. Scanlon, C. T. Gledhill, and M. Grace. 2000. Protection of fish spawning habitat for the conservation of warm-temperate reef-fish fisheries of shelf-edge reefs of Florida. Bull. Mar. Sci., 66(3):593–616.

Kondratieva, T. M., and E. Sosa. 1967. Productividad primaria de las aguas cubanas. Estudios, 2(2):21–44.

Konshina, Y. V. 1977. Biological data of the pomadasyid fishes (Familia Pomadasyidae) [in Russian]. Vopr. Ikhtiol., 17(4):621–633.

Koshelev, B. V. 1978. Ecomorphological investigations of gametogenesis, the development of sexual cells and the reproduction on fishes [in Russian]. In: B. V. Koshelev, ed. Ecological, Morphological and Ecological Physiological Investigations of Fish Development. Nauka, Moscow, pp.10–42.

Koshelev, B. V. 1981. Study of fish reproduction (gametogenesis, sexual maturity rate, sexual cycle, ovulation rate and spawning) [in Russian]. In: B. V. Koshelev, and M. V. Gulidov, eds. Investigations of Fish Reproduction and Growth. Methodological Manual. Nauka, Moscow, pp. 5–16.

Koshelev, B. V. 1984. Ecology of fish reproduction [in Russian]. Nauka, Moscow, 309 p.

Koslow, J. A., F. Hanley, and R. Wicklund. 1988. Effects of fishing on reef fish communities at Pedro Bank and Port Royal Cays, Jamaica. Mar. Ecol. Prog. Ser., 43:201–212.

Kozlov, A. N. 1978. Ecological-physiological characteristics of some Antarctic fish species [in Russian]. Tr. VNIRO, 120:75–84.

Lang, J. C., P. M. Alcolado, J. P. Carricart-Gavinet, M. Chiappone, H. A. Curran, P. Dustan, F. X. Geraldes, S. R. Gittings, S. R. Smith, J. W. Tunnell, and J. Weiner. 1998. Status of western Atlantic coral reefs in the northern areas of the Wider Caribbean. In: C. Wilkinson, ed. Status of Coral Reefs of the World: 1998. Global Reef Monitoring Network. Aust. Inst. Mar. Sci., pp. 123–134.

Lapin, V. I. 1976. Seasonal variation of physiological-biochemical indicators on different flatfish populations (*Platichthys flesus* L.) [in Russian]. Vopr. Ikhtiol., 16(4):703–714.

Lapina, N. N., and V. I. Lapin. 1982. Characteristics of the dynamics of physiological-biochemical indicators of the generative synthesis on females of different carp species [in Russian]. Vopr. Ikhtiol., 22(2):258–239.

Lapina, N. N., V. I. Lapin, and M. I. Shatunovsky. 1984. On the metabolic aspects of the intraspecies variability of fish reproduction [in Russian]. Dokl. AN S.S.S.R., 274(4):1005–1008.

Lapointe, B. E., M. M. Littler, and D. S. Littler. 1992. Modification of benthic community structure by natural eutrophication: The Belize barrier reef. Proc. 7th Int. Coral Reef Symp., Guam., 1:323–324.

Larkin, P. A. 1956. Interspecific competition and population control in freshwater fish. J. Fish. Res. Board Can., 13:327–342.

Larrañeta, M. G. 1967. Crecimiento de *Pagellus erythrinus* de las costas de Castellón. Invest. Pesq. (Barc.), 31(2):185–258.

Lasker, M. R. 1970. Utilization of zooplankton energy by a Pacific sardine population in the California current. Symp. Food Chain Stud., 4:265–284.

Lasker, H. R. 1985. Prey preferences and browsing pressure of the butterfly fish, *Chaetodon capistratus*, on Caribbean gorgonians. Mar. Ecol. Prog. Ser., 21:213–220.

Lee, T. N., and E. Williams. 1999. Mean distribution and seasonal variability of coastal currents and temperature in the Florida Keys with implications for larval recruitment. Bull. Mar. Sci., 64(1):35–56.

Leis, J. M. 1987. Review of the early life history of tropical groupers (Serranidae) and snappers (Lutjanidae). In: J. J. Polovina, and S. Ralston, eds. Tropical Snappers and Groupers: Biology and Fisheries Management. Westview Press, Boulder, CO, pp. 189–237.

Leis, J. M. 1991. The pelagic stage of reef fishes: The larval biology of coral reef fishes. In: P. F. Sale, ed. The Ecology of Fishes on Coral Reefs. Academic Press, San Diego, CA, pp. 183–230.

Leis, J. M., H. P. A. Sweatman, and S. E. Reader. 1996. What the pelagic stages of coral reefs are doing out in blue water: Daytime field observations of larval behavioral capabilities. Mar. Freshwat. Res., 47:401–411.

León. M. E., and M. Guardiola. 1984. Caracterización biológico-pesquera del género *Scomberomorus* de la zona suroriental de Cuba. Rev. Cub. Invest. Pesq., 9(3-4):1–26.

León, Y., E. Pugibet, and R. Sluka. In press. The abundance of fishes in shallow, algal/seagrass habitats in the waters surrounding Parque Nacional del Este, Dominican Republic. Proc. Gulf Carib. Fish. Inst. 48.

Leonce-Valencia, C. O., and C. Monroy. 1993. Growth parameter estimation based on length frequency data of red snapper (*Lutjanus campechanus*) on Campeche Bank. Paper presented at Int. Workshop on Tropical Groupers and Snapper, 28–29 Oct., 1993, Campeche, Mexico.

Lessios, H. A. 1988. Mass mortality of *Diadema antillarum* in the Carribean: What have we learned? Ann. Rev. Ecol. Sys., 19:371–393.

Lewis, J., G. Draper, C. Bourdon, C. Bowin, P. Mattson, F. Maurrasse, F. Nagle, and G. Pardo. 1990. Geology and tectonic evolution of the northern Caribbean margin. In: G. Dengo, and J. E. Case, eds. The Caribbean Region, vol. H, The Decade of North American Geology. Geol. Soc. Am., Boulder, CO, pp. 77–140.

Lewis, J. B. 1967. Food of the dolphin *Coryphaena hippurus* Linnaeus, and of the yellowfin tuna, *Thunnus albacares* (Lowe), from Barbados, West Indies. J. Fish. Res. Board Can., 24(3):683–686.

Ley, J. A., C. C. McIvor, and C. L. Montague. 1999. Fishes in mangrove prop-root habitats of northeastern Florida Bay: distinct assemblages across an estuarine gradient. Estuar. Coast. Shelf Sci., 48:701–723.

Lindeman, K. C. 1986. Development of larvae of the French grunt, *Haemulon flavolineatum,* and comparative development of twelve western Atlantic species of Haemulon. Bull. Mar. Sci., 39(3):673–716.

Lindeman, K. C. 1996. Review of *Ecologia de los Peces Marinos de Cuba,* R. Claro, ed. Bull. Mar. Sci., 58(3):568–570.

Lindeman, K. C. 1997. Comparative management of beach systems of Florida and the Antilles: Applications using ecological assessment and decision support procedures. In: G. Cambers, ed. Managing Beach Resources in the Smaller Caribbean Islands. UNESCO Coastal Region and Small Island Papers # 1, Paris, pp. 134–164.

Lindeman, K. C., and D. B. Snyder. 1999. Nearshore hardbottom fishes of southeast Florida and effects of habitat burial caused by dredging. Fish. Bull., 97(3):508–525.

Lindeman, K. C., G. A. Díaz, J. E. Serafy, and J. S. Ault. 1998. A spatial framework for assessing cross-shelf habitat use among newly settled grunts and snappers. Proc. Gulf Carib. Fish. Inst., 50:385–416.

Lindeman, K. C., R. Pugliese, G. T. Waugh, and J. S. Ault. 2000. Developmental patterns within a multispecies reef fishery: Management applications for essential habitats and protected areas. Bull. Mar. Sci., 66(3):929–956.

Lindeman, K. C., T. N. Lee, W. D. Wilson, R. Claro, and J. S. Ault. 2001. Transport of larvae originating in southwest Cuba and the Dry Tortugas: Evidence for partial retention in grunts and snappers. Proc. Gulf Carib. Fish. Inst. 52:732–747.

Lipskaia, N. 1975. Intensity of metabolism of juveniles of different tropical fish species [in Russian]. Vopr. Ikhtiol., 14(6):1076–1086.

Lirman, D., and P. Fong. 1995. The effects of Hurricane Andrew and Tropical Storm Gordon on Florida reefs. Coral Reefs, 14:172.

Liubimova, T. G., and A. M. Capote. 1971. Biological characteristics of some commercial fish species of the Campeche Bank [in Russian]. In: A. S. Bogdanov, ed. Joint Soviet-Cuban Fisheries Investigations Pishevaia Promishliennost, Moscow, vol. 3, pp. 82–88.

Lluis-Riera, M. 1972. Estudios hidrológicos del Golfo de Batabanó y de las aguas oceánicas adyacentes. Ser. Oceanol. Acad. Cienc. Cuba, 14:1–49.

Lluis-Riera, M. 1974. Estudios hidrológicos de un área semi-cerrada y de las aguas adyacentes en la costa norte de la provincia de La Habana. Ser. Oceanol. Acad. Cienc. Cuba, 19:1–45.

Lluis-Riera, M. 1977. Estudios hidrológicos de la plataforma suroriental de Cuba y aguas oceánicas adyacentes. Inf. Cient.-Téc., Inst. Oceanol. Acad. Cienc. Cuba, 16:1–29.

Lluis-Riera, M. 1981a. Condiciones hidrológicas de la plataforma nororiental de Cuba, durante febrero de 1976. Inf. Cient.-Téc., Inst. Oceanol. Acad. Cienc. Cuba, 161:1–32.

Lluis-Riera, M. 1981b. Informe de datos oceanográficos de la plataforma suroriental de Cuba y aguas oceánicas adyacentes. Editorial Academia, La Habana, 141 p., 42 figs.

Lluis-Riera, M. 1983a. Estudios hidrológicos de la plataforma noroccidental de Cuba. Rep. Invest. Inst. Oceanol. Acad. Cienc. Cuba, 13:1–48.

Lluis-Riera, M. 1983b. Physico-chemical characteristics of the Cuban shelf and its adjacent oceanic waters [in Russian]. Informatsionni Bulleten, Centro de Coordinación del CAME para el Problema Océano Mundial, Moscow, 15:29–45.

Lluis-Riera, M. 1983c. Régimen hidrológico de la plataforma insular de Cuba. Cienc. Tierra Espacio, 7:81–110.

Lluis-Riera, M. 1984. Estudios hidrológicos de la plataforma nororiental de Cuba (zona D). Editorial Academia, La Habana, 44 p.

Lo, N. C. H., J. Alheit, and B. Alegre. 1986. Fecundidad parcial de la sardina peruana (*Sardinops sagax*). Bol. Inst. Mar. Perd., 10(2):48–60.

Longley, W. H., and S. F. Hildebrand. 1941. Systematic catalogue of the fishes of Tortugas, Florida, with observation on color, habits and local distribution. Pap. Tortugas Lab., 34:1–331.

López-Baluja, L. 1978. Variaciones estacionales del fitoplancton en el Golfo de Batabanó. Cienc. Biol., 2:59–89.

López-Baluja, L., N. Borrero, and G. Popowski. 1980. Distribución cualitativa y cuantitativa del fitoplancton en la plataforma de Cuba. Unpublished, Inst. Oceanol., Acad. Cienc. Cuba, La Habana, 48 p., 13 tabs., 25 figs.

Losey, G. S., Jr. 1972. The ecological importance of cleaning symbiosis. Copeia, 4:820–833.

Love, R. R. 1980. The chemical biology of fishes. Academic Press, London, vol. 2, 943 p.

Lowe, R. M. 1971. Interspecific territoriality in a pomacentrid reef fish *Pomacentrus flavicauda* Witley. Ecology, 52(4):648–654.

Lowe-McConnell, R. M. 1962. The fishes of the British Guiana continental shelf, Atlantic coast of South America, with notes on their natural history. J. Linn. Soc. Lond. (Zool.), 44(301):669–700.

Lowe-McConnell, R. M. 1979. Ecological aspects of seasonality in fishes of tropical waters. Symp. Zool. Soc. Lond., 44:219–241.

Luckhurst, B. E. 1998. Site fidelity and return migration of tagged red hinds (*Epinephelus guttatus*) to a spawning aggregation site in Bermuda. Proc. Gulf Carib. Fish. Inst. 50:750–763.

Lugo, A. E., and S. C. Snedaker. 1974. The ecology of mangroves. Ann. Rev. Ecol. Syst., 5:39–64.

Mahon, R. 1993. Natural fishery management areas in the western central Atlantic region. Ocean and Coastal Management, 19:121–135.

Malfait, B., and M. Dinkelman. 1972. Circum-Caribbean tectonic and igneus activity and the evolution of the Caribbean plate. Bull. Geol. Soc. Amer., 83(2):251–272.

Man, A., R. Law, and N. V. C. Polunin. 1994. Role of marine reserves in recruitment to reef fisheries: A metapopulation model. Biol. Conserv., 7:197–204.

Manickchand-Dass, S. C. 1987. Reproduction, age and growth of the lane snapper, *Lutjanus synagris* (Linnaeus) in Trinidad, West Indies. Bull. Mar. Sci., 40(1):22–28.

Manickchand-Heileman, S. C., and D. A. T. Phillip. 1996. Reproduction, age and growth of the Caribbean red snapper (*Lutjanus purpureus*) in waters off Trinidad and Tobago. In: F. Arreguín-Sánchez, J. L. Munro, M. C. Balgos, and D. Pauly, eds. Biology, Fisheries and Culture of Tropical Groupers and Snappers. ICLARM Conf. Proc. 48, pp. 137–149.

Manooch, C. S., III. 1976. Age, growth and mortality of the white grunt, *Haemulon plumieri* Lacépède (Pisces: Pomadasyidae), from North Carolina and South Carolina. Proc. Ann. Conf. Southeast Assoc. Game Fish. Comm., 30:58–70.

Manooch, C. S., III. 1987. Age and growth of snappers and groupers. In: J. J. Polovina, and S. Ralston, eds. Tropical Snappers and Groupers. Biology and Fisheries Management. Westview Press, Boulder and London, pp. 329–373.

Manooch, C. S., III, and C. Barans. 1982. Distribution and abundance, and age and growth of the tomtate, *Haemulon aurolineatum* along the southeastern United States coast. Fish. Bull., 80(1):1–19.

Manooch, C. S., III, and C. Drennon. 1987. Age and growth of yellowtail snapper and queen triggerfish collected from US Virgin Islands and Puerto Rico. Fish. Res., 6:53–68.

Manooch, C. S., III, and M. Haimovicii. 1978. Age and growth of the gag, *Mycteroperca microlepis*, and size-age composition of the recreational catch of the southeastern United States. Trans. Am. Fish. Soc., 107(2):234–240.

Manooch, C. S., III, and W. T. Hogarth. 1983. Stomach contents and giant trematores from wahoo, *Acanthocybium solanderi*, collected adong the south Atlantic and Gulf coasts of the United States. Bull. Mar. Sci. 33(2):227–238.

Manooch, C. S., III, and D. L. Mason. 1984. Age, growth and mortality of lane snapper from southern Florida. Northeast Gulf Sci., 7(1):109–115.

Manooch, C. S., III, and D. L. Mason. 1985. Age and growth of mutton snapper along the east coast of Florida. Fish. Res. 3:93–104.

Manooch, C. S., III, and D. L. Mason. 1987. Age and growth of warsaw grouper from the southeast region of the United States. Northeast Gulf Sci., 9(2):65–75.

Manooch, C. S., III, and R. Matheson III. 1983. Age, growth and mortality of gray snapper collected from Florida waters. Proc. Annu. Conf. Southeast Assoc. Fish Wildl. Agencies, 35:331–344.

Manooch, C. S., III, and J. C. Potts. 1996. Growth characteristics of the spottail pinfish, *Diplodus holbrooki*, off North Carolina. J. Elish Mitchell Sci. Soc., 112(1):7–19.

Manooch, C. S., III, and J. C. Potts. 1997a. Age, growth and mortality of greater amberjack, Seriola dumerili, from the southeastern United States. Fish. Res., 30:229–240.

Manooch, C. S., III, and J. C. Potts. 1997b. Age, growth and mortality of greater amberjack, Seriola dumerili, from the U.S. Gulf of Mexico headboat fishery. Bull. Mar. Sci., 61(3): 671–683.

Manooch, C. S., III, and J. C. Potts. 1997c. Age and growth of red snapper, *Lutjanus campechanus*, Lutjanidae, collected along the southeastern United states from North Carolina through the east coast of Florida. J. Elisha Mitchell Sci. Soc., 113(3):111–122.

Manteifel, B. P. 1961. Vertical migrations of marine organisms. II. On the adaptive significance of vertical migrations of planktivorous fishes [in Russian]. Tr. Inst. Evoliutsionnoi Morfologii y Ecologii Zhivotnikh, Akad. Nauk SSSR, 39:5–46.

Manteifel, B. P. 1980. Ecology of animal behavior [in Russian]. Nauka, Moscow, 220 p.

Manteifel, B. P., I. I. Guirsa, T. S. Lesheva, and D. S. Pavlov. 1965. Ritmo diario de la alimentación y actividad de los movimientos de algunos peces depredadores de agua dulce [in Russian]. In: La Alimentación de los Peces Depredadores y su Interrelación con los Organismos Alimentarios. Nauka, Moscow, pp. 2–82.

Marshall. N. 1980. Fishery yield of coral reefs and adjacent shallow water environments. In: S. B. Saila and P. M. Roedel, eds. Stock Assessment for Tropical Small-Scale Fisheries. Int. Cent. Mar. Res. Dev., University of Rhode Island, Kingston, pp. 103–109.

Martínez, S., and E. D. Houde. 1975. Fecundity, sexual maturation and spawning of scaled sardine (*Harengula jaguana* Poey). Bull. Mar. Sci., 25(1):35–45.

Martínez-Estalella, N., and P. M. Alcolado. 1990. Caracteristicas de las comunidades de corales pétreos de la macrolaguna del Golfo de Batabanó. In: P. M. Alcolado, ed. El Bentos de la Macrolaguna del Golfo de Batabanó. Editorial Academia, La Habana, pp. 25–36.

Martínez-Iglesias, J. C., and P. M. Alcolado. 1990. Caracteristicas de la fauna de crustáceos decápodos de la macrolaguna del Golfo de Batabanó. In: P. M. Alcolado, ed. El Bentos de la Macrolaguna del Golfo de Batabanó. Editorial Academia, La Habana, pp. 75–84.

Maslennikova, N. V. 1978. Role of the liver in some marine fishes [in Russian]. Tr. VNIRO, 120:20–29.

Mason, D., and C. S. Manooch III. 1986. Growth characteristics of mutton snapper along the East coast of Florida. Poeyana, 311:1–15.

Matheson, R. H., and G. R. Huntsman. 1984. Growth, mortality and yield-recruit model for speckled hind and snowy grouper from United States South Atlantic Bight. Trans. Am. Fish. Soc., 113:607–616.

McCoy, E. D., and K. L. Heck. 1976. Biogeography of corals, seagrasses and mangroves: An alternative to the center of origin concept. Systematic Zool., 25:201–210.

McEachran, J. D., and J. D. Fechhelm. 1998. Fishes of the Gulf of Mexico. Vol 1: Myxiniformes to Gasterosteiformes. University of Texas Press, Austin. 1112 p.

McErlean, A. 1963. A study of the age and growth of *Mycteroperca microlepis* Goode and Bean, Pisces: Serranidae, on the west coast of Florida. Fla. State Board Conserv. Tech. Ser., 41:1–29.

McErlean, A., and C. L. Smith. 1964. The age of sexual succession in the protogynous hermaphrodite *Mycteroperca microlepis*. Trans. Am. Fish. Soc., 93(3):301–302.

McFarland, W. N., E. B. Brothers, J. C. Ogden, M. J. Shulman, E. L. Bermingham, and N. M. Kotchian-Prentis. 1985. Recruitment patterns in young French grunts, *Haemulon flavolineatum* (family Haemulidae), at St. Croix, Virgin Islands. Fish. Bull., 83:151–161.

McGovern, J. C., D. M. Wyanski, O. Pashuk, C. S. Manooch III, and G. R. Sedberry. 1998. Changes in the sex ratio and size at maturity of gag, *Mycteroperca microlepis*, from the Atlantic coast of the southeastern United States during 1976–1995. Fish. Bull., 96(4):797–807.

Medina-Quej, A., and M. Domínguez-Viveros. 1997. Edad y crecimiento del *Scomberomorus maculatus* (Scombriformes: Scombridae) en Quintana Roo, México. Rev. Biol. Trop., 45(3):1155–1161.

Melo, A. F. M. 1976. Aspectos biológico-pesqueros de *Epinephelus morio* (Val.) "mero." In: Memorias del Primer Simposio Nacional de Recursos Pesqueros Marinos de México, Inst. Nacl. Pesca, México, vol. 2, pp. 223–266.

Mendez, F. 1989. Contribución al estudio de la biología y la pesquería del pargo guanapo, *Lutjanus synagris* Linnaeus, 1758 (Pisces: Lutjanidae) en el Parque Nacional Archipiélago de los Roques, Venezuela. T. E. G. Univ. Central Venezuela, Caracas, 103 p.

Mendoza, A. 1968. Consideraciones sobre la biología pesquera de la sierra, *Scomberomorus maculatus* (Mitchill), en el Estado de Veracruz. Bios (Lisb.), 1(2):11–22.

Menge, B. A. 1972. Competition for food between two intertidal starfish species and its effect on body size and feeding. Ecology, 53(4):635–644.

Menon, M. D. 1950. The use of bones other than otoliths in determining the age and growth rate of fishes. J. Conserv. CIEM, 16(3):311–340.

Mester, R., R. M. Ros, and R. L. Mester. 1974. Estudio histológico de las gónadas de dos especies de sardinas *Harengula clupeola* (Cuvier) y *Harengula humeralis* (Cuvier). Ciencias, Ser. 8, Invest. Mar., 8:1–21.

Mexicano-Cíntora, G., and F. Arreguín-Sánchez. 1989a. Dinámica de las poblaciones de rubia (*Lutjanus synagris*) y canané (*Ocyurus chrysurus*) de las costas de Yucatán, México. Unpublished, Cent. Invest. y Estudios Avanzados del Inst. Politécnico Nacl., Unidad Mérida, México, 13 p., 5 tabs., 8 figs.

Mexicano-Cíntora, G., and F. Arreguín-Sánchez. 1989b. Estimación de edad y crecimiento del pargo canané (*Ocyurus chrysurus*) del litoral de Yucatán, México. Unpublished, Cent. Invest. y Estudios Avanzados del Instituto Politécnico Nacl., Unidad Mérida, México,12 p., 3 tabs., 1 fig.

Meyer, J. L., E. T. Schultz, and G. S. Helfman. 1983. Fish schools: An asset to corals. Science, 220:1047–1049.

Meyerhoff, A. A., and C. Hatten. 1968. Diapiric structure in Central Cuba. Am. Assoc. Petrol. Geol. Mem., 8:315–357.

Mijeev, V. N. 1983. Size of ingested prey and and feeding selectivity of juvenile fish [in Russian]. Vopr. Ikhtiol., 24(2):243–252.

Millares, N., M. Borrero, T. Damas, and E. González. 1979a. Desove inducido en la biajaiba Lutjanus synagris Linné, 1758. Rev. Cub. Invest. Pesq., 4(1):1–21.

Millares, N., M. Borrero, and E. González, 1979b. Desove inducido en Eugerres plumieri (patao) en condiciones de laboragtorio. Rev. Cub. Invest. Pesp., 4(1):67–87.

Mina, M. V. 1967. On the relationship between otolith (sagitta) weight and fish length on the cod populations of the White and Barents Seas [in Russian]. Biol. Nauk, 9:26–31.

Ministerio de la Industria Pesquera. 1980. Análisis general de las pesquerías de escama en Cuba. In: Reunión Nacional de Escama, Topes de Collantes, Cuba, 48 p.

Miranda, B. O. 1981. Daily lines of growth as components of the spawning bands in the otoliths in a tropical fish (Sparisoma aurofrenatum) (Cuvier et Valenciennes, 1839). Rev. Biol. Mar., Inst. Oceanol. Univ. Valparaiso, 17(2):253–265.

Mochek, A. D. 1977. Estudio comparativo de la conducta territorial de la chopita, Eupomacentrus fuscus (Cuvier et Valenciennes) en condiciones naturales y experimentales. Inf. Cient.-Téc., Inst. Oceanol. Acad. Cienc. Cuba, 22:1–11.

Mochek, A. D. 1980. Behaviour of fishes of the Black Sea coastal zone. [In Russian]. Zool. Zh., 59(7):1060–1066.

Mochek, A. D., and A. F. Silva Lee. 1975. Conducta social del género Haemulon. Ser. Oceanol. Acad. Cienc. Cuba, 27:1–27.

Mochek, A. D., and E. Valdés-Muñoz. 1983. Acerca de la conducta de los peces de las comunidades costeras en la plataforma cubana. Cienc. Biol., 9:87–106.

Mochek, A. D., and E. Valdés-Muñoz. 1985. Structure and functionality of fish communities of the Cuban shelf. In: V. E. Sokolov and G. García, eds. Ecology of fishes of the Cuban shelf [In Russian], Nauka, Moscow, pp. 232–283.

Moe, M. A., Jr. 1969. Biology of the red grouper, Epinephelus morio (Valenciennes) from the eastern Gulf of Mexico. Fla. Dep. Nat. Resour. Mar. Res. Lab., Prof. Pap. Ser., 10:1–95.

Moe, M. A., R. H. Lewis, and R. M. Ingle. 1968. Pompano mariculture: Preliminary data and basic consideration. Fla. State Board Conserv. Tech. Ser., 55:1–65.

Molnar, G., and J. Tolg. 1967. Experiments concerning gastric digestion of pikeperch (Lucioperca) in relation to water temperature. Acta Biol. Acad. Sci. Hung., 13:231–239.

Monte, S. 1964. Observacoes sobre a estructura histologica das gonadas da albacora, Thunnus atlanticus (Lesson), no nordeste do Brasil. Bol. Inst. Biol. Mar. Univ. Rio Grande do Norte, 1:17–31.

Montgomery, L. W. 1981. Mixed species schools and the significance of vertical territories of damselfishes. Copeia, 2:477–481.

Montolio, M., and M. Juárez. 1976. Estimado preliminar de la población en desove del bonito (Katsuwonus pelamis Lin., 1758) en el Caribe occidental. Rev. Invest. Cent. Invest. Pesq., La Habana, 2(3):267–275.

Moore, C. M., and R. F. Labinsky. 1984. Population parameters of a relatively unexploited stock of snowy grouper, Epinephelus niveatus, in the lower Florida Keys, USA. Trans. Am. Fish. Soc., 113:322–329.

Mota Alves, M. D., and G. S. Tomé. 1968. Observacoes sobre a desemvolvimento maturativo das gonadas da Serra, Scomberomorus maculatus (Mitchill, 1815). Arq. Est. Biol. Mar. Univ. Fed. Ceará, 8(1):25–30.

Moyer, J. T., and A. Nakazone. 1978. Population structure, reproductive behavior and protogynus hermaphroditism in the angelfishes Centropyge interruptus at Mujake-jima, Japan. Jap. J. Ichthyol., 25:25–39.

Moyer, J. T., and M. J. Zaiser. 1984. Early sex change: A possible mating strategy of Centropyge angelfishes (Pisces: Pomacanthidae). J. Ethol., 2:63–67.

Mueller, K. W., G. D. Dennis, D. B. Eggleston, and R. I. Wicklund. 1994. Size-specific social interactions and foraging styles in a shallow water population of mutton snapper, Lutjanus analis (Pisces: Lutjanidae), in the central Bahamas. Environ. Biol. Fish., 40:175–188.

Mullaney, M. D. Jr., and L. D. Gale. 1996. Morphological relationships in ontogeny: Anatomy and diet in gag, Mycteroperca microlepis (Pisces: Serranidae). Copeia, 1:167–180.

Munro, J. L. 1976. Aspects of the biology and ecology of Caribbean fishes: Mullidae (goat-fishes). J. Fish. Biol., 9(1):79–97.

Munro, J. L. 1977. Actual and potential production from the coralline shelves of the Caribbean Sea. FAO Fish. Rep., 200:301–321.

Munro, J. L. 1983a. Biological and ecological characteristics of Caribbean reef fishes. In: J. L. Munro, ed. Caribbean Coral Reef Fishery Resources. Intern. Center Living Aquatic Resources Management (ICLARM) Stud. Rev., 7:223–231.

Munro, J. L. 1983b. Coral reef fish and fisheries of the Caribbean Sea. In: J. L Munro, ed. Caribbean Coral Reef Fishery Resources. Intern. Center Living Aquatic Resources Management (ICLARM) Stud. Rev., 7:1–9.

Munro, J. L. 1983c. Epilogue: Progress in coral reef fisheries research, 1973–1982. In: J. L. Munro, ed. Caribbean Coral Reef Fishery Resources. Intern. Center Living Aquatic Resources Management (ICLARM) Stud. Rev., 7:249–265.

Munro, J. L., ed. 1983d. Caribbean Coral Reef Fishery Resources. Intern. Center Living Aquatic Resources Management (ICLARM) Stud. Rev., 7:1–276.

Munro, J. L. 1999. Marine protected areas and the management of coral reef fisheries. Tech. rep. Intern. Center Living Aquatic Resources Management (ICLARM) Caribbean and Eastern Pacific Office, 40 p.

Munro, J. L., and D. McB. Williams. 1985. Assessment and management of coral reef fisheries: Biological environmenttal and socio-economic aspects In: Proc. 5th Int. Coral Reef Congress, Tahiti, 27 May–1 June, 1985. Vol. 4 AntenneMuseum-EPHE, Moorea, French Polynesia, p. 543–578.

Munro, J. L., V. C. Gaut, R. Thompson, and P. H. Reeson. 1973. The spawning seasons of Caribbean reef fishes. J. Fish. Biol., 5:69–84.

Murdy, E. O., and C. J. Ferraris. 1980. The contribution of coral reef fisheries to Philippine fisheries production. ICLARM Newsl., 3(1):21–22.

Murina, V. V., V. D. Chukhchin, O. Gómez, and G. Suárez. 1969. Distribución cuantitativa de la macrofauna bentónica del sublitoral superior de la plataforma cubana (región noroccidental). Ser. Oceanol. Acad. Cienc. Cuba, 6:1–14.

Murray, P. A., and E. A. Moore 1992. Recruitment and explotation rate of Etelis oculatus Val. In: The St. Lucian Fishery. Abstract in Proc. Gulf Carib. Fish. Inst., 42:262.

Murray, P. A., L. E. Chinnery, and E. A. Moore. 1992. The recruitment of the queen snapper, Etelis oculatus Val. into the St. Lucian fishery: Recruitment of fish and recruitment of fishermen. Proc. Gulf Carib. Fish. Inst. 41:297–303.

Myrberg, A. A., Jr. and R. E. Thresher. 1978. Interspecific aggression and its relevance to the concept of territoriality in reef fishes. Am. Zool., 14:81–96.

Myrberg, A. A., Jr., B. D. Brahy, and A. R. Emery. 1967. Field observations on the reproduction of the damselfish, Chromis multilineatus (Pomacentridae) with additional notes on general behavior. Copeia, 4:819–827.

Nagelkerken, W. P. 1979. Biology of the graysby, Epinephelus cruentatus, of the coral reef of Curaçao. Stud. Fauna Curaçao, 60: 1–118.

Naranjo, A. 1956. Cordel y anzuelo. Editorial Cenit, La Habana, 251 p.

Naumov, V. M. 1956. Oogenesis and ecology of the sexual cycle of the Atlantic herring in Murmansk (*Clupea harengus harengus* L.) [in Russian]. Tr. PINRO, 9:176–224.

Naumov, V. M. 1968. Fecundity of the fishes of the Indian Ocean [in Russian]. Tr. VNIRO, 64(5, 28):431–436.

Nelson, J. S. 1994. Fishes of the World. John Wiley and Sons, New York, NY, 600 p.

Nelson, R. S., and C. S. Manooch III. 1982. Growth and mortality of red snapper in the West-Central Atlantic and Northern Gulf of Mexico. Trans. Am. Fish. Soc., 3:465–475.

Nelson, R. S., C. S. Manooch III, and D. L. Mason. 1985. Ecological effects of energy development on reef fish of the Flower Garden Banks: Reef fish bioprofiles. Final rep. Southeast Fish. Cent., Beaufort Lab. Natl. Mar. Fish. Serv., NOAA.,Beufort, N.C. 28515–9722.

Nikolsky, G. V. 1963. The Ecology of Fishes. Academic Press, London and New York, 329 p.

Nikolsky, G. V. 1974a. Theory of the Dynamics of Fish Populations [in Russian]. Nauka, Moscow, 447 p.

Nikolsky, G. V. 1974b. On the causes of the action of fish predators upon fish prey in the low latitudes [in Russian]. Zh. Obshch. Biol., 35(3):346–352.

Nomura, H. 1967. Dados biológicos sobre a serra, *Scomberomorus maculatus* (Mitchill), das aguas cearenses. Arq. Est. Biol. Mar. Univ. Fed. Ceará, 7(1):29–39.

Nomura, H., and M. S. Rodrigues. 1967. Biological notes on king mackerel, *Scomberomorus cavalla* (Cuvier), from northeastern Brazil. Arq. Est. Biol. Mar. Univ. Fed. Ceará, 7(1):79–85.

Nunes, M. L., F. J. Beserra, G. Hitzschky, F. Vieira, C. A. Sobreira Rocha, and J. W. Menezes da Nobrega. 1976. Composiçao química de algunos peixes marinhos do nordeste brazileiro. Arq. Cien. Mar., 16(1):23–26.

Núñez Jiménez, A. 1965. Geografia de Cuba. Editora Nacional de Cuba, La Habana, 141 p., 42 Illust.

Núñez Jiménez, A. 1982. Cuba: La Naturaleza y el Hombre: El Archipielago. Editorial Letras Cubanas, La Habana, 691 p.

Obando, E., and J. R. Leon. 1989. Reproducción del bolo, *Diplectrum formosum* (Linnaeus, 1766) (Pisces: Serranidae) en Punta Mosquito, Isla de Margarita, Venezuela. Sci. Mar., 53(4):771–777.

Obregón, M. H., and R. Pedroso. 1984. Resultados de la experiencia con la corona seleccionadora en barcos chinchorreros. Rev. Cub. Invest. Pesq., 9(3–4):64–98.

Obregón, M. H., E. Pozo, and S. Valle. 1990. Las pesquerías de biajaiba (*Lutjanus synagris*) en la plataforma norroriental de Cuba. Resúmenes, II Cong. Cienc. Mar, 18–22 junio, La Habana, Cuba, p. 151.

Ogden, J. C., and P. S. Lobel. 1978. The role of herbivorous fishes and urchins in coral reef communities. Environ. Biol. Fish., 3(1):49–65.

Okada, Y. K. 1965. Bisexuality in sparid fish. I and II. Proc. Jap. Acad., 41:294–304 [cited by Roede 1972].

Olaechea, A., and M. M. Quintana. 1970. Pre-evaluación sobre la determinación de la edad en la biajaiba, *Lutjanus synagris* (Linné), Cuba. In: Segunda Reunion de Balance del Centro de Investigaciones Pesqueras (mecanografiado) 2:50–61.

Olaechea, A., and M. A. Quintana. 1975. Desarrollo gonadal de la biajaiba en la plataforma suroccidental de Cuba. Resúmenes Invest., Cent. Invest. Pesq. [mimeograph], La Habana, 9 p., 4 tabs., 12 figs.

Olaechea, A., and U. Sauskan. 1974. Cartas de pesca del Banco de Campeche, 1972, Parte 1 Unpublished, Cent. Invest. Pesq., La Habana, 67 p.

Olaechea, A., C. Hernández, and M. de León. 1975. Evaluación de peces demersales en el Banco de Campeche. Unpublished, Cent. Invest. Pesq., La Habana, Cuba, 25 p.

Olsen, D., and J. A. LaPlace. 1979. A study of a Virgin Islands grouper fishery based on a breeding aggregation. Proc. Gulf Carib. Fish. Inst., 31:13–144.

Ortiz, M. 1976. Algunas características del bentos de Cuba. Ciencias, Ser. 8, Invest. Mar., 22:1–32.

Otsu, T., and R. N. Uchida. 1959. Sexual maturity and spawning of albacore in the Pacific Ocean. U.S. Fish. Wildl. Serv., Fish. Bull., 59(148):289–304.

Oven, L. S. 1967. On the reproduction of *Platichthys flesus luscus* (Pallas) at the Black Sea. Vopr. Ikhtiol., 7(1):96–100.

Oven, L. S. 1971. On the reproduction of *Trachurus trachurus trachurus* (Linné) at the Mediterranean Sea. (In Russian). Biol. Morya, 25:76–84.

Oven, L. S. 1974. On the multi-intermittent spawning of marine fishes [in Russian]. In: The Biological Productivity of the Southern Oceans. Naukova Dumka, Kiev, pp. 206–213.

Oven, L. S. 1976. Oogenesis and spawning in marine fishes [in Russian]. Naukova Dumka, Kiev, 132 p.

Oven, L. S. 1985. Spawning patterns of fishes in low latitude waters (in Russian). In: B.V. Koshelev, ed. Characteristics of the reproductive cycle of fishes of different geographic areas. Nauka, Moscow, pp. 35–45.

Oven, L. S., and L. P. Saliejova. 1970. Growth and reproduction of some tropical marine fishes [in Russian]. Biol. Morya, 21:245–266.

Oven, L. S., and L. P. Saliejova. 1971. Studies on the reproduction and growth of marine fishes [in Russian]. In: Issues of Marine Biology. Naukova, Dumka Kiev, pp. 95–99.

Overko, S. M. 1969. Reproductive biology of the genera *Trachurus* and *Decapterus* in the nearshore waters of northwestern Africa [in Russian]. Tr. Atlant-NIRO, 22:86–102.

Overko, S. M. 1971. Morphology, biology and exploitation of the jurel (*Trachurus trachurus trachurus* Linné) of the central-eastern Atlantic [in Russian]. Tr. Atlant-NIRO, 41:102–121.

Palazón, J. L., and L. W. González. 1986. Edad y crecimiento del pargo cebal, *Lutjanus analis* (Cuvier, 1828) (Teleostei: Lutjanidae) en la Isla Margarita y alrededores de Venezuela. Invest. Pesq. (Barc.), 50(2):151–165.

Panella, G. 1971. Fish otoliths: Daily growth layers and periodical patterns. Science, 173:1124–1127.

Panella, G. 1974. Otolith growth patterns: An aid in age determination in temperate and tropical fishes. In: T. B. Bagenal, ed. Proc. Int. Symp. on the Ageing of Fish. Unwin Brothers Ltd., Surrey, England, pp. 28–39.

Parrish, J. D. 1987. The trophic biology of snappers and groupers. In: J. J. Polovina, and S. Ralston, eds. Tropical Snappers and Groupers. Biology and Fisheries Management. Westview Press, Boulder and London, pp. 405–464.

Parrish, J. D. 1989. Fish communities of interacting shallow-water habitats in tropical oceanic regions. Mar. Ecol. Prog. Ser., 58:143–160.

Pastor, L. 1987. Utilización de diferentes tipos de alimento en los peces marinos ornamentales destinados a la exportación. In: Cong. Cienc. del Mar, "MARCUBA 87," La Habana, 1987, Resúmenes, p. 89.

Pauly, D. 1978. A preliminary compilation of fish length growth. Berichte des Instituts für Meereskunde an der Christian-Alberchts Universitat Kiel (55), 200 p.

Pauly, D. 1979. Theory and management of tropical multispecies stocks. ICLARM Stud. Rev., 1:1–35.

Pauly, D. 1980. A new methodology for rapidly acquiring basic information on tropical fish stocks: Growth, mortality and stock-recruitment relationships. In: S. B. Saila, and P. M. Roedel, eds. Stock Assessment for Tropical Small-Scale Fisheries. Int. Cent. Mar. Res. Dev., Univ. Rhode Island, Kingston, RI, pp. 154–172.

Pauly, D., and C. Binolhan. 1996. FishBase and AUXIM as tools for compararing the life history patterns, growth and natural mortality of fish:

Application to snappers and groupers. In: F. Arreguín-Sánchez, J. L. Munro, M. C. Balgos, and D. Pauly, eds. Biology, Fisheries and Culture of Tropical Grouper and Snappers. ICLARM Conf. Proc. 48, pp. 218–243, 449 p.

Pauly, D., and V. Christensen. 1996. Rehabilitating fished ecosystems: Insights from the past. NAGA, ICLARM Q. July, 1996:13–14.

Pauly, D., and N. David. 1981. ELEFAN I, a basic program for the objective extraction of growth parameters from length-frequency data. Meeresforschung, 28:205–211.

Pauly, D., and J. Ingles. 1981. Aspects of the growth and natural mortality of exploited coral reef fishes. Proc. IV Int. Coral Reef Symp., Manila, 1:89–98.

Pauly, D., and J. L. Munro. 1984. Once more on growth comparisons of fish and invertebrates. Fishbyte, 2(2):21.

Pauly, D., V. Christensen, J. Dalsgaard, R. Froese, and F. Torres, Jr. 1998. Fishing down marine food webs. Science, 279:860–863.

Peña, N., I. Alheit, and M. E. Nakama. 1986. Fecundidad parcial de la caballa del Perú (Scomber japonicus peruanus). Bol. Inst. Mar. Peru, 10(4):93–104.

Pérez Farfante, I., and B. Kensley. 1997. Penaeoid and sergestoid shrimps and prawns of the world. Keys and diagnoses for the families and genera. Mém. Mus. Natl. Hist. Nat., tome 2, 233 p.

Pérez-Pérez, D., and M. Alvarez-Conesa. 1983. Descripción de la pesca de escama con chinchorro de arrastre en la Empresa Combinado Pesquero-Industrial de Batabanó, durante 1982. Unpublished, Ministerio de la Industria Pesquera, La Habana, 5 p., 4 tabs.

Phillips, R. C., and E. G. Menez. 1988. Seagrasses. Smithsonian Contrib. Mar. Sci., No. 34, 104 p.

Pianka, E. 1970. On r- and K- selection. Am. Nat., 104:592–597.

Piedra Castañeda, G. 1965. Materials on the biology of the yellowtail snapper, Ocyurus chrysurus (Bloch) [in Russian; Spanish abstract]. In: A. S. Bogdanov, ed. Joint Soviet-Cuban Fisheries Investigations, Pishevaia Promishlennost, Moscow, vol. 1, pp. 267–283.

Pien, P. C., and I. C. Liao. 1975. Preliminary report of histological studies on the grey mullet gonad related to hormone treatment. Aquaculture, 5:31–39.

Pietsch, T. W., and D. B. Grobecker. 1980. Parental care as an alternative reproductive mode in an antennariid anglerfish. Copeia, 3:551–553.

Plan Development Team. 1990. The potential of marine fishery reserves for reef fish management in the U.S. Southern Atlantic. NOAA Tech. Memo. NMFS-SEFC-261, 45 p.

Poey Aloy, F. 1851–1861. Memorias sobre la historia natural de la Isla de Cuba, acompañadas de sumarios latinos y extractos en frances. La Habana. Imprenta de la Vda. de Barcina, 2 vol.

Poey Aloy, F. 1866. Ciguatera. Memoria sobre la enfermedad ocasionada por los peces venenosos. Repertorio Físico-Natural de la Isla de Cuba, La Habana, 2:1–39.

Poey Aloy, F. 2000. Ictiología Cubana. Transcripción, conjunción y edición cientifica de Dario Guitart Manday. Biblioteca Clásicos Cubanos. Imagen Contemporánea, La Habana, 3 vol., 975 pp, 572 illust.

Pol, J., L. Muñoz, S. Moreno, and A. Rodríguez. 1990. Evaluación de la abundancia de grandes pelágicos en la región noroccidental de Cuba. Rev. Invest. Marinas, 11(1):41–49.

Polunin, N. V. C. 1996. Trophodynamics of reef fisheries productivity. In: N. V. C. C. M. R. Polunin, ed. Reef Fisheries. Fish and Fisheries Ser. 20. Chapman, and Hall, Great Britain, pp. 113–135.

Polunin, W., and C. Roberts. 1996. Reef Fisheries. Chapman and Hall, Fish and Fisheries Ser. 20, 477 p.

Popova, O. A. 1967. The predatory-prey relationship among fish. In: Freshwater Fish Production. Blackwell Scientific Publication, Oxford and Edinburgh, pp. 359–376.

Popova, O. A. 1982. Reaction of predatory fishes to changes in habitat conditions produced by human activities. In: M. I. Shatunovsky, ed.

Changes in Fish Community Structure in an Eutrophic Lake [in Russian]. Nauka, Moscow, pp. 146–160.

Popova, O. A., and L. M. Sierra. 1983. Towards a methodology for studying the feeding habits of the fishes of the Cuban platform. Length-weight relationships of the main food objects of the bar jack, Caranx ruber (Bloch) [in Russian]. Vopr. Ikhtiol., 23(1): 159–162.

Posada, J. M., and R. S. Appeldoorn. 1996. The validity of length-based methods for estimating growth and mortality of groupers, as illustrated by comparative assessment of creole fish, Paranthias furcifer (Pisces: Serranidae). In: F. Arreguin-Sanchez, J. L. Munro, M. C. Balgos, and D. Pauly, eds. Biology, Fisheries and Culture of Tropical Grouper and Snappers. ICLARM Conf. Proc. 48:163–173, 449 p.

Potts, J. C., and C. S. Manooch III. 1995. Age and growth of red hind and rock hind collected from North Carolina through the Dry Tortugas, Florida. Bull. Mar. Sci., 56(3):784–794.

Potts, J. C., and C. S. Manooch III. 1999. Observations on the age and growth of Graysby and Coney from the Southeastern United States. Trans. Amer. Fish. Soc., 128(4):751–757.

Potts, J. C., and C. S. Manooch III. 2001. Differences in the age and growth of white grunt (Haemulon plumieri) from North Carolina and South Carolina compared with southeast Florida. Bull. Mar. Sci., 68(1):1–12.

Potts, J. C., C. S. Manooch III, and D. S. Vaughan. 1998. Age and growth of vermillion snapper from the southeastern United States. Trans. Amer. Fish. Soc., 127:787–795.

Powell, D. 1975. Age, growth and reproduction in Florida stocks of spanish mackerel, Scomberomorus maculatus. Fla. Mar. Res. Publ., 5:1–21.

Pozo, E. 1979. Edad y crecimiento del pargo criollo (Lutjanus analis [Cuvier, 1828]) en la plataforma noroccidental de Cuba. Rev. Cub. Invest. Pesq., 4(2):1–24.

Pozo, E., and L. Espinosa. 1982. Estudio de la edad y crecimiento del pargo del alto (Lutjanus vivanus) en la plataforma suroriental de Cuba. Rev. Cub. Invest. Pesq., 7(2):1–23.

Pozo, E., and L. Espinosa. 1983. Resultados de las investigaciones biológico-pesqueras del pargo del alto en la plataforma cubana. Unpublished, Cent. Invest. Pesq., La Habana, 27 p., 3 tabs., 5 figs.

Pozo, E., L. Espinosa, and R. Quevedo. 1983. Edad y crecimiento del pargo del alto (Lutjanus vivanus [Cuvier, 1828]) en el talud de la plataforma nororiental de Cuba. Rev. Cub. Invest. Pesq., 8(4):30–40.

Pozo, E., L. Espinosa, and M. Guardiola. 1984. Características biológicas del sesí (Lutjanus buccanella [Cuvier, 1828]) en la plataforma suroriental de Cuba. In: Cuarta Conferencia Cientifica de Educacion Superior, Univ. La Habana, Segundo Congr. Cienc. Biol. Resumenes, p. 90.

Prabhu, M. S. 1956. Maturation of intra ovarian eggs and spawning periodicities in some fishes. Indian J. Fish., 3(1–2):59–90.

Pratt, H. 1935. Remarques sur la faune et la flore associées aux sargasses flottantes. Nat. Can., 62:120–129.

Pratt, H., Jr., and J. Casey. 1983. Age and growth of the shortfin mako Isurus oxyrhinchus. NOAA Techn. Rep. NMFS 8:175–177.

Pressley, P. H. 1980. Lunar periodicity in the spawning of yellowtail damselfish, Microspathodon chrysurus. Environ. Biol. Fish., 5(2):153–159.

Puga, R., and S. Wong. 1978. Algunos aspectos biológicos de la mojarra blanca, Gerres cinereus (Walbaum, 1792). Cent. Invest. Pesq., La Habana, 10 p., 7 tabs., 5 figs.

Qasim, S. Z. 1955. Time and duration of spawning season of teleosts in relation to their distribution. J. Conserv. CIEM, 21:144–155.

Quevedo, R., and A. Aguilar. 1984. Algunos aspectos biológicos de las especies pelágico-oceánicas en la plataforma NW de Cuba. Unpublished,

Resúmenes de la Primera Jornada Científica de las BTJ, Cent. Invest. Pesq., La Habana, p. 8.

Radakov, D. V. 1972. Ecological significance of schooling in fishes [in Russian]. Nauka, Moscow, 137 p.

Radakov, D. V., A. D. Mochek, Y. N. Sbikin, R. Claro, and A. Silva. 1975. Acerca de la longitud de los peces comerciales en capturas de la zona noroccidental de Cuba. Ser. Oceanol. Acad. Cienc. Cuba, 28:1–9.

Radtke, R. L., and P. C. Hurley. 1983. Age estimation and growth of broadbill swordfish *Xiphias gladius*, from the North West Atlantic. Tech. rep. NOAA NMFS, 8:145–150.

Ramírez, E. A. 1974. Resultados parciales de investigaciones hidrológicas en los Cayos de la Enfermería. Resúmenes Invest. Cent. Invest. Pesq. Cuba, 1:18–21.

Ramos, I. 1988. Relaciones largo-peso de siete especies de peces. Cent. Invest. Pesq., MIP, Cuba, 3 p., 1 tab., 2 figs.

Ramos, I., and M. H. Obregón. 1983. Uso del arrastre y otras artes masivas en la plataforma y su influencia sobre la población. Unpublished, Cent. Invest. Pesq., MIP, La Habana, 20 p.

Ramos, I., and E. Pozo. 1984. Evaluación preliminar del ronco arará (*Haemulon plumieri* Lacépède, 1802) en la parte occidental de la plataforma nororiental de Cuba. Rev. Cub. Invest. Pesq., 9(3-4):45–63.

Ramos, J. 1983. Contribución al estudio de la oogénesis en el lenguado, *Solea solea* (Linnaeo, 1758) (Pisces: Soleidae). Invest. Pesq. (Barc.), 47(2):241–251.

Randall, J. E. 1961. Observations on the spawning of surgeonfishes (Acanthuridae) in the Society Islands. Copeia, 2:237–238.

Randall, J. E. 1962. Tagging reef fishes in the Virgin Islands. Proc. Gulf Carib. Fish. Inst., 14:201–241.

Randall, J. E. 1963a. Additional recoveries of tagged reef fishes from the Virgin Islands. Proc. Gulf Carib. Fish. Inst., 15:155–157.

Randall, J. E. 1963b. An analysis of the fish populations of artificial and natural reefs in the Virgin Islands. Carib. J. Sci., 3(1):31–47.

Randall, J. E. 1965. Grazing effect on sea grasses by herbivorous reef fishes in the West Indies. Ecology, 46(3):255–260.

Randall, J. E. 1967. Food habits of reef fishes of the West Indies. Stud. Trop. Oceanogr., 5, 84 p.

Randall, J. E. 1968. Caribbean Reef Fishes. T. F. H. Publications, Inc., Jersey City, NJ, 318 p.

Randall, J. E., and V. E. Brock. 1960. Observations on the ecology of epinepheline and lutjanid fishes of the Society Islands, with emphasis on food habits. Trans. Am. Fish. Soc., 89(1):9–16.

Randall, J. E., and H. A. Randall. 1963. The spawning and early development of the Atlantic parrotfish, *Sparisoma rubripinne*, with notes on others scarid and labrid fishes. Zoologia (N.Y.), 48(2): 49–60.

Reeson, P. H. 1983a. The biology, ecology and bionomics of the parrotfishes, Scaridae. In: J. L. Munro, ed. Caribbean Coral Reef Fishery Resources. Intern. Center Living Aquatic Resources Management (ICLARM) Stud. Rev., pp. 166–177.

Reeson, P. H. 1983b. The biology, ecology and bionomics of the surgeon-fishes, Acanthuridae. In: J. L. Munro, ed. Caribbean Coral Reef Fishery Resources. Intern. Center Living Aquatic Resources Management (ICLARM) Stud. Rev., pp. 178–190.

Reid, G. K. 1954. An ecological study of the Gulf of Mexico fishes, in the vicinity of Cedar Key, Florida. Bull. Mar. Sci. Gulf Carib., 4(1):1–94.

Reinboth, R. 1962. Morphologische und funktionelle zweigeschlechtlichkeit bei marinen Teleostiern (Serranidae, Sparidae, Centracanthidae, Labridae). Zool. J. Physiol., 69:405–480.

Reinboth, R. 1988. Physiological problems of teleost ambisexuality. Envir. Biol. Fishes, 22(4):249–259.

Reshetnikov, Y. S. 1976. The application of biochemical indices in the investigation of Salmonidae. Acta Biol. Iugosl. (E. Ichthyol.), 8(1):123–128.

Reshetnikov, Y. S. 1985. Aspects of the reproduction of tropical fishes [in Russian]. In: Aspects of the Reproductive Cycle of Fishes from Different Latitudes in Aquariums. Nauka, Moscow, pp. 12–35.

Reshetnikov, Y. S., R. Claro, and A. Silva. 1974. Ritmo alimentario y velocidad de digestión de algunos peces depredadores tropicales. Ser. Oceanol. Acad. Cienc. Cuba, 21:3–13.

Reshetnikov, Y. S., A. Silva, R. Claro, and O. A. Popova. 1975. Velocidad de digestión del alimento de los peces de aguas tropicales. Zool. Zh., 54(10):1506–1511.

Richards, C. E. 1967. Age, growth and fecundity of cobia *Rachycentron canadum*, from Chesapeake Bay and adjacent mid-Atlantic waters. Trans. Am. Fish. Soc., 96:343–350.

Richards, C. E., and M. Castagna. 1976. Distribution, growth and predation of juvenile white mullet (*Mugil curema*) in ocean side waters of Virginia's eastern shore. Chesapeake Sci., 17(4):308–309.

Ricker, W. E. 1975. Computation and interpretation of biological statistic of fish population. Bull. Fish. Res. Board Can., 191:1–382.

Rivera-Arriaga, E., A. L. Lara-Dominguez, J. Ramos-Miranda, P. Sanchez, and A. Yanez-Arencibia. 1996. Ecology and population dynamics of *Lutjanus synagris* on Campeche Bank. In: F. Arreguin-Sanchez, J. L. Munro, M. C. Balgos, and D. Pauly, eds. Biology, Fisheries and Culture of Tropical Grouper and Snappers. ICLARM Conf. Proc. 48, pp. 11–18.

Roberts, C. M. 1995. Rapid build-up of fish biomass in a Caribbean marine reserve. Conserv. Biol., 9(4):816–826.

Roberts, C. M. 1997. Connectivity and management of Caribbean coral reefs. Science, 278:1454–1457.

Roberts, C. M., N. Quinn, J. W. Tucker, Jr., and P. N. Woodward. 1995. Introduction of hatchery-reared Nassau grouper to a coral reef environment. N. Am. J. Fish. Manag, 15:159–164.

Robertson, D. R. 1972. Social control of sex reversal in a coral-reef fish. Science, 177:1007–1009.

Robertson, D. R. 1991. The role of adult biology in the timing of spawning of tropical reef fishes. In: P. F. Sale, ed. The Ecology of Fishes on Coral Reefs. Academic Press, San Diego, CA, pp. 356–386.

Robertson, D. R., and R. R. Warner. 1978. Sexual patterns in the labroid fishes of the Western Caribbean II. The parrot fishes (Scaridae). Smithson. Contrib. Zool., 255:1–26.

Robertson, D. R., H. P. A. Sweatman, E. A. Fletcher, and M. G. Cleland. 1976. Schooling as a mechanism for circumventing the territoriality of competitors. Ecology, 57:1208–1220.

Robins, C. R. 1971. Distributional patterns of fishes from coastal and shelf waters of the tropical western Atlantic. In: Symp.Investigations and Resources of the Caribbean Sea and Adjacent Regions. Pap. Fish. Resour. FAO, Rome, Italy, pp. 249–255.

Robins, C. R., R. M. Bailey, C. E. Bond, J. R. Brooker, E. A. Lachner, R. N. Lea, and W. B. Scott. 1980. A list of common and scientific names of fishes from the United States and Canada (4th ed.). Am. Fish. Soc. Spec. Publ. 20, Bethesda, MD.

Robins, C. R., R. M. Bailey, C. E. Bond, J. R. Brooker, E. A. Lachner, R. N. Lea, and W. B. Scott. 1991. Common and scientific names of fishes from the United States and Canada (5th ed.). Am. Fish. Soc. Spec. Publ. 20, Bethesda, MD.

Rodríguez, A., E. Valdés-Muñoz, and R. Valdés. 1984. Lista de nombres científicos y comunes de peces marinos cubanos [nomenclator]. Cent. Invest. Pesq., La Habana, 82 p.

Rodríguez, H. 1986. Estudio comparativo de dos estructuras rígidas (otolito y hueso mesopterigoide) para la estimación dey edad crecimiento del mero (*Epinephelus morio*) del Banco de Campeche. Tesis profesional, Univ. Auton. Nuevo León, México.

Rodríguez, M. M. 1974. Alimentaçao do ariacó, *Lutjanus synagris* (Linnaeus), do estado de Ceará (Brasil). Arq. Cienc. Mar., 14(1):60–62.

Rodríguez, R. B. 1984. Biología y cultivo de *Solea senegalensis* Kaup, 1858 en el Golfo de Cádiz. Tesis doctoral, Univ. Sevilla, 207 p.

Rodríguez Castro, J. 1992. Contribución al conocimiento de la biología pesquera del huachinango *Lutjanus campechanus* (Poey, 1860) en las costas del sur de Tamaulipas, México. Tesis de licenciatura, 82 p.

Rodríguez Pino, Z. 1962. Estudios estadísticos y biológicos sobre la biajaiba (*Lutjanus synagris*). Cent. Invest. Pesq., Notas Invest., 4: 1–92.

Rodríguez-Portal, J. P., and M. Nadal-Llosa. 1983. Consideraciones sobre el grado de alteración de las condiciones oceanográficas de las bahías de la Habana y Mariel. Rep. Invest. Inst. Oceanol., Acad. Cienc. Cuba, 18:1–15.

Roede, M. J. 1972. Color as related to size, sex and behavior in seven Caribbean labrid fish species (genera *Thalassoma*, *Halichoeres* and *Hemipteronotus*). Stud. Fauna CuraHao Other Carib. Isl., 42(73):1–264.

Roede, M. J. 1975. Reversal of sex in several labrid fish species. Coll. Papers. Carib. Mar. Biol. Inst. Curaçao, Neth. Antilles., 10(152):1–23.

Rojas, E. L. 1970. Estudios estadísticos y biológicos sobre el pargo criollo, *Lutjanus analis*. Cent. Invest. Pesq., Notas Invest., 2:1–16.

Román Cordero, A. M. 1991. Estudio sobre la dinamica reproductiva de la cachita blanca *Haemulon plumieri* (Lacepede, 1802) (Pisces: Pomadasyidae). M.S. thesis, Univ. Puerto Rico, Mayaguez, Puerto Rico.

Rooker, J. R. 1995. Feeding ecology of the schoolmaster snapper, *Lutjanus apodus* (Walbaum), from southwestern Puerto Rico. Bull. Mar. Sci., 56(3):881–894.

Rooker, J. R., and G. D. Dennis. 1991. Diel, lunar and seasonal changes in a mangrove fish assemblage off southwestern Puerto Rico. Bull. Mar. Sci., 49(3):684–698.

Ros, R. M., and C. M. Pérez. 1978. Contribución al conocimiento de la biología del pez sable, *Trichiurus lepturus* Linné, 1758. Ciencias, Ser. 8, Invest. Mar., 37:1–33.

Rosen, D. E. 1975. A vicariance model of Caribbean biogeography. Syst. Zool., 24(4):431–464.

Rosen, D. E. 1985. Geological hierarchies and biogeographic congruence in the Caribbean. Ann. Missouri Bot. Gard., 72:636–659.

Rossov, V. V., and H. Santana. 1966. Algunas características hidrológicas del Mediterráneo americano. Estudios, Inst. Oceanol., Acad. Cienc. Cuba, 1(1):47–77.

Roughgarden, J. 1995. Anolis Lizards of the Caribbean: Ecology, Evolution and Plate Tectonics. Oxford Univ. Press, New York, NY, 200 p.

Rubio, C. E. 1975. Crecimiento, sexualidad y desarrollo gonadal de la mojarra rayada *Eugerres plumieri* (Cuvier) de la Ciénaga Grande de Sta. Marta, con anotaciones sobre su biología. Divulg. Pesq. Inst. Desarr. Recurs. Nat. Renovables. Bogotá, 10(1):1–69.

Rubio, R., P. Salahange, and M. Betancourt. 1985. Relaciones de la edad con el largo, el peso y la fecundidad de la biajaiba (*Lutjanus synagris*) de la plataforma suroccidental de Cuba. Rev. Cub. Invest. Pesq., 10(3–4):77–90.

Ruiz, L. J., R. Figueroa-M., and A. Prieto-A. 1999. Ciclo reproductivo de *Lactophrys quadricornis* (Pisces: Ostraciidae) de la costa nororiental de Venezuela. Rev. Biol. Trop., 47(3):561–569.

Russ, G. R., and A. C. Alcala. 1996. Do marine reserves export adult fish biomass? Evidence from Apo Island, central Philippines. Mar. Ecol. Progr. Ser., 132:1–9.

Russ, G. R., and A. C. Alcala. 1998. Natural fishing experiments in marine reserves 1983–1993: Roles of life history and fishing intensity in family responses. Coral Reefs, 17:399–416.

Ryzhov, V., and M. Formoso. 1975. Ritmo de la alimentación de tres especies del género *Calamus* (Sparidae) en el banco de Campeche. Rev. Invest. Pesq., 2:113–114.

Sadovy, Y. 1994. Aggregation and spawning in the tiger grouper, *Mycteroperca tigris* (Pisces: Serranidae). Copeia, 2:511–516.

Sadovy, Y. 1996. Reproduction of reef fishery species. In: N. V. C. Polunin and C. M. Roberts, eds. Reef Fisheries, Chapman and Hall, London, pp. 15–59.

Sadovy, Y. 1997. The case of the disappearing grouper: *Epinephelus striatus*, the Nassau grouper in the Caribbean and western Atlantic. Proc. Gulf Carib. Fish. Inst., 45:5–22.

Sadovy, Y., and P. L. Colin. 1995. Sexual development and sexuality in the Nassau grouper. J. Fish. Biol., 46:961–976.

Sadovy, Y. and A. Eklund. 1999. Synopsis of biological data on the Nassau grouper, *Epinephelus striatus* (Bloch, 1792), and the jewfish, *E. itajara* (Lichtenstein, 1822). NOAA Tech. Rep. NMFS 146, 65 p.

Sadovy, Y., and D. Y. Shapiro. 1987. Criteria for the diagnosis of hermaphroditism in fishes. Copeia, 1:136–156.

Sadovy, Y., M. Figuerola, and A. Román. 1992. Age and growth of *Epinephelus guttatus* in Puerto Rico and St. Thomas. Fish. Bull., 90(3):516–528.

Sadovy, Y., A. Rosario, and A. Román. 1994. Reproduction in an aggregating grouper, the red hind, *Epinephelus guttatus*. Environ. Biol. Fish., 41:269–286.

Sakun, O. F., and N. A. Butskaia. 1963. Determination of maturity stages and study of the sexual cycle of fishes [in Russian]. Ribnoie Khoziatsvo, Moscow, 36 p.

Salazar, A. R. 1988. Contribución al conocimiento de la pesquería del mero (*Epinephelus morio*) de la flota menor de las costas de Yucatán Tesis profesional, ENEP-Iztacala, UNAM, Mexico.

Sale, P. F. 1980. The ecology of fishes on coral reefs. Oceanogr. Mar. Biol. Annu. Rev., 18:367–421.

Sale, P. F. 1991. Reef fish communities: Open nonequilibrial systems. In: P. F. Sale, ed. The Ecology of Fishes on Coral Reefs. Academic Press, San Diego, CA, pp. 564–596.

Saloman, C. H., and W. A. Fable, Jr. 1981. Length-frequency distribution of recreationally caught fishes from Panama City, Florida in 1978 and 1979. NOAA Tech. Memo. NMFS-SEFC-61.

San Martin, A. 1965. Informe biológico de la cherna criolla. Estadísticas de su captura en Cuba. Breve descripción sobre las pesquerías de la cherna criolla. Descripción del género *Epinephelus* y de la éspecie cherna criolla. Resumenes de su estudio sobre la biologia. Centro de Investigaciones Pesqueras, MIP, La Habana, 23 p.

Schaefer, M. B., and C. J. Orange. 1956. Estudios, mediante el examen de gónadas, del desarrollo sexual y desove del atún aleta amarilla (*Neothunnus macropterus*) y del barrilete (*Katsuwonus pelamis*) en tres regiones del Pacífico Oriental. Com. Interamericana Atun Trop., 1(6):283–349.

Scherbachev, Y. N. 1973. Biología y distribución de los corifénidos (Pisces: Coryphaenidae). Vopr. Ikhtiol., 13(2):219–230.

Schmidt, R. E., and M. C. Fabrizio. 1980. Daily growth increments on otoliths for ageing young-of-the-year largemouth bass from a wild population. Prog. Fish-Cult., 40:78–80.

Schwartz, F. 1983. Short ageing methods on age estimation of scalloped hammerhead, *Sphyrna lewini*, and dusty, *Carcharhinus obscurus* sharks based on vertebral ring counts. NOAA Tech. Rep. NMFS, 8:167–174.

Sedberry, G. R., and J. Carter. 1993. The fish community of a shallow tropical lagoon of Belize, Central America. Estuaries, 16(2):198–215.

Senta, T. 1962. Studies on floating sea-weeds in early summer around Oki Islands and larvae and juvenile of fishes accompanying them. Physiol. Ecol., 10(2):68–78.

Settle, L. R. 1993. Spatial and temporal variability in the distribution and abundance of larval and juvenile fishes associated with pelagic Sargassum. M.S. thesis, Dep. Biol., Univ. North Carolina, Wilmington, NC, 64 p.

Shapiro, D. Y. 1979. Social behaviour, group structure and the control of sex reversal in hermaphroditic fish. Adv. Stud. Behav., 10:43–102.

Shapiro, D. Y. 1981. Sequence of coloration changes during sex reversal in the tropical marine fish *Anthias squamipinnis* (Peters). Bull. Mar. Sci., 31(2):383–398.

Shapiro, D. Y. 1985. Behavioral influences on the initiation of adult sex change in coral reef fishes. In: B. Lofts and W. N. Holmes, eds. Current Trends in Comparative Endocrinology, Univ. Press, Hong Kong, pp. 583–585.

Shapiro, D. Y. 1987. Reproduction in groupers. In: J. J. Polovina, and S. Ralston, eds. Tropical Snappers and Groupers. Biology and Fisheries Management. Westview Press/Boulder and London, pp. 295–373.

Shapiro, D. Y. 1991. Intraespecific variability in social systems of coral reef fishes. In: P. F. Sale, ed. The ecology of fishes on coral reefs. Academic Press, San Diego, CA, pp. 331–355.

Shapiro, D. Y., D. A. Hensley, and R. A. Appeldoorn. 1988. Pelagic spawning and egg transport in coral-reef fishes: A skeptical overview. Environ. Biol. Fish., 22:3–14.

Shapiro, D. Y., Y. Sadovy, and M. A. McGehee. 1993a. Periodicity of sex change and reproduction in the red hind, *Epinephelus guttatus*, a protogynous grouper. Bull. Mar. Sci., 53(3):1151–1162.

Shapiro, D. Y., Y. Sadovy, and M. A. McGehee. 1993b. Size, composition and spatial structure of the annual spawning aggregation of the red hind, *Epinephelus guttatus* (Pisces: Serranidae). Copeia, 2:399–406.

Sharov, V. L. 1973. Fecundity of the big-eye tuna in the Atlantic Ocean [in Russian]. Tr. Atlant.-NIRO, 53:104–111.

Shatunovsky, M. I. 1976. Physiological-biochemical aspects of the population dynamics and the productivity of fishes [in Russian]. In: Third All-Union Conference on Fish Ecological Physiology, Summaries. Naukova Dumka, Kiev, pp. 9–11.

Shatunovsky, M. I. 1978. Significance of the study of physiology and biochemistry of fishes on fish management and culture [in Russian]. Tr. VNIRO, 120:7–12.

Shatunovsky, M. I. 1980. Ecological patterns of metabolism in marine fishes [in Russian]. Nauka, Moscow, 281 p.

Shatunovsky, M. I., and M. N. Krivobok. 1976. On some problems of ecological physiology of sea and anadromus fishes. Acta Biol. Iugosl. (E. Ichthyol.), 8(1):123–128.

Shilov, I. A. 1977. Ecological-physiological basis of animal population relationships [in Russian]. Editorial MGU, Moscow, 263 p.

Shoriguin, A. A. 1952. Feeding and feeding relationships among fishes of the Caspian Sea [in Russian]. Pishevaia Promishlennost, Moscow, 268 p.

Shorthouse, B. 1990. The Great Barrier Reef marine park—how does it work for fishermen? Aust. Fish., 49(12):16–17.

Shulman, G. E. 1972. Physiological-biochemical characteristics of biological cycles of fish [in Russian]. Pishevaia Promishlennost, Moscow, 368 p.

Shulman, G. E. 1985. Problems of ecological physiology and biochemistry of fishes [in Russian]. Gidrobiol. Zh., 21(6):49–56.

Siam Lahera, C. 1983. Corrientes superficiales alrededor de Cuba. Unpublished, Cent. Invest. Pesq., MIP, La Habana, 5 p., 4 figs.

Sierra, L. M. 1983. Características de la alimentación del jallao, *Haemulon album*, en la plataforma suroccidental de Cuba. Rep. Invest. Inst. Oceanol. Acad. Cienc. Cuba, 11:1–17.

Sierra, L. M. 1986. Ecología de la alimentación de algunos peces neríticos de Cuba. Ph.D. diss., Inst. Oceanol., Acad. Cienc. Cuba, La Habana, 116 p., 30 tabs., 41 figs.

Sierra, L. M. 1987. Peculiaridades de la alimentación de la sardina de ley, *Harengula humeralis* (Cuvier,1829), en la región oriental del Golfo de Batabanó. Rep. Invest. Inst. Oceanol. Acad. Cienc. Cuba, 69:1–19.

Sierra, L. M. 1996. Relaciones tróficas de los juveniles de cinco especies de pargos (Pisces: Lutjanidae) en Cuba. Rev. Biol. Trop., 44(3)/45(1):499–507.

Sierra, L. M., and R. Claro. 1979. Variación estacional de la velocidad de digestión en dos especies de peces lutiánidos, la biajaiba (*Lutjanus synagris*) y el caballerote (*Lutjanus griseus*). Cienc. Biol., 3:87–97.

Sierra, L. M., and J. Díaz-Zaballa. 1984. Alimentación de dos especies de sardinas, *Harengula humeralis* (Cuvier, 1829) y *Harengula clupeola* (Cuvier, 1829), en la costa N de la Ciudad de La Habana (rada del Instituto de Oceanología). Rep. Invest. Inst. Oceanol. Acad. Cienc. Cuba, 25:1–18.

Sierra, L. M., and O. A. Popova. 1982. Particularidades de la alimentación del civil (*Caranx ruber* Bloch) en la región suroccidental de la plataforma cubana. Rep. Invest. Inst. Oceanol. Acad. Cienc. Cuba, 3:1–19.

Sierra, L. M., and O. A. Popova. 1988. Velocidad de digestión, ritmo diario de alimentación y raciones alimenticias de algunas especies de peces del Golfo de Batabanó. Rep. Invest. Inst. Oceanol. Acad. Cienc. Cuba, 6:1–20.

Sierra, L. M., A. García-Cagide, and A. Hernández. 1986. Aspectos de la biología del cibí amarillo (*Caranx bartholomaei* [Cuvier, 1833]), en la región oriental del Golfo de Batabanó, Cuba. Rep. Invest. Inst. Oceanol., Acad. Cienc. Cuba, 55:1–36.

Sierra, L. M., R. Claro, J. P. García-Arteaga, and E. Valdés-Muñoz. 1990. Estructura trófica de las comunidades de peces que habitan en diferentes biótopos del Golfo de Batabanó. In: R. Claro, ed. Asociaciones de Peces en el Golfo de Batabanó Editorial Academia, La Habana, pp. 83–95.

Sierra, L. M., R. Claro, and O. A. Popova. 1994. Alimentación y relaciones tróficas. In: R. Claro, ed. Ecología de los Peces Marinos de Cuba. Inst. Oceanol. Acad. Cienc. Cuba y Cent. Invest. Quintana Roo, México, pp. 263–320.

Silva Lee, A. 1974a. Algunos datos sobre la biología de la sardina de ley, *Harengula humeralis* Cuvier y Valenciennes, en la costa noroccidental de Cuba. Ser. Oceanol. Acad. Cienc. Cuba, 22: 1–11.

Silva Lee, A. 1974b. Hábitos alimentarios de la cherna criolla, *Epinephelus striatus* Bloch, y algunos datos sobre su biología. Ser. Oceanol. Acad. Cienc. Cuba, 25:1–25.

Silva Lee, A. 1975a. Observaciones sobre arrecifes artificiales usados para pescar en Cuba. Ser. Oceanol. Acad. Cienc. Cuba, 26:1–13.

Silva Lee, A. 1975b. Acerca de la actividad de los depredadores y el arisqueo en una mancha de sardinas (*Harengula* spp.). Ser. Oceanol. Acad. Cienc. Cuba., 32:7–12.

Silva Lee, A. 1977. Notas sobre la coloración y la conducta de la cherna criolla (*Epinephelus striatus* Bloch). Inf. Cient.-Téc., Inst. Oceanol. Acad. Cienc. Cuba, 14:1–8.

Silva Lee, A., and E. Valdés–Muñoz. 1985. La ictiofauna de los arrecifes artificiales comerciales de la costa sur de Cuba. Rep. Invest. Inst. Oceanol. Acad. Cienc. Cuba., 39:1–29.

Simpson, J. G., and R. G. Griffiths. 1973. Edad, crecimiento y madurez del rabo amarillo (*Cetengraulis edentulus*), en el oriente de Venezuela, basados en estudios de frecuencia de longitud. Ser. Recursos y Explotación Pesq., 2(6):4–21.

Slobodkin, L. B., and L. Fishelson. 1974. The effect of the cleaner-fish *Labroides dimidiatus* on the point diversity of fishes on the reef front at Eilat. Am. Nat., 108(961):369–376.

Sluka, R., and K. M. Sullivan. 1996. The influence of habitat on the size distribution of groupers in the upper Florida Keys. Environ. Biol. Fish., 47:177–189.

Sluka, R., M. Chiappone, and K. M. Sullivan. 1994. Comparison of juvenile grouper populations in southern Florida and the central Bahamas. Bull. Mar. Sci. 54(3):871–880.

Smith, C. L. 1959. Hermaphroditism in some serranid fishes from Bermudas. Pap. Michigan Acad. Sci., 44:111–119.

Smith, C. L. 1961. Synopsis of biological data on groupers (Epinephelus and allied genera) of the Western North Atlantic. FAO Fish. Biol. Synop., 23(1):1–62.

Smith, C. L. 1965. The patterns of sexuality and the classification of serranid fishes. Am. Mus. Novit., 2207:1–20.

Smith, C. L. 1972. A spawning aggregation of Nassau grouper Epinephelus striatus (Bloch). Trans. Am. Fish. Soc., 101(2):257–261.

Smith, C. L., and J. C. Tyler. 1972. Space resource sharing in a coral reef fish community. In: B. B. Collette, and S. Earle, eds. Results of the Tektite: Program Ecology of Coral Reef Fishes, Los Angeles Mus. Sci. Bull. 14:125–170.

Smith, C. L., and J. C. Tyler. 1973a. Direct observations of resource sharing in coral reef fish. Helgol. Meeresunters, 24:264–275.

Smith, C. L., and J. C. Tyler. 1973b. Population ecology of Bahamian suprabenthic shore fish assemblage. Am. Mus. Novitates 2528:1–38.

Smith, H. 1967. Influence of temperature on the rate of gastric juice secretion in the brown bullhead, Ictalurus nebulosus. Comp. Biochem. Physiol., 21:125–132.

Smith-Vaniz, W. F., B. B. Collette, and B. E. Luckhurst. 1999. Fishes of Bermuda: History, zoogeography, annotated checklist, and identification keys. Am. Soc. Ichthyologists and Herpetologists, Spec. Publ. No. 4. Lawrence, KS.

Soemarto, T. 1958. Fish behaviour with special reference to pelagic schooling species: Lajang (Decapterus spp.). 8th Proc. Indo-Pacific Fish. Counc., 3:89–93.

Sokolov, V. E., and G. García, eds. 1985. Ecology of fishes of the Cuban shelf (In Russian). Nauka, Moscow, 298 p.

Sokolova, L. V. 1965. Distribution and biological aspects of the principal commercial fishes of the Bank of Campeche [in Russian]. In: Joint Soviet-Cuban Fisheries Investigations, Pishevaia Promishlennost, Moscow, vol. 1, pp. 223–240.

Soletchnik, P., M. Suquet, E. Thouard, and J. P. Mesdouze. 1988. Spawning of yellowtail snapper (Ocyurus chrysurus, Bloch, 1791) in captivity. Aquaculture, 77:287–289.

Speir, J., ed. 1999. Cuban Environmental Law. Tulane Law School, 99 p.

Springer, V. G. 1982. Pacific plate biogeography, with special reference to shorefishes. Smithsonian Contrib. Zool., 367:1–182.

Sponaugle, S., and R. K. Cowen. 1997. Early life history traits and recruitment patterns of Caribbean wrasses (Labridae). Ecol. Monogr., 67(2):177–202.

Stansby, M. E. 1963. Proximate composition of fish. In: E. Heen, and R. Kreuzer, eds. Fish in Nutrition. FAO Fishing News (Books), Ltd., pp. 55–66.

Starck, W. A. 1970. Biology of the gray snapper, Lutjanus griseus (Linnaeus) in the Florida Keys. Stud. Trop. Oceanogr., 10:11–150.

Starck, W. A., and W. P. Davis. 1966. Night habits of fishes of Alligator reef, Florida. Ichthyologica, Aquarium J., 38(4):313–356.

Steele, M. A. 1996. Effects of predators on reef fishes: Separating cage artifacts from effects of predation. J. Exp. Mar. Biol. Ecol., 198:249–267.

Steffensen, E. 1980. Daily growth increments observed in otoliths from juvenile East Baltic cod. Dana, 1:29–37.

Steimle, F., and R. B. Stone. 1973. Bibliography on artificial reefs. Coastal Plains Center for Marine Development Services, Wilmington, DL. 129 p.

Stevens, J. D. 1975. Vertebral rings as a means of age determination in the blue shark (Prionace glauca L.). J. Mar. Biol. Assoc. U.K., 55(3):657–665.

Stevenson, D. K. 1977. Management of a tropical fish pot fishery for maximum sustainable yield. Proc. Gulf Carib. Fish. Inst., 30:95–115.

Stiles, T. C., and M. L. Burton. 1994. Age, growth and mortality of red grouper, Epinephelus morio from the southeastern U. S. Gulf. Proc. Gulf Carib. Fish. Inst., 43:123–137.

Stilwell, C. E., and N. E. Kohler. 1982. Food, feeding habits and estimates of daily ration of the shortfin mako (Isurus oxyrhinchus) in the Northwest Atlantic. Can. J. Fish. Aquat. Sci., 39(3):407–414.

Stobutzki, I. C., and D. R. Bellwood. 1997. Sustained swimming abilities of the late pelagic stages of coral reef fishes. Mar. Ecol. Prog. Ser., 149:35–41.

Stobutzki, I. C., and D. R. Bellwood. 1998. Nocturnal orientation to reefs by late pelagic stage coral reef fishes. Coral Reefs, 17:103–110.

Stoner, A. W., and R. J. Livingston. 1984. Ontogenetic patterns in diet and feeding morphology in sympatric sparid fishes from seagrass meadows. Copeia, 1:174–87.

Storozhuk, A. Ya. 1975. Seasonal dynamics of the physiological-biochemical conditions of saida (Pollachius virens L.) of the North Sea [in Russian]. Tr. VNIRO, 96:114–120.

Struhsaker, P., and J. H. Uchiyama. 1976. Age and growth of the nehu Stolephorus purpureus (Pisces: Engraulidae) from Hawaiian Islands as indicated by daily growth increments of sagittae otoliths. Fish. Bull. U.S., 74(1):9–17.

Sturm, M. G. de L. 1978. Aspects of the biology of Scomberomorus maculatus (Mitchill) in Trinidad. J. Fish. Biol., 13:155–172.

Suárez, G., T. Romero, and G. Arencibia. 1983. Características contaminantes de los residuales de la industria azucarera. In: Cuarto Foro Cient. Cent. Invest. Pesq., Resúmenes de Investigaciones, La Habana, p. 89.

Suárez-Caabro, J. A., and P. P. Duarte-Bello. 1961. Biología pesquera del bonito (Katsuwonus pelamis) y la albacora (Thunnus atlanticus) en Cuba. I. Inst. Cubano Invest. Tec., Ser. Estudios de Trabajos de Investigación, La Habana, 15, 151 p., 40 tabs., 67 figs.

Suárez-Caabro, J. A., and P. P. Duarte-Bello. 1962. Hidrografía del Golfo de Batabanó. Inst. Cubano Invest. Tecnol., Bol. Informativo, 6(1):4–19.

Suárez-Caabro, J. A., P. P. Duarte-Bello, and J. Alvarez-Reguera. 1961. Biología y tecnología de las sardinas cubanas. I. Harengula pensacolae cubana Rivas y Harengula humeralis (Cuvier). Inst. Cubano Invest. Tec., Ser. Estudios de Trabajos de Investigación, La Habana, 19:1–87.

Sukhovey, V. F., G. K. Korotaev, and N. B. Shapiro. 1980. Hidrología del Mar Caribe y el Golfo de México [in Russian]. Gidrometeoizdat, Leningrado, 180 p.

Sullivan, K. M., and M. De Garine-Witchatitsky. 1994. Energetics of juvenile Epinephelus groupers; impact of summer temperatures and activilty patterns on growth rates. Proc. Gulf Carib. Fish. Inst., 43:148–167.

Sykes, L. R., W. R. McCann, and A. R. Kafka. 1982. Motion of Caribbean Plate during the last 7 million years and implications for earlier Cenozoic movements. J. Geophys. Res., 87:10656–10676.

Szedlmayer, S. T. 1998. Comparison of growth rate and formation of otolith increments in age-0 red snapper. J. Fish. Biol., 53:58–65.

Tallet, C. In press. Nuevos registros para la ictiofauna cubana, con notas sobre su ecología, coloración en vida y conducta en acuario. Rev. Oceanol. Ecol. Trop. Avicennia.

Tápanes, J. J. 1972a. Hidrología de la plataforma cubana. Juventud Técnica, 73:82–96.

Tápanes, J. J. 1972b. Hidrología de la plataforma cubana. Juventud Técnica, 74:73–88.

Targett, N. M., T. E. Targett, N. M. Vrolijic, and J. C. Ogden. 1986. Effect of macrophyte secondary metabolites on feeding preferences of the herbivorous parrotfish *Sparisoma radians*. Mar. Biol., 92:141–148.

Tariche, N. 1998. Peces capturados en las expediciones del Atlantis en aguas de Cuba en 1938 y 1939 y conservados en las colecciones del Centro de Investigaciones Marinas, Univ. la Habana. Tesis de Maestría, Univ. la Habana.

Tariche, N. In press a. Sobre la presencia de *Opsanus pardus* (Goode and Bean, 1879) (Pisces-Batrachoididae) en aguas de Cuba. Rev. Invest. Mar.

Tariche, N. In press b. Nuevo reporte para la ictiofauna cubana, *Aplatophus chauliodus* Böhlke, 1956 Ophichthidae. Rev. Invest. Mar.

Taubert, B. O., and D. N. Coble. 1978. Daily rings in otoliths of three species of *Lepomis* and *Tilapia mossambica*. J. Fish. Res. Board Can., 34:332–340.

Thayer, G. W., D. R. Colby, and W. F. Hettler. 1987. Utilization of the red mangrove prop root habitat by fishes in South Florida. Mar. Ecol. Prog. Ser., 35:25–28.

Thompson, R., and J. C. Munro. 1974. The biology, ecology and exploitation and management of Caribbean reef fishes. Part V. Carangidae (jacks). Res. Rep. Zool. Dep. Univ. West Indies, 3:1–43.

Thompson, R., and J. C. Munro. 1978. Aspects of the biology and ecology of Caribbean reef fishes: Serranidae (hinds and groupers). J. Fish. Biol., 12(2):115–146.

Thompson, R., and J. C. Munro. 1983. The biology, ecology and bionomics of the snappers, Lutjanidae. In: J. L. Munro, ed. Caribbean coral reef fishery resources ICLARM Stud. Rev., 17:94–109.

Thresher, R. E. 1984. Reproduction in reef fishes. TFH Publications, Neptune City, NJ, 399 p.

Tornes, E., P. George, and D. Sánchez. 1971. Variación en el contenido de grasa y sólidos no grasos en cuatro especies de peces de importancia industrial en Venezuela. Proyecto de Investigaciones y Desarrollo Pesquero. MAC/PNUD/FAO. Inf. Téc. 35, 17 p.

Torres, R., and E. Chávez. 1987. Evaluación y diagnóstico de la pesquería de rubia (*Lutjanus synagris*) en el estado de Yucatán. Cienc. Mar., 13(1):7–29.

Torres-Lara, R., and S. Salas-Márquez. 1990. Crecimiento y mortalidad de la rubia *Lutjanus synagris* de las costas de Yucatán durante las temporadas de pesca 1983–1985. Anales del Instituto de Ciencias del Mar y Limnología, UNAM, 17(2):205–214.

Townsend, D. W. 1980. Microstructural growth increments in some Antarctic fish otoliths. Cybium (3e Sér.), 8:17–22.

Tucker, J. W., Jr., P. G. Bush, and S. T. Slaybough. 1993. Reproductive patterns of Cayman Islands Nassau grouper (*Epinephelus striatus*) populations. Bull. Mar. Sci., 52(3):961–969.

Tucker, J. W., and P. N. Woodward. 1995. Egg-production and completion of the life-cycle of belted sandfish (*Serranus subligarius*) in captivity. Bull Mar. Sci., 56(2):701–707.

Tucker, J. W., P. N. Woodward, and D. G. Stennett. 1996. Voluntary spawning of captive Nassau groupers, *Epinephelus striatus*, in a concrete raceway. J. World Aquacult. Soc., 27(4):373–383.

Turingan, R. G., P. C. Wainwright, and D. A. Hensley. 1995. Interpopulation variation in prey use and feeding biomechanics in Caribbean triggerfishes. Oecologia, 102:296–304.

Tyler, A. V. 1970. Role of gastric emptying in young cod. J. Fish. Res. Board Can., 27:1177–1189.

Valdés, E., and M. Sotolongo. 1983. Algunos aspectos de la biología y pesquería del machuelo (*Opisthonema oglinum*) de la plataforma suroccidental cubana. Rev. Cub. Invest. Pesq., 8(1):66–96.

Valdés-Alonso, R., and D. Fuentes-Castellanos. 1987. Informe anual del programa de cruceros de investigaciones sobre el mero, *Epinephelus morio* (Valenciennes) en el Banco de Campeche. Unpublished, Cent. Invest. Pesq., La Habana, 10 p.

Valdés-Alonso, R., and G. Padrón. 1980. Pesquerías y palangre. Revista Cubana de Investigaciones Pesqueras 5(2):38–52.

Valdés-Muñoz, E. 1980. Toxicidad y alimentación de algunos peces sospechosos de provocar la ciguatera. Inf. Cient.-Téc., Inst. Oceanol. Acad. Cienc. Cuba, 123:1–25.

Valdés-Muñoz, E. 1981. Estructura y diversidad de la ictiofauna de los manglares de Punta del Este, Isla de la Juventud. Cienc. Biol., 6:111–124.

Valdés-Muñoz, E. 1982. Características ecológicas de la ictiofauna de la zona costera de Punta del Este, Isla de la Juventud. In: Sexta Jornada Cient. Inst. Oceanol., Acad. Cienc. Cuba, La Habana, 1982, Resúmenes, p. 36.

Valdés-Muñoz, E. 1987. Conducta diurna-nocturna de la ictiofauna de los manglares y zonas adyacentes. Rep. Invest. Inst. Oceanol. Acad. Cienc. Cuba, 60:1–16.

Valdés-Muñoz, E., and A. D. Mochek. 1994. Estructura de etológica de las communidades de peces. In: R. Claro, ed. Ecología de los Peces Marinos de Cuba. Inst. Oceanol. Cent. Invest. Quintana Roo, Mexico, pp. 13–54.

Valdés-Muñoz, E., and G. Padrón. 1980. Pesquerías de palangre. Rev. Cub. Invest. Pesq. 5(2):38–52.

Valdés-Muñoz, E., and A. Silva Lee. 1977. Alimentación de los peces de arrecifes artificiales en la plataforma suroccidental de Cuba. Inf. Cient.-Téc., Inst. Oceanol. Acad. Cienc. Cuba, 24:1–21.

Valdés-Muñoz, E., R. Claro, J. P. García-Arteaga, and L. M. Sierra. 1990. Características de las comunidades de peces en los manglares en el Golfo de Batabanó. In: R. Claro, ed. Asociaciones de Peces en el Golfo de Batabanó. Editorial Academia, La Habana, pp. 67–82.

Valdés-Muñoz, E., C. Tallet, and E. Gutierrez de los Reyes. 1999. Un género y ocho nuevos registros de peces marinos para Cuba. Unpublished, Inst. Oceanol., CITMA, La Habana.

Vales, M. A. Alvarez, L. Montes and A. Avila, eds. 1998. Estudio nacional sobre la diversidad biológica en la República de Cuba. Programa de Naciones Unidas para el Medio Ambiente/Centro Nacional de Biodiversidad/Instituto de Ecología y Sistemática/CITMA, La Habana, 480 p.

Valle, S. 1983. Evaluación de la albacora (*Thunnus atlanticus*) en la región suroccidental de Cuba. In: Cuarto Foro Cient. Cent. Invest. Pesq., Resumen de Investigaciones, La Habana, p. 51.

Valle, S., and C. Carles. 1983. Análisis de las pesquerías cubanas de bonito durante el período 1967–1982. In: Cuarto Foro Cien. Cent. Invest. Pesq., Resumen de Investigaciones, La Habana, p. 52.

Valle, S., J. P. García-Arteaga, and R. Claro. 1997. Growth parameters of marine fishes in Cuban waters. NAGA, ICLARM Q., 20(1): 34–37.

Vasconcelos Gesteira, T. C., and C. A. Sobreira Roche. 1976. Estudo sobre a fecundidade do ariacó, *Lutjanus synagris* (Linnaeus), da costa do estado do Ceará (Brasil). Arq. Cienc. Mar, 16(1):19–22.

Vasnetsov, V. V. 1953. Bony fish developmental stages [in Russian]. In: Essays on General Questions of Ichthyology. Akademy Nauk USSR, Moscow, pp. 207–217.

Vazzoler, A. 1962. Sobre a biología da corvina da costa sul do Brasil. Bol. Inst. Oceanogr. São Paulo, 12(1):53–102.

Vazzoler, A. 1969. Micropogonias furnieri: Fecundidade e tipos de desova. Bol. Inst. Oceanogr. São Paulo, 18:27–32.

Vergara, R. 1992. Principales características de la ictiofauna dulceacuicola Cubana. Información adicional I: Factores causales de su composición y diferenciación. Mem. Soc. Cienc. Nat. La Salle, 52(138):57–80.

Veron, J. E. N. 1995. Corals in Space and Time: The Biogeography and Evolution of the Scleractinia. Comstock, Cornell Univ. Press. Ithaca, NY, 321 p.

Victor, B. C. 1991. Settlement strategies and biogeography of reef fishes. In: P. F. Sale, ed. The Ecology of Fishes on Coral Reefs. Academic Press, San Diego, CA, pp. 231–260.

Victoria del Río, I., and I. Penié. 1998. Hidrología. In: M. Vales, A. Alvarez, L. Montes, and A. ávila, eds., Estudio Nacional sobre la Diversidad Biológica en la República de Cuba. PNUMA/CENBIO/IES/AMA/ CITMA, La Habana, pp. 117–125.

Victoria del Río, I., N. Melo, and R. Pérez. 1997. Results of Cuban oceanographic research related to global change studies. Proc. Conf./ Workshop in the Caribbean Countries and the Inter-American Institute for Global Change Research (IAI), pp. 131–141.

Volpe, A. V. 1959. Aspects of the biology of the common snook Centropomus undecimalis (Bloch) of the south-west Florida. Fla. State Board Conserv. Tech. Ser., 31:1–37.

von Götting, K. 1961. Beitrage zur Kenntnisl der Grundlagen der Fortpfllanzung und zur Fruchtberkeits-bestimmung bei marinen Tele-osteern. Helgol. Meeresunters. Band, 8(1):1–41.

Wainwright, P. C. 1988. Morphology and ecology: Functional basis of feeding constraints in Caribbean labrid fishes. Ecology, 69(3):635–645.

Wainwright, P. C. 1995. Predicting patterns of prey use from morphology of fishes. Environ. Biol. Fishes, 44(1–3):97–113.

Wakeman, J. M., C. R. Arnold, D. E. Wohlschlag, and S. C. Rabalais. 1979. Oxygen consumption, energy expenditure and growth of the red snapper (Lutjanus campechanus). Trans. Am. Fish. Soc., 108:288–292.

Wallace, R. A., and K. Selman. 1981. Cellular and dynamic aspects of oocyte growth in teleosts. Am. Zool., 21:325–343.

Walters, C. 2000. Impacts of dispersal, ecological interactions, and fishing effort dynamics on efficacy of marine protected areas: How large should protected areas be? Bull. Mar. Sci., 66(3):745–757.

Warner, R. R. 1978. The evolution of hermaphroditism and unisexuality in aquatic and terrestrial vertebrates. In: E. S. Reese and F Lighter, eds. Contrast in Behavior. Wiley Interscience, New York, pp. 77–101.

Warner, R. R. 1991. The use of phenotypic plasticity in coral reef fishes as tests of theory in evolutionary ecology. In: P. F. Sale, ed. The Ecology of Fishes on Coral Reefs. Academic Press, San Diego, pp. 387–398.

Warner, R. R., and D. R. Robertson. 1978. Sexual patterns in the labrid fishes of the western Caribbean. I. The Wrasses (Labridae). Smithson. Contrib. Zool., 254:1–27.

Wass, R. C. 1982. The shoreline fishery of American Samoa—past and present. In: J. L. Munro, ed. Ecological Aspects of Coastal Zone Management. Proc. Seminar on Marine and Coastal Processes in the Pacific. Motupore Is. Res. Centre, July 1980, UNESCO, Jakarta, pp. 51–83.

Watson, R. E. 2000. Sicydium from the Dominican Republic with description of a new species (Teleostei: Gobiidae). Stuttgarter Beitrage zur Naturkunde. Serie A (Biologie), no. 608, 31 p.

Weatherley, A. H., and H. S. Gill. 1987. The Biology of Fish Growth. Academic Press, London, 443 p.

Werner, E. E. 1974. The fish size, prey size, handling time relation in several sunfishes and some implications. J. Fish. Res. Board Can., 31:1531–1536.

Wijkstrom, V. N. 1974. Processing and marketing marine fish: Possible guidelines for the 1975–1979 period. In: A. S. Msangi, and J. J. Griffin, eds. Int. Conf. Marine Res. Dev. in Eastern Africa. Univ. Dar es Salaam and Int. Cent. Mar. Res. Dev., Univ. Rhode Island, Kingston, RI, pp. 55–67.

Williams, D. McB., S. English, and M. J. Milicich. 1994. Annual recruitment surveys of coral reef fishes are good indicators of patterns of settlement. Bull. Mar. Sci., 54(1):314–331.

Wilson, D. T., and M. I. McCormick. 1997. Spatial and temporal validation of settlement-marks in the otoliths of tropical reef fishes. Mar. Ecol. Prog. Ser., 153:259–271.

Wilson, K. H., and P. A. Larkin. 1980. Daily growth rings in the otoliths of juveniles sockeye salmon (Oncorhynchus nerka). Can. J. Fish. Aquat. Sci., 37(10):1495–1498.

Winn, H. E., and J. E. Bardach. 1960. Some aspects of the comparative biology of parrotfishes at Bermuda. Zoologica (N.Y.), 45:29–34.

Wittenberger, C., F. Coro, G. Suárez, and N. Portilla. 1969. Composition and bioelectrical activity of lateral muscles in Harengula humeralis. Mar. Biol., 3(1):24–27.

Woodley, J. D. 1992. The incidence of hurricanes on the north coast of Jamaica since 1870: Are the classic reef descriptions atypical? Hydro-biologica, 247:133–138

Woodley, J., P. M. Alcolado, T. Austin, J. Barnes, R. Claro-Madruga, G. Ebanks-Petric, R. Estrada, F. Geraldes, A. Glasspool, F. Homer, B. Luckhurst, E. Phillips, D. Shim, R. Smith, K. Sullivan-Sealey, M. Vega, J. Ward, and J. Wiener. 2000. Status of coral reefs in the northern Caribbean and western Atlantic. In: C. Wilkinson, ed. Status of Coral Reefs of the World, AIMS, Western Australia, pp. 261–286.

Wootton, R. J. 1990. Ecology of Teleost Fishes. Chapman and Hall, Fish and Fisheries Series, 404 p.

Wulff, J. L. 1995. Effects of a hurricane on survival and orientation of large erect coral reef sponges. Coral Reefs, 14:55–61.

Wyanski, D. M., and O. Pashuk. 1990. Processing and interpretation of fish reproductive tissue. Mar. Resour. Res. Inst., South Carolina Wildl. Mar. Resour. Dep. 16 p., 2 tabs., 18 figs.

Wyanski, D. M., D. B. White, and C. A. Barans. 2000. Growth, population age structure, and aspects of the reproductive biology of snowy grouper, Epinephelus niveatus, off North Carolina and South Carolina. Fish. Bull., 98(1):199–218.

Wyatt, J. R. 1983. The biology, ecology and bionomics of the squirrelfishes, Holocentridae. In: J. R. Munro, ed. Caribbean Coral Reef Fishery Resources ICLARM Stud. Rev., 125:50–58.

Ximenes-Carvalho, M. O., and A. A. Fonteles-Filho. 1996. Estudo de idade e crescimento da xira, Haemulon aurolineatum Cuvier (Pisces: Pomadasidae), da Ceará, Brazil. Arq. Cién. Mar, Fortaleza 30(1–2):73–77.

Yamamoto, K., and F. Yamazaki. 1961. Rhythm of development in the oocyte of the gold-fish, Carassius auratus. Bull. Fac. Fish. Hokkaido Univ., 12(2):93–110.

Zahnd, J. P., and G. J. Clavert. 1960. Etude comparative des modifications hépatiques liées á la vitellogenése chez quelques poissons. Comptes Rendus, 6(154):1317–1319.

Zhao, B. J., J. C. McGovern, and P. J. Harris. 1997. Age, growth and temporal change in size-at-age of the vermilion snapper from the South Atlantic Bight, based on otolith sections. Trans. Am. Fish. Soc., 126:181–193.

Zlatarsky, V., and N. Martínez-Estalella. 1980. Scleractinians of Cuba, with data on associated organisms [In Russian]. Bulgarian Academy of Sciences Press, Sofia. 312 p.

Zotin, A. I. 1961. Fish relative fecundity and egg size [in Russian]. Vopr. Ikhtiol., 1(2):307–313.

Zverkova, L. M. 1969. On the spawning of mintai (Theragra chalcogramma [Pallas]) in nearshore waters of the western coast of Kamchatka [in Russian]. Tr. Novorossisskoi Biologicheskoi Stantsii, 2:270–275.

Index

Page numbers followed by a t indicate pages with tables; numbers in italics refer to pages with figures.

Abudefduf (damselfish), 60, 71
  saxatilis (sergeant major)
    behavior, 60, *61*, 62, 70, *71*
    habitats and abundance, 22,
      25t, 26
Acanthostracion quadricornis, 27
Acanthuridae (acanthurids)
  behavior, 60, 62, 70
  fisheries, *204*
  habitat, 22
  trophic biology, 116, 118, 132
Acanthurus, 71, 216t
  bahianus (ocean surgeon), 25t, 62,
    63
  behavior, 60, *61*, 62, 70
  chirurgus (doctorfish), 25t, 26, *61*,
    62
  coeruleus (blue tang), 25t, 62, *63*
Aetobatus narinari (spotted eagle
  ray), 215
Africa, 77
age, *80*, 154–165, *166*
Agonostomus monticola (mountain
  mullet), 211
Albula vulpes, 28
algae
  drifting mats, 19, 29–30, 188
  as food source, 116, 116t, 119,
    120. *See also* macrophytes;
    phytoplankton
Alopias superciliosus (bigeye thresher
  shark), 215
Alopiidae, 29
Alphestes afer, 74t
Aluterus
  schoepfi, 28
  scriptus, 116
American Samoa, 217
anchoveta, Atlantic, 115, 163
Anchovia clupeoides, 28
anchovy, European, 179
angelfish, 98
Anguilliformes, 21

Anisotremus virginicus (porkfish), 69,
  117
Antennariidae, 29
anthropogenic effects, 18–20
  causeways, 11, 20, 28
  damming of rivers, 18, 28
  tourism, 19–20, 28
Apogon, 71
  affinis (bigtooth cardinalfish), 67
Apogonidae (apogonids), 67, 70, 116
aquaculture, 196
Archipiélago Jardines de la Reina
  community structure, 23, 23t
  physical characteristics, 3, 6, 11
  trophic structure, 131, 133t
Archipiélago Los Canarreos, 3, 23t
Archipiélago Los Colorados, 1, 15,
  99, 133t
Archipiélago Sabana-Camaguey
  fish assemblages, 22–23, 24–25t,
    26, 27, 28
  fisheries, 194, 199, 211, 213, 217
  habitats, 17, 18, 19
  physical characteristics, 1–3, 11,
    13
  productivity, 23, 23t
  trophic structure, 133t
Archosargus rhomboidalis, 28, 216t
Arctic Ocean, 180
artificial reefs, 13, 18, 116–117
  catch statistics in, 210
  for fisheries expansion, 27, 196,
    202, 218
  fish fauna/assemblages, 27, 29t,
    30–31
  fishing methods, 199
  materials, 27, 199
  trophic structure in, 131, 132
assemblage structure
  geographic variation, 6–7
  habitat characteristics, 13–14, *16*,
    20
Astrapogon stellatus (conchfish), 67

Atherinidae (atherinids), 70, 100
Atherinomorus stipes, 26
Atlantic Ocean, 179, 194, 214

back-calculation of size at age, 154,
  156
Bahamas
  fish behavior in, 68
  fisheries, 31, 194, 209
  physical characteristics, 1, 6
  spawning period, 88
Bahía Cienfuegos, 11, 93, 194, 209,
  216
Bahía de Buenavista, 3, 28, 206
Bahía de Cárdenas, *3*, 19
Bahía de la Habana, 1, 19
Bahía de Matanzas, 1, 7, 201
Bahía de Nazabal, 3
Bahía de Nipe, 19
Bahía de Sagua, 3
Bahía Guantánamo, 31
Bahía Honda, 1, 7
Bahía Jigüey, 3, 11, 28
Bahía La Gloria, 3, 28
Bahía Los Perros, 3, 11, 28
Bahía San Juan de los Remedios, 3,
  28, 206
Bahía Santa Clara, 3
Bahías de Nuevitas, 198
Bairdiella ronchus (ground croaker),
  91t, 120, 131t
Balistes vetula (queen triggerfish),
  116
Balistidae (balistids), 29, 30
Banco de Campeche, 11, 13, 161,
  163t, 166, 194
Banco de Jagua, 93
  barracuda. *See Sphyraena*
  great. *See Sphyraena barracuda*
bass, harlequin, 74t, 75
batfishes, 70
bathyal zone, fish fauna/
  assemblages, 30

Batrachoididae (batrachoidids), 27,
  117
beaugregory. *See Stegastes leucostictus*
behavior, 58–72, 77
Belonidae, 18, 58, 131t
benthic fishes, 64, 66, 70–71
benthophagous fish, 133, 133t, 134t
benthos, 12, 13, 70, 71
Bermuda, 29, 75, 79, 93
bigeyes, 116, 215
billfishes, 29, 32, 202, 206t, *214*, 214–
  215
biochemical indicators. *See*
  indicators
biogeographic patterns, 6–7
biomass
  gonadal, 183
  by habitat, 23t
  macroinfaunal, 13, 20
  marine macrophytes, 17, 20
  vs. fish density, *26*
biotoxicity, 19, 32
Blenniidae, 31, 116, 118
bluehead. *See Thalassoma bifasciatum*
Bodianus rufus (Spanish hogfish), 22,
  25t, 74t, 119, 129
body size. *See also* growth
  distribution, 31–32, 79, *80*
  minimum catch size, 207, 219
  of prey, 119, 120, 122, 130–131,
    131t, *132*
  relationships among different
    measures, 154
  relationship to habitat, 31–32
  sexual dimorphism, 77, *78*, 79, *80*
  at sexual maturity, 81
  at time of sex change, 75
  weight and food ration, 127
bonefish, 200
Bramidae, 30
Brazil, 155, *156*, 163, 180
butterflyfish
  banded, 64, *65*

fecundity, 98
foureye, 26, 64, 65, 69, 70
  spotfin, 64, 65, 69
bycatch, 195, 196t, 200, 201
  percent of total catch, 197
  statistics, 216–217, 218
Bythitidae, 30

Cabo Corrientes, 93, 99
Cabo Cruz, 3, 100
Cabo San Antonio, 1, 9, 11, 93, 99,
    194
Caibarién, 201, 202
Calamus, 216t
  bajonado (jolthead porgy), 74t, 77
  calamus, 91t
  nodosus (knobbed porgy), 77
  pennatula (pluma), 74t, 77
  proridens (littlehead porgy), 74t,
      77, 93
Cantherhines pullus, 116
Cantiles, 100
Carangidae (carangids), 18, 21
  behavior, 58
  as bycatch, 216t
  diurnal feeding patterns, 124
  ecophysiology, 180
  fisheries, 199, 200, 204
  food resources, 115, 118, 122
  growth patterns, 160
  habitats and abundance, 29, 29t
  predator-prey size relationship,
      131t
  reproduction, 91t, 96
Caranx, 71
  bartholomaei (yellow jack), 81
    feeding dynamics, 122, 123,
        124, 125, 127, 128
    fisheries, 212, 213
    food resources, 118, 120, 120t
    habitat, 22, 28
    length, 32, 98t
    predator-prey size
        relationship, 131t, 132
    reproduction, 88, 91t, 93–94,
        98t
  crysos (blue runner), 32, 212, 213
  hippos (crevalle jack), 28, 212
  latus (horse-eye jack), 212, 213,
      216t
  lugubris, 32
  ruber (bar jack), 78, 80, 81
    behavior, 59
    digestion rate, 126, 126
    ecophysiology, 180–182, 183,
        184, 185t, 186, 188–189,
        190, 191
    feeding dynamics, 122, 123,
        124, 125, 128, 129–130
    fisheries, 212, 213
    food consumption,
        interannual variations, 124
    food ration, 127, 129, 130
    food resources, 118, 120, 120t,
        130

gonadosomatic index, 89, 91t,
    92, 98t, 184
growth patterns, 149, 150,
    151, 152, 153, 160, 164
habitats and abundance, 22,
    27, 28
length, 32, 98t
predator-prey size
    relationship, 131t, 132
reproduction, 87, 88, 89, 93–
    94, 98t, 101
Carcharhinidae (carcharhinids), 21,
    29
Carcharhinus
  altimus (bignose shark), 215
  falciformis (silky shark), 215
  habitat, 22
  longimanus (whitetip shark), 215
  obscurus (dusky shark), 215
  signatus (night shark), 215
cardinalfish, 70
  bigtooth, 67
  dusky, 67
  food resources, 116
  sponge, 67
Caribbean, 1, 2, 23, 36, 77, 102, 157
Caribbean Current, 8
Caribbean Plate, 3–4, 5–6
carnivores. See feeding; piscivores;
    predator-prey interactions
Casilda, 3
catch per unit effort (CPUE). See
    under fisheries
catch statistics, 194, 196t, 197, 198,
    205–216, 217, 218–219
catfish, sea, 118
Cayman Trench, 3
Cayo Avalos, 100
Cayo Buenavista, 1
Cayo Cabeza del Este, 100
Cayo Coco, 1, 7, 11
Cayo Diego Pérez, 15, 99, 101
Cayo Francés, 17
Cayo Juan García, 19
Cayo Largo, 17, 19
Cayo Levisa, 15
Cayo Romano, 1, 11
Cayo Rosario, 15
Cayo Sabinal, 1, 3
Cayo Tablones, 29t, 30, 116, 118
Cayos de Doce Leguas, 131
Cayos de San Felipe, 8, 15, 100
Cayos Los Indios, 15
Centropomidae (centropomids), 70,
    216t
Centropomus undecimalis, 28
Cephalopholis
  cruentata (graysby)
    growth patterns, 152, 161
    habitat, 31
    reproduction, 74t, 79, 86, 91t
  fulva (coney), 24t, 31, 74t, 75, 91t,
      162
Cetengraulis edentulus (Atlantic
    anchoveta), 115, 163

Chaenopsidae, 31
Chaenopsis ocellata, 27
Chaetodon, 22, 71
  capistratus (foureye
      butterflyfish), 26, 64, 65, 69, 70
  lunula, 64
  ocellatus (spotfin butterflyfish),
      64, 65, 69
  striatus (banded butterflyfish), 64,
      65
Chaetodontidae (chaetodontids), 60,
    64, 65, 96
Chauliodontinae, 30
Chloroscombrus chrysurus, 216t
Chromis, 67, 70, 71
  cyanea (blue chromis), 25t, 60
  multilineata (brown chromis),
      25t, 60
chubs, 116
Ciénaga de Zapata, 13
Ciénaga Litoral del Sur, 28
ciguatera poisoning, 19, 32, 219
circadian rhythm, 149
circulation patterns, 1, 8–11, 20, 99–
    100, 219
Ciudad de la Habana, 19
clam, fisheries, 194, 196, 196t, 204
classification, 21–22
Clepticus parrae (creole wrasse), 25t,
    74t, 119
Clupea
  harengus harengus (Atlantic
      herring), 85, 133, 179
Clupeidae (clupeids), 18, 91t
  behavior, 67, 70, 72
  catch and bycatch statistics, 211,
      216t
  ecophysiology, 180, 192
  fishing gear, 199
  food resources, 115, 118–119, 134
  reproduction, 96, 98, 100
cobia, 32, 216
cod, 185–186
  Atlantic, 85, 179
Colombia, 9, 154, 162
Colorados reefs, 15
coloration
  changes at sexual maturity, 79
  predator-prey relationships, 62,
      64, 67, 69
  sexual dimorphism, 77–79, 101
common fish names, 33–57t
community differentiation, behavior
    mechanisms, 69–71
community structure, 23t, 28, 30–
    31, 115
  competition, 69, 117
conch, fisheries, 197–198, 196t,
    201
conchfish, 67
condition factor. See under indexes
coney. See Cephalopholis fulva
Congridae, 216t
conservation and management, 20,
    218, 219

coral
  bleaching, 19
  diseases, 19
  as food resource, 116, 119, 133
  species in Cuba, 17
coral reefs, 14–17
  distribution and habitat, 22–23,
      26
  diurnal changes in feeding
      intensity within, 122–125
  ecological subsystem and
      fisheries, 198
  fish fauna/fish assemblages, 22–
      23, 26, 60–64
  fishing activities, 201, 204
  patch reefs, 17, 18, 23t, 24–25t,
      132, 133t, 204
  reef crests, 15, 17, 23t, 24–25t,
      131, 132, 133t
  slope reefs, 15, 23t, 24–25t, 131,
      133t, 201, 204
  species diversity, 17, 32, 59
Corona San Carlos, 99, 100
Coryphaena
  equiselis, 29
  hippurus, 29
Coryphopterus, 25t, 116
CPUE. See fisheries
crab
  fisheries, 194
  as food source, 116, 116t, 117,
      118, 119, 120, 122, 127, 152
creole-fish, 75
creole wrasse, 25t, 74t, 119
crepuscular fishes, 69, 70, 71, 72
croaker, ground, 91t, 120, 131t
crustaceans
  bycatch statistics, 216t
  diurnal patterns, 70, 124
  fisheries, 3, 195–196
  as food source, 13, 115, 116, 116t,
      117, 118, 119, 120t, 121, 122,
      123, 135
Cryptotomus, 77
Cryptotomus roseus, 74t
Ctenogobius boleosoma, 27
Cuba
  climate, 8, 73
  fishing associations, 202, 203, 209,
      216
  geography, 1–3
  geologic history, 1, 3, 5–7
  geomorphology, 4–6, 7–8
  habitats, 6, 13–14, 16, 20
  hydrobiological characteristics,
      12–13
  hydrochemical characteristics,
      11–12
  hydrological characteristics, 8–
      11, 30
  legal infrastructure, 202–205, 218
  list of fish species, 33–57t
  tidal amplitude, 10–11
  trophic structure, 134t

Cuba. Ministry of Fishing Industries (Ministerio de la Industria Pesquera), 194, 202, 207, 218
Cuba. Ministry of Science, Technology and Environment, 20
Cuban Countercurrent, 8–9
Cuban fish names, 33–57t
Cuban National Aquarium, 129
Cuban shelf
  circulation over the, 1, 8–11, 20
  fisheries, 194, *204*, 217
  geographic description, 1–3, *4–5*
  spawning at the edge, 99–100, 102
  spawning on the inner, 100–101, 102
  water temperature, 181
currents, 8–11, *12*
cusk-eel, blackedge, 215, 216t
Cyprinodontidae (cyprinodontiforms), 22, 28
Cyprinodon variegatus, 28

daily increments. *See* otoliths
daily ration (DR). *See* feeding
damselfish
  behavior, 59–60, 64, *65*, 70, 71
  dusky, *63*, 66, 116
  yellowtail. *See Microspathodon chrysurus*
Dasyatis americana (southern stingray), 27, 28, 215
Decapterus, 118
deep-water fish, 30
defensive mechanisms, 60–62, 70, 71
detritivores, 115, 116t, 134t, 135
Diapterus, 217
  rhombeus, 28, 216t
dick, slippery. *See Halichoeres bivittatus*
digestion. *See* feeding, digestion rate
Diodon
  holocanthus, 27, 29
  hystrix, 27
diodontids, 27
Diplectrum, 216t
  formosum (sand perch), 74t, 75, 152, 161
dispersal. *See* circulation patterns
diurnal behavior, 58–66, 69–71
  feeding intensity, 27, 122–125, 135
  food resource changes, 117
doctorfish, 26, 61, 62
Dominican Republic, 31
drum, fisheries, 200

Echeneis naucrates (sharksucker), 117
ecological aspects
  competition, 69, 117
  latitudinal variation in trophic structure, 134t
  niches, 69–71, 116, 206

and patterns of spawning, 84–88, 102
subsystems and fisheries, 198
trophic groups, 115–117, 134–135
trophic structure with habitats and regions, 131–134, 135
ecophysiology, of fishes, 179–193
  fat content, 179–180, 184–188, 191–193
  lipid synthesis, 180, 192, 193
  liver weight, 179, *187*, *188*
  metabolism, 126, 179
  ontogenetic dynamics of indicators, 120–122, 186–191
  seasonal patterns of indicators, 180–186, *187*
education outreach, 20
El Niño, 23, 32
Elops saurus, 28
energy
  costs of predation, 189, 193
  and euryphagy, 133–134
  exchange between adjacent habitats, 22
  and food resources, 120, 122, 129, 130, 135
  hydrostatic and energetic fat storage, 180
  and reproduction, 90, 92, 94, 102, 120, 122
Engraulidae (engraulids), 18, 100, 115, 192
Engraulis, 98
  encrasicholus (European anchovy), 179
Ensenada de la Broa, 7, 11, 19, 198
Ensenada Siguanea, 7
Ephippidae, 22
epibenthic plankton feeders, behavior, 59–60, 67
Epigonidae, 30
Epinephelinae, 74t, 75
Epinephelus, 71, 75, 77, 79
  flavolimbatus (yellowedge grouper), 209
  food resources, 117
  guttatus (red hind)
    fisheries, 215
    growth patterns, 161, 162
    reproduction, 74t, 75, 79, 99
  itajara (jewfish), 74t, 75, 118, 131t, 161
  morio (red grouper)
    growth patterns, 151, *153*, 161
    reproduction, 74t, 75, 77, 79, 86, 88, 93
  mystacinus (misty grouper), 30, 209
  nigritus (Warsaw grouper), 152, 161
  niveatus, 74t
  striatus (Nassau grouper)
    behavior, 69
    catch statistics, 206t, 209, *209*
    diurnal feeding patterns, *125*

fisheries, 194
food ration, 127, 129
food resources, 117–118, *121*, 122
gonadosomatic index, 91t
growth patterns, 151, 152, 161–162, *164*
habitats and abundance, 22, 27, 31
length, 32
reproduction, 74t, 75, *76*, 77, 79, 82, 93
spawning, 86, 88, 92, 95, 99, 100
Equetus, 22
estuaries
  ecological subsystem, 3, 198
  fisheries, 3, 198, 211, 212, 217
  fish fauna/assemblages, 28–29
  species diversity, 32
Eucinostomus, 24t, 27, 28
  gula, 28
  jonesii, 28
Eugerres
  brasilianus (Brazilian mojarra)
    growth patterns, 151, 152, *153*, 162
    habitat, 28
    reproduction, 91t
    trophic biology, 120
    habitat, 28
  plumieri (striped mojarra), 91t, 162, 210
euryphagy, 116, 134
Euthynnus alletteratus (little tunny), 215
evolution, hermaphroditism, and gonochorism, 76, 77
Exclusive Economic Zone (EEZ), 2, 194, 195–198, 205, 218

fat content, ecophysiology, 179–180, 184–188, 191–193
fat reserves, 90, 92, 94, 153
  hydrostatic and energetic storage, 180, 192
  in the liver, 180, *187*, *188*
fecundity, 95, 98t
  "batch rate," 97, 97–98, 101, 102
  and gonadosomatic index, 89, 90
  and predation pressure, 101
  relative, definition, 95, 102
feeding
  behavior, 60, 61, 62, 64, 66, 67, 69, 70, 71, 72
  comparison among habitats and regions, 115–120
  daily ration (DR), 127, *128*, 129, 188
  diet and morhology, 119
  digestion rate, 126–127, 134t, 135
  diurnal feeding patterns, 122–125, 135
  food ration, 127–129, *130*, 134t, 193

major trophic groups, 115–117
ontogenetic variations, 120–122, *123–124*, *128*, 129–130
prey size, 130–131, *132*, 135
relationship to sexual maturity, 79, 122, 130
seasonal variations, 119–120, *128*, *129*, 130, *130*, 134t, 180, *189*
trophic structure with habitats and regions, 131–134
filefish, food resources, 116
finfish
  as bycatch, 217
  catch statistics, 205, *205*, 206t, 218
  fisheries, 195, 196t, 198
  percent of total catch, *197*
fish, as food source. *See* piscivores
fisheries, 194–219. *See also* overfishing
  associations, 202, *203*, 209, 216
  catch per unit effort (CPUE), 214
  catch statistics, 194, 196t, 198, 205–216, 217, 218–219
  conservation and management, 20, 218, 219
  effects, 89, 155, 201, 206, 211, 212
  historical background, 194, 199
  productivity, 217
  regulations, 31, 194, 196, 201, 206, 208, 218, 219
fishing
  effort, 194, 205, 208, 209, 210, 211, 212, 213, 217
  gear, 194, 198–202, *204*
  ghost, 201–202
  gill nets, 198, 200, *204*
  hook and line, 194
  illegal, 207
  lights and chumming, 209
  longlines, 198, 202, *202*, *204*
  recreational, 29, 207, 215
  rods, 198, 213
  seines, 198–199, *204*
  set nets, 198, 200, *200*, *204*, 205
  traps, 198, 200–202, *204*
  trawls, 198–200, *199*, *204*
fishing fleet, 194, 198, 199–200, 213, 217
fish names, scientific/common/Cuban, 33–57t
Flamenco-Traviesa, 29t
Florida
  Everglades, 160t
  fisheries, 194
  fish growth patterns in, 153, 154, 155, 156t, 157, 158t, 160t, 161t, 163t
  fish reproductive patterns in, 75, 79, 81
  food resources in, 120
  habitats and fish assemblages, 23t, 24–25t, 27
  Keys, 9, 23t, 31, 99, 131, 133t, 160t
  larval dispersal, 9

spawning periods, 95
Straits off, 1, 29–30
flounder, 85
food resources, 12, 13, 62, 69, 116t, 117–119
  availability of, 81, 119, 124, 128–130, 135, 156
  interannual variations in, 119, 120t, 122, 124
  and mouth morphology, 130–131, 189
  opportunistic responses to, 115, 116, 117, 133, 134
  relationship to growth, 160t
  seasonal variations in, 119–120, 129
  variations with fish growth, 120–122, 129–130, 152, 189
food web. See under ecological aspects
Fundulidae, 22

Gadus morhua (Atlantic cod), 85, 179
Galeocerdo cuvier (tiger shark), 29, 215
gametogenesis, 81–89
Gempylidae, 30
genetic aspects, of hermaphroditism, 75
Gerreidae (gerreids)
  fisheries, 216t, 217
  habitats and abundance, 18, 24t
  reproduction, 91t, 100
  trophic biology, 118
Gerres cinereus (yellowfin mojarra), 27, 28, 150, 152, 162, 210–211
Ginglymostoma cirratum (nurse shark), 215
goatfish, 22
  spotted, 119
  yellow, 119
Gobiidae (gobiids), 21, 27, 30, 31, 100
Golfo de Ana Maria
  bycatch, 216t
  habitat, 17, 19
  physical characteristics, 1, 3, 7–8, 10, 11
  spawning aggregation, 100
Golfo de Batabanó
  fish assemblages, 23, 24–25t, 26, 27, 29t
  fish ecophysiology, 185, 188, 189
  fisheries, 30, 199, 201, 202, 203, 217. See also overfishing
    catch statistics, 208, 210, 212–213
    productivity, 217
  fish growth patterns in, 153, 159, 160t
  food resources in, 117, 118, 119
  gonadosomatic index, 89, 92
  habitats, 14, 15, 17, 19
  larval dispersal, 100

physical characteristics, 1, 3, 7, 9, 10, 11, 13
size and age distribution, 80
spawning aggregations, 99–100, 101
spawning periods, 93, 94
trophic structure, 131
Golfo de Cazones, 3, 8, 9, 99–100
  fisheries, 206
  larval dispersal, 100
  spawning aggregation, 99
Golfo de Guacanayabo, 1, 3, 8, 10, 11, 18, 216t
Golfo de Guanahacahabibes, 1, 7, 17
gonadosomatic index (GSI). See under indexes
gonads. See also reproduction; vitellogenesis
  development and spawning patterns, 79–81, 81t, 92
  gametogenesis, 81–89
  growth marks and maturation of, 151
  oogenesis, oocyte development, 75, 76, 81t, 82–89, 183
  resorption of ovarian tissue, 76, 77, 81, 84, 86, 89, 90, 200
  spermatogenesis, 75, 76, 81t, 183
  weight, 179
gonochorism, 73, 75, 77, 102
gonostomatids, 32
Gramma loreto, 24t
Grammatidae, 30
Gran Banco de Buena Esperanza, 3
graysby. See Cephalopholis cruentata
Greater Antilles, 3–4, 5, 133, 215, 217
grouper(s), 71, 72, 85
  black. See Mycteroperca bonaci
  catch statistics, 209, 209–210
  diurnal feeding patterns, 123
  fisheries, 31, 201, 202, 204
  food resources, 117–118
  gag. See Mycteroperca microlepis
  misty, 30, 209
  Nassau. See Epinephelus striatus
  red. See Epinephelus morio
  tiger. See Mycteroperca tigris
  Warsaw, 152, 161
  yellowedge, 209
  yellowfin. See Mycteroperca venenosa
growth
  and age, 154–165, 166
  and hermaphroditism, 79
  juvenile marks, 152, 188
  latitudinal variations in, 134t, 159
  measurement methodology, 152, 154–155, 162, 165
  relationships among size measures, 154
  seasonal variations in, 130, 151, 152–154, 182, 189
  size at sexual maturity, 75

variations in feeding habits with, 120–122
von Bertalanffy function, 154
grunt(s), 3. See also Haemulidae
  behavior, 67, 69, 70, 71, 79
  bluestriped. See Haemulon sciurus
  catch statistics, 206t, 210
  ecophysiology, 180
  fisheries, 194
  fishing gear, 199, 201, 204
  food resources, 22, 115, 118, 122
  French. See Haemulon flavolineatum
  growth patterns, 159
  habitats and abundance, 26, 30–31
  white. See Haemulon plumieri
Guadeloupe, 23t, 24–25t, 26, 131, 133t
Gulf of Maine, 133, 134t
Gulf of Mexico, 1, 166, 194
Gymnothorax (moray eel), 22, 32, 72, 201

habitat. See also Cuba
  diurnal patterns in spatial distribution, 70–72
  formation, 6
  fragmentation and degradation, 30, 201, 206, 211, 212, 219
  nursery areas, 18, 27, 102
  structure, 13–14, 16, 20
  trophic structure with, 131–134
  types, fish/fish assemblages of, 22–30
  usage patterns, and reproductive cycles, 95
Haemulidae (haemulids), 21, 79
  behavior, 67–68
  diurnal feeding patterns, 125–126
  food resources, 115, 130
  as food source, 117, 118
  growth patterns, 163t
  predator-prey size relationship, 131t
  reproduction, 91t, 92, 94, 96
Haemulon, 67, 71, 82, 88, 120, 150, 216t
  album (margate), 78, 80, 81
    digestion rate, 126, 126
    ecophysiology, 180–182, 183, 184, 185t, 186, 190
    feeding dynamics, 122, 123, 124–125, 125, 128
    fisheries, 210
    food resources, 118, 120t
    gonadosomatic index, 89–90, 91t, 92, 98t, 184
    growth patterns, 150, 151, 152, 153, 154, 159, 162, 163
    mean length, 98t
    predator-prey relationships, 131t, 132
    reproduction, 87, 88, 95, 97, 98t, 100

aurolineatum (tomtate), 91t
  ecophysiology, 180
  effects of lane snapper overfishing, 206
  fisheries, 210
  growth patterns, 159
  habitats and abundance, 24t, 29t, 30
  length, 31, 98t
  reproduction, 98t
carbonarium, 91t
flavolineatum (French grunt)
  behavior, 65, 67, 70, 71
  growth patterns, 159
  habitats and abundance, 24t, 26, 31
parra (sailor's choice), 69, 91t, 210
plumieri (white grunt), 78, 80, 81
  behavior, 69, 71
  ecophysiology, 180
  fisheries, 210
  food resources, 118
  gonadosomatic index, 91t, 94, 98t
  growth patterns, 151, 152, 159, 163, 164
  habitats and abundance, 25t, 29t
  length, 32
  mean length, 98t
  reproduction, 87, 88, 89, 94–95, 98t
sciurus (bluestriped grunt), 78, 80, 81
  behavior, 67, 68
  ecophysiology, 180
  fisheries, 210
  food resources, 118
  gonadosomatic index, 91t, 92, 94, 98t
  growth patterns, 151, 152, 153, 154, 159, 163, 164
  habitats and abundance, 25t, 26, 29t
  length, 32
  mean length, 98t
  reproduction, 87, 88, 89, 94–95, 98t
halfbeaks, 116
halibut, 179
Halichoeres
  bivittatus (slippery dick)
    behavior, 63, 64, 70, 71, 74t, 77
    habitats and abundance, 25t
  food resources, 119
  as food source, 118
  garnoti (yellowhead wrasse), 25t, 74t, 77
  maculipinna, 74t
  pictus, 74t
  poeyi (blackear wrasse), 63, 64, 70, 74t
  radiatus, 74t

hamlet
　barred, 74t, 75
　butter, 74t, 75
　yellowtail, 74t, 75
Harengula
　clupeola (false pilchard), 67, 91t
　　diurnal feeding patterns, 125,
　　　125
　　fisheries, 211
　　food resources, 115, 119, 122,
　　　123
　　habitats and abundance, 26, 27
　humeralis (redear sardine)
　　behavior, 67
　　diurnal feeding patterns, 125,
　　　125
　　ecophysiology, 180, *184*, 186,
　　　189–190, *190*, 191, 193
　　fecundity, 97, 98t, 101
　　fisheries, 211
　　food resources, 115, 118–119,
　　　120t, 122, *123*, 133
　　as food source, *127*
　　gonadosomatic index, 89, 91t,
　　　92, 98t, *184*
　　growth patterns, 149, *150*, *153*
　　mean length, 98t
　　reproduction, 78, 80, 81, 82,
　　　92, 95, 97, 98t, 101
　　spawning pattern, *87*, 88
　jaguana (scaled sardine), 67, 91t,
　　163, 211
Havana. *See* La Habana
Hemiramphus (halfbeak), 116
hepatosomatic index (HSI). *See*
　indexes
herbivores and herbivory, 131–134,
　133t, 134–135, 134t
hermaphroditism, 31, 73–77, 101–
　102
　behavior factors, 77
　diandric, 74t, 75t, 76, 77
　monandric, 74t, 75t, 77
　protandric, 73, 74t, 75t
　protogynous, 73, 74t, 75t, 76, 77,
　　79, 101
　synchronous, 73, 74t, 75
herring(s)
　Atlantic, 85, 132, 179
　Atlantic thread, 115, 206t, 211
　catch statistics, 206t
　dwarf. *See Jenkinsia lamprotaenia*
　fisheries, *204*, 211
Hexanchidae, 29
Hexanchus griseus (sixgill shark), 30,
　215
hind, red. *See Epinephelus guttatus*
Hippocampus, 216t
Hispaniola, 217
Histrio histrio, 29
hogfish. *See also Lachnolaimus*
　maximus
　Spanish (*Bodianus rufus*), 22, 74t,
　　119, 129
Holacanthus, 117

tricolor (rock beauty), 74t, 79
Holocentridae (holocentrids), 67, 69,
　71, 116
Holocentrus, 71
　rufus (longspine squirrelfish),
　　24t, *68*, 69
home range, 58, 71
horse mackerel, Mediterranean, 179
hypersaline habitats, 11, 18, 22
Hypoplectrus
　chlorurus (yellowtail hamlet), 74t,
　　75
　nigricans, 74t
　puella (barred hamlet), 74t, 75
　unicolor (butter hamlet), 74t, 75

ichthyosarcotoxins, 19, 32
indexes
　condition factor (K), 179, 180t,
　　182t, 188
　gonadosomatic index (GSI), 89–
　　90, 91t, 179, 180t, 182t, *184*
　hepatosomatic index (HSI), 179,
　　180t, 186, *187*
　index of gastric repletion (IGR),
　　124, 125, *125*, 128, 130
　mesenteric fat index (MFI), 179,
　　180t, 182t, *183*, 189, *190*
indicators
　behavioral, 69
　growth marks, 149, 151, 188
　ontogenetic dynamics, 120–122,
　　123–124, 128, 129–130, 186–
　　191
　otoliths. *See* otoliths
　physiological and biochemical,
　　179, 180t. *See also* indexes
　　digestion rate, 126
　　fecundity, 95
　　scales. *See* scales
　seasonal patterns in, 180–186, *187*
　spines, 152
invertebrates
　benthic, 7, 13, 70, 119, 122
　fisheries, 3
　as food source, 116t, 117, 118,
　　119, 122, 157
　in mangroves, 18
　in seagrass beds, 18
Isabela de Sagua, 3
Isla de la Juventud
　diversity, 58
　habitats and fish assemblages, 17,
　　26
　physical characteristics, *3*, 8, 11,
　　13
　turtle fisheries, 198
Istiophoridae, 29
Istiophorus platypterus (sailfish), 91t,
　202, *214*, 214–215
Isurus oxyrinchus (shortfin mako
　shark), 215

jack(s), 3, 71, 79. *See also* Carangidae
　bar. *See Caranx ruber*

　behavior, 58–59
　catch statistics, 206t, *212*, 212–213
　crevalle, 212
　habitats and abundance, 28, 29,
　　29t, 30
　horse-eye, 212, 213, 216t
　yellow. *See Caranx bartholomaei*
Jagua Trough, 8
Jamaica, 79, 81
　fisheries, 31, 213, 217
　fish growth patterns in, 160,
　　163t, 165
　juvenile mark study, 152
　spawning periods, 93, 94, 95
　water temperature, 151
Jenkinsia lamprotaenia (dwarf
　herring), 26, 67
　ecophysiology, *184*, 186
　food resources, 119
　as food source or bait, 119, *127*,
　　198, 211, 213
　growth patterns, 163
　habitats and abundance, 24t, 26,
　　27
　mean length, 98t
　reproduction, 91t, 94–95, 98t, *184*
jewfish. *See Epinephelus itajara*
Joturus pichardi (bobo mullet), 211
juveniles. *See also* nursery areas;
　recruitment
　growth marks, 152, 166, 188
　seasonal patterns of indicators,
　　180–182, 187–189, 193

Katsuwonus pelamis (skipjack tuna),
　29, 198, 206t, 211, 213–214
Kyphosidae, 116
Kyphosus sectatrix, 29

Labridae (labrids)
　behavior, 60, 62–64, 70
　food resources, 22, 116, 119
　as food source, 118
　reproduction, 76, 77, 79, 91t, 96
Labrisomidae (labrisomids), 31, 100
Labroides dimidiatus, 77
Lachnolaimus maximus (hogfish)
　abundance, 29t
　behavior, 62
　diurnal feeding patterns, 125, *125*
　fisheries, 215–216
　food resources, 119
　growth patterns, 149, *150*, 151,
　　162, *165*
　mean length, 98t
　reproduction, 74t, *76*, 77, 79, 91t,
　　98t
Lactophrys
　bicaudalis, 27
　trigonus, 27
lagoons
　euhaline, 18
　fisheries, 211, 212
　fish fauna/fish assemblages,
　　27–28

variability of food supply, 120
low-salinity, fish fauna/
　assemblages, 28–29
La Habana, 10, 11, 17
La Isabela, 7
Lamnidae, 29
Langmuir circulation, 30
larvae, 9, 98–99, 120, 219
length. *See* body size
Lepophidium brevibarbe (blackedge
　cusk-eel), 215, 216t
Lesser Antilles, 6, 26, 133
life cycle
　annual reproduction cycles, 90,
　　92–95, 96, 97, 102
　ontogenetic variations in feeding,
　　120–122, *123–124*, 128, 129–
　　130
Limia vittata, 28
liza. *See Mugil liza*
lobster
　fisheries, 3, 18, 195–196, 198, 202,
　　204, 217, 218
　as food source or bait, 117, 118,
　　201
　percent of total catch, *197*
　spiny, 18, 195–196
longevity, 79, 134t, 159
Los Indios, 100
lunar cycle and spawning, 90, 94, 95,
　102
Lutjanidae (lutjanids), 21, 67–68
　catch statistics, 205–209
　diurnal feeding patterns, 124
　fisheries, 199, 201, *204*, 205–209
　food resources, 115
　growth patterns, 152–154, 155–
　　161
　habitats and abundance, 22, 24t,
　　28, 30
　predator-prey size relationship,
　　131t
　reproduction, 91t, 92, *96*
Lutjanus, 71
　analis (mutton snapper)
　　behavior, 68, 69
　　catch statistics, 206–207, 206t,
　　　207
　　ecophysiology, 182, *184*, 186,
　　　190
　　feeding intensity, 124, *128*
　　food ration, *127*, 129, 130, *130*
　　food resources, 68, 69, 117,
　　　121, 122
　　gonadosomatic index, 89, 91t,
　　　92, *184*
　　growth patterns, 149, 150,
　　　150, 152, *153*, 156–157, *158*,
　　　166
　　habitat and abundance, 28, 31
　　length, 31
　　predator-prey size
　　　relationship, 131t
　　reproduction, 78, 80, 81, 82,
　　　92, 93, 95

spawning, 86, 88, 99, 100, *100*
apodus (schoolmaster)
  behavior, 68
  habitats and abundance, 22,
    24t, 26
  length, 31
  reproduction, 91t
  trophic biology, 120, 124, 127,
    131t
buccanella (blackfin snapper)
  fisheries, 209
  growth patterns, *150*, 151,
    152, *153*, 158–159
  habitat, 30
  reproduction, 88, 93
cyanopterus (cubera snapper)
  behavior, 69, 70
  catch statistics, 206t, 207–209,
    208
  diurnal feeding patterns, 124
  habitat, 22
  spawning aggregation, 99, 100
griseus (gray snapper), 18
  behavior, 68, 69, 70
  catch statistics, 206t, 207–209,
    208
  digestion rate, 126, *126*
  ecophysiology, *184*, *186*, *190*
  effects of lane snapper
    overfishing, 206
  feeding dynamics, 124, *128*,
    131t
  fisheries, 194, 201
  food resources, 117, 120, *121*
  gonadosomatic index, 89, 91t,
    92, *184*
  growth patterns, 149, *150*,
    151, 152, *153*, 157–158, *159*,
    160t, *165*, *166*, 180
  habitats and abundance, 22,
    24t, 26, 28, 29t, 31
  length, 31
  reproduction, 78, 79, *80*, 81,
    82, 93
  spawning, 86, 88, 99, 100
jocu (dog snapper)
  behavior, 68–69, 70, 82
  diurnal feeding patterns, 124
  food resources, 117
  growth patterns, 151, 152,
    157, *162*
  habitats and abundance, 26
  reproduction, 88, 89, 91t, 100
synagris (lane snapper)
  behavior, 68, 69
  as bycatch, 216t, 217
  catch statistics, 205–206, 206t
  digestion rate, 126, *126*, 127
  ecophysiology, 182–183, *184*,
    185t, *186*, *187*, *190*
  feeding intensity, 124, *128*
  fisheries, 30–31, 200, *200*
  food ration, 127
  food resources, 117, 120, *121*,
    121–122

gonadosomatic index, 89, 91t,
  92, *94*, *184*
growth patterns, 149, *150*,
  152, *153*, 155–156, *157*, *166*,
  180
habitats and abundance, 22,
  28, 29t, 30
length, 31
reproduction, 78, 79, *80*, 81,
  82, 89, 93, 95
spawning pattern, 82, 86, *87*,
  88
vivanus (silk snapper)
  catch statistics, 209
  growth patterns, *150*, 151,
    152, 158–159, 164
  habitat, *30*
  reproduction, 88, 93

mackerel(s)
  catch statistics, 206t
  cero, 131t, 214
  chub, 98, 116
  fisheries, 200, *204*, 214, *214*
  king, *153*, 163, 214
  Mediterranean horse, 179
  Spanish, *153*, 163, 214
macrophytes
  benthic, 12, 13, 70
  floating, 18, 116
  as food source, 116, 116t
  marine reef, 17
Makaira nigricans (blue marlin), 29,
  214, 215
Malacosteidae, 30
mangroves, 1, 12, 13–14, *16*
  ecological subsystem and
    fisheries, 198
  euhaline, 27, 59, 70
  fish fauna/fish assemblages, 18,
    23t, 24–25t, 26–27, 28–29
  species diversity, 32, *59*
  trophic structure, 131, 132, 133t
margate. *See Haemulon album*
marlin
  blue, 29, *214*, 215
  white, 29, *214*, 215
Marshall Islands, 133, 134t
Martinique, 23t, 24–25t, 26, 131,
  133t
maximum sustainable yield, 194,
  206, 207
Megalops atlanticus, 28
Melanostomiinae, 30
mesenteric fat index (MSI). *See*
  under indexes
metabolic expenditure, 182, 189, 192
Mexico. *See* Yucatán
Micropogonias furnieri, 91t, 216t
Microspathodon chrysurus (yellowtail
  damselfish)
  behavior, 62, 65, 66, *71*
  habitats and abundance, 25t
  reproduction, 101
  trophic biology, 117

migration. *See* spawning, migration;
  vertical migration
Ministry of Fishing Industries
  (Ministerio de la Industria
  Pesquera), 195, 202, 207, 215, 218
Ministry of Science, Technology and
  Environment, 20
minnow, sheepshead, 28
mojarra(s), 18, 79, 117. *See also*
  Gerreidae
  Brazilian. *See Eugerres brasilianus*
  catch statistics, 206t, 210–211, *211*
  fisheries, 200, *204*
  habitats and abundance, 27, 28
  striped, 91t, 162, 210
  yellowfin. *See Gerres cinereus*
mollusks, 13, 18, 70, 216t
  bycatch statistics, 216t
  fisheries, 195, 196–197, 196t
  as food source, 116t, 117, 118,
    119, 120t, *121*, 122, *123*, 135
Monacanthidae, 29, 116
Monacanthus ciliatus, 28, 116
moray eel, 22, 32, 72, 201
morralla. *See* bycatch
mortality, 200, 201–202. *See also*
  bycatch; predator-prey
  interactions
Mugil, 212
  curema (white mullet)
    fisheries, 211
    growth patterns, 149, *150*, *153*,
      162
    habitat, 28
    reproduction, 82, 88, 89, 91t,
      102
  hospes (hospe mullet), 91t, 162,
    211
  incilis (Parassi mullet), 28, 211
  liza (liza)
    fisheries, 211
    growth patterns, 150, *150*, *153*,
      162
    habitat, 28
    reproduction, 82, 88, 89, 91t
  longicauda (rabúa), 211
  trichodon (fantail mullet), 28, 91t,
    149, *150*, *153*, 211
Mugilidae (mugilids)
  catch statistics, 206t
  fisheries, 194, 200, *204*, 211–212
  food resources, 115, 134, 135
  growth patterns, 152, *153*
  reproduction, 91t, 96
mullet(s), 28
  bobo, 211
  fantail. *See Mugil trichodon*
  hospe, 91t, 162, 211
  mountain, 211
  Parassi, 211
  white. *See Mugil curema*
Mullidae, 22
Mulloidichthys martinicus (yellow
  goatfish), 119
Mycteroperca, 32, 75, 79

bonaci (black grouper)
  fisheries, 215–216
  growth patterns, 152, 161
  predator-prey size
    relationship, 131t
  reproduction, 74t, 82, 88, 91t
  toxicity, 32
  food resources, 117
  habitat, 22
interstitialis, 74t
microlepis (gag), 74t, 75, 79, 91t,
  152, 161
tigris (tiger grouper)
  food resources, 118
  growth patterns, 151, 152,
    161, 162, *165*
  predator-prey size
    relationship, 131t
  reproduction, 74t, 75, 82, 88,
    91t
  toxicity, 32
venenosa (yellowfin grouper), 152
  predator-prey size
    relationship, 131t
  reproduction, 74t, 79, 82, 88,
    91t
Myctophidae (myctophids), 30, 32
Myrichthys breviceps, 27
Myripristis jacobus, 24t

Nazabal, 28
needlefishes, 22, 58, 59
nests, 101
niche, 69–71, 116
nocturnal fishes, 64, 67–69, 70–71
  spawning migration, 99
nomenclature of fishes, 33–57t
nursery areas, 18, 27, 102
nutrient levels
  in brackish lagoons, 18
  calcium and scale development,
    151, 166
  on coral reefs, 17
  on the Cuban shelf, 11–12, 20
nutrient transfer, between habitats,
  70

oceanic water complex, 29–30, 32,
  198
Oculina, 17
Ocyurus chrysurus (yellowtail
  snapper)
  behavior, 68, 69, 70
  catch statistics, 206t, *208*, 209
  ecophysiology, *184*, *186*, *190*
  feeding intensity, 124, *128*
  food resources, 116, 117, 120,
    121, *121*
  gonadosomatic index, 89, 91t, *92*,
    184
  growth patterns, *150*, 151, 152,
    153, 158, *159*, 161t, *165*, *166*
  habitats and abundance, 24t, 31
  reproduction, *80*, 81, 82, 88, 93,
    100

Ogcocephalidae, 70, 216t
omnivores, 133t, 134t
Onchorhynchus mykiss (rainbow trout), 186
oogenesis, 75, *76*, 81t, 82–89, 183
ophichthids, 27
Ophichthus
   gomesii, 27
   ophis, 27
ophiuroids, 13, 116, 119
Opisthonema oglinum (Atlantic thread herring), 115, 206t, 211
Opsanus, 124
   beta, 28
   phobetron (scarecrow toadfish), 27, 28, 117
Oreochromis aureus (tilapia), 21, 28
otoliths
   back-calculation of age from, 154, 156
   changes in pattern with life stage, 151–152
   daily increments, 149
   environmental conditions and first formation of, 149, 152
   methodology, 152, 154–155, 162, 165
   periods of formation, 149–151
   winter mark, 149–151
overfishing, 19–20, 219
   altered sex ratios due to, 77, 79, 102
   crab, 219
   grouper, 209–210, 219
   herring, 211
   historical background, 30–31, 194
   mangrove oyster, 196
   mullet, 212, 219
   productivity statistics, 217
   sexual maturity and, 81
   shrimp, 196, 219
   signs of recovery, 206
   snapper, 30–31, 200, 207, 219
oyster, fisheries, 194, 196t, *204*

Pacific Ocean, 194
Pagellus erythrinus, 150
Panama, 9
Paralepididae, 30
Paranthias furcifer (creole-fish), 75
parasite cleaners, 30, 62, 117, 119
parental behavior, 101
parrotfish
   bucktooth. *See Sparisoma radians*
   fisheries, *204*
   food resources, 116
   habitats and abundance, 27, 29t
   redtail, 75t, 116
   spotlight, 60, *61*, 75t
   striped. *See Scarus iserti*
pelagic fishes, 18
   fat content, 180, 192
   fecundity, 98
   fisheries, 214–215
   fish fauna/assemblages, 29–30

fishing gear, 202
   inshore, behavior, 58–59, 67, 71–72
   length, 32
Peninsula de Hicacos, 99
Peninsula de Zapata, 11
Perca fluviatilis, 85
perch, sand. *See Diplectrum formosum*
Percophidae, 30
Phaeoptyx
   pigmentaria (dusky cardinalfish), 67
   xenus (sponge cardinalfish), 67
photoperiod, 90, 95, 149
physiology. *See* ecophysiology
phytoplankton and phytoplanktivores, 12, 115, 116, 116t, 127
Pico Turquino, 3
pilchard
   European, 88, 179
   false. *See Harengula clupeola*
Pinar del Río, 11, 19, 99, 202
piscivores
   behavior, 69, 70, 72
   food resources of, 115, 116, 116t, 117, 118
   by habitat and region, 133t, 134t
   ontogenic variation of food composition, *121*, *123*
   percent of population, 135
   seasonal variation of food composition, *129*
planktivores, 70, 115, 118–119, 133t, 134t
plasticity of behavior, 69, 117
Platichthys flesus, 85, 179
Playa del Este, *3*
Playa Habana, *3*
pluma. *See Calamus pennatula*
Pollachius virens (cod), 185–186
pollution
   corals near, 17
   sources of, 19–20
Pomacanthidae (pomacanthids), 22, 74t, 77, 79, 116
Pomacentridae (pomacentrids), 26, 59–60, 64, *65*, 70, 96, 116
pompano, 118, 212, 216
population
   density, 77
   genetics, 75
   structure, 30–31
porgy, 3, 29t, *204*
   jolthead, 74t, 77
   knobbed, 77
   littlehead, 74t, 77, 93
porkfish, 69, 117
predator-prey interactions. *See also* coloration
   behavior, 59, 67, 69, 70, 72
   diurnal changes in feeding intensity, 122–125, *123*
   in drifting algal mats, 30
   energy costs, 189, 193

fecundity, 101
   relative size, 130–131, *132*, 135
   timing of spawning, 98
Priacanthidae, 32, 116
primary production, 12–13, 20
Prionace glauca (blue shark), 215
Prionotus, 216t
Pristipomoides macrophthalmus (cardinal snapper), 209
productivity, 23, 180, 217, 219. *See also* upwelling
Pseudupeneus maculatus (spotted goatfish), 119
Puerto de Sagua, 28
Puerto Rico, 27, 75, 79, 118, 119, 120, 133, 161, 217
puffers, spiny, 70
Punta del Este, *3*, 26, 58
Punta de Maisí, 1, 3, 194
Punta Francés, 19
Punta Hicacos, 1, *3*, 194
Puntalón de Cayo Guano, 99

rabúa, 211
Rachycentron canadum (cobia), 32, 215
ray, spotted eagle, 215
razorfish, green, 74t, 119
recruitment, 9, 22, 98, 99, 185, 188–189. *See also* bycatch; circulation patterns; nursery areas; overfishing
reefs. *See* coral reefs
Remora remora (remora), 117
reproduction, 73–102. *See also* gonads; spawning
   annual cycle, 92–95, *96*, 97, 102
   attainment of sexual maturity, 75, 81, 81t
   duration, 88
   ecophysiology, 183, 193
   fecundity, 95, 97–98, 98t, 101
   sex ratio, 77, *78*, 79
   sexuality, 73–79
   strategy, 98–101. *See also* spawning
resorption of ovarian tissue, *76*, 77, 81, 84, *86*, 89, *90*, 200
resource availability. *See* food resources, availability of
Rhomboplites aurorubens (vermilion snapper), 151, 159, 209
Río Cauto, 28
Río Mosquito, 19
Río Sagua la Grande, *3*
Río Zaza, 28
robustness, 179
rock beauty, 74t, 79
runner, blue, 212, 213
Rypticus, 216t
   saponaceus, 74t

sailfish. *See Istiophorus platypterus*
sailor's choice, 69, 91t, 210
salinity, 11, 18, 20, 22, 28

sandfish, belted, 75
Santa Cruz del Norte, *3*, 19
Santa Cruz del Sur, 202
Sardina pilchardus (European pilchard), 88, 179
sardine(s), 3, 67, 70. *See also* Clupeidae
   Peruvian, 98
   redear. *See Harengula humeralis*
   scaled. *See Harengula jaguana*
   Spanish, 211
Sardinella
   anchovia (round sardinella), 180
   aurita (Spanish sardine), 211
Sardinops sagax (Peruvian sardine), 98
Sargocentron, 71
   vexillarium (dusky squirrelfish), *68*, 69
scad, bigeye, 212, 216, 216t
scales
   effects on digestion rate of prey, 127
   periods of mark formation, *150*, 151
   use in aging, 149–152
Scaridae (scarids), 22
   behavior, 60–62, 70
   food resources, 116, 117, 134
   as food source, 117, 118, 120t
   habitats and abundance, 29t
   reproduction, 74–75t, 76, 77, 79, 96
Scarinae, 74–75t, 77
Scarus, 77
   coelestinus, 74t, 215
   coeruleus, 75t
   croicensis. *See Scarus iserti*
   guacamaia, 75t, 215
   iserti (striped parrotfish), 25t, 60, *61*, 69, 70, 75t
   taeniopterus, 75t
   vetula, 75t
schooling, 22, 58–59, 60, 62, 66, 67, 71–72, 135
schoolmaster. *See Lutjanus apodus*
Sciaenidae, 91t, 96, 131t
Scomber japonicus (chub mackerel), 98, 116
Scomberomorus, 32, 214, 216t
   cavalla (king mackerel), *153*, 163, 214
   maculatus (Spanish mackerel), 22, 153, 163, 214
   regalis (cero), 22, 131t, 214
Scombridae (scombrids), 29, 96, 131t, 192
Scorpaenidae, 69, 72
scorpionfishes, 72
Scyliorhinidae, 29
sea bream, 28
seagrass beds
   distribution, 12, 13, *15*, *16*, 17–18
   ecological subsystem and fisheries, 198, *204*

fish fauna/fish assemblages, 18, 20, 27–28, 59, 218
  as food source, 116t
  movement patterns, 18, 20, 59, 61–62
  species diversity, 32, *59*
seagrass-reef complex, 14, 117, 198, 210
seasonal aspects
  annual cycle of reproduction, 90, 92–95, *96*, *97*, 102
  digestion rate, 126–127
  feeding habits, 119–120, *128*, *129*, 130, *130*, 134t, 135, *189*
  gonadosomatic index, 89, *92*, 94
  growth mark formation, *150*
  growth rate, 130, 151, 152–154, 166, 182, *189*
  photoperiod, 90, 95, 149
  physiological and biochemical indicators, 180–186, *187*, *188*
sediment types, 7–8, *15*, 18
Selar, 118
  crumenophthalmus (bigeye scad), 28, 32, 212, 216, 216t
self-fertilization, 75
sergeant major. *See Abudefduf saxatilis*
Seriola, 118
Serranidae (serranids), 21
  behavior, 69, 72
  as bycatch, 216t
  growth patterns, 152, 160–162, *164*, *165*
  habitats, 22, 30
  predator-prey size relationship, 131t
  reproduction, 74t, 77, 91t, 92, *96*
Serraninae, 74t, 75
Serranus
  phoebe (tattler), 74t, 75
  subligarius (belted sandfish), 75
  tabacarius (tobaccofish), 74t, 75, 129
  tigrinus (harlequin bass), 74t, 75
sex ratio, 31, 77, 78, 79, 101–102
sexual differentiation, 79, 81, 102
sexual dimorphism, 77–79, *80*, 101–102
sexual maturation, 75, 102
  age at, 134t
  fat reserves at, 182, 182t, 186
  stages and spawning, 97
  timeframe, 92, 93
shark(s), 32, 59
  bigeye thresher, 215
  bignose, 215
  blue, 215
  dusky, 215
  fisheries, 194, 200, 202, *204*, 206t, 215
  habitat and abundance, 29
  hammerhead, 215
  night, 215

nurse, 215
shortfin mako, 215
silky, 215
sixgill, 30, 215
tiger, 215
whitetip, 215
sharksucker, 117
shelter site, 67–69, 71
shrimp
  fisheries, 3, 196, 196t, 198, *204*, 216
  as food source, 116, 116t, 117, 118, 119, 120t, 122
  seasonal availability, *129*
silverside, 70
  hardhead, 26, 27
similarity index. *See* indexes
slope reefs. *See* coral reefs
snapper(s), 85, 379. *See also* Lutjanidae
  behavior, 67, 68–69, 70, 71
  blackfin. *See Lutjanus buccanella*
  cubera. *See Lutjanus cyanopterus*
  dog. *See Lutjanus jocu*
  food resources, 115, 120–121
  gray. *See Lutjanus griseus*
  lane. *See Lutjanus synagris*
  mutton. *See Lutjanus analis*
  silk. *See Lutjanus vivanus*
  vermilion. *See Rhomboplites aurorubens*
  yellowtail. *See Ocyurus chrysurus*
snooks, 70
Spain, 150
Sparidae (sparids)
  fisheries, 206t
  habitats and abundance, 29t
  reproduction, 74t, 75, 77, 91t, 96
  trophic biology, 118
Sparisoma, 71, 77
  atomarium, 75t
  aurofrenatum, 75t
  chrysopterum (redtail parrotfish), 25t, 75t, 115
  as food source, 117, 118, 216
  radians (bucktooth parrotfish), *61*, 61–62, 75t, 115
  rubripinne, 75t
  viride (spotlight parrotfish), 25t, *60*, *61*, 75t
Sparisomatinae, 75t, 77
spatial distribution, diurnal patterns, 70–71
spawning
  aggregation, 92, 94, 99, 100
  categories, 82, 102
  ecological patterns, 82, 84–88, *85*
  and fishing strategy, 31, 205, 206, 207, 208, 209. *See also* overfishing
  induced, 86
  on the inner shelf, 101, 102
  migration, 95, 99, 206
  periodicity, and calculation of fecundity, 95, 97, 98

seasonal aspects, 90, 92–95, *96*, *97*
  at the shelf edge, 99–100, 102
  strategies, 18, 98–102
spearfish, longbill, 215
specialization, ecological niches, 69–71, 116, 206
species diversity, 23, 23t, 32, 33–57t, *59*
species richness, 23, 23t
spermatogenesis, 75, *76*, 81t, 183
Sphoeroides, 27, 216t
Sphyraena, 58, 59, 71
  barracuda (great barracuda), 59, *93*, 131t
  habitats and abundance, 22, 26, 27, 32
  toxicity, 32
  guachancho, 131t
Sphyraenidae (sphyraenids), 29t, 131t
Sphyrna (hammerhead shark), 22, 215
Sphyrnidae (sphyrnids), 29
sponges, 13, 17
  bycatch statistics, 216t
  commensalism with, 67
  fisheries, 196t, 198, *204*
  as food source, 116, 116t
Sprattus sprattus (European sprat), 179
Squalidae, 29
squirrelfish(es)
  behavior, 67, 70, 71
  dusky, 68, 69
  food resources, 22, 116
  longspine, 68, 69
Stegastes, 22, 71, *71*
  adustus (dusky damselfish), 25t, *63*, 66, 116
  leucostictus (beaugregory), 26, *65*, 66, 70, 116, 129
  partitus, 25t
  planifrons, 116
  variabilis, 116
stenophagy, 117, 132, 135
Stenotomus chrysops, 75
Stephanolepis setifer, 28
Sternoptychidae, 30
stingray(s)
  fisheries, 194, 200, *204*, 206t, 215
  habitats, 27, 28, 32
  southern, 215
  spotted eagle ray, 215
stress
  and growth rate, 164
  oocyte resorption due to, 89, 200
Stromateidae, 30
Strongylura notata, 22, 27
suprabenthic fishes, behavior, 60–64, 67–69, 71
surgeon, ocean, 62, *63*
swordfish. *See Xiphias gladius*
symbiotic relationships, 59, 62, 117, 119
Symphysanodontidae, 30

Syngnathidae, 216t
Syngnathus pelagicus, 29
Synodontidae, 118, 216t
tang, blue, 62, *62*
tattler, 74t, 75
taxonomy, 21–22
territoriality
  behavior, 64, 66, 67–68, 70, 71–72
  and diurnal changes in feeding intensity, 122
  and spawning strategy, 101
Tetraodontidae (tetraodontids), 21, 27
Tetrapturus
  albidus (white marlin), 29, *214*, 215
  pfluegeri (longbill spearfish), 215
Thalassoma bifasciatum (bluehead)
  behavior, 62, *63*, 64, 71
  food resources, 117, 119
  habitats and abundance, 25t
  reproduction, 74t, 77
Theragra chalcogramma, 85
thermocline, 29
thigmotropism, 30
Thunnus
  albacares (yellowfin tuna), 213
  atlanticus (blackfin tuna), 29, 198, 206t, 211, 213
  obesus, 86
tilapia, 21, 28, 151
Tilapia esculenta (tilapia), 151
toadfish, scarecrow, 117
tobaccofish, 74t, 75, 129
tomtate. *See Haemulon aurolineatum*
topographic complexity, 70, 71
Trachinotus (pompano), 118, 212, 216
Trachipteridae, 30
Trachurus mediterraneus (Mediterranean horse mackerel), 179
Trichiuridae, 91t
Trichiurus lepturus, 91t
triggerfish, 30, 116
Trinidad, 155–156, 156t, 163, 180
Tripterygiidae, 31
trout, rainbow, 186
trunkfish, 116
tuna, 29, 30, 98
  blackfin. *See Thunnus atlanticus*
  fisheries, 202, *204*, 213, 213–214
  skipjack. *See Katsuwonus pelamis*
  yellowfin. *See Thunnus albacares*
Tunas de Zaza, 28, 120, 163, 208, 210, 211, 212
tunny, little, 216
turtle
  fisheries, 118, *195*, 196t, 198, 200
  as food source, 118
twilight changeover. *See* diurnal behavior
Tylosurus, 27
  acus acus, 22, 131t
  crocodilus crocodilus, 22

upwelling, 11, 32, 73, 180
urohyal bone, use in aging, 149–151, 152

Venezuela, 81, 134t
  fish ecophysiology, 180
  fish growth patterns in, 155, 157, 158t, 160t, 162, 163
  productivity, 180
vertical migration, 30
vessels. *See* fishing fleet
Virgin Islands
  fish growth patterns in, 155, 158, 161t

food resources in, 118, 119, 133
  trophic structure, 134t
visual signals, 62
vitellogenesis, *76*, 82, *83*, 84t, 85–89, 102, 183, 186
von Bertalanffy function, 154

water depth, on the shelf, 3, 10
water temperature, 11, *12*, 18, 20, 181
  and digestion rate, 126–127, 134t, 135
  and ecophysiology, *189*

and growth marks, 149, 166
  and spawning period, 95, 102
  stable, 73
wenchman, 209
West Cayo Sigua, 29t
wrasse(s)
  behavior, 60, 62–64, 70, 71
  blackear. *See Halichoeres poeyi*
  food resources, 119
  as food source, 120t
  yellowhead, 74t, 77

Xiphias gladius (swordfish), 29, 32, 202, 206t, *214*, 214–215

Xyrichtys, 119
  martinicensis, 74t
  splendens (green razorfish), 74t, 119

Yucatán, 9, 10, 75, 155, 156t

zooplankton and zooplanktivores, 26, 70, 115–116, 116t, 117, 118, 120, 120t, 121, *121*, 122, *123*
  percent of population, 134
  seasonal availability, *129*, 135